Student's Solutions Manual

for use with

Introductory Algebra

A Real-World Approach

Second Edition

Ignacio Bello
Hillsborough Community College

Boston Burr Ridge, IL Dubuque, IA Madison, WI New York San Francisco St. Louis
Bangkok Bogotá Caracas Kuala Lumpur Lisbon London Madrid Mexico City
Milan Montreal New Delhi Santiago Seoul Singapore Sydney Taipei Toronto

The McGraw-Hill Companies

Student's Solutions Manual for use with
INTRODUCTORY ALGEBRA: A REAL-WORLD APPROACH, SECOND EDITION
Ignacio Bello

Published by McGraw-Hill Higher Education, an imprint of The McGraw-Hill Companies, Inc., 1221 Avenue of the Americas, New York, NY 10020.

This book is printed on acid-free paper

1 2 3 4 5 6 7 8 9 0 QSR/QSR 0 9 8 7 6 5

ISBN 0-07-294557-5

www.mhhe.com

Contents

Preface

Dear Student,

This Student's Solution Manual for use with Bello, *Introductory Algebra,* 2nd Edition, provides complete, worked-out solutions to the following:

- Odd-numbered items in each section's set of exercises.
- All problems paired with examples.
- Odd- and even-numbered items in the chapter Review Exercises.
- Odd- and even-numbered items in the Cumulative Review exercises.

How to use this manual:

If you want to get good grades on your tests, then this manual can help you. Treat every homework assignment as a practice test. Resist the urge to look up all the solutions just to get it over with, because you're not allowed to do that on a test.

So, first try every exercise on your own, then check your work against the solution. If you're stumped, then do as much of the exercise as you can. Next, look up the rest of the steps one at a time, and try to figure out each step before you look at it. (It's handy to keep a piece of paper covering the solution steps so you don't see them all at once.) This is a study habit that will help you improve your test scores and build your confidence in math.

Seeing some solutions will give you that "aha!" sense that you knew how to do it all along and just needed a refresher. If you see other solutions that still appear too complex to follow, then it will help you to review the textbook's explanations and examples to see how similar problems were discussed and solved.

You can also go to the text's MathZone website to view lecture demonstrations, practice problems, and videos by the authors explaining how to work through sample exercises. With all these resources, you'll find that this is a course that you can do well in.

Chapter R Prealgebra Review

R.1 Fractions

Problems R.1

1. a. $18 = \frac{18}{1}$

b. $-24 = \frac{-24}{1}$

2. $\frac{4}{7} = \frac{?}{21}$
7 was multiplied by 3 to get 21, so 4 must be multiplied by 3.
$\frac{4}{7} = \frac{4 \times 3}{7 \times 3} = \frac{12}{21}$
The equivalent fraction is $\frac{12}{21}$.

3. $\frac{24}{30} = \frac{?}{5}$
30 was divided by 6 to get 5, so 24 must be divided by 6.
$\frac{24}{30} = \frac{24 \div 6}{30 \div 6} = \frac{4}{5}$
The equivalent fraction is $\frac{4}{5}$.

4. a. The largest natural number exactly dividing 30 and 50 is 10.
$\frac{30}{50} = \frac{30 \div 10}{50 \div 10} = \frac{3}{5}$

b. The largest natural number exactly dividing 45 and 60 is 15.
$\frac{45}{60} = \frac{45 \div 15}{60 \div 15} = \frac{3}{4}$

c. The largest natural number exactly dividing 84 and 72 is 12.
$\frac{84}{72} = \frac{84 \div 12}{72 \div 12} = \frac{7}{6}$

Exercises R.1

1. $28 = \frac{28}{1}$

3. $-42 = \frac{-42}{1}$

5. $0 = \frac{0}{1}$

7. $-1 = \frac{-1}{1}$

9. $\frac{1}{8} = \frac{?}{24}$
8 was multiplied by 3 to get 24, so 1 must be multiplied by 3.
$\frac{1}{8} = \frac{1 \times 3}{8 \times 3} = \frac{3}{24}$
The missing number is 3.

11. $\frac{7}{1} = \frac{?}{6}$
1 was multiplied by 6 to get 6, so 7 must be multiplied by 6.
$\frac{7}{1} = \frac{7 \times 6}{1 \times 6} = \frac{42}{6}$
The missing number is 42.

13. $\frac{5}{3} = \frac{?}{15}$
3 was multiplied by 5 to get 15, so 5 must be multiplied by 5.
$\frac{5}{3} = \frac{5 \times 5}{3 \times 5} = \frac{25}{15}$
The missing number is 25.

15. $\frac{7}{11} = \frac{?}{33}$
11 was multiplied by 3 to get 33, so 7 must be multiplied by 3.
$\frac{7}{11} = \frac{7 \times 3}{11 \times 3} = \frac{21}{33}$
The missing number is 21.

17. $\frac{1}{8}=\frac{4}{?}$

1 was multiplied by 4 to get 4, so 8 must be multiplied by 4.

$\frac{1}{8}=\frac{1\times 4}{8\times 4}=\frac{4}{32}$

The missing number is 32.

19. $\frac{5}{6}=\frac{5}{?}$

5 was multiplied by 1 to get 5, so 6 must be multiplied by 1.

$\frac{5}{6}=\frac{5\times 1}{6\times 1}=\frac{5}{6}$

The missing number is 6.

21. $\frac{8}{7}=\frac{16}{?}$

8 was multiplied by 2 to get 16, so 7 must be multiplied by 2.

$\frac{8}{7}=\frac{8\times 2}{7\times 2}=\frac{16}{14}$

The missing number is 14.

23. $\frac{6}{5}=\frac{36}{?}$

6 was multiplied by 6 to get 36, so 5 must be multiplied by 6.

$\frac{6}{5}=\frac{6\times 6}{5\times 6}=\frac{36}{30}$

The missing number is 30.

25. $\frac{21}{56}=\frac{?}{8}$

56 was divided by 7 to get 8, so 21 must be divided by 7.

$\frac{21}{56}=\frac{21\div 7}{56\div 7}=\frac{3}{8}$

The missing number is 3.

27. $\frac{36}{180}=\frac{?}{5}$

180 was divided by 36 to get 5, so 36 must be divided by 36.

$\frac{36}{180}=\frac{36\div 36}{180\div 36}=\frac{1}{5}$

The missing number is 1.

29. $\frac{18}{12}=\frac{3}{?}$

18 was divided by 6 to get 3, so 12 must be divided by 6.

$\frac{18}{12}=\frac{18\div 6}{12\div 6}=\frac{3}{2}$

The missing number is 2.

31. $\frac{15}{12}=\frac{3\cdot 5}{2\cdot 2\cdot 3}=\frac{5}{4}$

33. $\frac{13}{52}=\frac{13}{2\cdot 2\cdot 13}=\frac{1}{4}$

35. $\frac{56}{24}=\frac{2\cdot 2\cdot 2\cdot 7}{2\cdot 2\cdot 2\cdot 3}=\frac{7}{3}$

37. $\frac{22}{33}=\frac{2\cdot 11}{3\cdot 11}=\frac{2}{3}$

39. $\frac{100}{25}=\frac{2\cdot 2\cdot 5\cdot 5}{5\cdot 5}=\frac{4}{1}=4$

41. **a.** 1000 hours for 45 days $=\frac{1000 \text{ hours}}{45 \text{ days}}$
$=\frac{1000}{45}$

b. $\frac{1000}{45}=\frac{1000\div 5}{45\div 5}=\frac{200}{9}$

c. $\frac{1000}{45}=22\frac{2}{9}$

d. $\frac{1000}{24}=\frac{1000\div 8}{24\div 8}=\frac{125}{3}=41\frac{2}{3}$

The 1000 free hours would last $41\frac{2}{3}$ days.

43. 41 out of 100 $=\frac{41}{100}$

45. a. $\frac{1}{3}=\frac{?}{6}$

3 was multiplied by 2 to get 6, so 1 must be multiplied by 2.

$\frac{1}{3}=\frac{1\times 2}{3\times 2}=\frac{2}{6}$

b. $\frac{2}{6}$ of the pizza or 2 pieces

c. $\frac{2}{6}=\frac{2\div 2}{6\div 2}=\frac{1}{3}$

She ate $\frac{1}{3}$ of the pizza.

d. Each ate $\frac{1}{3}$, so they ate the same amount.

47–49. Count the number of sections from E to the line described. Divide that number by the total number of equal sections, 8.

47. $\frac{1}{8}$

49. $\frac{5}{8}$

51. a. 48 minutes × 60 seconds/minute
= 2880 seconds

b. Each team shoots 60 shots.
60 × 2 = 120 total shots

c. $\frac{2880 \text{ seconds}}{120 \text{ shots}}=\frac{2880\div 120}{120\div 120}=\frac{24}{1}$ or 24

53. Two segments of 4 minutes each are devoted to Sports.

$\frac{8}{60}=\frac{8\div 4}{60\div 4}=\frac{2}{15}$

55. From the graph, National and International News occupies the most segments, two segments of 15 minutes each.

$\frac{30}{60}=\frac{30\div 30}{60\div 30}=\frac{1}{2}$

57. 3 + 3 + 1 + 3 + 3 + 1 = 14 minutes are devoted to Weather.

$\frac{14}{60}=\frac{14\div 2}{60\div 2}=\frac{7}{30}$

59. News & Beyond the Bay use the most time: $\frac{13}{30}$

Traffic uses the least time: $\frac{1}{10}$

61. Answers may vary.

R.2 Operations with Fractions

Problems R.2

1. $\frac{9}{7}\cdot\frac{3}{5}=\frac{9\cdot 3}{7\cdot 5}=\frac{27}{35}$

2. a. $\frac{4}{\cancel{9}_1}\cdot\frac{\cancel{9}^1}{11}=\frac{4\cdot 1}{1\cdot 11}=\frac{4}{11}$

b. $\frac{\cancel{7}^1}{\cancel{14}_2}\cdot\frac{\cancel{5}^1}{\cancel{20}_4}=\frac{1\cdot 1}{2\cdot 4}=\frac{1}{8}$

3. First convert $2\frac{3}{4}$ to a fraction.

$2\frac{3}{4}=\frac{2\cdot 4+3}{4}=\frac{11}{4}$

$2\frac{3}{4}\cdot\frac{8}{11}=\frac{\cancel{11}^1}{\cancel{4}_1}\cdot\frac{\cancel{8}^2}{\cancel{11}_1}=\frac{1\cdot 2}{1\cdot 1}=2$

4. a. $\frac{3}{4}\div\frac{5}{7}=\frac{3}{4}\cdot\frac{7}{5}=\frac{21}{20}$

b. $\frac{3}{7} \div 5 = \frac{3}{7} \cdot \frac{1}{5} = \frac{3}{35}$

5. a. $2\frac{1}{5} \div \frac{7}{10} = \frac{11}{5} \div \frac{7}{10} = \frac{11}{\cancel{5}_1} \cdot \frac{\cancel{10}^2}{7} = \frac{22}{7}$

b. $\frac{11}{15} \div 7\frac{1}{3} = \frac{11}{15} \div \frac{22}{3} = \frac{\cancel{11}^1}{\cancel{15}_5} \cdot \frac{\cancel{3}^1}{\cancel{22}_2} = \frac{1}{10}$

6. $20 = 2 \cdot 2 \cdot 5$
$16 = 2 \cdot 2 \cdot 2 \cdot 2$
$\text{LCD} = 2 \cdot 2 \cdot 2 \cdot 2 \cdot 5 = 80$
$\frac{3}{20} = \frac{3 \cdot 4}{20 \cdot 4} = \frac{12}{80}$
$1\frac{1}{16} = \frac{17}{16} = \frac{17 \cdot 5}{16 \cdot 5} = \frac{85}{80}$
$\frac{3}{20} + 1\frac{1}{16} = \frac{12}{80} + \frac{85}{80} = \frac{97}{80}$ or $1\frac{17}{80}$

7. $18 = 2 \cdot 3 \cdot 3$
$12 = 2 \cdot 2 \cdot 3$
$\text{LCD} = 2 \cdot 2 \cdot 3 \cdot 3 = 36$
$\frac{7}{18} = \frac{7 \cdot 2}{18 \cdot 2} = \frac{14}{36}$
$\frac{1}{12} = \frac{1 \cdot 3}{12 \cdot 3} = \frac{3}{36}$
$\frac{7}{18} - \frac{1}{12} = \frac{14}{36} - \frac{3}{36} = \frac{11}{36}$

8. $8 = 2 \cdot 2 \cdot 2$
$6 = 2 \cdot 3$
$\text{LCD} = 2 \cdot 2 \cdot 2 \cdot 3 = 24$

$$4\frac{1}{8} + 2\frac{5}{6} = \frac{33}{8} - \frac{17}{6} = \frac{33 \cdot 3}{8 \cdot 3} - \frac{17 \cdot 4}{6 \cdot 4} = \frac{99}{24} - \frac{68}{24} = \frac{31}{24} \text{ or } 1\frac{7}{24}$$

9. $8 = 2 \cdot 2 \cdot 2$
$21 = 3 \cdot 7$
$28 = 2 \cdot 2 \cdot 7$
$\text{LCD} = 2 \cdot 2 \cdot 2 \cdot 3 \cdot 7 = 168$
$1\frac{1}{8} = \frac{9}{8} = \frac{9 \cdot 21}{8 \cdot 21} = \frac{189}{168}$
$\frac{7}{21} = \frac{7 \cdot 8}{21 \cdot 8} = \frac{56}{168}$
$\frac{3}{28} = \frac{3 \cdot 6}{28 \cdot 6} = \frac{18}{168}$

$$1\frac{1}{8} + \frac{7}{21} - \frac{3}{28} = \frac{189}{168} + \frac{56}{168} - \frac{18}{168} = \frac{189 + 56 - 18}{168} = \frac{227}{168} \text{ or } 1\frac{59}{168}$$

10. We need $1\frac{1}{4}$ cups water and $\frac{1}{2}$ cup oat bran.

$$1\frac{1}{4} + \frac{1}{2} = \frac{5}{4} + \frac{1}{2} = \frac{5}{4} + \frac{1 \cdot 2}{2 \cdot 2} = \frac{5}{4} + \frac{2}{4} = \frac{7}{4} = 1\frac{3}{4}$$

A total of $1\frac{3}{4}$ cups is needed.

Exercises R.2

1. $\frac{2}{3} \cdot \frac{7}{3} = \frac{2 \cdot 7}{3 \cdot 3} = \frac{14}{9}$

3. $\frac{\cancel{6}^1}{5} \cdot \frac{7}{\cancel{6}_1} = \frac{1 \cdot 7}{5 \cdot 1} = \frac{7}{5}$

5. $\frac{\cancel{7}^1}{\cancel{3}_1} \cdot \frac{\cancel{6}^2}{\cancel{7}_1} = \frac{1 \cdot 2}{1 \cdot 1} = \frac{2}{1}$ or 2

7. $7\cdot\frac{8}{7}=\frac{\overset{1}{\cancel{7}}}{1}\cdot\frac{8}{\underset{1}{\cancel{7}}}=\frac{1\cdot 8}{1\cdot 1}=\frac{8}{1}$ or 8

9. $2\frac{3}{5}\cdot 2\frac{1}{7}=\frac{13}{\underset{1}{\cancel{5}}}\cdot\frac{\overset{3}{\cancel{15}}}{7}=\frac{13\cdot 3}{1\cdot 7}=\frac{39}{7}$

11. $7\div\frac{3}{5}=\frac{7}{1}\div\frac{3}{5}=\frac{7}{1}\cdot\frac{5}{3}=\frac{35}{3}$

13. $\frac{3}{5}\div\frac{9}{10}=\frac{\overset{1}{\cancel{3}}}{\underset{1}{\cancel{5}}}\cdot\frac{\overset{2}{\cancel{10}}}{\underset{3}{\cancel{9}}}=\frac{2}{3}$

15. $\frac{9}{10}\div\frac{3}{5}=\frac{\overset{3}{\cancel{9}}}{\underset{2}{\cancel{10}}}\cdot\frac{\overset{1}{\cancel{5}}}{\underset{1}{\cancel{3}}}=\frac{3}{2}$

17. $1\frac{1}{5}\div\frac{3}{8}=\frac{6}{5}\div\frac{3}{8}=\frac{\overset{2}{\cancel{6}}}{5}\cdot\frac{8}{\underset{1}{\cancel{3}}}=\frac{16}{5}$

19. $2\frac{1}{2}\div 6\frac{1}{4}=\frac{5}{2}\div\frac{25}{4}=\frac{\overset{1}{\cancel{5}}}{\underset{1}{\cancel{2}}}\cdot\frac{\overset{2}{\cancel{4}}}{\underset{5}{\cancel{25}}}=\frac{2}{5}$

21. $\frac{1}{5}+\frac{2}{5}=\frac{1+2}{5}=\frac{3}{5}$

23. $\frac{3}{8}+\frac{5}{8}=\frac{3+5}{8}=\frac{8}{8}=\frac{1}{1}$ or 1

25. $8=2\cdot 2\cdot 2$
$4=2\cdot 2$
$\text{LCD}=2\cdot 2\cdot 2=8$
$\frac{3}{4}=\frac{3\cdot 2}{4\cdot 2}=\frac{6}{8}$
$\frac{7}{8}+\frac{3}{4}=\frac{7}{8}+\frac{6}{8}=\frac{13}{8}$

27. $6=2\cdot 3$
$10=2\cdot 5$
$\text{LCD}=2\cdot 3\cdot 5=30$
$\frac{5}{6}=\frac{5\cdot 5}{6\cdot 5}=\frac{25}{30}$
$\frac{3}{10}=\frac{3\cdot 3}{10\cdot 3}=\frac{9}{30}$
$\frac{5}{6}+\frac{3}{10}=\frac{25}{30}+\frac{9}{30}=\frac{34}{30}=\frac{17}{15}$

29. $\text{LCD}=3\cdot 2=6$
$2\frac{1}{3}=\frac{7}{3}=\frac{7\cdot 2}{3\cdot 2}=\frac{14}{6}$
$1\frac{1}{2}=\frac{3}{2}=\frac{3\cdot 3}{2\cdot 3}=\frac{9}{6}$
$2\frac{1}{3}+1\frac{1}{2}=\frac{14}{6}+\frac{9}{6}=\frac{23}{6}$

31. $5=5$
$10=2\cdot 5$
$\text{LCD}=2\cdot 5=10$
$\frac{1}{5}=\frac{1\cdot 2}{5\cdot 2}=\frac{2}{10}$
$2=\frac{2}{1}=\frac{2\cdot 10}{1\cdot 10}=\frac{20}{10}$
$\frac{1}{5}+2+\frac{9}{10}=\frac{2}{10}+\frac{20}{10}+\frac{9}{10}=\frac{31}{10}$

33. $2=2$
$7=7$
$4=2\cdot 2$
$\text{LCD}=2\cdot 2\cdot 7=28$
$3\frac{1}{2}=\frac{7}{2}=\frac{7\cdot 14}{2\cdot 14}=\frac{98}{28}$
$1\frac{1}{7}=\frac{8}{7}=\frac{8\cdot 4}{7\cdot 4}=\frac{32}{28}$
$2\frac{1}{4}=\frac{9}{4}=\frac{9\cdot 7}{4\cdot 7}=\frac{63}{28}$
$3\frac{1}{2}+1\frac{1}{7}+2\frac{1}{4}=\frac{98}{28}+\frac{32}{28}+\frac{63}{28}=\frac{193}{28}$

35. $\frac{5}{8}-\frac{2}{8}=\frac{5-2}{8}=\frac{3}{8}$

37. $3 = 3$
$6 = 2 \cdot 3$
$\text{LCD} = 2 \cdot 3 = 6$
$\frac{1}{3} = \frac{1 \cdot 2}{3 \cdot 2} = \frac{2}{6}$
$\frac{1}{3} - \frac{1}{6} = \frac{2}{6} - \frac{1}{6} = \frac{1}{6}$

39. $10 = 2 \cdot 5$
$20 = 2 \cdot 2 \cdot 5$
$\text{LCD} = 2 \cdot 2 \cdot 5 = 20$
$\frac{7}{10} = \frac{7 \cdot 2}{10 \cdot 2} = \frac{14}{20}$
$\frac{7}{10} - \frac{3}{20} = \frac{14}{20} - \frac{3}{20} = \frac{11}{20}$

41. $15 = 3 \cdot 5$
$25 = 5 \cdot 5$
$\text{LCD} = 3 \cdot 5 \cdot 5 = 75$
$\frac{8}{15} = \frac{8 \cdot 5}{15 \cdot 5} = \frac{40}{75}$
$\frac{2}{25} = \frac{2 \cdot 3}{25 \cdot 3} = \frac{6}{75}$
$\frac{8}{15} - \frac{2}{25} = \frac{40}{75} - \frac{6}{75} = \frac{34}{75}$

43. $\text{LCD} = 5 \cdot 4 = 20$
$2\frac{1}{5} = \frac{11}{5} = \frac{11 \cdot 4}{5 \cdot 4} = \frac{44}{20}$
$1\frac{3}{4} = \frac{7}{4} = \frac{7 \cdot 5}{4 \cdot 5} = \frac{35}{20}$
$2\frac{1}{5} - 1\frac{3}{4} = \frac{44}{20} - \frac{35}{20} = \frac{9}{20}$

45. $\text{LCD} = 4$
$3 = \frac{3}{1} = \frac{3 \cdot 4}{1 \cdot 4} = \frac{12}{4}$
$3 - 1\frac{3}{4} = \frac{12}{4} - \frac{7}{4} = \frac{5}{4}$

47. $6 = 2 \cdot 3$
$9 = 3 \cdot 3$
$3 = 3$
$\text{LCD} = 2 \cdot 3 \cdot 3 = 18$
$\frac{5}{6} = \frac{5 \cdot 3}{6 \cdot 3} = \frac{15}{18}$
$\frac{1}{9} = \frac{1 \cdot 2}{9 \cdot 2} = \frac{2}{18}$
$\frac{1}{3} = \frac{1 \cdot 6}{3 \cdot 6} = \frac{6}{18}$
$\frac{5}{6} + \frac{1}{9} - \frac{1}{3} = \frac{15}{18} + \frac{2}{18} - \frac{6}{18}$
$= \frac{15 + 2 - 6}{18}$
$= \frac{11}{18}$

49. $\text{LCD} = 3 \cdot 5 = 15$
$1\frac{1}{3} = \frac{4}{3} = \frac{4 \cdot 5}{3 \cdot 5} = \frac{20}{15}$
$2\frac{1}{3} = \frac{7}{3} = \frac{7 \cdot 5}{3 \cdot 5} = \frac{35}{15}$
$1\frac{1}{5} = \frac{6}{5} = \frac{6 \cdot 3}{5 \cdot 3} = \frac{18}{15}$
$1\frac{1}{3} + 2\frac{1}{3} - 1\frac{1}{5} = \frac{20}{15} + \frac{35}{15} - \frac{18}{15}$
$= \frac{20 + 35 - 18}{15}$
$= \frac{37}{15}$

51. $450 \cdot \frac{1}{6} = \frac{\overset{75}{\cancel{450}}}{1} \cdot \frac{1}{\underset{1}{\cancel{6}}} = \frac{75}{1}$ or 75

The Lunar Rover weighed 75 pounds on the moon.

53. $28{,}000 \cdot \frac{1}{5} = \frac{\overset{5600}{\cancel{28{,}000}}}{1} \cdot \frac{1}{\underset{1}{\cancel{5}}} = \frac{5600}{1}$ or 5600

The number of working actors is 5600.

55. $66\frac{1}{2} \div 4 = \frac{133}{2} \div \frac{4}{1} = \frac{133}{2} \cdot \frac{1}{4} = \frac{133}{8}$ or $16\frac{5}{8}$

It removed $16\frac{5}{8}$ cubic yards in one hour.

57. $\frac{9}{50} + \frac{3}{25}$

LCD = 50

$\frac{9}{50} + \frac{3}{25} = \frac{9}{50} + \frac{3 \cdot 2}{25 \cdot 2} = \frac{9}{50} + \frac{6}{50} = \frac{15}{50} = \frac{3}{10}$

$\frac{3}{10}$ of American households make more than \$20,000.

59. $46\frac{3}{5} - 38\frac{9}{10}$

LCD = 10

$46\frac{3}{5} = \frac{233}{5} = \frac{233 \cdot 2}{5 \cdot 2} = \frac{466}{10}$

$38\frac{9}{10} = \frac{389}{10}$

$46\frac{3}{5} - 38\frac{9}{10} = \frac{466}{10} - \frac{389}{10} = \frac{77}{10}$ or $7\frac{7}{10}$

U.S. workers work $7\frac{7}{10}$ hours more per week.

61. Find the LCM of 10 and 8.

$10 = 2 \cdot 5$

$8 = 2 \cdot 2 \cdot 2$

$\text{LCM} = 2 \cdot 2 \cdot 2 \cdot 5 = 40$

$10 \cdot 4 = 40$

$8 \cdot 5 = 40$

Buy 4 packages of hot dogs and 5 packages of buns.

63. Answers may vary.

65. a. Multiply the numerators to get the new numerator. Multiply the denominators to get the new denominator.

b. Multiply the first fraction by the reciprocal of the second fraction.

67. $\frac{128}{16} = 8$

69. The largest natural number exactly dividing 20 and 100 is 20.

$\frac{20}{100} = \frac{20 \div 20}{100 \div 20} = \frac{1}{5}$

R.3 Decimals and Percents

Problems R.3

1. $46.325 = 40 + 6 + \frac{3}{10} + \frac{2}{100} + \frac{5}{1000}$

2. a. $0.045 = \frac{45}{1000} = \frac{9}{200}$

b. $0.0375 = \frac{375}{10{,}000} = \frac{3}{80}$

3. a. $3.13 = \frac{313}{100}$

b. $5.25 = \frac{525}{100} = \frac{21}{4}$

4.

$$\begin{array}{r} \scriptstyle 1 \\ 6.00 \\ +\,17.25 \\ \hline 23.25 \end{array}$$

5.

$$\begin{array}{r} \scriptstyle 6\,12\,13 \\ 2\not{7}\not{3}.\not{3}2 \\ -14.50 \\ \hline 258.82 \end{array}$$

6.

$$\begin{array}{r} \scriptstyle 9 \\ \scriptstyle 3\,15\;\;5\,\not{10}\,10 \\ \not{4}\not{5}8.\not{6}\,\not{0}\,\not{0} \\ -\,193.3\,4\,1 \\ \hline 265.2\,5\,9 \end{array}$$

7. a.

$$\begin{array}{rl} 6.203 & \text{3 decimal digits} \\ \times\ 31.03 & \text{2 decimal digits} \\ \hline 18609 & \\ 62030 & \\ 18609 & \\ \hline 192.47909 & 2 + 3 = 5 \text{ decimal digits} \end{array}$$

b.
$$\begin{array}{rl} 3.123 & \text{3 decimal digits} \\ \times\ 0.0015 & \text{4 decimal digits} \\ \hline 15615 & \\ 3123 & \\ \hline 0.0046845 & 3+4=7 \text{ decimal digits} \end{array}$$

8. $0045.\overline{)1800.}$ quotient $40.$; 180; 00

$$\frac{1.8}{0.045}=40$$

9. a. $6\overline{)5.000}$ quotient $0.833\ldots$; 48, 20, 18, 20, 18, 2

$$\frac{5}{6}=0.8\overline{3}$$

b. $4\overline{)1.00}$ quotient 0.25; 8, 20, 20, 0

$$\frac{1}{4}=0.25$$

10. a. $86\% = 86. = 0.86$

b. $48.9\% = 48.9 = 0.489$

c. $8.3\% = 08.3 = 0.083$

11. a. $0.49 = 0.49\% = 49\%$

b. $4.17 = 4.17\% = 417\%$

c. $89.2 = 89.20\% = 8920\%$

12. a. $37\% = \frac{37}{100}$

b. $25\% = \frac{25}{100} = \frac{1}{4}$

13. $\frac{3}{5} = \frac{3\times 20}{5\times 20} = \frac{60}{100} = 60\%$

14. $8\overline{)3.00}$ quotient 0.37; $2\,4$, 60, 56, 4

The remainder is 4; $\frac{4}{8}=\frac{1}{2}$.

$$\frac{3}{8} = 0.37\frac{1}{2} = 37\frac{1}{2}\%$$

15. a. $0.\overline{6} = 0.6\underline{6}66\ldots = 0.67$

Underline 6. The digit to the right of 6 is $6 \geq 5$, so add 1 to the underlined digit and drop all the numbers to the right of 7.

b. $0.\overline{18} = 0.1\underline{8}18\ldots = 0.18$

Underline 8. The digit to the right of 8 is $1 < 5$, so do not change the underlined digit and drop all the numbers to the right of 8.

Exercises R.3

1. $4.7 = 4 + \frac{7}{10}$

3. $5.62 = 5 + \frac{6}{10} + \frac{2}{100}$

5. $16.123 = 10 + 6 + \frac{1}{10} + \frac{2}{100} + \frac{3}{1000}$

7. $49.012 = 40 + 9 + \frac{1}{100} + \frac{2}{1000}$

9. $57.104 = 50 + 7 + \dfrac{1}{10} + \dfrac{4}{1000}$

11. $0.9 = \dfrac{9}{10}$

13. $0.06 = \dfrac{6}{100} = \dfrac{3}{50}$

15. $0.12 = \dfrac{12}{100} = \dfrac{3}{25}$

17. $0.054 = \dfrac{54}{1000} = \dfrac{27}{500}$

19. $2.13 = \dfrac{213}{100}$

21.
$$\begin{array}{r} \$648.01 \\ +\ \$341.06 \\ \hline \$989.07 \end{array}$$

23.
$$\begin{array}{r} {}^{1} \\ 72.030 \\ +\ 847.124 \\ \hline 919.154 \end{array}$$

25.
$$\begin{array}{r} {}^{1} \\ 104.000 \\ +\ 78.103 \\ \hline 182.103 \end{array}$$

27.
$$\begin{array}{r} {}^{1} \\ 0.350 \\ 3.600 \\ +\ 0.127 \\ \hline 4.077 \end{array}$$

29.
$$\begin{array}{r} 6\ 11\ 10 \\ 2\cancel{7}.\cancel{2}\,\cancel{0} \\ -\ 0.\ 3\ \ 5 \\ \hline 26.\ 8\ \ 5 \end{array}$$

31.
$$\begin{array}{r} 9 \\ 8\ \cancel{10}\ 10 \\ \$1\cancel{9}.\cancel{0}\ \cancel{0} \\ -\ \$16.6\ \ 2 \\ \hline \$2.3\ \ 8 \end{array}$$

33.
$$\begin{array}{r} 2\ \ 10 \\ 9.4\cancel{3}\ \cancel{0} \\ -\ 6.40\ \ 6 \\ \hline 3.02\ \ 4 \end{array}$$

35.
$$\begin{array}{r} 9 \\ 7\ 11\ \cancel{10}\ 10 \\ \cancel{8}.\cancel{2}\ \cancel{0}\ \cancel{0} \\ -\ 1.3\ \ 5\ \ 6 \\ \hline 6.8\ \ 4\ \ 4 \end{array}$$

37.
$$\begin{array}{r} 6.0900 \\ +\ 3.0046 \\ \hline 9.0946 \end{array}$$

39.
$$\begin{array}{r} 4.0700 \\ +\ 8.0035 \\ \hline 12.0735 \end{array}$$

41.
$$\begin{array}{rl} 0.613 & \text{3 decimal digits} \\ \times\ 9.2 & \text{1 decimal digit} \\ \hline 1226 & \\ 5517 & \\ \hline 5.6396 & 3 + 1 = 4 \text{ decimal digits} \end{array}$$

43.
$$\begin{array}{rl} 8.7 & \text{1 decimal digit} \\ \times\ 11 & \\ \hline 87 & \\ 87 & \\ \hline 95.7 & \text{1 decimal digit} \end{array}$$

45.
$$\begin{array}{rl} 7.03 & \text{2 decimal digits} \\ \times\ 0.0035 & \text{4 decimal digits} \\ \hline 3515 & \\ 2109 & \\ \hline 0.024605 & 2 + 4 = 6 \text{ decimal digits} \end{array}$$

47.
$$\begin{array}{rl} 3.0012 & \text{4 decimal digits} \\ \times\ 4.3 & \text{1 decimal digit} \\ \hline 90036 & \\ 120048 & \\ \hline 12.90516 & 4 + 1 = 5 \text{ decimal digits} \end{array}$$

49.
$$\begin{array}{rl} 0.0031 & \text{4 decimal digits} \\ \times\ 0.82 & \text{2 decimal digits} \\ \hline 62 & \\ 248 & \\ \hline 0.002542 & 4 + 2 = 6 \text{ decimal digits} \end{array}$$

51.
$$\begin{array}{r} 0.6 \\ 15\overline{)9.0} \\ 9\ 0 \\ \hline 0 \end{array}$$

53.
$$\begin{array}{r} 6.4 \\ 5\overline{)32.0} \\ \underline{30} \\ 2\,0 \\ \underline{2\,0} \\ 0 \end{array}$$

55.
$$\begin{array}{r} 1700. \\ 0005.\overline{)8500.} \\ \underline{5} \\ 35 \\ \underline{35} \\ 0 \end{array}$$

57.
$$\begin{array}{r} 80. \\ 005.\overline{)400.} \\ \underline{40} \\ 0 \end{array}$$

59.
$$\begin{array}{r} 0.046 \\ 60\overline{)2.760} \\ \underline{2\,40} \\ 360 \\ \underline{360} \\ 0 \end{array}$$

61.
$$\begin{array}{r} 0.2 \\ 5\overline{)1.0} \\ \underline{1\,0} \\ 0 \end{array}$$

$\frac{1}{5} = 0.2$

63.
$$\begin{array}{r} 0.875 \\ 8\overline{)7.000} \\ \underline{6\,4} \\ 60 \\ \underline{56} \\ 40 \\ \underline{40} \\ 0 \end{array}$$

$\frac{7}{8} = 0.875$

65.
$$\begin{array}{r} 0.1875 \\ 1\overline{)\ 3.0000} \\ \underline{1\,6} \\ 1\,40 \\ \underline{1\,28} \\ 120 \\ \underline{112} \\ 80 \\ \underline{80} \\ 0 \end{array}$$

$\frac{3}{16} = 0.1875$

67.
$$\begin{array}{r} 0.222... \\ 9\overline{)2.000} \\ \underline{1\,8} \\ 20 \\ \underline{18} \\ 20 \\ \underline{18} \\ 2 \end{array}$$

$\frac{2}{9} = 0.\overline{2}$

69.
$$\begin{array}{r} 0.5454... \\ 11\overline{)6.0000} \\ \underline{5\,5} \\ 50 \\ \underline{44} \\ 60 \\ \underline{55} \\ 50 \\ \underline{44} \\ 6 \end{array}$$

$\frac{6}{11} = 0.\overline{54}$

71.
$$\begin{array}{r} 0.2727... \\ 11\overline{)3.0000} \\ \underline{2\,2} \\ 80 \\ \underline{77} \\ 30 \\ \underline{22} \\ 80 \\ \underline{77} \\ 3 \end{array}$$

$\frac{3}{11} = 0.\overline{27}$

73. $$\begin{array}{r} 0.166\ldots \\ 6\overline{)1.000} \\ \underline{6} \\ 40 \\ \underline{36} \\ 40 \\ \underline{36} \\ 4 \end{array}$$

$\frac{1}{6} = 0.1\overline{6}$

75. $$\begin{array}{r} 1.11\ldots \\ 9\overline{)10.00} \\ \underline{9} \\ 10 \\ \underline{9} \\ 10 \\ \underline{9} \\ 1 \end{array}$$

$\frac{10}{9} = 1.\overline{1}$

77. $33\% = 33. = 0.33$

79. $5\% = 05. = 0.05$

81. $300\% = 300. = 3$

83. $11.8\% = 11.8 = 0.118$

85. $0.5\% = 00.5 = 0.005$

87. $0.05 = 0.05\% = 5\%$

89. $0.39 = 0.39\% = 39\%$

91. $0.416 = 0.416\% = 41.6\%$

93. $0.003 = 0.003\% = 0.3\%$

95. $1.00 = 1.00\% = 100\%$

97. $30\% = \frac{30}{100} = \frac{3}{10}$

99. $6\% = \frac{6}{100} = \frac{3}{50}$

101. $7\% = \frac{7}{100}$

103. $4\frac{1}{2}\% = \frac{4\frac{1}{2}}{100} = \frac{9}{200}$

105. $1\frac{1}{3}\% = \frac{1\frac{1}{3}}{100} = \frac{4}{300} = \frac{1}{75}$

107. $\frac{3}{5} = \frac{3 \times 20}{5 \times 20} = \frac{60}{100} = 60\%$

109. $\frac{1}{2} = \frac{1 \times 50}{2 \times 50} = \frac{50}{100} = 50\%$

111. $$\begin{array}{r} 0.83 \\ 6\overline{)5.00} \\ \underline{4\,8} \\ 20 \\ \underline{18} \\ 2 \end{array}$$

The remainder is 2; $\frac{2}{6} = \frac{1}{3}$.

$\frac{5}{6} = 0.83\frac{1}{3} = 83\frac{1}{3}\%$

113. $\frac{4}{8} = \frac{1}{2} = \frac{1 \times 50}{2 \times 50} = \frac{50}{100} = 50\%$

115. $$\begin{array}{r} 1.33 \\ 3\overline{)4.00} \\ \underline{3} \\ 1\,0 \\ \underline{9} \\ 10 \\ \underline{9} \\ 1 \end{array}$$

The remainder is 1; $\frac{1}{3}$.

$\frac{4}{3} = 1.33\frac{1}{3} = 133\frac{1}{3}\%$

117. Underline 6. The digit to the right of 6 is $2 < 5$, so do not change the underlined digit and drop all the numbers to the right of 6. $27.\underline{6}263 = 27.6$

119. Underline 7. The digit to the right of 7 is $0 < 5$, so do not change the underlined digit and drop all the numbers to the right of 7.
$26.746\underline{7}06 = 26.7467$

121. Underline 9. The digit to the right of 9 is $8 \geq 5$, so add 1 to the underlined digit. Notice when you add 1 to 9, 49 becomes 50. Drop all numbers to the right of 50.
$35.24\underline{9}86 = 35.250$

123. Underline 7. The digit to the right of 7 is $8 \geq 5$, so add 1 to the underlined digit and drop all numbers to the right of 8.
$52.3\underline{7}8 = 52.38$

125. Underline 0. The digit to the right of 0 is $8 \geq 5$, so add 1 to the underlined digit and drop all numbers to the right of 1.
$74.8460\underline{0}8 = 74.84601$

127. $\dfrac{36}{100} = \dfrac{36 \div 4}{100 \div 4} = \dfrac{9}{25}$

129. $\dfrac{480}{1000} = \dfrac{480 \div 40}{1000 \div 40} = \dfrac{12}{25}$

131. $\dfrac{7}{10} = \dfrac{7 \times 10}{10 \times 10} = \dfrac{70}{100} = 70\%$

133. **a.** $0.40 = 0.\underset{\smile\smile}{40}\% = 40\%$

b. $0.40 = \dfrac{40}{100} = \dfrac{4}{10} = \dfrac{2}{5}$

135. **a.** $49\% = \dfrac{49}{100}$

b. $49\% = .\underset{\smile\smile}{49} = 0.49$

137. $\dfrac{2600}{52{,}000} = \dfrac{5}{100} = 5\%$

139. $\dfrac{520}{52{,}000} = \dfrac{1}{100} = 1\%$

Chapter 1 Real Numbers and Their Properties

1.1 Introduction to Algebra

Problems 1.1

1. **a.** The sum of p and q is written as $p+q$.

   b. q minus p is written as $q-p$.

   c. $3q$ plus $5y$ minus 2 is written as $3q+5y-2$.

2. **a.** -3 times x is written as $-3x$.

   b. a times b times c is written as abc.

   c. 5 times a times a is written as $5aa$.

   d. $\frac{1}{2}$ of a is written as $\frac{1}{2}a$.

   e. The product of 5 and a is written as $5a$.

3. **a.** The quotient of a and b is written as $\frac{a}{b}$.

   b. The quotient of b and a is written as $\frac{b}{a}$.

   c. The quotient of $(x-y)$ and z is written as $\frac{x-y}{z}$.

   d. The difference of x and y, divided by the sum of x and y is written as $\frac{x-y}{x+y}$.

4. **a.** Substitute 22 for a and 3 for b in $a+b$.
   $a+b=22+3=25$

   b. $a-b=22-3=19$

   c. $5b=5(3)=15$

   d. $\frac{2a}{b}=\frac{2(22)}{3}=\frac{44}{3}$

   e. $$\begin{aligned}2a-3b&=2(22)-3(3)\\&=44-9\\&=35\end{aligned}$$

5. **a.** Refer to Example 5c of the text. The algebraic expression is $A-S-3050E$.

   b. Write an arithmetic expression for $A-S-3050E$ where $A=\$20{,}000$, $S=\$4750$, and $E=6$.
   $$\begin{aligned}A-S-3050E&=20{,}000-4750-3050(6)\\&=20{,}000-4750-18{,}300\end{aligned}$$

   c. Line 27 directs you to subtract line 26 from line 25 unless line 26 is more than line 25. Line 25 directs you to line 24 (S) from line 22 (A).
   $A-S=20{,}000-4750=\$15{,}250$
   Line 26 directs you to multiply \$3050 by the total number of exceptions E.
   $3050E=3050(6)=\$18{,}300$
   Since line 26 (\$18,300) is more than line 25 (\$15,250), the number you should enter on line 27 is 0.

Exercises 1.1

1. The sum of a and c is written as $a+c$.

3. The sum of $3x$ and y is written as $3x+y$.

5. $9x$ plus $17y$ is written as $9x+17y$.

7. The difference of $3a$ and $2b$ is written as $3a-2b$.

9. $-2x$ less 5 is written as $-2x-5$.

11. 7 times a is written as $7a$.

13. $\frac{1}{7}$ of a is written as $\frac{1}{7}a$.

15. The product of b and d is written as bd.

17. xy multiplied by z is written as xyz.

19. $-b$ times $(c+d)$ is written as $-b(c+d)$.

21. $(a-b)$ times x is written as $(a-b)x$.

23. The product of $(x-3y)$ and $(x+7y)$ is written as $(x-3y)(x+7y)$.

25. $(c-4d)$ times $(x+y)$ is written as $(c-4d)(x+y)$.

27. y divided by $3x$ is written as $\frac{y}{3x}$.

29. The quotient of $2b$ and a is written as $\frac{2b}{a}$.

31. The quotient of a and the sum of x and y is written as $\frac{a}{x+y}$.

33. The quotient of the difference of a and b, and c is written as $\frac{a-b}{c}$.

35. The quotient when x is divided into y is written as $\frac{y}{x}$.

37. The quotient when the sum of p and q is divided into the difference of p and q is written as $\frac{p-q}{p+q}$.

39. The quotient obtained when the sum of x and $2y$ is divided by the difference of x and $2y$ is written as $\frac{x+2y}{x-2y}$.

41. Substitute 7 for a and 9 for c in $a+c$.
$a+c=7+9=16$

43. Substitute 3 for x and 2 for y in $9x+17y$.
$$\begin{aligned}9x+17y &= 9(3)+17(2)\\ &= 27+34\\ &= 61\end{aligned}$$

45. Substitute 5 for a and 3 for b in $3a-2b$.
$$\begin{aligned}3a-2b &= 3(5)-2(3)\\ &= 15-6\\ &= 9\end{aligned}$$

47. Substitute 4 for x in $2x-5$.
$2x-5=2(4)-5=8-5=3$

49. Substitute 2 for a and 4 for b in $7ab$.
$7ab=7(2)(4)=14(4)=56$

51. Substitute 3 for b and 2 for d in bd.
$bd=(3)(2)=6$

53. Substitute 10 for x, 5 for y, and 1 for z in xyz.
$xyz=(10)(5)(1)=50(1)=50$

55. Substitute 3 for a, 1 for x, and 2 for y in $\frac{a}{x+y}$.
$$\frac{a}{x+y}=\frac{3}{1+2}=\frac{3}{3}=1$$

57. Substitute 2 for x and 8 for y in $\frac{y}{x}$.
$$\frac{y}{x}=\frac{8}{2}=4$$

59. Substitute 2 for x and 8 for y in $\frac{x}{y}$.
$$\frac{x}{y}=\frac{2}{8}=\frac{1}{4}$$

61. $V=IR$

63. $P=P_A+P_B+P_C$

65. $D=RT$

67. $E=mc^2$

69. $c^2=a^2+b^2$

71. The word "of" signifies multiplication.

73. Sample answer: "x divided by y" is written as $\frac{x}{y}$, while "x divided into y" is written as $\frac{y}{x}$.

75. The product of 3 and xy is written as $3xy$.

77. The quotient of $3x$ and $2y$ is written as $\frac{3x}{2y}$.

79. The difference of b and c divided by the sum of b and c is written as $\frac{b-c}{b+c}$.

81. Substitute 9 for p and 3 for q in $\frac{p-q}{3}$.

$$\frac{p-q}{3} = \frac{9-3}{3} = \frac{6}{3} = 2$$

1.2 The Real Numbers

Problems 1.2

1. a. The additive inverse of -8 is $-(-8) = 8$.

b. The additive inverse of 9 is -9.

c. The additive inverse of -3 is $-(-3) = 3$.

2. a. The additive inverse of $\frac{3}{11}$ is $-\frac{3}{11}$.

b. The additive inverse of -7.4 is $-(-7.4) = 7.4$.

c. The additive inverse of $-9\frac{8}{13}$ is $-\left(-9\frac{8}{13}\right) = 9\frac{8}{13}$.

d. The additive inverse of 3.4 is -3.4.

3. a. $|-5| = 5$

b. $|10| = 10$

c. $|2| = 2$

d. $-|-5| = -5$

4. a. $\left|-\frac{5}{7}\right| = \frac{5}{7}$

b. $|3.4| = 3.4$

c. $\left|-3\frac{1}{8}\right| = 3\frac{1}{8}$

d. $|-3.8| = 3.8$

e. $-\left|-\frac{1}{5}\right| = -\frac{1}{5}$

5. a. -9 is an integer, a rational number, and a real number.

b. 200 is a whole number, an integer, a rational number, and a real number.

c. $\sqrt{7}$ is an irrational number and a real number.

d. 0.9 is a terminating decimal, so it is a rational number and a real number.

e. 0.010010001... never terminates and never repeats, so it is an irrational number and a real number.

f. 0.010010001 terminates, so it is a rational number and a real number.

6. a. $\frac{1}{2}$

b. $-80°$F

c. -1850

7. a. 20,320 feet above sea level

b. 1349 feet below sea level

c. 36,198 feet below sea level

Exercises 1.2

1. -4

3. $-(-49) = 49$

5. $-\frac{7}{3}$

7. $-(-6.4) = 6.4$

9. $-3\frac{1}{7}$

11. -0.34

13. $-(-0.\overline{5}) = 0.\overline{5}$

15. $-\sqrt{7}$

17. $-\pi$

19. $|-2| = 2$

21. $|48| = 48$

23. $|-(-3)| = |3| = 3$

25. $\left|-\frac{4}{5}\right| = \frac{4}{5}$

27. $|-3.4| = 3.4$

29. $\left|-1\frac{1}{2}\right| = 1\frac{1}{2}$

31. $-\left|-\frac{3}{4}\right| = -\frac{3}{4}$

33. $-|-0.\overline{5}| = -0.\overline{5}$

35. $-|-\sqrt{3}| = -\sqrt{3}$

37. $-|-\pi| = -\pi$

39. 17 is a natural number, a whole number, an integer, a rational number, and a real number.

41. $-\frac{4}{5}$ is a rational number and a real number.

43. 0 is a whole number, an integer, a rational number, and a real number.

45. 3.76 is a terminating decimal, so it is a rational number and a real number.

47. $17.\overline{28}$ is a repeating decimal, so it is a rational number and a real number.

49. $-\sqrt{3}$ is an irrational number and a real number.

51. −0.888... is a repeating decimal, so it is a rational number and a real number.

53. 0.202002000 is a terminating decimal, so it is a rational number and a real number.

55. Natural numbers: 8

57. Positive integers: 8

59. Nonnegative integers: 0, 8

61. Rational numbers: −5, $\frac{1}{5}$, 0, 8, $0.\overline{1}$, 3.666...

63. True

65. False; $|0| = 0$

67. True

69. False; sample answer: $\frac{3}{5}$ is a rational number, but is not an integer.

71. False; sample answer: 0.12345... is not rational. Every nonterminating but repeating decimal is rational.

73. True

75. 20 yards

77. −1312 feet

79. −4°F to 45°F

81.
$$\begin{array}{r} 8.6 \\ -\,3.4 \\ \hline 5.2 \end{array}$$

83. Find the LCD.
LCD $= 2\cdot 2\cdot 2\cdot 5 = 40$
Write each fraction with 40 as the denominator.
$\frac{5}{8}=\frac{5\cdot 5}{8\cdot 5}=\frac{25}{40}$
$\frac{2}{5}=\frac{2\cdot 8}{5\cdot 8}=\frac{16}{40}$
Thus,
$\frac{5}{8}-\frac{2}{5}=\frac{25}{40}-\frac{16}{40}=\frac{9}{40}$

85.
$$\begin{array}{r} 3.1 \\ \times\, 4.2 \\ \hline 62 \\ 124 \\ \hline 13.02 \end{array}$$

87. $\frac{3}{4}\cdot\frac{5}{2}=\frac{3\cdot 5}{4\cdot 2}=\frac{15}{8}$

89. Possible answer: −43°F

91. Possible answer: −40°F

93. a. Sample answer: The additive inverse of a number is a number whose distance from the origin is the same as the given number but in the opposite direction.

b. Sample answer: The absolute value of a number is the distance of the given number from the origin.

95. Sample answer: All numbers with either terminating or repeating decimal representations.

97. Answers may vary.

99. $-(-\sqrt{19})=\sqrt{19}$

101. $-\left(-8\frac{1}{4}\right)=8\frac{1}{4}$

103. $|-0.\overline{4}|=0.\overline{4}$

105. $-\left|-\frac{1}{2}\right|=-\frac{1}{2}$

107. −2 is an integer, a rational number, and a real number.

109. $4\frac{1}{3}$ is a rational number and a real number.

111. 0 is a whole number, an integer, a rational number, and a real number.

1.3 Adding and Subtracting Real Numbers

Problems 1.3

1. Start at zero. Move 4 units to the right and then 2 units to the left. The result is 2. Thus $4+(-2)=2$, as shown below.

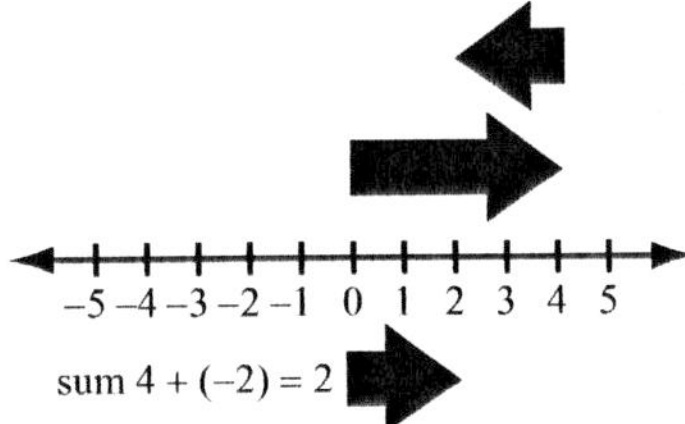

2. Start at zero. Move 2 units to the left and then 1 more unit to the left. The result is −3. Thus $-2+(-1)=-3$, as shown below.

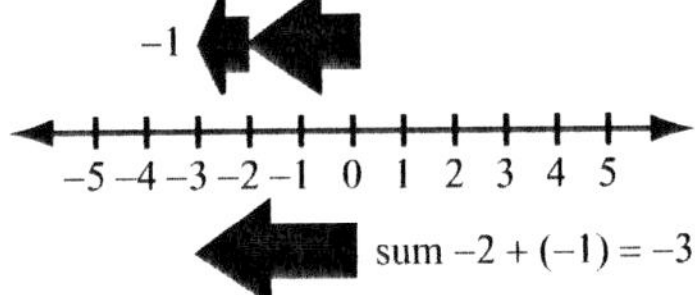

3. a. $-10+4=-(10-4)=-6$

b. $10+(-4)=+(10-4)=6$

4. a. $-7.8+2.5=-(7.8-2.5)=-5.3$

b. $5.4+(-7.8)=-(7.8-5.4)=-2.4$

c. $-3.4+(-5.1)=-(3.4+5.1)=-8.5$

5. a. $-\frac{4}{9}+\frac{5}{9}=+\left(\frac{5}{9}-\frac{4}{9}\right)=\frac{1}{9}$

b. Find the LCM of 5 and 4.
LCM = 20

Write $\frac{2}{5}=\frac{8}{20}$ and $-\frac{3}{4}=-\frac{15}{20}$.

$$\begin{aligned}\frac{2}{5}+\left(-\frac{3}{4}\right)&=\frac{8}{20}+\left(-\frac{15}{20}\right)\\&=-\left(\frac{15}{20}-\frac{8}{20}\right)\\&=-\frac{7}{20}\end{aligned}$$

6. a. $15-8=15+(-8)=7$

b. $-23-5=-23+(-5)=-28$

c. $-12-(-4)=-12+4=-8$

d. $-5-(-7)=-5+7=2$

7. a. $3.2-(-2.1)=3.2+2.1=5.3$

b. $-3.4-(-6.9)=-3.4+6.9=3.5$

c. $\frac{2}{7}-\left(-\frac{3}{7}\right)=\frac{2}{7}+\frac{3}{7}=\frac{5}{7}$

d. LCD is 8.

Write $\frac{9}{4}=\frac{18}{8}$.

$$\begin{aligned}-\frac{5}{8}-\frac{9}{4}&=-\frac{5}{8}-\frac{18}{8}\\&=-\frac{5}{8}+\left(-\frac{18}{8}\right)\\&=-\frac{23}{8}\end{aligned}$$

8. $$\begin{aligned}&14-(-15)+10-23-15\\&=14+15+10+(-23)+(-15)\\&=14+10+(-23)\\&=24+(-23)\\&=1\end{aligned}$$

9. We have to find the difference between 63 and −21; that is, we have to find $63-(-21)$.
$63-(-21)=63+21=84$
The temperature fell 84°F.

Exercises 1.3

1. $3+3=6$

3. $-5+1=-4$

5. $6+(-5)=6-5=1$

7. $-2+(-5)=-(2+5)=-7$

9. $3+(-3)=3-3=0$

11. $-18+21=21-18=3$

13. $19+(-6)=19-6=13$

15. $-9+11=11-9=2$

17. $-18+9=-(18-9)=-9$

19. $-17+(+5)=-(17-5)=-12$

21. $-3.8+6.9=6.9-3.8=3.1$

23. $-7.8+(3.1)=-(7.8-3.1)=-4.7$

25. $3.2+(-8.6)=-(8.6-3.2)=-5.4$

27. $-3.4+(-5.2)=-(3.4+5.2)=-8.6$

29. $-\frac{2}{7}+\frac{5}{7}=\frac{5}{7}-\frac{2}{7}=\frac{3}{7}$

31. $-\frac{3}{4}+\frac{1}{4}=-\left(\frac{3}{4}-\frac{1}{4}\right)=-\frac{2}{4}=-\frac{1}{2}$

33. LCD is 12.

$\frac{3}{4}=\frac{9}{12}$ and $-\frac{5}{6}=-\frac{10}{12}$

$\frac{3}{4}+\left(-\frac{5}{6}\right)=\frac{9}{12}+\left(-\frac{10}{12}\right)=-\left(\frac{10}{12}-\frac{9}{12}\right)=-\frac{1}{12}$

35. LCD is 12.

$-\frac{1}{6}=-\frac{2}{12}$ and $\frac{3}{4}=\frac{9}{12}$

$-\frac{1}{6}+\frac{3}{4}=-\frac{2}{12}+\frac{9}{12}=\frac{9}{12}-\frac{2}{12}=\frac{7}{12}$

37. LCD is 21.

$-\frac{1}{3}=-\frac{7}{21}$ and $-\frac{2}{7}=-\frac{6}{21}$

$$-\frac{1}{3}+\left(-\frac{2}{7}\right)=-\frac{7}{21}+\left(-\frac{6}{21}\right)$$
$$=-\left(\frac{7}{21}+\frac{6}{21}\right)$$
$$=-\frac{13}{21}$$

39. LCD is 18.

$-\frac{5}{6}=-\frac{15}{18}$ and $-\frac{8}{9}=-\frac{16}{18}$

$$-\frac{5}{6}+\left(-\frac{8}{9}\right)=-\frac{15}{18}+\left(-\frac{16}{18}\right)$$
$$=-\left(\frac{15}{18}+\frac{16}{18}\right)$$
$$=-\frac{31}{18}$$

41. $-5-11=-5+(-11)=-16$

43. $-4-16=-4+(-16)=-20$

45. $7-13=7+(-13)=-6$

47. $9-(-7)=9+7=16$

49. $0-4=0+(-4)=-4$

51. $-3.8-(-1.2)=-3.8+1.2=-2.6$

53. $-3.5-(-8.7)=-3.5+8.7=5.2$

55. $4.5-8.2=4.5+(-8.2)=-3.7$

57. $\frac{3}{7}-\left(-\frac{1}{7}\right)=\frac{3}{7}+\frac{1}{7}=\frac{4}{7}$

59. LCD is 12.

$-\frac{5}{4}=-\frac{15}{12}$ and $\frac{7}{6}=\frac{14}{12}$

$$-\frac{5}{4}-\frac{7}{6}=-\frac{15}{12}-\frac{14}{12}$$
$$=-\frac{15}{12}+\left(-\frac{14}{12}\right)$$
$$=-\frac{29}{12}$$

61. $8-(-10)+5-20-10$
$$=8+10+5+(-20)+(-10)$$
$$=8+5+(-20)$$
$$=13+(-20)$$
$$=-7$$

63. $-15+12-8-(-15)+5$
$$=-15+12+(-8)+15+5$$
$$=12+(-8)+5$$
$$=4+5$$
$$=9$$

65. $-10+9-14-3-(-14)$
$$=-10+9+(-14)+(-3)+14$$
$$=-10+9+(-3)$$
$$=-1+(-3)$$
$$=-4$$

67. $5000-1500=3500$
The difference in temperature is 3500°C.

69. $47+(+1)+(+2)+(-1)+(-2)+(-1)$
$$=47+(-1)$$
$$=46$$
The price of the stock at the end of the week is $46.

71. $15+(+2)+(+1)+(-1)+(-3)$
$$=15+(+2)+(-3)$$
$$=17+(-3)$$
$$=14$$
The temperature is 14°C.

73. $6\frac{1}{6}\cdot 5\frac{7}{10}=\frac{37}{\cancel{6}_{2}}\cdot\frac{\cancel{57}^{19}}{10}=\frac{703}{20}$ or $35\frac{3}{20}$

75. $\frac{18}{11} \div 5\frac{1}{2} = \frac{18}{11} \div \frac{11}{2} = \frac{18}{11} \cdot \frac{2}{11} = \frac{36}{121}$

77. $476 - (-323) = 476 + 323 = 799$
799 years

79. $1776 - 1492 = 284$
284 years

81. $1939 - (-323) = 1939 + 323 = 2262$
2262 years

83. Answers may vary.

85. Answers may vary.

87. $3.8 - 6.9 = -(6.9 - 3.8) = -3.1$

89. $-3.5 - 4.2 = -(3.5 + 4.2) = -7.7$

91. $-5 - 15 = -(5 + 15) = -20$

93. $-3.4 - (-4.6) = -3.4 + 4.6 = 1.2$

95. $3.9 + (-4.2) = -(4.2 - 3.9) = -0.3$

97. $-3.2 + (-2.5) = -(3.2 + 2.5) = -5.7$

99. LCD is 20.
$\frac{3}{4} = \frac{15}{20}, \frac{2}{5} = \frac{8}{20}, \frac{4}{5} = \frac{16}{20}$, and $\frac{5}{4} = \frac{25}{20}$

$$\begin{aligned}
&\frac{3}{4} - \left(-\frac{2}{5}\right) + \frac{4}{5} - \frac{2}{5} + \frac{5}{4} \\
&= \frac{15}{20} - \left(-\frac{8}{20}\right) + \frac{16}{20} - \frac{8}{20} + \frac{25}{20} \\
&= \frac{15}{20} + \frac{8}{20} + \frac{16}{20} + \left(-\frac{8}{20}\right) + \frac{25}{20} \\
&= \frac{15}{20} + \frac{16}{20} + \frac{25}{20} \\
&= \frac{31}{20} + \frac{25}{20} \\
&= \frac{56}{20} \\
&= \frac{14}{5}
\end{aligned}$$

1.4 Multiplying and Dividing Real Numbers

Problems 1.4

1. a. 9 and 6 have the same sign; the product is positive.
$9 \cdot 6 = 54$

b. -7 and 8 have different signs; the product is negative.
$-7 \cdot 8 = -56$

c. 5 and -4 have different signs; the product is negative.
$5 \cdot (-4) = -20$

d. -5 and -6 have the same sign; the product is positive.
$-5 \cdot (-6) = 30$

2. a. -4.1 and 3.2 have different signs; the product is negative.
$-4.1 \cdot 3.2 = -13.12$

b. -1.3 and -4.2 have the same sign; the product is positive.
$-1.3(-4.2) = 5.46$

c. $-\frac{3}{5}$ and $-\frac{7}{10}$ have the same sign; the product is positive.
$-\frac{3}{5} \cdot \left(-\frac{7}{10}\right) = \frac{21}{50}$

d. $\frac{5}{9}$ and $-\frac{6}{7}$ have different signs; the product is negative.
$\frac{5}{9} \cdot \left(-\frac{6}{7}\right) = -\frac{30}{63} = -\frac{10}{21}$

3. a. The base is −7.
$(-7)^2 = (-7)(-7) = 49$

b. The base is 7.
$-7^2 = -(7 \cdot 7) = -49$

c. The base is $-\frac{1}{4}$.
$\left(-\frac{1}{4}\right)^2 = \left(-\frac{1}{4}\right)\left(-\frac{1}{4}\right) = \frac{1}{16}$

d. The base is $\frac{1}{4}$.
$-\left(\frac{1}{4}\right)^2 = -\left(\frac{1}{4} \cdot \frac{1}{4}\right) = -\frac{1}{16}$

4. a. The base is −4.
$(-4)^3 = (-4)(-4)(-4) = 16(-4) = -64$

b. The base is 4.
$-4^3 = -(4 \cdot 4 \cdot 4) = -(64) = -64$

5. a. 56 and 8 have the same sign; the quotient is positive.
$56 \div 8 = 7$

b. 36 and −4 have different signs; the quotient is negative.
$\frac{36}{-4} = -9$

c. −49 and −7 have the same sign; the quotient is positive.
$\frac{-49}{-7} = 7$

d. −24 and 6 have different signs; the quotient is negative.
$-24 \div 6 = -4$

e. $-7 \div 0$ is not defined.

6. a. $\frac{4}{5}$ and $-\frac{3}{4}$ have different signs; the quotient is negative.
$\frac{4}{5} \div \left(-\frac{3}{4}\right) = \frac{4}{5} \cdot \left(-\frac{4}{3}\right) = -\frac{16}{15}$

b. $-\frac{5}{6}$ and $-\frac{7}{4}$ have the same sign; the quotient is positive.
$-\frac{5}{6} \div \left(-\frac{7}{4}\right) = -\frac{5}{6} \cdot \left(-\frac{4}{7}\right) = \frac{20}{42} = \frac{10}{21}$

c. $-\frac{4}{7}$ and $\frac{8}{7}$ have different signs; the quotient is negative.
$-\frac{4}{7} \div \frac{8}{7} = -\frac{4}{7} \cdot \frac{7}{8} = -\frac{28}{56} = -\frac{1}{2}$

7. a. Your starting speed is $s = 55$ miles per hour, your final speed is $f = 70$ miles per hour, and the time period is $t = 5$ seconds.
$a = \frac{(70-55)\text{ mi/hr}}{5\text{ sec}} = \frac{15\text{ mi/hr}}{5\text{ sec}} = 3\ \frac{\text{mi/hr}}{\text{sec}}$
Thus your acceleration is 3 miles per hour each second.

b. Your starting speed is $s = 50$ miles per hour, your final speed is $f = 40$ miles per hour, and the time period is $t = 5$ seconds.
$a = \frac{(40-50)\text{ mi/hr}}{5\text{ sec}} = \frac{-10\text{ mi/hr}}{5\text{ sec}} = -2\ \frac{\text{mi/hr}}{\text{sec}}$
Thus your acceleration is −2 miles per hour each second.

Exercises 1.4

1. $4 \cdot 9 = 36$

3. $-10 \cdot 4 = -40$

5. $-9 \cdot 9 = -81$

7. $-6 \cdot (-3)(-2) = 18 \cdot (-2) = -36$

9. $-9 \cdot (-2)(-3) = 18 \cdot (-3) = -54$

11. $-2.2(3.3) = -7.26$

13. $-1.3(-2.2) = 2.86$

15. $\frac{5}{6}\left(-\frac{5}{7}\right) = -\frac{25}{42}$

17. $-\frac{3}{5}\left(-\frac{5}{12}\right) = \frac{15}{60} = \frac{1}{4}$

19. $-\frac{6}{7}\left(\frac{35}{8}\right) = -\frac{210}{56} = -\frac{15}{4}$

21. $-4^2 = -4 \cdot 4 = -16$

23. $(-5)^2 = (-5)(-5) = 25$

25. $-5^3 = -5 \cdot 5 \cdot 5 = -25 \cdot 5 = -125$

27. $(-6)^4 = (-6)(-6)(-6)(-6)$
$= 36(-6)(-6)$
$= -216(-6)$
$= 1296$

29. $-\left(\frac{1}{2}\right)^5 = -\left(\frac{1}{2} \cdot \frac{1}{2} \cdot \frac{1}{2} \cdot \frac{1}{2} \cdot \frac{1}{2}\right)$
$= -\left(\frac{1}{32}\right)$
$= -\frac{1}{32}$

31. $\frac{14}{2} = 7$

33. $-50 \div 10 = -5$

35. $\frac{-30}{10} = -3$

37. $\frac{-0}{3} = 0$

39. $-5 \div 0$ is undefined.

41. $\frac{0}{7} = 0$

43. $-15 \div (-3) = 5$

45. $\frac{-25}{-5} = 5$

47. $\frac{18}{-9} = -2$

49. $30 \div (-5) = -6$

51. $\frac{3}{5} \div \left(-\frac{4}{7}\right) = \frac{3}{5} \cdot \left(-\frac{7}{4}\right) = -\frac{21}{20}$

53. $-\frac{2}{3} \div \left(-\frac{7}{6}\right) = -\frac{2}{3} \cdot \left(-\frac{6}{7}\right) = \frac{12}{21} = \frac{4}{7}$

55. $-\frac{5}{8} \div \frac{7}{8} = -\frac{5}{8} \cdot \frac{8}{7} = -\frac{40}{56} = -\frac{5}{7}$

57. $\frac{-3.1}{6.2} = -\frac{31}{62} = -\frac{1}{2}$ or -0.5

59. $\frac{-1.6}{-9.6} = \frac{16}{96} = \frac{1}{6}$ or $0.1\overline{6}$

61. $2(+45) + 5(-15) = 90 + (-75) = 15$
The person gains 15 calories.

63. $2(+45) + 2(+65) + 15(-15)$
$= 90 + 130 + (-225)$
$= 220 + (-225)$
$= -5$
The person loses 5 calories.

65. First find the number of calories the person gains in eating 2 beef franks.
$2(+45) = 90$
The person must *spend* -90 calories. Divide -90 by -15.
$\frac{-90}{-15} = 6$
The person has to run for 6 minutes.

67. $s = 0$ mi/hr, $f = 60$ mi/hr, $t = 3.89$ sec

$$a = \frac{(60-0)\text{ mi/hr}}{3.89\text{ sec}} = \frac{60\text{ mi/hr}}{3.89\text{ sec}} \approx 15.4\ \frac{\text{mi/hr}}{\text{sec}}$$

69. $\$3.05 + 2(\$0.70) = \$3.05 + \$1.40 = \$4.45$

71. $3 \cdot \frac{1}{3} = \frac{3}{3} = 1$

73. $7 + (-7) = 0$

75. $(0.54) \cdot (-2.5) = -1.35$

77. $(1.45) \cdot (-2.1) = -3.045$

79. $(1.25) \cdot (3.1) = 3.875$

81. Sample answer: $(-a)^2 = (-a)(-a)$ is the product of two numbers with the same sign. Therefore it is positive.

83. Sample answer: Suppose $\frac{a}{0} = b$. Then $a = b \cdot 0 = 0$ for any a, which is not possible. Therefore $\frac{a}{0}$ is not defined.

85. $\frac{-3.1}{-12.4} = \frac{3.1}{12.4} = \frac{31}{124} = \frac{1}{4}$ or 0.25

87. $-3 \cdot 11 = -33$

89. $-5 \cdot (-11) = 55$

91. $(-8)^2 = (-8)(-8) = 64$

93. $-2.2(3.2) = -7.04$

95. $-\frac{3}{7}\left(-\frac{4}{5}\right) = \frac{12}{35}$

97. $\frac{3}{5} \div \left(-\frac{4}{7}\right) = \frac{3}{5} \cdot \left(-\frac{7}{4}\right) = -\frac{21}{20}$

99. $-\frac{4}{5} \div \frac{8}{5} = -\frac{4}{5} \cdot \frac{5}{8} = -\frac{20}{40} = -\frac{1}{2}$

1.5 Order of Operations

Problems 1.5

1. a.

$$\begin{aligned} & 3 \cdot 8 - 9 \\ &= 24 - 9 \\ &= 15 \end{aligned}$$

b.

$$\begin{aligned} & 22 + 4 \cdot 9 \\ &= 22 + 36 \\ &= 58 \end{aligned}$$

2. a.

$$\begin{aligned} & 56 \div 8 - (3+1) \\ &= 56 \div 8 - 4 \\ &= 7 - 4 \\ &= 3 \end{aligned}$$

b.

$$\begin{aligned} & 54 \div 3^3 + 5 - 2 \\ &= 54 \div 27 + 5 - 2 \\ &= 2 + 5 - 2 \\ &= 7 - 2 \\ &= 5 \end{aligned}$$

3.

$$\begin{aligned} & 10 \div 5 \cdot 2 + 4(5-3) - 4 \cdot 2 \\ &= 10 \div 5 \cdot 2 + 4(2) - 4 \cdot 2 \\ &= 2 \cdot 2 + 4(2) - 4 \cdot 2 \\ &= 4 + 4(2) - 4 \cdot 2 \\ &= 4 + 8 - 4 \cdot 2 \\ &= 4 + 8 - 8 \\ &= 12 - 8 \\ &= 4 \end{aligned}$$

4.

$$\begin{aligned} & 50 \div 5 + \{2 \cdot 7 - [4 + (5-2)]\} \\ &= 50 \div 5 + \{2 \cdot 7 - [4+3]\} \\ &= 50 \div 5 + \{2 \cdot 7 - 7\} \\ &= 50 \div 5 + \{14 - 7\} \\ &= 50 \div 5 + 7 \\ &= 10 + 7 \\ &= 17 \end{aligned}$$

5. $-3^2+\dfrac{2(3-6)}{3}+20\div 5$

$=-9+\dfrac{2(3-6)}{3}+20\div 5$

$=-9+\dfrac{2(-3)}{3}+20\div 5$

$=-9+\dfrac{-6}{3}+20\div 5$

$=-9+(-2)+20\div 5$

$=-9+(-2)+4$

$=-11+4$

$=-7$

6. Ideal rate $=[(205-A)\cdot 7]\div 10$

$=[(205-20)\cdot 7]\div 10$

$=[185\cdot 7]\div 10$

$=1295\div 10$

$=129.5$

Rounded to the nearest whole number, the ideal heart rate is 130.

Exercises 1.5

1. $4\cdot 5+6=20+6=26$

3. $7+3\cdot 2=7+6=13$

5. $7\cdot 8-3=56-3=53$

7. $20-3\cdot 5=20-15=5$

9. $48\div 6-(3+2)=48\div 6-5$

$=8-5$

$=3$

11. $3\cdot 4\div 2+(6-2)=3\cdot 4\div 2+4$

$=12\div 2+4$

$=6+4$

$=10$

13. $36\div 3^2+4-1=36\div 9+4-1$

$=4+4-1$

$=8-1$

$=7$

15. $8\div 2^3-3+5=8\div 8-3+5$

$=1-3+5$

$=-2+5$

$=3$

17. $10\div 5\cdot 2+8\cdot(6-4)-3\cdot 4$

$=10\div 5\cdot 2+8\cdot 2-3\cdot 4$

$=2\cdot 2+8\cdot 2-3\cdot 4$

$=4+8\cdot 2-3\cdot 4$

$=4+16-3\cdot 4$

$=4+16-12$

$=20-12$

$=8$

19. $4\cdot 8\div 2-3(4-1)+9\div 3$

$=4\cdot 8\div 2-3(3)+9\div 3$

$=32\div 2-3(3)+9\div 3$

$=16-3(3)+9\div 3$

$=16-9+9\div 3$

$=16-9+3$

$=7+3$

$=10$

21. $20\div 5+\{3\cdot 4-[4+(5-3)]\}$

$=20\div 5+\{3\cdot 4-[4+2]\}$

$=20\div 5+\{3\cdot 4-6\}$

$=20\div 5+\{12-6\}$

$=20\div 5+6$

$=4+6$

$=10$

23. $(20-15)\cdot[20\div 2-(2\cdot 2+2)]$

$=5\cdot[20\div 2-(2\cdot 2+2)]$

$=5\cdot[20\div 2-(4+2)]$

$=5\cdot[20\div 2-6]$

$=5\cdot[10-6]$

$=5\cdot 4$

$=20$

25. $\{4\div 2\cdot 6-(3+2\cdot 3)+[5(3+2)-1]\}$
$=\{4\div 2\cdot 6-(3+6)+[5(3+2)-1]\}$
$=\{4\div 2\cdot 6-9+[5(3+2)-1]\}$
$=\{4\div 2\cdot 6-9+[5(5)-1]\}$
$=\{4\div 2\cdot 6-9+[25-1]\}$
$=\{4\div 2\cdot 6-9+24\}$
$=\{2\cdot 6-9+24\}$
$=\{12-9+24\}$
$=\{3+24\}$
$=27$

27. $-7^2+\frac{3(8-4)}{4}+10\div 2\cdot 3$
$=-49+\frac{3(8-4)}{4}+10\div 2\cdot 3$
$=-49+\frac{3(4)}{4}+10\div 2\cdot 3$
$=-49+\frac{12}{4}+10\div 2\cdot 3$
$=-49+3+10\div 2\cdot 3$
$=-49+3+5\cdot 3$
$=-49+3+15$
$=-46+15$
$=-31$

29. $(-6)^2\cdot 4\div 4-\frac{3(7-9)}{2}-4\cdot 3\div 2^2$
$=(-6)^2\cdot 4\div 4-\frac{3(-2)}{2}-4\cdot 3\div 2^2$
$=(-6)^2\cdot 4\div 4-\frac{-6}{2}-4\cdot 3\div 2^2$
$=36\cdot 4\div 4-\frac{-6}{2}-4\cdot 3\div 2^2$
$=36\cdot 4\div 4-\frac{-6}{2}-4\cdot 3\div 4$
$=144\div 4-\frac{-6}{2}-4\cdot 3\div 4$
$=36-\frac{-6}{2}-4\cdot 3\div 4$
$=36-(-3)-4\cdot 3\div 4$
$=36-(-3)-12\div 4$
$=36-(-3)-3$
$=39-3$
$=36$

31. $\left[\frac{7-(-3)}{8-6}\right]\left[\frac{3+(-8)}{7-2}\right]=\left[\frac{10}{8-6}\right]\left[\frac{3+(-8)}{7-2}\right]$
$=\left[\frac{10}{2}\right]\left[\frac{3+(-8)}{7-2}\right]$
$=[5]\left[\frac{3+(-8)}{7-2}\right]$
$=[5]\left[\frac{-5}{7-2}\right]$
$=[5]\left[\frac{-5}{5}\right]$
$=[5][-1]$
$=-5$

33. $\frac{(-3)(-2)(-4)}{3(-2)}-(-3)^2=\frac{6(-4)}{3(-2)}-(-3)^2$
$=\frac{-24}{3(-2)}-(-3)^2$
$=\frac{-24}{-6}-(-3)^2$
$=\frac{-24}{-6}-9$
$=4-9$
$=-5$

35. $\frac{(-10)(-6)}{(-2)(-3)(-5)}-(-3)^3$
$=\frac{60}{(-2)(-3)(-5)}-(-3)^3$
$=\frac{60}{6(-5)}-(-3)^3$
$=\frac{60}{-30}-(-3)^3$
$=\frac{60}{-30}-(-27)$
$=-2-(-27)$
$=25$

37. $\frac{(-2)^3(-3)(-9)}{(-2)^2(-3)^2}-3^2=\frac{-8(-3)(-9)}{(-2)^2(-3)^2}-3^2$

$=\frac{24(-9)}{(-2)^2(-3)^2}-3^2$

$=\frac{-216}{(-2)^2(-3)^2}-3^2$

$=\frac{-216}{4(-3)^2}-3^2$

$=\frac{-216}{4(9)}-3^2$

$=\frac{-216}{36}-3^2$

$=\frac{-216}{36}-9$

$=-6-9$

$=-15$

39. $-5^2-\frac{(-2)(-3)(-4)}{(-12)(-2)}=-5^2-\frac{6(-4)}{(-12)(-2)}$

$=-5^2-\frac{-24}{(-12)(-2)}$

$=-5^2-\frac{-24}{24}$

$=-25-\frac{-24}{24}$

$=-25-(-1)$

$=-24$

41. $\frac{R+M}{2}=\frac{92+82}{2}=\frac{174}{2}=87$

43. a. $0.72(220-A)=0.72(220-20)$

$=0.72(200)$

$=144$

b. $0.72(220-A)=0.72(220-45)$

$=0.72(175)$

$=126$

45. $11\cdot\frac{1}{11}=1$

47. $-2.3+2.3=0$

49. $-2.5\cdot\frac{1}{-2.5}=1$

51. (Age in months · Adult dose) ÷ 150

$=(10\cdot 75)\div 150$

$=750\div 150$

$=5$

Child's dose is 5 milligrams.

53. (Age · Adult dose) ÷ (Age + 12)

$=(6\cdot 4)\div(6+12)$

$=24\div 18$

$=\frac{24}{18}$

$=\frac{4}{3}$ or $1\frac{1}{3}$

Child's dose is $1\frac{1}{3}$ tablets every 12 hours.

55. a. Sample answer: Parentheses are not needed because multiplication is done before addition.

$2+(3\cdot 4)=2+3\cdot 4$

b. Sample answer: Parentheses are needed because multiplication is done before addition.

$2\cdot(3+4)\neq 2\cdot 3+4$

57. $63\div 9-(2+5)=63\div 9-7=7-7=0$

59. $16+2^3+3-9=16+8+3-9$

$=24+3-9$

$=27-9$

$=18$

61. $18+4\cdot 5=18+20=38$

63. $15\div 3+\{2\cdot 4-[6+(2-5)]\}$

$=15\div 3+\{2\cdot 4-[6+(-3)]\}$

$=15\div 3+\{2\cdot 4-3\}$

$=15\div 3+\{8-3\}$

$=15\div 3+5$

$=5+5$

$=10$

65. $[(205-A)\cdot 7]\div 10=[(205-35)\cdot 7]\div 10$

$=[170\cdot 7]\div 10$

$=1190\div 10$

$=119$

The ideal heart rate is 119.

1.6 Properties of Real Numbers

Problems 1.6

1. a. We changed the *grouping* of the numbers. The associative property of multiplication was used.

b. We changed the *order* of multiplication. The commutative property of multiplication was used.

c. We changed the *grouping* of the numbers. The associative property of addition was used.

d. We changed the *order* of 3 and 4 within the parentheses. The commutative property of addition was used.

2. a. Commutative property of multiplication:
$5\cdot\frac{1}{3}=\frac{1}{3}\cdot\underline{5}$

b. Associative property of multiplication:
$\frac{1}{3}\cdot(4\cdot\underline{7})=\left(\frac{1}{3}\cdot 4\right)\cdot 7$

c. Commutative property of addition:
$\underline{\frac{1}{3}}+\frac{1}{5}=\frac{1}{5}+\frac{1}{3}$

d. Associative property of addition:
$(-8+\underline{4})+3.1=-8+(4+3.1)$

3. $$\begin{aligned}3+5x+8&=(3+5x)+8\\&=8+(3+5x)\\&=(8+3)+5x\\&=11+5x \text{ or } 5x+11\end{aligned}$$

4. a. Identity element for multiplication

b. Multiplicative inverse

c. Identity element for addition

d. Additive inverse

5. a. Identity element for addition:
$\frac{7}{3}+\underline{0}=\frac{7}{3}$

b. Multiplicative inverse property:
$\frac{4}{9}\cdot\underline{\frac{9}{4}}=1$

c. Additive inverse property:
$\frac{5}{8}+\underline{\left(-\frac{5}{8}\right)}=0$

d. Identity element for multiplication:
$2.4\cdot\underline{1}=2.4$

e. Additive inverse property:
$9.1+\underline{(-9.1)}=0$

f. Identity element for multiplication:
$0.2\cdot\underline{1}=0.2$

6. $$\begin{aligned}-5x+7+5x&=(-5x+7)+5x\\&=5x+(-5x+7)\\&=[5x+(-5x)]+7\\&=0+7\\&=7\end{aligned}$$

7. a. $(a+b)c=ac+bc$

b. $5(a+b+c)=5a+5b+5c$

c. $$\begin{aligned}4(a+7)&=4a+4\cdot 7\\&=4a+28\end{aligned}$$

d. $$\begin{aligned}3(2+9)&=3\cdot 2+3\cdot 9\\&=6+27\\&=33\end{aligned}$$

8. a. $$\begin{aligned}-3(x+5)&=-3x+(-3)\cdot 5\\&=-3x+(-15)\\&=-3x-15\end{aligned}$$

b. $$\begin{aligned}-5(a-2)&=-5a+(-5)(-2)\\&=-5a+10\end{aligned}$$

c. $$\begin{aligned}-(x-4)&=-1\cdot(x-4)\\&=-1\cdot x+(-1)(-4)\\&=-x+4\end{aligned}$$

Exercises 1.6

1. Commutative property of addition

3. Commutative property of multiplication

5. Distributive property

7. Commutative property of multiplication

9. Associative property of addition

11. Associative property of multiplication:
$-3+(5+1.5)=(-3+\underline{5})+1.5$

13. Commutative property of multiplication:
$7\cdot\frac{1}{8}=\frac{1}{8}\cdot\underline{7}$

15. Commutative property of addition:
$\underline{6.5}+\frac{3}{4}=\frac{3}{4}+6.5$

17. Associative property of multiplication:
$\left(\frac{3}{7}\cdot 5\right)\cdot\underline{2}=\frac{3}{7}\cdot(5\cdot 2)$

19. $$\begin{aligned}5+2x+8&=(5+2x)+8\\&=8+(5+2x)\\&=(8+5)+2x\\&=13+2x\end{aligned}$$

21. $$\begin{aligned}\frac{1}{5}a-3+3-\frac{1}{5}a&=\left(\frac{1}{5}a-3\right)+3-\frac{1}{5}a\\&=\frac{1}{5}a+(-3+3)-\frac{1}{5}a\\&=\frac{1}{5}a+0-\frac{1}{5}a\\&=\frac{1}{5}a-\frac{1}{5}a\\&=0\end{aligned}$$

23. $$\begin{aligned}5c+(-5c)+\frac{1}{5}\cdot 5&=5c+(-5c)+1\\&=[5c+(-5c)]+1\\&=0+1\\&=1\end{aligned}$$

25. Multiplicative inverse property

27. Identity property for addition

29. $8(9+\underline{3})=8\cdot 9+8\cdot 3$

31. $\underline{5}(3+b)=15+5b$

33. $\underline{a}(b+c)=ab+ac$

35. $-4(x+\underline{-2})=-4x+8$

37. $-2(5+c)=-2\cdot 5+\underline{(-2)}\cdot c$

39. $$\begin{aligned}6(4+x)&=6\cdot 4+6x\\&=24+6x\end{aligned}$$

41. $8(x+y+z)=8x+8y+8z$

43. $$\begin{aligned}6(x+7)&=6x+6\cdot 7\\&=6x+42\end{aligned}$$

45. $(a+5)b=ab+5b$

47. $$\begin{aligned}6(5+b)&=6\cdot 5+6b\\&=30+6b\end{aligned}$$

49. $$\begin{aligned}-4(x+y)&=-4x+(-4y)\\&=-4x-4y\end{aligned}$$

51. $$\begin{aligned}-9(a+b)&=-9a+(-9b)\\&=-9a-9b\end{aligned}$$

53. $$\begin{aligned}-3(4x+2)&=-3\cdot 4x+(-3\cdot 2)\\&=-12x+(-6)\\&=-12x-6\end{aligned}$$

55. $$\begin{aligned}-\left(\frac{3a}{2}-\frac{6}{7}\right)&=-1\cdot\left(\frac{3a}{2}-\frac{6}{7}\right)\\&=-1\cdot\frac{3a}{2}+(-1)\left(-\frac{6}{7}\right)\\&=-\frac{3a}{2}+\frac{6}{7}\end{aligned}$$

57. $$\begin{aligned}-(2x-6y)&=-1\cdot(2x-6y)\\&=-1\cdot 2x+(-1)(-6y)\\&=-2x+6y\end{aligned}$$

59. $-(2.1+3y)=-1\cdot(2.1+3y)$
$=-1\cdot 2.1+(-1)(3y)$
$=-2.1-3y$

61. $-4(a+5)=-4a+(-4)(5)$
$=-4a+(-20)$
$=-4a-20$

63. $-x(6+y)=-x\cdot 6+(-x)(y)$
$=-6x+(-xy)$
$=-6x-xy$

65. $-8(x-y)=-8x+(-8)(-y)$
$=-8x+8y$

67. $-3(2a-7b)=-3(2a)+(-3)(-7b)$
$=-6a+21b$

69. $0.5(x+y-2)=0.5x+0.5y+(0.5)(-2)$
$=0.5x+0.5y+(-1)$
$=0.5x+0.5y-1$

71. $\frac{6}{5}(a-b+5)=\frac{6}{5}a-\frac{6}{5}b+\frac{6}{5}(5)$
$=\frac{6}{5}a-\frac{6}{5}b+6$

73. $-2(x-y+4)=-2x+(-2)(-y)+(-2)(4)$
$=-2x+2y+(-8)$
$=-2x+2y-8$

75. $-0.3(x+y-6)$
$=-0.3x+(-0.3)y+(-0.3)(-6)$
$=-0.3x-0.3y+1.8$

77. $-\frac{5}{2}(a-2b+c-1)$
$=-\frac{5}{2}a+\left(-\frac{5}{2}\right)(-2b)+\left(-\frac{5}{2}c\right)+\left(-\frac{5}{2}\right)(-1)$
$=-\frac{5}{2}a+5b-\frac{5}{2}c+\frac{5}{2}$

79. $3^4=3\cdot 3\cdot 3\cdot 3=81$

81. $(-3)^4=(-3)(-3)(-3)(-3)=81$

83. $(-3)^3=(-3)(-3)(-3)=-27$

85. $(-2)^6=(-2)(-2)(-2)(-2)(-2)(-2)=64$

87. $7(38)=7(30+8)$
$=210+56$
$=266$

89. $6(46)=6(40+6)$
$=240+36$
$=276$

91. Sample answer: Suppose a is the reciprocal of 0. Then $a\cdot 0=1$, but $a\cdot 0=0$ for all a. Thus zero has no reciprocal.

93. No. Sample answer: for example, let $a=1$, $b=1$, and $c=1$.
$a-(b-c)=1-(1-1)=1-0=1$
$(a-b)-c=(1-1)-1=0-1=-1$

95. Yes. Sample answer:
$a\cdot(b-c)=a\cdot[b+(-c)]$
$=a\cdot b+a\cdot(-c)$
$=a\cdot b+(-a\cdot c)$
$=a\cdot b-a\cdot c$

97. $-4(a-5)=-4a+(-4)(-5)$
$=-4a+20$

99. $4(a+6)=4a+4\cdot 6$
$=4a+24$

101. $-2x+7+2x=(-2x+7)+2x$
$=2x+(-2x+7)$
$=[2x+(-2x)]+7$
$=0+7$
$=7$

103. $7+4x+9=(7+4x)+9$
$=9+(7+4x)$
$=(9+7)+4x$
$=16+4x$

105. $\frac{2}{5}+\underline{0}=\frac{2}{5}$

107. $-3.2\cdot\underline{1}=-3.2$

109. $3(x+\underline{2})=3x+6$

111. Identity property for multiplication

113. Identity property for addition

115. Associative property for multiplication

117. Associative property for addition

119. Associative property for multiplication:
$\left(\frac{1}{5}\cdot 4\right)\cdot 2=\frac{1}{5}\cdot(4\cdot \underline{2})$

121. Commutative property for multiplication:
$\frac{2}{5}\cdot \underline{1}=1\cdot\frac{2}{5}$

123. Associative property for addition:
$-2+(3+1.4)=(-2+\underline{3})+1.4$

1.7 Simplifying Expressions

Problems 1.7

1. a. $-9x+3x=(-9+3)x=-6x$

b. $-6x+8x=(-6+8)x=2x$

c. $-3x+(-7x)=[-3+(-7)]x=-10x$

d. $x+(-6x)=1x+(-6)x$
$=[1+(-6)]x$
$=-5x$

2. a. $3xy-8xy=3xy+(-8xy)$
$=[3+(-8)]xy$
$=-5xy$

b. $7x^2-5x^2=7x^2+(-5x^2)$
$=[7+(-5)]x^2$
$=2x^2$

c. $-3xy^2-(-7xy^2)=-3xy^2+7xy^2$
$=(-3+7)xy^2$
$=4xy^2$

d. $-5x^2y-(-4x^2y)=-5x^2y+4x^2y$
$=(-5+4)x^2y$
$=(-1)x^2y$
$=-x^2y$

3. a. $(7a-5)+(8a-2)=7a-5+8a-2$
$=15a-7$

b. $(4x+6y)+(3x-8y)$
$=4x+6y+3x-8y$
$=7x-2y$

4. $9a-3(a-2)-(a+4)$
$=9a-3a+6-(a+4)$
$=9a-3a+6-a-4$
$=(9a-3a-a)+(6-4)$
$=5a+2$

5. $[(a^2-2)+(3a+4)]+[(a-5)-(4a^2+2)]$
$=[a^2-2+3a+4]+[a-5-4a^2-2]$
$=[a^2+3a+2]+[-4a^2+a-7]$
$=a^2+3a+2-4a^2+a-7$
$=-3a^2+4a-5$

6. a.

The surface area S of a sphere of radius r	is obtained by	multiplying 4π by the square of the radius r.
S	$=$	$4\pi r^2$

b.

The perimeter P of a triangle	is obtained by	adding the lengths a, b, and c of each of the sides.
P	$=$	$a+b+c$

c.

The principal P	is	the quotient of the interest I and the rate r.
P	$=$	$\frac{I}{r}$

7. Pick a number. n
Add 3 to it. $n+3$
Double the result. $2(n+3)$
Subtract twice the original number. $2(n+3)-2n$
Divide the result by 2. $\frac{2(n+3)-2n}{2}$
What is the answer? Simplify!
Distributive law. $\frac{2n+6-2n}{2}$
Simplify numerator. $\frac{6}{2}$
Reduce. 3
Thus, you always get 3.

8. The temperature T_K in degrees Kelvin is the sum of the temperature T_C in degrees Celsius and 273.15.

Exercises 1.7

1. $19a+(-8a)=[19+(-8)]a=11a$

3. $-8c+3c=(-8+3)c=-5c$

5. $4n^2+8n^2=(4+8)n^2=12n^2$

7. $-3ab^2+(-4ab^2)=[-3+(-4)]ab^2=-7ab^2$

9. $-4abc+7abc=(-4+7)abc=3abc$

11. $0.7ab+(-0.3ab)+0.9ab$
$=[0.7+(-0.3)+0.9]ab$
$=(0.4+0.9)ab$
$=1.3ab$

13. $-0.3xy^2+0.2x^2y+(-0.6)xy^2$
$=0.2x^2y+[-0.3+(-0.6)]xy^2$
$=0.2x^2y+(-0.9)xy^2$
$=0.2x^2y-0.9xy^2$

15. $8abc^2+3ab^2c+(-8abc^2)$
$=[8+(-8)]abc^2+3ab^2c$
$=0abc^2+3ab^2c$
$=3ab^2c$

17. $-8ab+9xy+2ab+(-2xy)$
$=(-8+2)ab+[9+(-2)]xy$
$=-6ab+7xy$

19. $\frac{1}{5}a+\frac{3}{7}a^2b+\frac{2}{5}a+\frac{1}{7}a^2b$
$=\left(\frac{1}{5}+\frac{2}{5}\right)a+\left(\frac{3}{7}+\frac{1}{7}\right)a^2b$
$=\frac{3}{5}a+\frac{4}{7}a^2b$

21. $13x-2x=13x+(-2x)$
$=[13+(-2)]x$
$=11x$

23. $6ab-(-2ab)=6ab+2ab$
$=(6+2)ab$
$=8ab$

25. $-4a^2b-(3a^2b)=-4a^2b+(-3a^2b)$
$=[-4+(-3)]a^2b$
$=-7a^2b$

27. $3.1t^2-3.1t^2=3.1t^2+(-3.1t^2)$
$=[3.1+(-3.1)]t^2$
$=0t^2$
$=0$

29. $0.3x^2-0.3x^2=0.3x^2+(-0.3x^2)$
$=[0.3+(-0.3)]x^2$
$=0x^2$
$=0$

31. $(3xy+5)+(7xy-9)=3xy+5+7xy-9$
$=10xy-4$

33. $(7R-2)+(8-9R)=7R-2+8-9R$
$=-2R+6$

35. $(5L-3W)+(W-6L)=5L-3W+W-6L$
$=-L-2W$

37. $5x-(8x+1)=5x-8x-1$
$=-3x-1$

39. $\frac{2x}{9}-\left(\frac{x}{9}-2\right)=\frac{2x}{9}-\frac{x}{9}+2$
$=\frac{x}{9}+2$

41. $4a-(a+b)+3(b+a)$
$=4a-a-b+3(b+a)$
$=4a-a-b+3b+3a$
$=(4a-a+3a)+(-b+3b)$
$=6a+2b$

43. $7x-3(x+y)-(x+y)$
$=7x-3x-3y-(x+y)$
$=7x-3x-3y-x-y$
$=(7x-3x-x)+(-3y-y)$
$=3x+(-4y)$
$=3x-4y$

45. $-(x+y-2)+3(x-y+6)-(x+y-16)$
$=-x-y+2+3(x-y+6)-(x+y-16)$
$=-x-y+2+3x-3y+18-(x+y-16)$
$=-x-y+2+3x-3y+18-x-y+16$
$=(-x+3x-x)+(-y-3y-y)+(2+18+16)$
$=x+(-5y)+36$
$=x-5y+36$

47. $(x^2+7-x)+[-2x^3+(8x^2-2x)+5]$
$=(x^2+7-x)+[-2x^3+8x^2-2x+5]$
$=x^2+7-x-2x^3+8x^2-2x+5$
$=-2x^3+(x^2+8x^2)+(-x-2x)+(7+5)$
$=-2x^3+9x^2+(-3x)+12$
$=-2x^3+9x^2-3x+12$

49. $\left[\left(\frac{1}{4}x^2+\frac{1}{5}x\right)-\frac{1}{8}\right]+\left[\left(\frac{3}{4}x^2-\frac{3}{5}x\right)+\frac{5}{8}\right]$
$=\left[\frac{1}{4}x^2+\frac{1}{5}x-\frac{1}{8}\right]+\left[\frac{3}{4}x^2-\frac{3}{5}x+\frac{5}{8}\right]$
$=\frac{1}{4}x^2+\frac{1}{5}x-\frac{1}{8}+\frac{3}{4}x^2-\frac{3}{5}x+\frac{5}{8}$
$=\left(\frac{1}{4}x^2+\frac{3}{4}x^2\right)+\left(\frac{1}{5}x-\frac{3}{5}x\right)+\left(-\frac{1}{8}+\frac{5}{8}\right)$
$=\frac{4}{4}x^2+\left(-\frac{2}{5}x\right)+\frac{4}{8}$
$=x^2-\frac{2}{5}x+\frac{1}{2}$

51. $2[3(2a-4)+5]-[2(a-1)+6]$
$=2[6a-12+5]-[2a-2+6]$
$=2[6a-7]-[2a+4]$
$=12a-14-2a-4$
$=10a-18$

53. $-3[4a-(3+2b)]-[6(a-2b)+5a]$
$=-3[4a-3-2b]-[6a-12b+5a]$
$=-3[4a-3-2b]-[11a-12b]$
$=-12a+9+6b-11a+12b$
$=(-12a-11a)+(6b+12b)+9$
$=-23a+18b+9$

55. $-[-(0.2x+y)+3(x-y)]-[2(x+0.3y)-5]$
$=-[-0.2x-y+3x-3y]-[2x+0.6y-5]$
$=-[2.8x-4y]-[2x+0.6y-5]$
$=-2.8x+4y-2x-0.6y+5$
$=(-2.8x-2x)+(4y-0.6y)+5$
$=-4.8x+3.4y+5$

57. $\frac{1}{f}=\frac{1}{u}+\frac{1}{v}$; 2 terms

59. $F=\frac{n}{4}+40$; 2 terms

61. $I=Prt$; 1 term

63. $P=a+b+c$; 3 terms

65. $A=\frac{bh}{2}$; 1 term

67. $K = Cmv^2$; 1 term

69. $d = 16t^2$; 1 term

71. $\frac{1}{4} + \frac{3}{8} = \frac{2}{8} + \frac{3}{8} = \frac{5}{8}$

73. $\frac{3}{4} - \frac{1}{5} = \frac{15}{20} - \frac{4}{20} = \frac{11}{20}$

75. $P = S + S + S + S$
$= 4S$
The perimeter P of a square is the product of 4 and side S.

77. $P = (3 \text{ ft} + 1 \text{ in.}) + (6 \text{ ft} + 2 \text{ in.}) + (3 \text{ ft} + 1 \text{ in.}) + (6 \text{ ft} + 2 \text{ in.})$
$= 18 \text{ ft} + 6 \text{ in.}$
The perimeter is 18 ft, 6 in.

79. Girth $= h + w + h + w = 2h + 2w$
The sum of the length and girth is $2h + 2w + L$.

81. The length of the metal plate is
$6x \text{ in.} + 2 \text{ in.} + x \text{ in.} = 7x \text{ in.} + 2 \text{ in.}$
$= (7x + 2) \text{ in.}$

83. Answers may vary.

85. Sample answer: when a minus sign precedes an expression within parentheses, multiply each term within the parentheses by -1 and remove the parentheses.

87. $A = LW$; 1 term

89. $(8y - 7) + 2(3y - 2) = 8y - 7 + 6y - 4$
$= 14y - 11$

91. $-6ab^3 - (-3ab^3) = -6ab^3 + 3ab^3$
$= (-6 + 3)ab^3$
$= -3ab^3$

93. $3[(5 - 2x^2) + (2x - 1)] - 3[(5 - 2x) - (3 + x^2)]$
$= 3[5 - 2x^2 + 2x - 1] - 3[5 - 2x - 3 - x^2]$
$= 3[4 - 2x^2 + 2x] - 3[2 - 2x - x^2]$
$= 12 - 6x^2 + 6x - 6 + 6x + 3x^2$
$= (-6x^2 + 3x^2) + (6x + 6x) + (12 - 6)$
$= -3x^2 + 12x + 6$

Review Exercises

1. a. $a + b$

b. $a - b$

c. $7a + 2b - 8$

2. a. $3m$

b. $-mnr$

c. $\frac{1}{7}m$

d. $8m$

3. a. $\frac{m}{9}$

b. $\frac{9}{n}$

4. a. $\frac{m + n}{r}$

b. $\frac{m + n}{m - n}$

5. a. $m + n = 9 + 3 = 12$

b. $m - n = 9 - 3 = 6$

c. $4m = 4(9) = 36$

6. a. $\frac{m}{n} = \frac{9}{3} = 3$

b. $2m-3n=2(9)-3(3)$
$=18-3(3)$
$=18-9$
$=9$

c. $\frac{2m+n}{n}=\frac{2(9)+3}{3}=\frac{18+3}{3}=\frac{21}{3}=7$

7. a. 5

b. $-\frac{2}{3}$

c. -0.37

8. a. $|-8|=8$

b. $\left|3\frac{1}{2}\right|=3\frac{1}{2}$

c. $-|0.76|=-0.76$

9. a. Integer, rational, real

b. Whole, integer, rational, real

c. Rational, real

d. Irrational real

e. Irrational, real

f. Rational, real

10. a. $7+(-5)=7-5=2$

b. $(-0.3)+(-0.5)=-(0.3+0.5)=-0.8$

c. $-\frac{3}{4}+\frac{1}{5}=-\frac{15}{20}+\frac{4}{20}$
$=-\left(\frac{15}{20}-\frac{4}{20}\right)$
$=-\frac{11}{20}$

d. $3.6+(-5.8)=3.6-5.8$
$=-(5.8-3.6)$
$=-2.2$

e. $\frac{3}{4}+\left(-\frac{1}{2}\right)=\frac{3}{4}+\left(-\frac{2}{4}\right)=\frac{3}{4}-\frac{2}{4}=\frac{1}{4}$

11. a. $-16-4=-(16+4)=-20$

b. $-7.6-(-5.2)=-7.6+5.2$
$=-(7.6-5.2)$
$=-2.4$

c. $\frac{5}{6}-\frac{9}{4}=\frac{10}{12}-\frac{27}{12}=-\left(\frac{27}{12}-\frac{10}{12}\right)=-\frac{17}{12}$

12. a. $20-(-12)+15-12-5$
$=20+12+15-12-5$
$=20+15-5$
$=35-5$
$=30$

b. $-17+(-7)+10-(-7)-8$
$=-17+(-7)+10+7-8$
$=-17+10-8$
$=-7-8$
$=-15$

13. a. $-5\cdot 7=-35$

b. $8(-2.3)=-(8\cdot 2.3)=-18.4$

c. $-6(3.2)=-19.2$

d. $-\frac{3}{7}\left(-\frac{4}{5}\right)=\frac{12}{35}$

14. a. $(-4)^2=(-4)(-4)=16$

b. $-3^2=-3\cdot 3=-9$

c. $\left(-\frac{1}{3}\right)^3=\left(-\frac{1}{3}\right)\left(-\frac{1}{3}\right)\left(-\frac{1}{3}\right)$
$=\frac{1}{9}\left(-\frac{1}{3}\right)$
$=-\frac{1}{27}$

d. $-\left(-\frac{1}{3}\right)^3 = -\left(-\frac{1}{3}\right)\left(-\frac{1}{3}\right)\left(-\frac{1}{3}\right)$
$= -\frac{1}{9}\left(-\frac{1}{3}\right)$
$= \frac{1}{27}$

15. a. $\frac{-40}{10} = -4$

b. $-8 \div (-4) = 2$

c. $\frac{3}{8} \div \left(-\frac{9}{16}\right) = \frac{3}{8}\left(-\frac{16}{9}\right) = -\frac{48}{72} = -\frac{2}{3}$

d. $-\frac{3}{4} \div \left(-\frac{1}{2}\right) = -\frac{3}{4}\left(-\frac{2}{1}\right) = \frac{6}{4} = \frac{3}{2}$

16. a. $64 \div 8 - (3-5) = 64 \div 8 - (-2)$
$= 64 \div 8 + 2$
$= 8 + 2$
$= 10$

b. $27 \div 3^2 + 5 - 8 = 27 \div 9 + 5 - 8$
$= 3 + 5 - 8$
$= 8 - 8$
$= 0$

17. a. $20 \div 5 + \{2 \cdot 3 - [4 + (7-9)]\}$
$= 20 \div 5 + \{2 \cdot 3 - [4 + (-2)]\}$
$= 20 \div 5 + \{2 \cdot 3 - 2\}$
$= 20 \div 5 + \{6 - 2\}$
$= 20 \div 5 + 4$
$= 4 + 4$
$= 8$

b. $-6^2 + \frac{2(3-6)}{2} + 8 \div (-4)$
$= -6^2 + \frac{2(-3)}{2} + 8 \div (-4)$
$= -6^2 + \frac{-6}{2} + 8 \div (-4)$
$= -36 + \frac{-6}{2} + 8 \div (-4)$
$= -36 + (-3) + 8 \div (-4)$
$= -36 + (-3) + (-2)$
$= -39 + (-2)$
$= -41$

18. $0.80(220 - A) = 0.80(220 - 30)$
$= 0.80(190)$
$= 152$

19. a. Commutative property of addition

b. Associative property of multiplication

c. Associative property of addition

20. a. $6 + 4x - 10 = 4x + 6 - 10 = 4x - 4$

b. $-8 + 7x + 10 - 15x$
$= (7x - 15x) + (-8 + 10)$
$= -8x + 2$

21. a. Identity element for multiplication:
$\underline{1} \cdot \frac{1}{5} = \frac{1}{5}$

b. Additive inverse property:
$\frac{3}{4} + \underline{-\frac{3}{4}} = 0$

c. Additive inverse property:
$\underline{-3.7} + 3.7 = 0$

d. Multiplicative inverse property:
$\underline{\frac{2}{3}} \cdot \frac{3}{2} = 1$

e. Identity element for addition:
$\underline{0} + \frac{2}{7} = \frac{2}{7}$

f. Distributive property:
$3(x+\underline{-5})=3x+(-15)$

22. a. $-3(a+8)=-3a+(-3)(8)=-3a-24$

b. $-4(x-5)=-4[x+(-5)]$
$=-4x+(-4)(-5)$
$=-4x+20$

c. $-(x-4)=-1(x-4)$
$=-1[x+(-4)]$
$=-1x+(-1)(-4)$
$=-x+4$

23. a. $-7x+2x=(-7+2)x=-5x$

b. $-2x+(-8x)=[-2+(-8)]x=-10x$

c. $x+(-9x)=[1+(-9)]x=-8x$

d. $-6a^2b-(-9a^2b)=-6a^2b+9a^2b$
$=(-6+9)a^2b$
$=3a^2b$

24. a. $(4a-7)+(8a+2)=4a-7+8a+2$
$=12a-5$

b. $9x-3(x+2)-(x+3)$
$=9x-3x-6-(x+3)$
$=9x-3x-6-x-3$
$=(9x-3x-x)+(-6-3)$
$=5x+(-9)$
$=5x-9$

25. a. $m=\frac{p}{n}(2.75)$; 1 term

b. $W=\frac{11}{2}h-220$; 2 terms

Chapter 2 Equations, Problem Solving, and Inequalities

2.1 The Addition and Subtraction Properties of Equality

Problems 2.1

1. a. If x is 7, then $x - 5 = 3$ becomes $7 - 5 = 3$, which is a *false* statement. Hence, 7 is not a solution of the equation.

b. If y is 4, then $1 = 5 - y$ becomes $1 = 5 - 4$, which is a *true* statement. Thus, 4 is a solution of the equation.

c. If z is 6, $\frac{1}{3}z - 2 = 0$ becomes

$\frac{1}{3}(6) - 2 = 0$, which is a *true* statement.

Thus, 6 is a solution of the equation.

2. a.

$$\begin{aligned} x - 5 &= 7 \\ x - 5 + 5 &= 7 + 5 \\ x &= 12 \end{aligned}$$

The solution is 12.

CHECK

$x - 5 \stackrel{?}{=} 7$	
$12 - 5$	7
7	

b.

$$\begin{aligned} x - \frac{1}{5} &= \frac{3}{5} \\ x - \frac{1}{5} + \frac{1}{5} &= \frac{3}{5} + \frac{1}{5} \\ x &= \frac{4}{5} \end{aligned}$$

The solution is $\frac{4}{5}$.

CHECK

$x - \frac{1}{5} \stackrel{?}{=} \frac{3}{5}$	
$\frac{4}{5} - \frac{1}{5}$	$\frac{3}{5}$
$\frac{3}{5}$	

3. a.

$$\begin{aligned} 5y + 2 - 4y + 3 &= 17 \\ y + 5 &= 17 \\ y + 5 - 5 &= 17 - 5 \\ y &= 12 \end{aligned}$$

Thus 12 is the solution of the equation.

CHECK

$5y + 2 - 4y + 3 \stackrel{?}{=} 17$	
$5(12) + 2 - 4(12) + 3$	17
$60 + 2 - 48 + 3$	
17	

b.

$$\begin{aligned} -5z + \frac{3}{8} + 6z - \frac{1}{8} &= \frac{5}{8} \\ z + \frac{2}{8} &= \frac{5}{8} \\ z + \frac{2}{8} - \frac{2}{8} &= \frac{5}{8} - \frac{2}{8} \\ z &= \frac{3}{8} \end{aligned}$$

Thus, $\frac{3}{8}$ is the solution of the equation.

CHECK

$-5z + \frac{3}{8} + 6z - \frac{1}{8} \stackrel{?}{=} \frac{5}{8}$	
$-5\left(\frac{3}{8}\right) + \frac{3}{8} + 6\left(\frac{3}{8}\right) - \frac{1}{8}$	$\frac{5}{8}$
$-\frac{15}{8} + \frac{3}{8} + \frac{18}{8} - \frac{1}{8}$	
$\frac{5}{8}$	

4. a.

$$\begin{aligned} 5 &= 7 + x \\ 5 - 7 &= 7 - 7 + x \\ -2 &= x \\ x &= -2 \end{aligned}$$

The solution is −2.

CHECK

$5 \stackrel{?}{=}$	$7 + x$
5	$7 + (-2)$
	5

b.
$$\begin{aligned}5x-2&=4x+3\\5x-2+2&=4x+3+2\\5x&=4x+5\\5x-4x&=4x-4x+5\\x&=5\end{aligned}$$
The solution is 5.

CHECK $5x - 2 \stackrel{?}{=} 4x + 3$

5(5) − 2	4(5) + 3
25 − 2	20 + 3
23	23

c.
$$\begin{aligned}0&=3(y-3)+7-2y\\0&=3y-9+7-2y\\0&=y-2\\0+2&=y-2+2\\2&=y\end{aligned}$$
The solution is 2.

CHECK $0 \stackrel{?}{=} 3(y-3)+7-2y$

0	3(2 − 3) + 7 − 2(2)
	3(−1) + 7 − 4
	−3 + 7 − 4
	0

d.
$$\begin{aligned}3(z+1)&=4z+8\\3z+3&=4z+8\\3z+3-3&=4z+8-3\\3z&=4z+5\\3z-4z&=4z-4z+5\\-z&=5\\z&=-5\end{aligned}$$
The solution is −5.

CHECK $3(z+1) \stackrel{?}{=} 4z+8$

3(−5 + 1)	4(−5) + 8
3(−4)	−20 + 8
−12	−12

5.
$$\begin{aligned}6y+5&=7y+2\\6y+5-5&=7y+2-5\\6y&=7y-3\\6y-7y&=7y-7y-3\\-y&=-3\\y&=3\end{aligned}$$
The solution is 3.

CHECK $6y+5 \stackrel{?}{=} 7y+2$

6(3) + 5	7(3) + 2
18 + 5	21 + 2
23	23

6.
$$\begin{aligned}5+3(z-1)&=4+3z\\5+3z-3&=4+3z\\2+3z&=4+3z\\2-4+3z&=4-4+3z\\-2+3z&=3z\\-2+3z-3z&=3z-3z\\-2&=0\end{aligned}$$
Since this statement is *false*, the equation has no solution.

7.
$$\begin{aligned}3+4(y+2)&=11+4y\\3+4y+8&=11+4y\\11+4y&=11+4y\end{aligned}$$
Since both sides are *identical*, this equation is an identity. The solution is all real numbers.

8. a. Let $x = 2003 - 1998 = 5$
$$\begin{aligned}400x+7400&=400(5)+7400\\&=2000+7400\\&=9400\end{aligned}$$
The estimated tuition and fees in 2003 are $9400.

b.
$$\begin{aligned}400x+7400&=9000\\400x&=9000-7400\\400x&=1600\\x&=4\end{aligned}$$
This means 4 years after 1998 or in 1998 + 4 = 2002, tuition and fees are estimated to be $9000.

Exercises 2.1

1. If x is 3, then $x - 1 = 2$ becomes $3 - 1 = 2$, which is a *true* statement. Thus, 3 is a solution of the equation.

3. If y is -2, then $3y + 6 = 0$ becomes $3(-2) + 6 = 0$, which is a *true* statement. Thus, -2 is a solution of the equation.

5. If n is 2, then $12 - 3n = 6$ becomes $12 - 3(2) = 6$, which is a *true* statement. Thus, 2 is a solution of the equation.

7. If d is 10, then $\frac{2}{5}d + 1 = 3$ becomes

   $\frac{2}{5}(10) + 1 = 3$, which is a *false* statement. Hence, 10 is not a solution of the equation.

9. If a is 2.1, then $4.6 = 11.9 - 3a$ becomes $4.6 = 11.9 - 3(2.1)$, which is a *false* statement. Hence 2.1 is not a solution of the equation.

11.
$$\begin{aligned} x - 5 &= 9 \\ x - 5 + 5 &= 9 + 5 \\ x &= 14 \end{aligned}$$
The solution is 14.

CHECK

$x - 5 \stackrel{?}{=} 9$	
$14 - 5$	9
9	

13.
$$\begin{aligned} 11 &= m - 8 \\ 11 + 8 &= m - 8 + 8 \\ 19 &= m \end{aligned}$$
The solution is 19.

CHECK

$11 \stackrel{?}{=} m - 8$	
11	$19 - 8$
	11

15.
$$\begin{aligned} y - \frac{2}{3} &= \frac{8}{3} \\ y - \frac{2}{3} + \frac{2}{3} &= \frac{8}{3} + \frac{2}{3} \\ y &= \frac{10}{3} \end{aligned}$$
The solution is $\frac{10}{3}$.

CHECK

$x - \frac{2}{3} \stackrel{?}{=} \frac{8}{3}$	
$\frac{10}{3} - \frac{2}{3}$	$\frac{8}{3}$
$\frac{8}{3}$	

17.
$$\begin{aligned} 2k - 6 - k - 10 &= 5 \\ k - 16 &= 5 \\ k - 16 + 16 &= 5 + 16 \\ k &= 21 \end{aligned}$$
The solution is 21.

CHECK

$2k - 6 - k - 10 \stackrel{?}{=} 5$	
$2(21) - 6 - 21 - 10$	5
$42 - 6 - 21 - 10$	
5	

19.
$$\begin{aligned} \frac{1}{4} &= 2z - \frac{2}{3} - z \\ \frac{1}{4} &= z - \frac{2}{3} \\ \frac{1}{4} + \frac{2}{3} &= z - \frac{2}{3} + \frac{2}{3} \\ \frac{3}{12} + \frac{8}{12} &= z \\ \frac{11}{12} &= z \end{aligned}$$
The solution is $\frac{11}{12}$.

CHECK $\frac{1}{4} \stackrel{?}{=} 2z-\frac{2}{3}-z$

$\frac{1}{4}$	$2\left(\frac{11}{12}\right)-\frac{2}{3}-\frac{11}{12}$
	$\frac{22}{12}-\frac{8}{12}-\frac{11}{12}$
	$\frac{3}{12}$
	$\frac{1}{4}$

21.
$$0=2x-\frac{3}{2}-x-2$$
$$0=x-\frac{3}{2}-\frac{4}{2}$$
$$0=x-\frac{7}{2}$$
$$0+\frac{7}{2}=x-\frac{7}{2}+\frac{7}{2}$$
$$\frac{7}{2}=x$$
The solution is $\frac{7}{2}$.

CHECK $0 \stackrel{?}{=} 2x-\frac{3}{2}-x-2$

0	$2\left(\frac{7}{2}\right)-\frac{3}{2}-\frac{7}{2}-2$
	$7-\frac{3}{2}-\frac{7}{2}-2$
	$5-\frac{10}{2}$
	$5-5$
	0

23.
$$\frac{1}{5}=4c+\frac{1}{5}-3c$$
$$\frac{1}{5}=c+\frac{1}{5}$$
$$\frac{1}{5}-\frac{1}{5}=c+\frac{1}{5}-\frac{1}{5}$$
$$0=c$$
The solution is 0.

CHECK $\frac{1}{5} \stackrel{?}{=} 4c+\frac{1}{5}-3c$

$\frac{1}{5}$	$4(0)+\frac{1}{5}-3(0)$
	$0+\frac{1}{5}-0$
	$\frac{1}{5}$

25.
$$-3x+3+4x=0$$
$$3+x=0$$
$$3-3+x=0-3$$
$$x=-3$$
The solution is –3.

CHECK $-3x+3+4x \stackrel{?}{=} 0$

$-3(-3)+3+4(-3)$	0
$9+3-12$	
0	

27.
$$\frac{3}{4}y+\frac{1}{4}+\frac{1}{4}y=\frac{1}{4}$$
$$y+\frac{1}{4}=\frac{1}{4}$$
$$y+\frac{1}{4}-\frac{1}{4}=\frac{1}{4}-\frac{1}{4}$$
$$y=0$$
The solution is 0.

CHECK $\frac{3}{4}y+\frac{1}{4}+\frac{1}{4}y \stackrel{?}{=} \frac{1}{4}$

$\frac{3}{4}(0)+\frac{1}{4}+\frac{1}{4}(0)$	$\frac{1}{4}$
$0+\frac{1}{4}+0$	
$\frac{1}{4}$	

29.
$$3.4=-3c+0.8+2c+0.1$$
$$3.4=-c+0.9$$
$$3.4-0.9=-c+0.9-0.9$$
$$2.5=-c$$
$$-2.5=c$$
The solution is –2.5.

CHECK

$3.4 \stackrel{?}{=} -3c + 0.8 + 2c + 0.1$	
3.4	$-3(-2.5) + 0.8 + 2(-2.5) + 0.1$
	$7.5 + 0.8 - 5 + 0.1$
	3.4

31.
$$\begin{aligned} 6p+9&=5p \\ 6p-6p+9&=5p-6p \\ 9&=-p \\ -9&=p \end{aligned}$$
The solution is −9.

CHECK

$6p + 9$	$\stackrel{?}{=} 5p$
$6(-9) + 9$	$5(-9)$
$-54 + 9$	-45
-45	

33.
$$\begin{aligned} 3x+3+2x&=4x \\ 5x+3&=4x \\ 5x-5x+3&=4x-5x \\ 3&=-x \\ -3&=x \end{aligned}$$
The solution is −3.

CHECK

$3x + 3 + 2x$	$\stackrel{?}{=} 4x$
$3(-3) + 3 + 2(-3)$	$4(-3)$
$-9 + 3 - 6$	-12
-12	

35.
$$\begin{aligned} 4(m-2)+2-3m&=0 \\ 4m-8+2-3m&=0 \\ m-6&=0 \\ m-6+6&=0+6 \\ m&=6 \end{aligned}$$
The solution is 6.

CHECK

$4(m - 2) + 2 - 3m$	$\stackrel{?}{=} 0$
$4(6 - 2) + 2 - 3(6)$	0
$4(4) + 2 - 18$	
$16 + 2 - 18$	
0	

37.
$$\begin{aligned} 5(y-2)&=4y+8 \\ 5y-10&=4y+8 \\ 5y-10+10&=4y+8+10 \\ 5y&=4y+18 \\ 5y-4y&=4y-4y+18 \\ y&=18 \end{aligned}$$
The solution is 18.

CHECK

$5(y - 2)$	$\stackrel{?}{=} 4y + 8$
$5(18 - 2)$	$4(18) + 8$
$5(16)$	$72 + 8$
80	80

39.
$$\begin{aligned} 3a-1&=2(a-4) \\ 3a-1&=2a-8 \\ 3a-1+1&=2a-8+1 \\ 3a&=2a-7 \\ 3a-2a&=2a-2a-7 \\ a&=-7 \end{aligned}$$
The solution is −7.

CHECK

$3a - 1$	$\stackrel{?}{=} 2(a - 4)$
$3(-7) - 1$	$2(-7 - 4)$
$-21 - 1$	$2(-11)$
-22	-22

41.
$$\begin{aligned} 5(c-2) &= 6c-2 \\ 5c-10 &= 6c-2 \\ 5c-10+10 &= 6c-2+10 \\ 5c &= 6c+8 \\ 5c-6c &= 6c-6c+8 \\ -c &= 8 \\ c &= -8 \end{aligned}$$

The solution is –8.

CHECK

$5(c-2) \stackrel{?}{=} 6c-2$	
$5(-8-2)$	$6(-8)-2$
$5(-10)$	$-48-2$
-50	-50

43.
$$\begin{aligned} 3x+5-2x+1 &= 6x+4-6x \\ x+6 &= 4 \\ x+6-6 &= 4-6 \\ x &= -2 \end{aligned}$$

The solution is –2.

CHECK

$3x+5-2x+1 \stackrel{?}{=} 6x+4-6x$	
$3(-2)+5-2(-2)+1$	$6(-2)+4-6(-2)$
$-6+5+4+1$	$-12+4+12$
4	4

45.
$$\begin{aligned} -2g+4-5g &= 6g+1-14g \\ -7g+4 &= -8g+1 \\ -7g+4-4 &= -8g+1-4 \\ -7g &= -8g-3 \\ -7g+8g &= -8g+8g-3 \\ g &= -3 \end{aligned}$$

The solution is –3.

CHECK

$-2g+4-5g \stackrel{?}{=} 6g+1-14g$	
$-2(-3)+4-5(-3)$	$6(-3)+1-14(-3)$
$6+4+15$	$-18+1+42$
25	25

47.
$$\begin{aligned} 6(x+4)+4-2x &= 4x \\ 6x+24+4-2x &= 4x \\ 4x+28 &= 4x \\ 4x-4x+28 &= 4x-4x \\ 28 &= 0 \end{aligned}$$

Since this statement is *false*, the equation has no solution.

49.
$$\begin{aligned} 10(z-2)+10-2z &= 8(z+1)-18 \\ 10z-20+10-2z &= 8z+8-18 \\ 8z-10 &= 8z-10 \end{aligned}$$

Since both sides are *identical*, this equation is an identity. The solution is all real numbers.

51.
$$\begin{aligned} 3b+6-2b &= 2(b-2)+4 \\ b+6 &= 2b-4+4 \\ b+6 &= 2b \\ b-b+6 &= 2b-b \\ 6 &= b \end{aligned}$$

The solution is 6.

CHECK

$3b+6-2b \stackrel{?}{=} 2(b-2)+4$	
$3(6)+6-2(6)$	$2(6-2)+4$
$18+6-12$	$2(4)+4$
12	$8+4$
	12

53. $2p+\frac{2}{3}-5p=-4p+7\frac{1}{3}$

$$\frac{2}{3}-3p=-4p+7\frac{1}{3}$$
$$\frac{2}{3}-\frac{2}{3}-3p=-4p+7\frac{1}{3}-\frac{2}{3}$$
$$-3p=-4p+\frac{22}{3}-\frac{2}{3}$$
$$-3p=-4p+\frac{20}{3}$$
$$-3p+4p=-4p+4p+\frac{20}{3}$$
$$p=\frac{20}{3}$$

The solution is $\frac{20}{3}$.

CHECK

$2p+\frac{2}{3}-5p$	$\stackrel{?}{=}$ $-4p+7\frac{1}{3}$
$2\left(\frac{20}{3}\right)+\frac{2}{3}-5\left(\frac{20}{3}\right)$	$-4\left(\frac{20}{3}\right)+7\frac{1}{3}$
$\frac{40}{3}+\frac{2}{3}-\frac{100}{3}$	$-\frac{80}{3}+\frac{22}{3}$
$-\frac{58}{3}$	$-\frac{58}{3}$

55. $5r+\frac{3}{8}-9r=-5r+1\frac{1}{2}$

$$-4r+\frac{3}{8}=-5r+\frac{3}{2}$$
$$-4r+\frac{3}{8}-\frac{3}{8}=-5r+\frac{3}{2}-\frac{3}{8}$$
$$-4r=-5r+\frac{9}{8}$$
$$-4r+5r=-5r+5r+\frac{9}{8}$$
$$r=\frac{9}{8}$$

The solution is $\frac{9}{8}$.

CHECK

$5r+\frac{3}{8}-9r$	$\stackrel{?}{=}$ $-5r+1\frac{1}{2}$
$5\left(\frac{9}{8}\right)+\frac{3}{8}-9\left(\frac{9}{8}\right)$	$-5\left(\frac{9}{8}\right)+1\frac{1}{2}$
$\frac{45}{8}+\frac{3}{8}-\frac{81}{8}$	$-\frac{45}{8}+\frac{12}{8}$
$-\frac{33}{8}$	$-\frac{33}{8}$

57. Let h = average hourly earnings the previous year.

$$9.81=h+0.40$$
$$9.81-0.40=h+0.40-0.40$$
$$9.41=h$$

The average hourly earnings the previous year were $9.41.

59. Let y = cost 6 years ago.

$$y+142.2=326.9$$
$$y+142.2-142.2=326.9-142.2$$
$$y=184.7$$

The cost of medical care was 184.7 points 6 years ago.

61. Let w = waste generated in 1960.

$$195.7=w+107.9$$
$$195.7-107.9=w+107.9-107.9$$
$$87.8=w$$

87.8 million tons were generated in 1960.
Let i = increase from 1960.

$$87.8+i=229.9$$
$$87.8-87.8+i=229.9-87.8$$
$$i=142.1$$

The increase from 1960 was 142.1 million tons.

63. Let m = percent of males.

$$38=17+m$$
$$38-17=17-17+m$$
$$21=m$$

21% of males engage in exercise-walking.

65. Let p = percent that attended art museums.

$$35=8+p$$
$$35-8=8-8+p$$
$$27=p$$

27% attended art museums.

67. $4(-5) = -20$

69. $-\frac{2}{3}\left(\frac{3}{4}\right) = -\frac{6}{12} = -\frac{1}{2}$

71. The reciprocal of $\frac{3}{2}$ is $\frac{2}{3}$.

73. $6 = 2 \cdot 3$
$16 = 2 \cdot 2 \cdot 2 \cdot 2$
$\text{LCM} = 2 \cdot 2 \cdot 2 \cdot 2 \cdot 3 = 48$

75. $10 = 2 \cdot 5$
$8 = 2 \cdot 2 \cdot 2$
$\text{LCM} = 2 \cdot 2 \cdot 2 \cdot 5 = 40$

77. If $f = 40$ and $H = 120$, then $H = 1.95f + 72.85$ becomes $120 = 1.95(40) + 72.85$ which is a *false* statement. Therefore, the bone cannot belong to the missing female.

79. If $L_t = 36,\ L_a = 144,\ V_t = 30$ and $V_a = 50$, then $V_a^2 = \frac{L_a V_t^2}{L_t}$ becomes $50^2 = \frac{144(30^2)}{36}$, which is a *false* statement. Therefore, you cannot believe him.

81. Answers may vary.

83. Answers may vary.

85.
$$\begin{aligned} 5 + 4(x+1) &= 3 + 4x \\ 5 + 4x + 4 &= 3 + 4x \\ 9 + 4x &= 3 + 4x \\ 9 + 4x - 4x &= 3 + 4x - 4x \\ 9 &= 3 \end{aligned}$$
Since this statement is *false*, there is no solution.

87.
$$\begin{aligned} x - 5 &= 4 \\ x - 5 + 5 &= 4 + 5 \\ x &= 9 \end{aligned}$$
The solution is 9.

89.
$$\begin{aligned} x - 2.3 &= 3.4 \\ x - 2.3 + 2.3 &= 3.4 + 2.3 \\ x &= 5.7 \end{aligned}$$
The solution is 5.7.

91.
$$\begin{aligned} 2x + 6 - x + 2 &= 12 \\ x + 8 &= 12 \\ x + 8 - 8 &= 12 - 8 \\ x &= 4 \end{aligned}$$
The solution is 4.

93.
$$\begin{aligned} -5x + \frac{2}{9} + 6x - \frac{4}{9} &= \frac{5}{9} \\ x - \frac{2}{9} &= \frac{5}{9} \\ x - \frac{2}{9} + \frac{2}{9} &= \frac{5}{9} + \frac{2}{9} \\ x &= \frac{7}{9} \end{aligned}$$
The solution is $\frac{7}{9}$.

95.
$$\begin{aligned} 0 &= 4(z-3) + 5 - 3z \\ 0 &= 4z - 12 + 5 - 3x \\ 0 &= z - 7 \\ 0 + 7 &= z - 7 + 7 \\ 7 &= z \end{aligned}$$
The solution is 7.

97.
$$\begin{aligned} 3(x+2) + 3 &= 2 - (1 - 3x) \\ 3x + 6 + 3 &= 2 - 1 + 3x \\ 3x + 9 &= 1 + 3x \\ 3x - 3x + 9 &= 1 + 3x - 3x \\ 9 &= 1 \end{aligned}$$
This statement is *false*, so there is no solution.

99. If $z = -7$, then $\frac{1}{7}z - 1 = 0$ becomes $\frac{1}{7}(-7) - 1 = 0$, which is a *false* statement. Hence, -7 is not a solution of the equation.

2.2 The Multiplication and Division Properties of Equality

Problems 2.2

1. a. $\frac{x}{5}=3$

$\overset{1}{\cancel{5}}\cdot\frac{x}{\cancel{5}}=5\cdot 3$

$x=15$

The solution is 15.

CHECK

$\frac{x}{5} \overset{?}{=} 3$	
$\frac{15}{5}$	3
3	

b. $\frac{y}{4}=-5$

$\overset{1}{\cancel{4}}\cdot\frac{y}{\cancel{4}}=4(-5)$

$y=-20$

The solution is −20.

CHECK

$\frac{y}{4} \overset{?}{=} -5$	
$\frac{-20}{4}$	−5
−5	

2. a. $3x=12$

$\frac{\overset{1}{\cancel{3}}x}{\cancel{3}}=\frac{12}{3}$

$x=4$

The solution is 4.

CHECK

$3x \overset{?}{=} 12$	
$3\cdot 4$	12
12	

b. $7x=-21$

$\frac{\overset{1}{\cancel{7}}x}{\cancel{7}}=\frac{-21}{7}$

$x=-3$

The solution is −3.

CHECK

$7x \overset{?}{=} -21$	
$7(-3)$	−21
−21	

c. $-5x=20$

$\frac{\overset{1}{\cancel{-5}}x}{\cancel{-5}}=\frac{20}{-5}$

$x=-4$

The solution is −4.

CHECK

$-5x \overset{?}{=} 20$	
$-5(-4)$	20
20	

3. a. $\frac{3}{5}x=12$

$\frac{5}{3}\left(\frac{3}{5}x\right)=\frac{5}{3}(12)$

$1\cdot x=\frac{5}{\underset{1}{\cancel{3}}}\cdot\frac{\overset{4}{\cancel{12}}}{1}=\frac{20}{1}$

$x=20$

The solution is 20.

CHECK

$\frac{3}{5}x \stackrel{?}{=} 12$	
$\frac{3}{5}(20)$	12
$\frac{3}{\not{5}_1}\cdot\frac{\not{20}^4}{1}$	
12	

b. $$-\frac{2}{5}x = 6$$

$$-\frac{5}{2}\left(-\frac{2}{5}x\right) = -\frac{5}{2}(6)$$

$$1\cdot x = -\frac{5}{\not{2}_1}\cdot\frac{\not{6}^3}{1} = -\frac{15}{1}$$

The solution is −15.

CHECK

$-\frac{2}{5}x \stackrel{?}{=} 6$	
$-\frac{2}{5}(-15)$	6
$-\frac{2}{\not{5}_1}\left(\frac{\not{-15}^{-3}}{1}\right)$	
6	

c. $$-\frac{4}{5}x = -8$$

$$-\frac{5}{4}\left(-\frac{4}{5}x\right) = -\frac{5}{4}(-8)$$

$$1\cdot x = \frac{-5(\not{-8}^{-2})}{\not{4}_1} = 10$$

The solution is 10.

CHECK

$-\frac{4}{5}x \stackrel{?}{=} -8$	
$-\frac{4}{5}(10)$	−8
$-\frac{4}{\not{5}_1}(\not{10}^2)$	
−8	

4. a. $$\begin{array}{r|rr} 2 & 10 & 6 \\ \hline & 5 & 3 \end{array}$$

LCM of 10 and 6 is 2 · 5 · 3 = 30.

$$\frac{x}{10} + \frac{x}{6} = 8$$

$$30\cdot\frac{x}{10} + 30\cdot\frac{x}{6} = 30\cdot 8$$

$$3x + 5x = 30\cdot 8$$

$$8x = 30\cdot 8$$

$$x = 30$$

The solution is 30.

CHECK

$\frac{x}{10} + \frac{x}{6} \stackrel{?}{=} 8$	
$\frac{30}{10} + \frac{30}{6}$	8
3 + 5	
8	

b. The LCM of 4 and 5 is 4 · 5 = 20.

$$\frac{x}{4} - \frac{x}{5} = 1$$

$$20\cdot\frac{x}{4} - 20\cdot\frac{x}{5} = 20\cdot 1$$

$$5x - 4x = 20$$

$$x = 20$$

The solution is 20.

CHECK

$\frac{x}{4} - \frac{x}{5} \stackrel{?}{=} 1$	
$\frac{20}{4} - \frac{20}{5}$	1
5 − 4	
1	

5. a. The LCM of 3 and 4 is 3 · 4 = 12.

$$\frac{x+2}{3}+\frac{x-2}{4}=6$$

$$\overset{4}{\cancel{12}}\cdot\frac{x+2}{\cancel{3}}+\overset{3}{\cancel{12}}\cdot\frac{x-2}{\cancel{4}}=12\cdot 6$$

$$4(x+2)+3(x-2)=72$$

$$4x+8+3x-6=72$$

$$7x+2=72$$

$$7x=70$$

$$x=10$$

The solution is 10.

CHECK

$\frac{x+2}{3}+\frac{x-2}{4}$	$\stackrel{?}{=}$ 6
$\frac{10+2}{3}+\frac{10-2}{4}$	6
$\frac{12}{3}+\frac{8}{4}$	
4 + 2	
6	

b. The LCM of 5 and 3 is 5 · 3 = 15.

$$\frac{x+2}{5}-\frac{x-2}{3}=0$$

$$15\left(\frac{x+2}{5}\right)-15\left(\frac{x-2}{3}\right)=15(0)$$

$$3(x+2)-5(x-2)=0$$

$$3x+6-5x+10=0$$

$$-2x+16=0$$

$$-2x=-16$$

$$x=8$$

The solution is 8.

6. Forty percent of 30 is what number?

$$\frac{40}{100}\cdot 30=n$$

$$\frac{2}{5}\cdot 30=n$$

$$\frac{60}{5}=n$$

$$12=n$$

Thus, 40% of 30 is 12.

7. What percent of 30 is 6?

$$x\cdot 30=6$$

$$\frac{x\cdot 30}{30}=\frac{6}{30}$$

$$x=\frac{1}{5}=\frac{20}{100}$$

Thus, 6 is 20% of 30.

8. 20 is 40% of what number?

$$20=\frac{40}{100}\cdot n$$

$$\frac{40}{100}\cdot n=20$$

$$\frac{2}{5}\cdot n=20$$

$$\frac{\cancel{5}}{\cancel{2}}\cdot\frac{\cancel{2}}{\cancel{5}}\cdot n=\frac{5}{\underset{1}{\cancel{2}}}\cdot\overset{10}{\cancel{20}}$$

$$n=50$$

Thus, 20 is 40% of 50.

9. a. 5.3% of 225 means 0.053 · 225 = 11.925 or 12 when rounded to the nearest whole number.

b. 10.6% of 226 means 0.106 · 226 = 23.956 or 24 when rounded to the nearest whole number.

Exercises 2.2

Checks are left to the student.

1.

$$\frac{x}{7}=5$$

$$7\cdot\frac{x}{7}=7\cdot 5$$

$$x=35$$

The solution is 35.

3.

$$-4=\frac{x}{2}$$

$$2\cdot(-4)=2\cdot\frac{x}{2}$$

$$-8=x$$

The solution is –8.

5. $\frac{b}{-3}=5$

$-3\cdot\frac{b}{-3}=-3\cdot 5$

$b=-15$

The solution is –15.

7. $-3=\frac{f}{-2}$

$-2(-3)=-2\left(\frac{f}{-2}\right)$

$6=f$

The solution is 6.

9. $\frac{v}{4}=\frac{1}{3}$

$4\cdot\frac{v}{4}=4\cdot\frac{1}{3}$

$v=\frac{4}{3}$

The solution is $\frac{4}{3}$.

11. $\frac{x}{5}=\frac{-3}{4}$

$5\cdot\frac{x}{5}=5\cdot\left(\frac{-3}{4}\right)$

$x=-\frac{15}{4}$

The solution is $-\frac{15}{4}$.

13. $3z=33$

$\frac{3z}{3}=\frac{33}{3}$

$z=11$

The solution is 11.

15. $-42=6x$

$\frac{-42}{6}=\frac{6x}{6}$

$-7=x$

The solution is –7.

17. $-8c=56$

$\frac{-8c}{-8}=\frac{56}{-8}$

$c=-7$

The solution is –7.

19. $-5x=-35$

$\frac{-5x}{-5}=\frac{-35}{-5}$

$x=7$

The solution is 7.

21. $-3y=11$

$\frac{-3y}{-3}=\frac{11}{-3}$

$y=-\frac{11}{3}$

The solution is $-\frac{11}{3}$.

23. $-2a=1.2$

$\frac{-2a}{-2}=\frac{1.2}{-2}$

$a=-0.6$

The solution is –0.6.

25. $3t=4\frac{1}{2}$

$3t=\frac{9}{2}$

$\frac{1}{3}\cdot 3t=\frac{1}{\cancel{3}_1}\cdot\frac{\cancel{9}^3}{2}$

$t=\frac{3}{2}$

The solution is $\frac{3}{2}$.

27. $\frac{1}{3}x=-0.75$

$3\cdot\frac{1}{3}x=3(-0.75)$

$x=-2.25$

The solution is –2.25.

29. $-6 = \frac{3}{4}C$

$\frac{4}{3} \cdot (-6) = \frac{4}{3} \cdot \frac{3}{4}C$

$-8 = C$

The solution is –8.

31. $\frac{5}{6}a = 10$

$\frac{6}{5} \cdot \frac{5}{6}a = \frac{6}{5} \cdot 10$

$a = 12$

The solution is 12.

33. $-\frac{4}{5}y = 0.4$

$-\frac{5}{4} \cdot \left(-\frac{4}{5}y\right) = -\frac{5}{4} \cdot 0.4$

$y = -0.5$

The solution is –0.5.

35. $\frac{-2}{11}p = 0$

$\frac{11}{-2} \cdot \left(\frac{-2}{11}p\right) = \frac{11}{-2} \cdot 0$

$p = 0$

The solution is 0.

37. $-18 = \frac{3}{5}t$

$\frac{5}{3} \cdot (-18) = \frac{5}{3} \cdot \frac{3}{5}t$

$-30 = t$

The solution is –30.

39. $\frac{7x}{0.02} = -7$

$\frac{0.02}{7}\left(\frac{7x}{0.02}\right) = \frac{0.02}{7} \cdot (-7)$

$x = -0.02$

The solution is –0.02.

41. The LCM of 2 and 3 is 2 · 3 = 6.

$\frac{y}{2} + \frac{y}{3} = 10$

$6 \cdot \frac{y}{2} + 6 \cdot \frac{y}{3} = 6 \cdot 10$

$3y + 2y = 60$

$5y = 60$

$y = 12$

The solution is 12.

43. The LCM of 7 and 3 is 7 · 3 = 21.

$\frac{x}{7} + \frac{x}{3} = 10$

$21 \cdot \frac{x}{7} + 21 \cdot \frac{x}{3} = 21 \cdot 10$

$3x + 7x = 210$

$10x = 210$

$x = 21$

The solution is 21.

45. The LCM of 5 and 10 is 10.

$\frac{x}{5} + \frac{x}{10} = 6$

$10 \cdot \frac{x}{5} + 10 \cdot \frac{x}{10} = 10 \cdot 6$

$2x + x = 60$

$3x = 60$

$x = 20$

The solution is 20.

47. $2\,\underline{|\,6 \quad 8}$
$\quad\;\; 3 \quad 4$

The LCM of 6 and 8 is 2 · 3 · 4 = 24.

$$\frac{t}{6}+\frac{t}{8}=7$$
$$24\cdot\frac{t}{6}+24\cdot\frac{t}{8}=24\cdot 7$$
$$4t+3t=168$$
$$7t=168$$
$$t=24$$

The solution is 24.

49. The LCM of 2 and 5 is 2 · 5 = 10.

$$\frac{x}{2}+\frac{x}{5}=\frac{7}{10}$$
$$10\cdot\frac{x}{2}+10\cdot\frac{x}{5}=10\cdot\frac{7}{10}$$
$$5x+2x=7$$
$$7x=7$$
$$x=1$$

The solution is 1.

51. The LCM of 3 and 5 is 3 · 5 = 15.

$$\frac{c}{3}-\frac{c}{5}=2$$
$$15\cdot\frac{c}{3}-15\cdot\frac{c}{5}=15\cdot 2$$
$$5c-3c=30$$
$$2c=30$$
$$c=15$$

The solution is 15.

53. 6 = 2 · 3
8 = 2 ·2 · 2
12 = 2 · 2 · 3
LCM = 2 · 2 · 2 · 3 = 24

$$\frac{W}{6}-\frac{W}{8}=\frac{5}{12}$$
$$24\cdot\frac{W}{6}-24\cdot\frac{W}{8}=24\cdot\frac{5}{12}$$
$$4W-3W=10$$
$$W=10$$

The solution is 10.

55. The LCM of 5, 10, and 2 is 10.

$$\frac{x}{5}-\frac{3}{10}=\frac{1}{2}$$
$$10\cdot\frac{x}{5}-10\cdot\frac{3}{10}=10\cdot\frac{1}{2}$$
$$2x-3=5$$
$$2x=8$$
$$x=4$$

The solution is 4.

57. The LCM of 4, 3, and 2 is 4 · 3 = 12.

$$\frac{x+4}{4}-\frac{x+2}{3}=-\frac{1}{2}$$
$$12\cdot\frac{x+4}{4}-12\cdot\frac{x+2}{3}=12\cdot\left(-\frac{1}{2}\right)$$
$$3(x+4)-4(x+2)=-6$$
$$3x+12-4x-8=-6$$
$$-x+4=-6$$
$$-x=-10$$
$$x=10$$

The solution is 10.

59. $2\,\underline{|\,6 \quad 4}$
$\quad\;\; 3 \quad 2$

The LCM of 6 and 4 is 2 · 3 · 2 = 12.

$$\frac{x}{6}+\frac{3}{4}=x-\frac{7}{4}$$
$$12\cdot\frac{x}{6}+12\cdot\frac{3}{4}=12\cdot x-12\cdot\frac{7}{4}$$
$$2x+9=12x-21$$
$$2x=12x-30$$
$$-10x=-30$$
$$x=3$$

The solution is 3.

61. 30% of 40 is what number?

$$\frac{30}{100}\cdot 40=n$$
$$\frac{3}{10}\cdot 40=n$$
$$12=n$$

Thus, 30% of 40 is 12.

63. 40% of 70 is what number?

$$\frac{40}{100}\cdot 70 = n$$

$$\frac{2}{\not{5}_{1}}\cdot \not{70}^{14} = n$$

$$28 = n$$

Thus, 40% of 70 is 28.

65. What percent of 30 is 15?

$$x\cdot 30 = 15$$

$$\frac{x\cdot 30}{30} = \frac{15}{30}$$

$$x = \frac{1}{2} = \frac{50}{10}$$

Thus, 15 is 50% of 30.

67. 30 is 20% of what number?

$$30 = \frac{20}{100}\cdot n$$

$$30 = \frac{1}{5}\cdot n$$

$$5\cdot 30 = 5\cdot\frac{1}{5}\cdot n$$

$$150 = n$$

Thus, 30 is 20% of 150.

69. 12 is 60% of what number?

$$12 = \frac{60}{100}\cdot n$$

$$12 = \frac{3}{5}\cdot n$$

$$\frac{5}{3}\cdot 12 = \frac{5}{3}\cdot\frac{3}{5}\cdot n$$

$$20 = n$$

Thus, 12 is 60% of 20.

71. 16 is what percent of 211?

$$16 = n\cdot 211$$

$$\frac{16}{211} = \frac{n\cdot 211}{211}$$

$$0.076 \approx n$$

Thus, 7.6% of 211 is 16.

73. What percent of 211 is 20?

$$n\cdot 211 = 20$$

$$\frac{2\cdot 211}{211} = \frac{20}{211}$$

$$n \approx 0.095$$

Thus, 9.5% of the 211 patients had heart attacks.

75. What percent of 211 is 8?

$$n\cdot 211 = 8$$

$$\frac{n\cdot 211}{211} = \frac{8}{211}$$

$$n \approx 0.038$$

Thus, 3.8% of the 211 patients had a stroke.

77. What percent of 171 is 9?

$$n\cdot 171 = 9$$

$$\frac{n\cdot 171}{171} = \frac{9}{171}$$

$$n \approx 0.053$$

Thus, 5.3% of the 171 patients died.

79. What percent of 211 is 8?

$$n\cdot 211 = 8$$

$$\frac{n\cdot 211}{211} = \frac{8}{211}$$

$$n = 0.038$$

Thus, 3.8% of the 211 patients had a stroke.

81. 50% of what price is $12?

$$\frac{50}{100}\cdot x = 12$$

$$\frac{1}{2}\cdot x = 12$$

$$2\cdot\frac{1}{2}\cdot x = 2\cdot 12$$

$$x = 24$$

The original price was $24.

83. If the item is $\frac{1}{3}$ off, then it is selling for $\frac{2}{3}$ of the original price.

$\frac{2}{3}$ of what price is $8?

$$\frac{2}{3}\cdot x=8$$

$$\frac{3}{2}\cdot\frac{2}{3}\cdot x=\frac{3}{\not{2}_1}\cdot\not{8}^4$$

$$x=12$$

The original price was $12.

85. If the beer has $\frac{1}{3}$ fewer calories, the it has $\frac{2}{3}$ of the calories of the regular beer. $\frac{2}{3}$ of how many calories is 100?

$$\frac{2}{3}\cdot x=100$$

$$\frac{3}{2}\cdot\frac{2}{3}\cdot x=\frac{3}{2}\cdot 100$$

$$x=150$$

The regular beer has 150 calories.

87. $5(8-y)=5\cdot 8-5\cdot y=40-5y$

89. $9(6-3y)=9\cdot 6-9\cdot 3y=54-27y$

91. $-5(3x-4)=-5\cdot 3x-5\cdot(-4)=-15x+20$

93. $24\cdot\frac{1}{6}=\frac{\not{24}^4}{1}\cdot\frac{1}{\not{6}_1}=\frac{4}{1}=4$

95. $-7\cdot\left(-\frac{7}{3}\right)=\frac{\not{-7}^1}{1}\cdot\left(\frac{3}{\not{-7}_1}\right)=\frac{3}{1}=3$

97. $8\times 100\times 20=16{,}000$
Her lost wages are $16,000.

99. a. Answers may vary.

b. Answers may vary.

101. Answers may vary. Sample answer: It would be easier to multiply by the reciprocal of $-\frac{3}{4}$.

$$-\frac{3}{4}x=15$$

$$-\frac{4}{3}\cdot\left(-\frac{3}{4}x\right)=-\frac{4}{\not{3}_1}\cdot\not{15}^5$$

$$x=-20$$

103. What percent of 45 is 9?

$$x\cdot 45=9$$

$$\frac{x\cdot 45}{45}=\frac{9}{45}$$

$$x=\frac{1}{5}=\frac{20}{100}$$

Thus, 9 is 20% of 45.

105. The LCM of 4 and 5 is $4\cdot 5=20$.

$$\frac{x+2}{4}+\frac{x-1}{5}=3$$

$$20\cdot\frac{x+2}{4}+20\cdot\frac{x-1}{5}=20\cdot 3$$

$$5(x+2)+4(x-1)=60$$

$$5x+10+4x-4=60$$

$$9x+6=60$$

$$9x=54$$

$$x=6$$

The solution is 6.

107. The LCM of 6 and 10 is $2\cdot 3\cdot 5=30$.

$$\frac{y}{6}+\frac{y}{10}=8$$

$$30\cdot\frac{y}{6}+30\cdot\frac{y}{10}=30\cdot 8$$

$$5y+3y=240$$

$$8y=240$$

$$y=30$$

109. $\frac{4}{5}y = 8$

$\frac{5}{4}\cdot\frac{4}{5}y = \frac{5}{\cancel{4}_1}\cdot\cancel{8}^2$

$y = 10$

The solution is 10.

111. $-\frac{2}{7}y = -4$

$-\frac{7}{2}\cdot\left(-\frac{2}{7}y\right) = -\frac{7}{\cancel{2}_1}\cdot(\cancel{-4}^2)$

$y = 14$

The solution is 14.

113. $\frac{x}{2} = -7$

$2\cdot\frac{x}{2} = 2(-7)$

$x = -14$

2.3 Linear Equations

Problems 2.3

1. a.

$$\begin{aligned} 4x+5 &= 17 \\ 4x+5-5 &= 17-5 \\ 4x &= 12 \\ \frac{4x}{4} &= \frac{12}{4} \\ x &= 3 \end{aligned}$$

The solution is 3.

CHECK

$4x + 5$	$\stackrel{?}{=}$ 17
$4(3) + 5$	17
$12 + 5$	
17	

b.

$$\begin{aligned} -3x-5 &= 2 \\ -3x-5+5 &= 2+5 \\ -3x &= 7 \\ \frac{-3x}{-3} &= \frac{7}{-3} \\ x &= -\frac{7}{3} \end{aligned}$$

The solution is $-\frac{7}{3}$.

CHECK

$-3x - 5$	$\stackrel{?}{=}$ 2
$-3\left(-\frac{7}{3}\right)-5$	2
$7 - 5$	
2	

2.

$$\begin{aligned} 7(x+1) &= 4(x+2)+5 \\ 7x+7 &= 4x+8+5 \\ 7x+7 &= 4x+13 \\ 7x+7-7 &= 4x+13-7 \\ 7x &= 4x+6 \\ 7x-4x &= 4x-4x+6 \\ 3x &= 6 \\ \frac{3x}{3} &= \frac{6}{3} \\ x &= 2 \end{aligned}$$

The solution is 2.

CHECK

$7(x + 1)$	$\stackrel{?}{=}$ $4(x + 2) + 5$
$7(2 + 1)$	$4(2 + 2) + 5$
$7(3)$	$4(4) + 5$
21	$16 + 5$
	21

3. a. The LCM of 3, 7, and 21 is 21.

$$\frac{20}{21}=\frac{x}{7}+\frac{x}{3}$$

$$\cancel{21}\cdot\frac{20}{\cancel{21}}=\overset{3}{\cancel{21}}\cdot\frac{x}{\cancel{7}}+\overset{7}{\cancel{21}}\cdot\frac{x}{\cancel{3}}$$

$$20=3x+7x$$
$$20=10x$$
$$\frac{20}{10}=\frac{10x}{10}$$
$$2=x$$

The solution is 2.

CHECK

$\frac{20}{21} \stackrel{?}{=} \frac{x}{7}+\frac{x}{3}$	
$\frac{20}{21}$	$\frac{2}{7}+\frac{2}{3}$
	$\frac{6}{21}+\frac{14}{21}$
	$\frac{20}{21}$

b. The LCM of 4, 5, and 20 is 20.

$$\frac{1}{4}-\frac{x}{5}=\frac{17(x+4)}{20}$$

$$\overset{5}{\cancel{20}}\cdot\frac{1}{\cancel{4}}-\overset{4}{\cancel{20}}\cdot\frac{x}{\cancel{5}}=\cancel{20}\cdot\frac{17(x+4)}{\cancel{20}}$$

$$5-4x=17(x+4)$$
$$5-4x=17x+68$$
$$5-5-4x=17x+68-5$$
$$-4x=17x+63$$
$$-4x-17x=17x-17x+63$$
$$-21x=63$$
$$\frac{-21x}{-21}=\frac{63}{-21}$$
$$x=-3$$

The solution is –3.

CHECK

$\frac{1}{4}-\frac{x}{5} \stackrel{?}{=}$	$\frac{17(x+4)}{20}$
$\frac{1}{4}-\frac{-3}{5}$	$\frac{17(-3+4)}{20}$
$\frac{5}{20}+\frac{12}{20}$	$\frac{17(1)}{20}$
$\frac{17}{20}$	$\frac{17}{20}$

4.

$$S=\frac{1}{6}\left(\boxed{C}-4\right)$$
$$6\cdot S=6\cdot\frac{1}{6}\left(\boxed{C}-4\right)$$
$$6S=\boxed{C}-4$$
$$6S+4=\boxed{C}-4+4$$
$$6S+4=\boxed{C}$$
$$C=6S+4$$

5. a. Let $S = 12$.

$$S=3L-24$$
$$12=3L-24$$
$$12+24=3L-24+24$$
$$36=3L$$
$$\frac{36}{3}=\frac{3L}{3}$$
$$12=L$$

The length of their foot is 12 inches.

b.

$$S=3\boxed{L}-24$$
$$S+24=3\boxed{L}-24+24$$
$$S+24=3\boxed{L}$$
$$\frac{S+24}{3}=\frac{3\boxed{L}}{3}$$
$$\frac{S+24}{3}=\boxed{L}$$
$$L=\frac{S+24}{3}$$

6.
$$\begin{aligned}3x+4y&=7\\3x+4y-4y&=7-4y\\3x&=7-4y\\\frac{3x}{3}&=\frac{7-4y}{3}\\x&=\frac{7-4y}{3}\end{aligned}$$

Exercises 2.3

Checks are left to the student.

1.
$$\begin{aligned}3x-12&=0\\3x-12+12&=0+12\\3x&=12\\\frac{3x}{3}&=\frac{12}{3}\\x&=4\end{aligned}$$
The solution is 4.

3.
$$\begin{aligned}2y+6&=8\\2y+6-6&=8-6\\2y&=2\\\frac{2y}{2}&=\frac{2}{2}\\y&=1\end{aligned}$$
The solution is 1.

5.
$$\begin{aligned}-3z-4&=-10\\-3z-4+4&=-10+4\\-3z&=-6\\\frac{-3z}{-3}&=\frac{-6}{-3}\\z&=2\end{aligned}$$
The solution is 2.

7.
$$\begin{aligned}-5y+1&=-13\\-5y+1-1&=-13-1\\-5y&=-14\\\frac{-5y}{-5}&=\frac{-14}{-5}\\y&=\frac{14}{5}\end{aligned}$$
The solution is $\frac{14}{5}$.

9.
$$\begin{aligned}3x+4&=x+10\\3x+4-4&=x+10-4\\3x&=x+6\\3x-x&=x-x+6\\2x&=6\\\frac{2x}{2}&=\frac{6}{2}\\x&=3\end{aligned}$$
The solution is 3.

11.
$$\begin{aligned}5x-12&=6x-8\\5x-12+12&=6x-8+12\\5x&=6x+4\\5x-6x&=6x-6x+4\\-x&=4\\x&=-4\end{aligned}$$
The solution is –4.

13.
$$\begin{aligned}4v-7&=6v+9\\4v-7+7&=6v+9+7\\4v&=6v+16\\4v-6v&=6v-6v+16\\-2v&=16\\\frac{-2v}{-2}&=\frac{16}{-2}\\v&=-8\end{aligned}$$
The solution is –8.

15.
$$\begin{aligned}6m-3m+12&=0\\3m+12&=0\\3m+12-12&=0-12\\3m&=-12\\\frac{3m}{3}&=\frac{-12}{3}\\m&=-4\end{aligned}$$
The solution is –4.

17.

$$\begin{aligned}10-3z&=8-6z\\10-10-3z&=8-10-6z\\-3z&=-2-6z\\-3z+6z&=-2-6z+6z\\3z&=-2\\\frac{3z}{3}&=\frac{-2}{3}\\z&=-\frac{2}{3}\end{aligned}$$

The solution is $-\frac{2}{3}$.

19.

$$\begin{aligned}5(x+2)&=3(x+3)+1\\5x+10&=3x+9+1\\5x+10&=3x+10\\5x+10-10&=3x+10-10\\5x&=3x\\5x-3x&=3x-3x\\2x&=0\\\frac{2x}{2}&=\frac{0}{2}\\x&=0\end{aligned}$$

The solution is 0.

21.

$$\begin{aligned}5(4-3a)&=7(3-4a)\\20-15a&=21-28a\\20-20-15a&=21-20-28a\\-15a&=1-28a\\-15a+28a&=1-28a+28a\\13a&=1\\\frac{13a}{13}&=\frac{1}{13}\\a&=\frac{1}{13}\end{aligned}$$

The solution is $\frac{1}{13}$.

23.

$$\begin{aligned}-\frac{7}{8}c+5.6&=-\frac{5}{8}c-3.3\\-\frac{7}{8}c+5.6-5.6&=-\frac{5}{8}c-3.3-5.6\\-\frac{7}{8}c&=-\frac{5}{8}c-8.9\\-\frac{7}{8}c+\frac{5}{8}c&=-\frac{5}{8}c+\frac{5}{8}c-8.9\\-\frac{2}{8}c&=-8.9\\-\frac{1}{4}c&=-8.9\\-4\cdot\left(-\frac{1}{4}c\right)&=-4\cdot(-8.9)\\c&=35.6\end{aligned}$$

The solution is 35.6.

25.

$$\begin{aligned}-2x+\frac{1}{4}&=2x+\frac{4}{5}\\-2x+\frac{1}{4}-\frac{1}{4}&=2x+\frac{4}{5}-\frac{1}{4}\\-2x&=2x+\frac{16}{20}-\frac{5}{20}\\-2x&=2x+\frac{11}{20}\\-2x-2x&=2x-2x+\frac{11}{20}\\-4x&=\frac{11}{20}\\-\frac{1}{4}\cdot(-4x)&=-\frac{1}{4}\cdot\frac{11}{20}\\x&=-\frac{11}{80}\end{aligned}$$

The solution is $-\frac{11}{80}$.

27.
$$\frac{x-1}{2}+\frac{x-2}{2}=3$$
$$2\cdot\frac{x-1}{2}+2\cdot\frac{x-2}{2}=2\cdot3$$
$$x-1+x-2=6$$
$$2x-3=6$$
$$2x-3+3=6+3$$
$$2x=9$$
$$\frac{2x}{2}=\frac{9}{2}$$
$$x=\frac{9}{2}$$
The solution is $\frac{9}{2}$.

29.
$$\frac{x}{5}-\frac{x}{4}=1$$
$$20\cdot\frac{x}{5}-20\cdot\frac{x}{4}=20\cdot1$$
$$4x-5x=20$$
$$-x=20$$
$$x=-20$$
The solution is –20.

31.
$$\frac{x+1}{4}-\frac{2x-2}{3}=3$$
$$12\cdot\frac{x+1}{4}-12\cdot\frac{2x-2}{3}=12\cdot3$$
$$3(x+1)-4(2x-2)=36$$
$$3x+3-8x+8=36$$
$$-5x+11=36$$
$$-5x+11-11=36-11$$
$$-5x=25$$
$$\frac{-5x}{-5}=\frac{25}{-5}$$
$$x=-5$$
The solution is –5.

33.
$$\frac{2h-1}{3}=\frac{h-4}{12}$$
$$12\cdot\frac{2h-1}{3}=12\cdot\frac{h-4}{12}$$
$$4(2h-1)=h-4$$
$$8h-4=h-4$$
$$8h-4+4=h-4+4$$
$$8h=h$$
$$8h-h=h-h$$
$$7h=0$$
$$\frac{7h}{7}=\frac{0}{7}$$
$$h=0$$
The solution is 0.

35.
$$\frac{2w+3}{2}-\frac{3w+1}{4}=1$$
$$4\cdot\frac{2w+3}{2}-4\cdot\frac{3w+1}{4}=4\cdot1$$
$$2(2w+3)-(3w+1)=4$$
$$4w+6-3w-1=4$$
$$w+5=4$$
$$w+5-5=4-5$$
$$w=-1$$
The solution is –1.

37.
$$\frac{8x-23}{6}+\frac{1}{3}=\frac{5}{2}x$$
$$6\cdot\frac{8x-23}{6}+6\cdot\frac{1}{3}=6\cdot\frac{5}{2}x$$
$$8x-23+2=15x$$
$$8x-21=15x$$
$$8x-8x-21=15x-8x$$
$$-21=7x$$
$$\frac{-21}{7}=\frac{7x}{7}$$
$$-3=x$$
The solution is –3.

39.
$$\frac{x-5}{2}-\frac{x-4}{3}=\frac{x-3}{2}-(x-2)$$
$$6\cdot\frac{x-5}{2}-6\cdot\frac{x-4}{3}=6\cdot\frac{x-3}{2}-6\cdot(x-2)$$
$$3(x-5)-2(x-4)=3(x-3)-6x+12$$
$$3x-15-2x+8=3x-9-6x+12$$
$$x-7=-3x+3$$
$$x-7+7=-3x+3+7$$
$$x=-3x+10$$
$$x+3x=-3x+3x+10$$
$$4x=10$$
$$\frac{4x}{4}=\frac{10}{4}$$
$$x=\frac{5}{2}$$
The solution is $\frac{5}{2}$.

41. $-4x+\frac{1}{2}=4\left(\frac{1}{8}-x\right)$
$$-4x+\frac{1}{2}=\frac{1}{2}-4x$$
This is an identity. The solution is all real numbers.

43. $\frac{1}{2}(8x+4)-5=\frac{1}{4}(4x+8)+1$
$$4x+2-5=x+2+1$$
$$4x-3=x+3$$
$$4x-3+3=x+3+3$$
$$4x=x+6$$
$$4x-x=x-x+6$$
$$3x=6$$
$$\frac{3x}{3}=\frac{6}{3}$$
$$x=2$$
The solution is 2.

45.
$$x+\frac{x}{2}-\frac{3x}{5}=9$$
$$10\cdot x+10\cdot\frac{x}{2}-10\cdot\frac{3x}{5}=10\cdot 9$$
$$10x+5x-6x=90$$
$$9x=90$$
$$\frac{9x}{9}=\frac{90}{9}$$
$$x=10$$
The solution is 10.

47.
$$\frac{4x}{9}-\frac{3}{2}=\frac{5x}{6}-\frac{3x}{2}$$
$$18\cdot\frac{4x}{9}-18\cdot\frac{3}{2}=18\cdot\frac{5x}{6}-18\cdot\frac{3x}{2}$$
$$8x-27=15x-27x$$
$$8x-27=-12x$$
$$8x-8x-27=-12x-8x$$
$$-27=-20x$$
$$\frac{-27}{-20}=\frac{-20x}{-20}$$
$$\frac{27}{20}=x$$
The solution is $\frac{27}{20}$.

49.

$$\frac{3x+4}{2}-\frac{1}{8}(19x-3)=1-\frac{7x+18}{12}$$
$$24\cdot\frac{3x+4}{2}-24\cdot\frac{1}{8}(19x-3)=24\cdot 1-24\cdot\frac{7x+18}{12}$$
$$12(3x+4)-3(19x-3)=24-2(7x+18)$$
$$36x+48-57x+9=24-14x-36$$
$$-21x+57=-14x-12$$
$$-21x+57-57=-14x-12-57$$
$$-21x=-14x-69$$
$$-21x+14x=-14x+14x-69$$
$$-7x=-69$$
$$\frac{-7x}{-7}=\frac{-69}{-7}$$
$$x=\frac{69}{7}$$

The solution is $\frac{69}{7}$.

51.

$$C=2\pi\boxed{r}$$
$$\frac{C}{2\pi}=\frac{2\pi\boxed{r}}{2\pi}$$
$$\frac{C}{2\pi}=\boxed{r}$$
$$r=\frac{C}{2\pi}$$

53.

$$3x+2y=6$$
$$3x-3x+2y=6-3x$$
$$2y=6-3x$$
$$\frac{2y}{2}=\frac{6-3x}{2}$$
$$y=\frac{6-3x}{2}$$

55.

$$A=\pi(r^2+rs)$$
$$A=\pi r^2+\pi rs$$
$$A-\pi r^2=\pi r^2-\pi r^2+\pi rs$$
$$A-\pi r^2=\pi rs$$
$$\frac{A-\pi r^2}{\pi r}=\frac{\pi rs}{\pi r}$$
$$\frac{A-\pi r^2}{\pi r}=s$$
$$s=\frac{A-\pi r^2}{\pi r}$$

57.
$$\frac{V_2}{V_1} = \frac{P_1}{P_2}$$
$$V_1 \cdot \frac{V_2}{V_1} = V_1 \cdot \frac{P_1}{P_2}$$
$$V_2 = \frac{P_1 V_1}{P_2}$$

59.
$$S = \frac{f}{H-h}$$
$$(H-h) \cdot S = (H-h) \cdot \frac{f}{H-h}$$
$$HS - hS = f$$
$$HS - hS + hS = f + hS$$
$$HS = f + hS$$
$$\frac{HS}{S} = \frac{f+hS}{S}$$
$$H = \frac{f+Sh}{S}$$

61. a.
$$S = 3L - 22$$
$$S + 22 = 3L - 22 + 22$$
$$S + 22 = 3L$$
$$\frac{S+22}{3} = \frac{3L}{3}$$
$$\frac{S+22}{3} = L$$
$$L = \frac{S+22}{3}$$

b.
$$S = 3L - 21$$
$$S + 21 = 3L - 21 + 21$$
$$S + 21 = 3L$$
$$\frac{S+21}{3} = \frac{3L}{3}$$
$$\frac{S+21}{3} = L$$
$$L = \frac{S+21}{3}$$

63.
$$W = 5H - 200$$
$$W + 200 = 5H - 200 + 200$$
$$W + 200 = 5H$$
$$\frac{W+200}{5} = \frac{5H}{5}$$
$$\frac{W+200}{5} = H$$
$$H = \frac{W+200}{5}$$

65. $S = 3L - 22$

or $L = \frac{S+22}{3}$

$$L = \frac{11+22}{3}$$
$$L = \frac{33}{3}$$
$$L = 11$$

Tyrone's foot is 11 inches long.

67. First, find the length of a man's foot for size 7.

$$L = \frac{S+22}{3}$$
$$L = \frac{7+22}{3}$$
$$L = \frac{29}{3}$$

Then, solve $S = 3L - 21$ for women with $L = \frac{29}{3}$.

$$S = 3\left(\frac{29}{3}\right) - 21$$
$$S = 29 - 21$$
$$S = 8$$

Sue wears a size 8.

69. Let $C = 11.50$.

$$\begin{aligned} C &= 1 + 0.75(h-1) \\ 11.50 &= 1 + 0.75(h-1) \\ 11.50 &= 1 + 0.75h - 0.75 \\ 11.50 &= 0.25 + 0.75h \\ 11.50 - 0.25 &= 0.25 - 0.25 + 0.75h \\ 11.25 &= 0.75h \\ \frac{11.25}{0.75} &= \frac{0.75h}{0.75} \\ 15 &= h \end{aligned}$$

After 15 hours, the cost is \$11.50.

71. Quotient means divide.

$\frac{a+b}{c}$

73. Product means multiply and sum means add.

$a(b + c)$

75. Difference means subtract and product means multiply.

$a - bc$

77. Let $C = 40$ and solve for m.

$$\begin{aligned} C &= 30 + 0.15m \\ 40 &= 30 + 0.15m \\ 40 - 30 &= 30 - 30 + 0.15m \\ 10 &= 0.15m \\ \frac{10}{0.15} &= \frac{0.15m}{0.15} \\ 66\frac{2}{3} &= m \end{aligned}$$

The mileage rate and flat rate are the same at $66\frac{2}{3}$ miles.

79. a.

$$\begin{aligned} S &= C + M \\ S &= C + 0.2C \\ S &= 1.2C \end{aligned}$$

b. Let $C = 8$.

$S = 1.2C$

$S = 1.2(8)$

$S = 9.6$

The selling price is \$9.60.

81. Answers may vary.

83. The solution to the equation is $x = 0$. If $x = 0$, then dividing by x (or 0) is undefined.

85.

$$\begin{aligned} 50 &= 40 + 0.2(m - 100) \\ 50 &= 40 + 0.2m - 20 \\ 50 &= 20 + 0.2m \\ 50 - 20 &= 20 - 20 + 0.2m \\ 30 &= 0.2m \\ \frac{30}{0.2} &= \frac{0.2m}{0.2} \\ 150 &= m \\ m &= 150 \end{aligned}$$

87.

$$\begin{aligned} \frac{1}{3} - \frac{x}{5} &= \frac{8(x+2)}{15} \\ 15\cdot\frac{1}{3} - 15\cdot\frac{x}{5} &= 15\cdot\frac{8(x+2)}{15} \\ 5 - 3x &= 8(x+2) \\ 5 - 3x &= 8x + 16 \\ 5 - 5 - 3x &= 8x + 16 - 5 \\ -3x &= 8x + 11 \\ -3x - 8x &= 8x - 8x + 11 \\ -11x &= 11 \\ \frac{-11x}{-11} &= \frac{11}{-11} \\ x &= -1 \end{aligned}$$

The solution is -1.

89.

$$\begin{aligned} -5(x+2) &= -3(x+1) - 9 \\ -5x - 10 &= -3x - 3 - 9 \\ -5x - 10 &= -3x - 12 \\ -5x - 10 + 10 &= -3x - 12 + 10 \\ -5x &= -3x - 2 \\ -5x + 3x &= -3x + 3x - 2 \\ -2x &= -2 \\ \frac{-2x}{-2} &= \frac{-2}{-2} \\ x &= 1 \end{aligned}$$

The solution is 1.

91.

$$\begin{aligned} 3x + 8 &= 11 \\ 3x + 8 - 8 &= 11 - 8 \\ 3x &= 3 \\ \frac{3x}{3} &= \frac{3}{3} \\ x &= 1 \end{aligned}$$

The solution is 1.

2.4 Problem Solving: Integer, General, and Geometry Problems

Problems 2.4

1. 1. **Read the problem.**
We are asked to find three consecutive *odd* integers.

2. **Select the unknown.**
Let n = first odd integer.
Then $n + 2$ = 2nd odd integer.
and $n + 4$ = 3rd odd integer.

3. **Think of a plan.**
Translate the sentence into an equation.
$n + (n + 2) + (n + 4) = 129$

4. **Use algebra to solve the resulting equation.**

$$\begin{aligned} n+(n+2)+(n+4)&=129 \\ n+n+2+n+4&=129 \\ 3n+6&=129 \\ 3n+6-6&=129-6 \\ 3n&=123 \\ \frac{3n}{3}&=\frac{123}{3} \\ n&=41 \end{aligned}$$

Thus, the three consecutive odd integers are 41, $41 + 2 = 43$, and $41 + 4 = 45$.

5. **Verify the solution.**
Since $41 + 43 + 45 = 129$, our result is correct.

2. 1. **Read the problem.**
We are asked to find a certain number.

2. **Select the unknown.**
Let n represent the number.

3. **Think of a plan.**
Translate the sentence into an equation.
$3n - 14 = n + 2$

4. **Use algebra to solve the resulting equation.**

$$\begin{aligned} 3n-14&=n+2 \\ 3n-14+14&=n+2+14 \\ 3n&=n+16 \\ 3n-n&=n-n+16 \\ 2n&=16 \\ \frac{2n}{2}&=\frac{16}{2} \\ n&=8 \end{aligned}$$

The number is 8.

5. **Verify the solution.**
Is $3(8) - 14 = 8 + 2$? Yes, since $24 - 14 = 10$ is true.

3. 1. **Read the problem.**
We are asked to find the number of calories in the cheeseburger and in the fries.

2. **Select the unknown.**
Let f = number of calories in the fries.
Then $f + 120$ = number of calories in the cheeseburger.

3. **Think of a plan.**
Translate the problem.
$f + (f + 120) = 540$

4. **Use algebra to solve the equation.**

$$\begin{aligned} f+(f+120)&=540 \\ f+f+120&=540 \\ 2f+120&=540 \\ 2f+120-120&=540-120 \\ 2f&=420 \\ \frac{2f}{2}&=\frac{420}{2} \\ f&=210 \end{aligned}$$

Thus, the fries have 210 calories and the cheeseburger has $210 + 120 = 330$ calories.

5. **Verify the solution.**
Does $210 + 330 = 540$? Yes, the equation is true.

4. 1. **Read the problem.**
We are asked to find the measure of an angle.

2. **Select the unknown.**
   Let m = measure of the angle.
   $90 - m$ is its complement.
   $180 - m$ is its supplement.

3. **Think of a plan.**
   Translate the problem statement into an equation.
   $180 - m = 4(90 - m) - 45$

4. **Use algebra to solve the equation.**
$$\begin{aligned} 180 - m &= 4(90 - m) - 45 \\ 180 - m &= 360 - 4m - 45 \\ 180 - m &= 315 - 4m \\ 180 - 180 - m &= 315 - 180 - 4m \\ -m &= 135 - 4m \\ -m + 4m &= 135 - 4m + 4m \\ 3m &= 135 \\ \frac{3m}{3} &= \frac{135}{3} \\ m &= 45 \end{aligned}$$
   The measure of the angle is 45°.

5. **Verify the solution.**
   Is the supplement (180° – 45° = 135°) 45° less than 4 times its complement (4 · (90° – 45°) = 180°)? Yes, since 135° = 180° – 45°.

Exercises 2.4

Students should use the RSTUV method to solve. An outline of each solution is given.

1. Let n = first even integer.
   Then $n + 2$ = 2nd even integer, and $n + 4$ = 3rd even integer.
$$\begin{aligned} n + (n+2) + (n+4) &= 138 \\ n + n + 2 + n + 4 &= 138 \\ 3n + 6 &= 138 \\ 3n + 6 - 6 &= 138 - 6 \\ 3n &= 132 \\ \frac{3n}{3} &= \frac{132}{3} \\ n &= 44 \end{aligned}$$
   Thus, the three integers are 44, 44 + 2 = 46, and 44 + 4 = 48.

3. Let n = first even integer.
   Then $n + 2$ = 2nd even integer, and $n + 4$ = 3rd even integer.
$$\begin{aligned} n + (n+2) + (n+4) &= -24 \\ n + n + 2 + n + 4 &= -24 \\ 3n + 6 &= -24 \\ 3n + 6 - 6 &= -24 - 6 \\ 3n &= -30 \\ \frac{3n}{3} &= \frac{-30}{3} \\ n &= -10 \end{aligned}$$
   Thus, the integers are –10, –10 + 2 = –8, and –10 + 4 = –6.

5. Let n = first integer.
   Then $n + 1$ = next consecutive integer.
$$\begin{aligned} n + (n+1) &= -25 \\ n + n + 1 &= -25 \\ 2n + 1 &= -25 \\ 2n + 1 - 1 &= -25 - 1 \\ 2n &= -26 \\ \frac{2n}{2} &= \frac{-26}{2} \\ n &= -13 \end{aligned}$$
   Thus, the integers are –13 and –13 + 1 = –12.

7. Let n = first integer,
   $n + 1$ = second integer,
   $n + 2$ = last integer.
$$\begin{aligned} n + 2 + 2n &= 23 \\ 3n + 2 &= 23 \\ 3n + 2 - 2 &= 23 - 2 \\ 3n &= 21 \\ \frac{3n}{3} &= \frac{21}{3} \\ n &= 7 \end{aligned}$$
   Thus, the integers are 7, 7 + 1 = 8, and 7 + 2 = 9.

9. Let n = first locker number,
   $n + 1$ = middle locker number,
   $n + 2$ = last locker number.
$$\begin{aligned} n + (n+2) &= 2(n+1) \\ n + n + 2 &= 2n + 2 \\ 2n + 2 &= 2n + 2 \end{aligned}$$
   This is an identity, so the solution is any consecutive integers.

11. Let p = price of cheaper book.
Then $p + 24$ = price of more expensive book.

$$p+(p+24)=64$$
$$p+p+24=64$$
$$2p+24=64$$
$$2p+24-24=64-24$$
$$2p=40$$
$$\frac{2p}{2}=\frac{40}{2}$$
$$p=20$$

Thus, the books cost \$20 and \$20 + \$24 = \$44.

13. Let x = points scored by losing team.
Then $x + 55$ = points scored by winning team.

$$x+(x+55)=133$$
$$x+x+55=133$$
$$2x+55=133$$
$$2x+55-55=133-55$$
$$2x=78$$
$$\frac{2x}{2}=\frac{78}{2}$$
$$x=39$$

Thus, the losing team scored 39 points and the winning team scored 39 + 55 = 94 points.

15. Let x charges on other card.
Then $\frac{3}{8}x+4=$ charges on one card.

$$x+\left(\frac{3}{8}x+4\right)=147$$
$$x+\frac{3}{8}x+4=147$$
$$\frac{8}{8}x+\frac{3}{8}x+4=147$$
$$\frac{11}{8}x+4=147$$
$$\frac{11}{8}x+4-4=147-4$$
$$\frac{11}{8}x=143$$
$$\frac{8}{11}\cdot\frac{11}{8}x=\frac{8}{11}\cdot 143$$
$$x=8\cdot 13$$
$$x=104$$

Thus, the charges on the cards are \$104 and $\frac{3}{8}(104)+4=3\cdot 13+4=39+4=\$43$.

17. Let n = first number.
Then $3n$ = second number,
and $3n - 5$ = third number.

$$n+3n+(3n-5)=254$$
$$n+3n+3n-5=254$$
$$7n-5=254$$
$$7n-5+5=254+5$$
$$7n=259$$
$$\frac{7n}{7}=\frac{259}{7}$$
$$n=37$$

Thus, the three numbers are 37, 3(37) = 111, and 111 − 5 = 106.

19. Let n = smaller number.
Then $6n$ = larger number.

$$n+6n=147$$
$$7n=147$$
$$\frac{7n}{7}=\frac{147}{7}$$
$$n=21$$

Thus, the integers are 21 and 6(21) = 126.

21. Let w = percent accessing from work.
Then $w + 25$ = percent accessing from home.

$$\begin{aligned} w+(w+25)+15&=66\\ w+w+25+15&=66\\ 2w+40&=66\\ 2w+40-40&=66-40\\ 2w&=26\\ \frac{2w}{2}&=\frac{26}{2}\\ w&=13\\ w+25&=38 \end{aligned}$$

Thus 13% access the Internet from work and 38% from home.

23. Let x = salary per week for men 16–24 years old.
Then $x + 330$ = salary per week for men over 25 years old.

$$\begin{aligned} x+330&=722\\ x+330-330&=722-330\\ x&=392 \end{aligned}$$

Men 16–24 years old make $392 per week.

25. Let a = height of antenna.

$$\begin{aligned} 1250+a&=1472\\ 1250-1250+a&=1472-1250\\ a&=222 \end{aligned}$$

The antenna is 222 feet high.

27.

$$\begin{aligned} v&=96-32t\\ 0&=96-32t\\ 0-96&=96-96-32t\\ -96&=-32t\\ \frac{-96}{-32}&=\frac{-32t}{-32}\\ 3&=t \end{aligned}$$

The rocket will reach its highest point at 3 seconds.

29. Let r votes received by loser.
Then $r + 372$ = votes received by winner.

$$\begin{aligned} r+(r+372)&=980\\ r+r+372&=980\\ 2r+372&=980\\ 2r+372-372&=980-372\\ 2r&=608\\ \frac{2r}{2}&=\frac{608}{2}\\ r&=304\\ r+372&=676 \end{aligned}$$

Thus, the candidates received 304 votes and 676 votes.

31. Let m = number of minutes after the first 3 minutes.

$$\begin{aligned} 3.05+0.7m&=7.95\\ 3.05-3.05+0.7m&=7.95-3.05\\ 0.7m&=4.90\\ \frac{0.7m}{0.7}&=\frac{4.90}{0.7}\\ m&=7 \end{aligned}$$

He talked for 7 minutes after the first 3 minutes. The call was 10 minutes long.

33. **a.** Let m = number of miles.

$$\begin{aligned} 0.95+1.25m&=28.45\\ 0.95-0.95+1.25m&=28.45-0.95\\ 1.25m&=27.50\\ \frac{1.25m}{1.25}&=\frac{27.50}{1.25}\\ m&=22 \end{aligned}$$

The ride was 22 miles long.

b. Find the cost of a 12-mile ride using the equation from **a.**

$$\begin{aligned} 0.95+1.25m&=0.95+1.25(12)\\ &=0.95+15\\ &=15.95 \end{aligned}$$

The limo is cheaper, since $15 < $15.95.

35. Let m = measure of the angle.
$90 - m$ = complement and
$180 - m$ = supplement.

$$\begin{aligned}180 - m &= 2(90 - m) + 20\\ 180 - m &= 180 - 2m + 20\\ 180 - m &= 200 - 2m\\ 180 - 180 - m &= 200 - 180 - 2m\\ -m &= 20 - 2m\\ -m + 2m &= 20 - 2m + 2m\\ m &= 20\end{aligned}$$

The measure of the angle is 20°.

37. Let m = measure of the angle.
$90 - m$ = complement and
$180 - m$ = supplement.

$$\begin{aligned}180 - m &= 3(90 - m)\\ 180 - m &= 270 - 3m\\ 180 - 180 - m &= 270 - 180 - 3m\\ -m &= 90 - 3m\\ -m + 3m &= 90 - 3m + 3m\\ 2m &= 90\\ \frac{2m}{2} &= \frac{90}{2}\\ m &= 45\end{aligned}$$

The measure of the angle is 45°.

39. Let m = measure of the angle.
$90 - m$ = complement and
$180 - m$ = supplement.

$$\begin{aligned}180 - m &= 4(90 - m) - 54\\ 180 - m &= 360 - 4m - 54\\ 180 - m &= 306 - 4m\\ 180 - 180 - m &= 306 - 180 - 4m\\ -m &= 126 - 4m\\ -m + 4m &= 126 - 4m + 4m\\ 3m &= 126\\ \frac{3m}{3} &= \frac{126}{3}\\ m &= 42\end{aligned}$$

The measure of the angle is 42°.

41.

$$\begin{aligned}55T &= 100\\ \frac{55T}{55} &= \frac{100}{55}\\ T &= \frac{20}{11}\end{aligned}$$

43.

$$\begin{aligned}15T &= 120\\ \frac{15T}{15} &= \frac{120}{15}\\ T &= 8\end{aligned}$$

45.

$$\begin{aligned}-75x &= -600\\ \frac{-75x}{-75} &= \frac{-600}{-75}\\ x &= 8\end{aligned}$$

47.

$$\begin{aligned}-0.02P &= -70\\ \frac{-0.02P}{-0.02} &= \frac{-70}{-0.02}\\ P &= 3500\end{aligned}$$

49.

$$\begin{aligned}-0.04x &= -40\\ \frac{-0.04x}{-0.04} &= \frac{-40}{-0.04}\\ x &= 1000\end{aligned}$$

51. Let x = the number of years Diophantus lived.

$\frac{1}{6}x$ = youth years

$\frac{1}{12}x$ = grew beard

$\frac{1}{7}x$ = married

$\frac{1}{2}x$ = one-half life span of father's age.

$$\begin{aligned}\frac{1}{6}x + \frac{1}{12}x + \frac{1}{7}x + 5 + \frac{1}{2}x + 4 &= x\\ \frac{14}{84}x + \frac{7}{84}x + \frac{12}{84}x + \frac{42}{84}x + 9 &= x\\ \frac{75}{84}x + 9 &= x\\ \frac{75}{84}x - \frac{75}{84}x + 9 &= \frac{84}{84}x - \frac{75}{84}x\\ 9 &= \frac{9}{84}x\\ \frac{84}{9} \cdot 9 &= \frac{84}{9} \cdot \frac{9}{84}x\\ 84 &= x\end{aligned}$$

Diophantus lived 84 years.

53. Answers may vary.

55. Let n = first odd integer.
Then $n + 2$ = 2nd odd integer,
and $n + 4$ = 3rd odd integer.

$$n+(n+2)+(n+4)=249$$
$$n+n+2+n+4=249$$
$$3n+6=249$$
$$3n+6-6=249-6$$
$$3n=243$$
$$\frac{3n}{3}=\frac{243}{3}$$
$$n=81$$
$$n+2=83$$
$$n+4=85$$

The integers are 81, 83, and 85.

57. Let p = calories in the pizza.
Then $p + 70$ = calories in the shake.

$$p+(p+70)=530$$
$$p+p+70=530$$
$$2p+70=530$$
$$2p+70-70=530-70$$
$$2p=460$$
$$\frac{2p}{2}=\frac{460}{2}$$
$$p=230$$
$$p+70=300$$

The pizza had 230 calories and the shake had 300 calories.

2.5 Problem Solving: Motion, Mixture, and Investment Problems

Problems 2.5

1. **1. Read the problem.**
We are asked to find the average speed of the bus.

2. Select the unknown.
Let R = average speed.

3. Think of a plan.
$T = 214 - 12 - 24 = 178$
$$R \times T = D$$
$$R \times 178 = 6000$$

4. Use algebra to solve the equation.
$$178R = 6000$$
$$\frac{178R}{178}=\frac{6000}{178}$$
$$R \approx 33.7$$
The average speed of the bus is 33.7 miles per hour.

5. Verify the solution.
$$R \times T = D$$
$$33.7 \times 178 = 5998.6$$

2. **1. Read the problem.**
We are asked to find the number of hours if the car is going 60 mph.

2. Select the unknown.
Let T = number of hours.

3. Think of a plan.

	R	×	T	=	D
car	60		T		$60T$
bus	40		$T+3$		$40(T+3)$

$60T = 40(T + 3)$

4. Use algebra to solve the equation.
$$60T = 40(T+3)$$
$$60T = 40T + 120$$
$$60T - 40T = 40T - 40T + 120$$
$$20T = 120$$
$$\frac{20T}{20}=\frac{120}{20}$$
$$T = 6$$
It takes the car 6 hours to overtake the bus.

5. Verify the solution.
car: $60 \times 6 = 360$
bus: $40 \times 9 = 360$

3. **1. Read the problem.**
We are asked to find how many hours to meet if the faster bus travels at 50 mph.

2. Select the unknown.
Let T = hours until they meet.

3. Think of a plan.

	R	×	T	=	D
Supercruiser	40		T		$40T$
bus	50		T		$50T$

$40T + 50T = 3240$

4. Use algebra to solve the equation.

$$40T + 50T = 3240$$
$$90T = 3240$$
$$\frac{90T}{90} = \frac{3240}{90}$$
$$T = 36$$

Each bus traveled 36 hours before they met.

5. Verify the solution.

$40 \times 36 = 1440$

$50 \times 36 = 1800$

$1440 + 1800 = 3240$

4. 1. Read the problem.

We are asked to find the number of ounces of the 50% solution that should be added to make a 30% solution.

2. Select the unknown.

Let x = number of ounces of 50% solution to added.

3. Think of a plan.

	%	×	ounces	=	Amount of Pure acid in Final Mixture
50% solution	0.50		x		$0.50x$
5% solution	0.05		32		1.60
30% solution	0.30		$x + 32$		$0.30(x + 32)$

$0.50x + 1.60 = 0.30(x + 32)$

4. **Use algebra to solve the equation.**

$$10 \cdot 0.50x + 10 \cdot 1.60 = 10 \cdot [0.30(x+32)]$$
$$5x + 16 = 3(x+32)$$
$$5x + 16 = 3x + 96$$
$$5x + 16 - 16 = 3x + 96 - 16$$
$$5x = 3x + 80$$
$$5x - 3x = 3x - 3x + 80$$
$$2x = 80$$
$$\frac{2x}{2} = \frac{80}{2}$$
$$x = 40$$

The photographer must add 40 ounces of the 30% solution.

5. **Verify the solution.**
0.50(40) + 1.60 = 21.6
0.30(40 + 32) = 21.6

5. 1. **Read the problem.**
We are asked to find how much is invested if her income is only $400.

2. **Select the unknown.**
Let s = amount invested in stocks.
Then $6000 - s$ = amount invested in bonds.

3. **Think of a plan.**

	P	×	r	=	I
stocks	s		0.05		$0.05s$
bonds	$6000 - s$		0.10		$0.10(6000 - s)$

$0.05s + 0.10(6000 - s) = 400$

4. **Use algebra to solve the equation.**

$$0.05s + 0.10(6000 - s) = 400$$
$$0.05s + 600 - 0.10s = 400$$
$$600 - 0.05s = 400$$
$$600 - 600 - 0.05s = 400 - 600$$
$$-0.05s = -200$$
$$\frac{-0.05s}{-0.05} = \frac{-200}{-0.05}$$
$$s = 4000$$

The woman invested $4000 in stocks and $2000 in bonds.

5. **Verify the solution.**
5% of $4000 is $200 and 10% of $2000 is $200, so the total interest is $400.

Exercises 2.5

Students should use the RSTUV method to solve. An outline of each solution is given.

1. Let $D = 400$ and $T = 8$.

$$R \times T = D$$
$$R \times 8 = 400$$
$$\frac{8R}{8} = \frac{400}{8}$$
$$R = 50$$

The bus' average speed is 50 miles per hour.

3. Let $D = 246$ and $T = 120$.

$$R \times T = D$$
$$R \times 120 = 246$$
$$\frac{120R}{120} = \frac{246}{120}$$
$$R = 2.05 \approx 2$$

The rate of play of the tape is 2 meters per minute.

5. Let $D = 200$ and $R = 400$.

$$R \times T = D$$
$$400 \times T = 200$$
$$\frac{400T}{400} = \frac{200}{400}$$
$$T = \frac{1}{2}$$

It takes $\frac{1}{2}$ hour to go from Tampa to Miami.

7.

	R	$\times$	T	$=$	D
bus	60		$T+2$		$60(T+2)$
car	90		T		$90T$

$$60(T+2) = 90T$$
$$60T + 120 = 90T$$
$$60T - 60T + 120 = 90T - 60T$$
$$120 = 30T$$
$$\frac{120}{30} = \frac{30T}{30}$$
$$4 = T$$

It will take her 4 hours to catch the bus.

9.

	R	$\times$	T	$=$	D
bicycle	15		T		$15T$
car	60		$T-\frac{1}{2}$		$60\left(T+\frac{1}{2}\right)$

$$15T = 60\left(T - \frac{1}{2}\right)$$
$$15T = 60T - 30$$
$$15T - 60T = 60T - 60T - 30$$
$$-45T = -30$$
$$\frac{-45T}{-45} = \frac{-30}{-45}$$
$$T = \frac{2}{3}$$

Find either the bicycle's distance or the car's distance.

$$R \times T = D$$
$$15 \times \frac{2}{3} = 10$$

Their house is 10 miles from the school.

11.

	R	$\times$	T	$=$	D
car	50		T		$50T$
another car	55		T		$55T$

$$50T + 55T = 630$$
$$105T = 630$$
$$\frac{105T}{105} = \frac{630}{105}$$
$$T = 6$$

The cars will meet in 6 hours.

13.

	R	$\times$	T	$=$	D
out	480		T		$480T$
back	640		$7 - T$		$640(7 - T)$

$$480T = 640(7 - T)$$
$$480T = 4480 - 640T$$
$$480T + 640T = 4480 - 640T + 640T$$
$$1120T = 4480$$
$$\frac{1120T}{1120} = \frac{4480}{1120}$$
$$T = 4$$

Find the distance out or the distance back.

$R \times T = D$

$480 \times 4 = 1920$

The base is 1920 miles from the target.

15.

	R	$\times$	T	$=$	D
first seconds	$20R$		180		$20R(180)$
last seconds	R		60		$R(60)$

$$20R(180) + R(60) = 366$$
$$3600R + 60R = 366$$
$$3660R = 366$$
$$\frac{3660R}{3660} = \frac{366}{3660}$$
$$R = 0.1$$

During the first 180 seconds, the shuttle was traveling at $20(0.1) = 2$ feet per second or 24 inches per second. During the last 60 seconds, the shuttle was traveling at 0.1 feet per second or 1.2 inches per second.

17.

	%	×	liters	=	Amount of glycerin in final mixture
40% solution	0.40		x		$0.40x$
80% solution	0.80		10		8
65% solution	0.65		$x + 10$		$0.65(x + 10)$

$$\begin{aligned} 0.40x + 8 &= 0.65(x+10) \\ 0.40x + 8 &= 0.65x + 6.5 \\ 0.40x &= 0.65x - 1.5 \\ -0.25x &= -1.5 \\ x &= 6 \end{aligned}$$

Mix 6 liters of the 40% glycerin solution.

19.

	price per pound	×	pounds	=	price
copper	0.65		x		$0.65x$
zinc	0.30		$70 - x$		$0.30(70 - x)$
brass	0.45		70		$0.45(70)$

$$\begin{aligned} 0.65x + 0.30(70 - x) &= 0.45(70) \\ 100 \cdot 0.65x + 100 \cdot 0.30(70 - x) &= 100 \cdot 0.45(70) \\ 65x + 30(70 - x) &= 45(70) \\ 65x + 2100 - 30x &= 3150 \\ 35x + 2100 &= 3150 \\ 35x &= 1050 \\ x &= 30 \\ 70 - x &= 40 \end{aligned}$$

Mix 30 pounds of copper with 40 pounds of zinc.

21.

	price per pound	×	pounds	=	price
Blue Jamaican	5		x		$5x$
Regular	2		80		$2(80)$
Mixture	2.60		$x + 80$		$2.6(x + 80)$

$$\begin{aligned} 5x + 2(80) &= 2.6(x+80) \\ 5x + 160 &= 2.6x + 208 \\ 5x &= 2.6x + 48 \\ 2.4x &= 48 \\ x &= 20 \end{aligned}$$

Mix 20 pounds of Blue Jamaican coffee.

23.

	%	×	ounces	=	amount of alcohol in final mixture
40%	0.4		x		$0.4x$
20%	0.2		$64-x$		$0.2(64-x)$
30%	0.3		64		$0.3(64)$

$$\begin{aligned} 0.4x+0.2(64-x)&=0.3(64) \\ 10\cdot 0.4x+10\cdot 0.2(64-x)&=10\cdot 0.3(64) \\ 4x+2(64-x)&=3(64) \\ 4x+128-2x&=192 \\ 2x+128&=192 \\ 2x&=64 \\ x&=32 \\ 64-x&=32 \end{aligned}$$

Mix 32 ounces of each.

25.

	%	×	quarts	=	Amount of antifreeze in final solution
50%	0.5		$30-x$		$0.5(30-x)$
0%	0		x		0
30%	0.3		30		$0.3(30)$

$$\begin{aligned} 0.5(30-x)+0&=0.3(30) \\ 15-0.5x&=9 \\ -0.5x&=-6 \\ x&=12 \end{aligned}$$

Drain 12 quarts of 50% solution and replace with 12 quarts of water.

27.

	%	×	ounces	=	Amount of juice in mixture
Welch's	x		12		$12x$
Water	0		$3(12)=36$		$0\cdot 36=0$
30% juice	0.3		$12+3(12)=48$		$0.3(48)$

$$\begin{aligned} 12x+0&=0.3(48) \\ 12x&=14.4 \\ x&=1.2 \end{aligned}$$

This is impossible, since it would need 120% of juice in the concentrate.

29.

	P	$\times$ r	$=$ I
6%	x	0.06	$0.06x$
8%	$20{,}000-x$	0.08	$0.08(20{,}000-x)$

$$\begin{aligned} 0.06x+0.08(20{,}000-x)&=1500\\ 100\cdot 0.06x+100\cdot 0.08(20{,}000-x)&=100\cdot 1500\\ 6x+8(20{,}000-x)&=150{,}000\\ 6x+160{,}000-8x&=150{,}000\\ 160{,}000-2x&=150{,}000\\ -2x&=-10{,}000\\ x&=5000\\ 20{,}000-x&=15{,}000 \end{aligned}$$

\$5000 is invested at 6% and \$15,000 is invested at 8%.

31.

	P	$\times$ r	$=$ I
5%	x	0.05	$0.05x$
7%	$18{,}000-x$	0.07	$0.07(18{,}000-x)$

$$\begin{aligned} 0.05x+0.07(18{,}000-x)&=1100\\ 0.05x+1260-0.07x&=1100\\ 1260-0.02x&=1100\\ -0.02x&=-160\\ x&=8000 \end{aligned}$$

\$8000 is invested at 5% in the savings account.

33.

	P	$\times$ r	$=$ I
5%	x	0.05	$0.05x$
6%	$10{,}000-x$	0.06	$0.06(10{,}000-x)$

$$\begin{aligned} 0.05x&=0.06(10{,}000-x)+60\\ 0.05x&=600-0.06x+60\\ 0.05x&=660-0.06x\\ 0.11x&=660\\ x&=6000\\ 10{,}000-x&=4000 \end{aligned}$$

\$6000 is invested at 5% and \$4000 is invested at 6%.

35.

	P	$\times$ r	$=$ I
6%	x	0.06	$0.06x$
10%	$40{,}000-x$	0.10	$0.10(40{,}000-x)$

$$0.06x = 0.10(40{,}000 - x)$$
$$0.06x = 4000 - 0.1x$$
$$0.16x = 4000$$
$$x = 25{,}000$$
$$40{,}000 - x = 15{,}000$$

\$25,000 is invested at 6% and \$15,000 is invested at 10%.

37. $-16(2)^2 + 118 = -16(4) + 118 = -64 + 118 = 54$

39. $-4 \cdot 8 \div 2 + 20 = -32 \div 2 + 20 = -16 + 20 = 4$

41. Answers may vary.

43. Answers may vary.

45.

	%	× gallons	= Amount of salt in mixture
10%	0.1	x	$0.1x$
20%	0.2	15	$0.2(15)$
16%	0.16	$x + 15$	$0.16(x + 15)$

$$0.1x + 0.2(15) = 0.16(x + 15)$$
$$0.1x + 3 = 0.16x + 2.4$$
$$0.1x = 0.16x - 0.6$$
$$-0.06x = -0.6$$
$$x = 10$$

Thus 10 gallons of 10% salt solution should be added.

47.

	R	× T	= D
bus	50	$T + 1$	$50(T + 1)$
car	60	T	$60T$

$$50(T + 1) = 60T$$
$$50T + 50 = 60T$$
$$50 = 10T$$
$$5 = T$$

The car overtakes the bus in 5 hours.

2.6 Formulas and Geometry Applications

Problems 2.6

1. a. Let $h = 15$.

$$\begin{aligned} H &= 2.8h + 28.1 \\ &= 2.8(15) + 28.1 \\ &= 42 + 28.1 \\ &= 70.1 \end{aligned}$$

Thus a woman with a 15-inch humerus should be about 70.1 inches tall.

b.

$$\begin{aligned} H &= 2.8\boxed{h} + 28.1 \\ H - 28.1 &= 2.8\boxed{h} \\ \frac{H-28.1}{2.8} &= \frac{2.8\boxed{h}}{2.8} \\ \frac{H-28.1}{2.8} &= \boxed{h} \\ h &= \frac{H-28.1}{2.8} \end{aligned}$$

c. Let $H = 61.7$.

$$h = \frac{61.7 - 28.1}{2.8} = 12$$

Thus the length of the humerus of a 61.7-inch-tall woman is 12 inches.

2. a. Let $C = 22$.

$$\begin{aligned} S &= \frac{1}{6}(C-4) \\ &= \frac{1}{6}(22-4) \\ &= \frac{1}{6}(18) \\ &= 3 \end{aligned}$$

The ant's speed is 3 cm/sec.

b.

$$\begin{aligned} S &= \frac{1}{6}\left(\boxed{C}-4\right) \\ 6 \cdot S &= 6 \cdot \frac{1}{6}\left(\boxed{C}-4\right) \\ 6S &= \boxed{C} - 4 \\ 6S + 4 &= \boxed{C} - 4 + 4 \\ 6S + 4 &= \boxed{C} \\ C &= 6S + 4 \end{aligned}$$

c. Let $S = 2$.

$$\begin{aligned} C &= 6(2) + 4 \\ &= 12 + 4 \\ &= 16 \end{aligned}$$

Thus, the temperature is 16°C.

3. a. The final cost is obtained by adding the original cost C and the tax T.

$$F = C + T$$

b. Let $C = 10$.

$$\begin{aligned} F &= C + T \\ F &= 10 + T \\ F - 10 &= T \\ T &= F - 10 \end{aligned}$$

4. a.

$$\begin{aligned} A &= LW \\ &= (2.4 \text{ in.}) \cdot (1.2 \text{ in.}) \\ &= 2.88 \text{ in.}^2 \end{aligned}$$

b.

$$\begin{aligned} P &= 2L + 2W \\ &= 2(2.4 \text{ in.}) + 2(1.2 \text{ in.}) \\ &= 4.8 \text{ in.} + 2.4 \text{ in.} \\ &= 7.2 \text{ in.} \end{aligned}$$

c. Let $P = 100$ and $L = 20$.

$$\begin{aligned} P &= 2L + 2W \\ 100 &= 2(20) + 2W \\ 100 &= 40 + 2W \\ 100 - 40 &= 2W \\ 60 &= 2W \\ 30 &= W \end{aligned}$$

The width is 30 inches.

5. a. $A = \frac{1}{2}bh$
$= \frac{1}{2}(30 \text{ in.})(15 \text{ in.})$
$= 225 \text{ in.}^2$

b. Let $A = 300$ and $b = 20$.
$A = \frac{1}{2}bh$
$300 = \frac{1}{2}(20)h$
$300 = 10h$
$30 = h$
Thus, the height of the sail is 30 feet.

6. a. Let $C = 9\pi$.
$C = 2\pi r$
$9\pi = 2\pi r$
$\frac{9}{2} = r$
$r = 4.5$
The radius is 4.5 inches.

b. Find the area of the entire CD.
$A = \pi r^2 = \pi(4.5)^2 = 20.25\pi$
Find the area of the nonrecorded part.
$A = \pi r^2 = \pi(0.75)^2 = 0.5625\pi$
The area of the recorded region is
$20.25\pi - 0.5625\pi = 19.6875\pi \text{ in.}^2$

7. a. $(2x - 10) + 8x = 180$
$10x - 10 = 180$
$10x = 190$
$x = 19$
$2(19) - 10 = 28$ and $8(19) = 152$
The angles measure 28° and 152°.

b. $8x - 15 = 4x - 3$
$8x = 4x + 12$
$4x = 12$
$x = 3$
$8(3) - 15 = 9$ and $4(3) - 3 = 9$
Both measure 9°.

c. $(3x - 3) + (x + 17) = 90$
$4x + 14 = 90$
$4x = 76$
$x = 19$
$3(19) - 3 = 54$ and $19 + 17 = 36$
The angles measure 54° and 36°.

8. a. Let $S = 2$.
$L = \frac{22 + S}{3}$
$= \frac{22 + 2}{3}$
$= \frac{24}{3} = 8$
The foot is 8 inches long.

b. Let $S = 3$.
$L = \frac{21 + S}{3}$
$= \frac{21 + 3}{3}$
$= \frac{24}{3} = 8$
The foot is 8 inches long.

c. Let $L = 17$.
$S = 3L - 22$
$= 3(17) - 22$
$= 51 - 22$
$= 29$
His right foot needs a size 29.

Exercises 2.6

1. a. Let $R = 30$ and $T = 4$.
$D = RT = 30(4) = 120$

b. Let $R = 55$ and $T = 5$.
$D = RT = 55(5) = 275$
The car traveled 275 miles.

c. $D = RT$
$\frac{D}{T} = \frac{RT}{T}$
$\frac{D}{T} = R$ or $R = \frac{D}{T}$

d. Let $D = 180$ and $T = 3$.

$R = \frac{D}{T} = \frac{180}{3} = 60$

The rate is 60 miles per hour.

3. a. Let $H = 60$.

$$\begin{aligned} W &= 5H - 190 \\ &= 5(60) - 190 \\ &= 300 - 190 \\ &= 110 \end{aligned}$$

He should weigh 110 pounds.

b.

$$\begin{aligned} W &= 5H - 190 \\ W + 190 &= 5H \\ \frac{W+190}{5} &= H \\ H &= \frac{W+190}{5} \end{aligned}$$

c. Let $W = 200$.

$$\begin{aligned} H &= \frac{200+190}{5} \\ &= \frac{390}{5} \\ &= 78 \text{ in.} \\ &= 6\frac{1}{2} \text{ ft} \end{aligned}$$

He should be 78 inches or 6 feet 6 inches tall.

5. a. Let $C = 15$.

$$\begin{aligned} F &= \frac{9}{5}C + 32 \\ &= \frac{9}{5}(15) + 32 \\ &= 27 + 32 \\ &= 59 \end{aligned}$$

The corresponding temperature is 59°F.

b.

$$\begin{aligned} F &= \frac{9}{5}C + 32 \\ F - 32 &= \frac{9}{5}C \\ \frac{5}{9}\cdot(F-32) &= \frac{5}{9}\cdot\frac{9}{5}C \\ \frac{5}{9}(F-32) &= C \text{ or } C = \frac{5}{9}(F-32) \end{aligned}$$

c. Let $F = 50$.

$$\begin{aligned} C &= \frac{5}{9}(F-32) \\ &= \frac{5}{9}(50-32) \\ &= \frac{5}{9}(18) \\ &= 10 \end{aligned}$$

The corresponding temperature is 10°C.

7. a. $\text{EER} = \frac{\text{Btu}}{w}$

b. Let Btu = 9000 and $w = 1000$.

$\text{EER} = \frac{\text{Btu}}{w} = \frac{9000}{1000} = 9$

The EER is 9.

c.

$$\begin{aligned} \text{EER} &= \frac{\text{Btu}}{w} \\ w(\text{EER}) &= w\left(\frac{\text{Btu}}{w}\right) \\ (\text{EER})(w) &= \text{Btu} \\ \text{Btu} &= (\text{EER})(w) \end{aligned}$$

d. Let $w = 2000$ and EER = 10.

$\text{Btu} = (\text{EER})(w) = 10(2000) = 20{,}000$

It produces 20,000 Btu per hour.

9. a. $S = C + M$

b. Let $M = 15$ and $C = 52$.

$S = C + M = 52 + 15 = 67$

The selling price should be $67.

c.

$$\begin{aligned} S &= C + M \\ S - C &= M \\ M &= S - C \end{aligned}$$

d. Let $S = 18.75$ and $C = 10.50$.

$M = S - C = 18.75 - 10.50 = 8.25$

The markup is $8.25.

11. a. Let $W = 10$ and $L = 20$.

$$\begin{aligned} P &= 2L + 2W \\ &= 2(20) + 2(10) \\ &= 40 + 20 \\ &= 60 \end{aligned}$$

The perimeter is 60 cm.

b. Let $P = 220$ and $W = 20$.
$$\begin{aligned} P &= 2L + 2W \\ 220 &= 2L + 2(20) \\ 220 &= 2L + 40 \\ 180 &= 2L \\ 90 &= L \end{aligned}$$
The length is 90 cm.

13. a. Let $r = 10$.
$C = 2\pi r = 2(3.14)(10) = 62.8$
The circumference is 62.8 inches.

b.
$$\begin{aligned} C &= 2\pi r \\ \frac{C}{2\pi} &= \frac{2\pi r}{2\pi} \\ \frac{C}{2\pi} &= r \\ r &= \frac{C}{2\pi} \end{aligned}$$

c. Let $C = 20\pi$.
$$r = \frac{C}{2\pi} = \frac{20\pi}{2\pi} = 10$$
The radius is 10 inches.

15. a. Let $L = 4.2$ and $W = 3.1$.
$A = LW = (4.2)(3.1) = 13.02$
The area is 13.02 m^2.

b.
$$\begin{aligned} A &= LW \\ \frac{A}{L} &= \frac{LW}{L} \\ \frac{A}{L} &= W \text{ or } W = \frac{A}{L} \end{aligned}$$

c. Let $A = 60$ and $L = 10$.
$$W = \frac{A}{L} = \frac{60}{10} = 6$$
The width is 6 meters.

17. Vertical angles must be equal.
$$\begin{aligned} 15 - 4x &= 25 - 2x \\ -4x &= 10 - 2x \\ -2x &= 10 \\ x &= -5 \end{aligned}$$
$15 - 4(-5) = 35$ and $25 - 2(-5) = 35$
The angles each measure 35°.

19. Vertical angles must be equal.
$$\begin{aligned} 80 + 3x &= 40 + 5x \\ 3x &= -40 + 5x \\ -2x &= -40 \\ x &= 20 \end{aligned}$$
$80 + 3(20) = 140$ and $40 + 5(20) = 140$
The angles each measure 140°.

21. Vertical angles must be equal.
$$\begin{aligned} 6x - 5 &= 5x + 25 \\ 6x &= 5x + 30 \\ x &= 30 \end{aligned}$$
$6(30) - 5 = 175$ and $5(30) + 25 = 175$
The angles each measure 175°.

23. The angles are complementary.
$$\begin{aligned} 7x + (5x - 30) &= 90 \\ 12x - 30 &= 90 \\ 12x &= 120 \\ x &= 10 \end{aligned}$$
$7(10) = 70$ and $5(10) - 30 = 20$
The angles measure 70° and 20°.

25. The angles are complementary.
$$\begin{aligned} (4x - 25) + x &= 90 \\ 5x - 25 &= 90 \\ 5x &= 115 \\ x &= 23 \end{aligned}$$
$4(23) - 25 = 67$ and 23
The angles measure 67° and 23°.

27. The angles are supplementary.
$$\begin{aligned} (8x + 30) + (2x + 10) &= 180 \\ 10x + 40 &= 180 \\ 10x &= 140 \\ x &= 14 \end{aligned}$$
$8(14) + 30 = 142$ and $2(14) + 10 = 38$
The angles measure 142° and 38°.

29. The angles are supplementary.
$$\begin{aligned} (3x + 6) + (5x + 6) &= 180 \\ 8x + 12 &= 180 \\ 8x &= 168 \\ x &= 21 \end{aligned}$$
$3(21) + 6 = 69$ and $5(21) + 6 = 111$
The angles measure 69° and 111°.

31. Let $P = 1110$ and $L = 480$.

$$\begin{aligned} P &= 2L + 2W \\ 1110 &= 2(480) + 2W \\ 1110 &= 960 + 2W \\ 150 &= 2W \\ 75 &= W \end{aligned}$$

The pool is 75 meters wide.

33. Let $C = 14.13$.

$$\begin{aligned} C &= 2\pi r \\ 14.13 &= 2(3.14)r \\ 2.25 &= r \end{aligned}$$

The diameter is $2(2.25) = 4.5$ inches.

35. **a.**

$$\begin{aligned} C &= A + 10 \\ C - 10 &= A \\ A &= C - 10 \end{aligned}$$

b. Let $C = 50$.

$A = C - 10 = 50 - 10 = 40$

The American size is 40.

37. **a.** Let $t = 1985 - 1975 = 10$.

$$\begin{aligned} N &= 9.74 + 0.40t \\ &= 9.74 + 0.40(10) \\ &= 9.74 + 4 \\ &= 13.74 \end{aligned}$$

In 1985, there were 13.74 million recreational boats.

b.

$$\begin{aligned} N &= 9.74 + 0.40t \\ N - 9.74 &= 0.40t \\ \frac{N - 9.74}{0.40} &= t \end{aligned}$$

$$t = \frac{N - 9.74}{0.40} \text{ or } t = \frac{5(N - 9.74)}{2}$$

c. Let $N = 17.74$.

$$t = \frac{N - 9.74}{0.40} = \frac{17.74 - 9.74}{0.40} = \frac{8}{0.40} = 20$$

$1975 + 20 = 1995$

In 1995, the expected number of recreational boats is 17.74 million.

39.

$$\begin{aligned} 2x - 1 &= x + 3 \\ 2x - 1 + 1 &= x + 3 + 1 \\ 2x &= x + 4 \\ 2x - x &= x - x + 4 \\ x &= 4 \end{aligned}$$

41.

$$\begin{aligned} 4(x + 1) &= 3x + 7 \\ 4x + 4 &= 3x + 7 \\ 4x + 4 - 4 &= 3x + 7 - 4 \\ 4x &= 3x + 3 \\ 4x - 3x &= 3x - 3x + 3 \\ x &= 3 \end{aligned}$$

43.

$$\begin{aligned} \frac{x}{3} - \frac{x}{2} &= 1 \\ 6 \cdot \frac{x}{3} - 6 \cdot \frac{x}{2} &= 6 \cdot 1 \\ 2x - 3x &= 6 \\ -x &= 6 \\ x &= -6 \end{aligned}$$

45. Let $A = 5400$ and $W = 60$.

$$\begin{aligned} A &= LW \\ 5400 &= L(60) \\ 90 &= L \end{aligned}$$

The sod area can be 90 feet long.

47. Answers may vary.

49. Answers may vary.

51. Answers may vary. Sample answer: $C = 2\pi r$

Let $r = \frac{4}{2} = 2$ and $r = \frac{6}{2} = 3$ and compare.

$$\frac{2\pi(3)}{2\pi(2)} = \frac{3}{2} = 1.5 \text{ or } 150\%$$

Using this method, the Monster burger is 50% bigger than the Lite burger.

53. The angles are complementary.

$$\begin{aligned} (3x - 25) + 2x &= 90 \\ 5x - 25 &= 90 \\ 5x &= 115 \\ x &= 23 \end{aligned}$$

$3(23) - 25 = 44$ and $2(23) = 46$

The angles measure 44° and 46°.

55. The angles are supplementary.

$$\begin{aligned} (7x + 30) + (3x + 10) &= 180 \\ 10x + 40 &= 180 \\ 10x &= 140 \\ x &= 14 \end{aligned}$$

$7(14) + 30 = 128$ and $3(14) + 10 = 52$

The angles measure 128° and 52°.

57. The angles are supplementary.

$$(2x+6)+(6x+14)=180$$
$$8x+20=180$$
$$8x=160$$
$$x=20$$

$2(20) + 6 = 46$ and $6(20) + 14 = 134$

The angles measure 46° and 134°.

59. a. Let $h = 20$.

$$\begin{aligned} H &= 2.75h + 71.48 \\ &= 2.75(20) + 71.48 \\ &= 126.48 \end{aligned}$$

The woman is 126.48 centimeters.

b.

$$H = 2.75h + 71.48$$
$$H - 71.48 = 2.75h$$
$$\frac{H-71.48}{2.75} = h$$
$$h = \frac{H-71.48}{2.75}$$

c. Let $H = 140.23$.

$$h = \frac{H-71.48}{2.75} = \frac{140.23-71.48}{2.75} = 25$$

The woman's humerus is 25 centimeters.

61. a. Let $t = 1985 - 1975 = 10$.

$$\begin{aligned} N &= 15 + 0.60t \\ &= 15 + 0.60(10) \\ &= 15 + 6 \\ &= 21 \end{aligned}$$

There were 21 thousand theaters in 1985.

b.

$$N = 15 + 0.60t$$
$$N - 15 = 0.60t$$
$$\frac{N-15}{0.60} = t$$
$$t = \frac{N-15}{0.60} \text{ or } t = \frac{5(N-15)}{3}$$

c. Let $N = 27$.

$$t = \frac{27-15}{0.60}$$
$$t = 20$$

$1975 + 20 = 1995$

In 1995, the number of theaters totaled 27,000.

63. a. Let $F = 41$.

$$\begin{aligned} C &= \frac{5}{9}F - \frac{160}{9} \\ &= \frac{5}{9}(41) - \frac{160}{9} \\ &= \frac{205}{9} - \frac{160}{9} \\ &= \frac{45}{9} \\ &= 5 \end{aligned}$$

The temperature is 5°C.

b.

$$C = \frac{5}{9}F - \frac{160}{9}$$
$$\frac{9}{5}\cdot C = \frac{9}{5}\cdot\frac{5}{9}F - \frac{9}{5}\cdot\frac{160}{9}$$
$$\frac{9}{5}C = F - 32$$
$$\frac{9}{5}C + 32 = F \text{ or } F = \frac{9}{5}C + 32$$

c. Let $C = 20$.

$$\begin{aligned} F &= \frac{9}{5}C + 32 \\ &= \frac{9}{5}(20) + 32 \\ &= 36 + 32 \\ &= 68 \end{aligned}$$

The temperature is 68°F.

2.7 Properties of Inequalities

Problems 2.7

1. a. Since 5 is to the right of 3, $5 > 3$.

b. Since -1 is to the right of -4, $-1 > -4$.

c. Since -5 is to the left of -4, $-5 < -4$.

2. a. For $x \le -2$, any number less than or equal to -2 is a solution.

−4 −3 −2 −1 0 1 2 3 4

b. For $x > -3$, any number greater than -3 is a solution.

−4 −3 −2 −1 0 1 2 3 4

3. a.

$$\begin{aligned} 4x-3 &< 3(x-2) \\ 4x-3 &< 3x-6 \\ 4x-3+3 &< 3x-6+3 \\ 4x &< 3x-3 \\ 4x-3x &< 3x-3x-3 \\ x &< -3 \end{aligned}$$

Any number less than −3 is a solution.

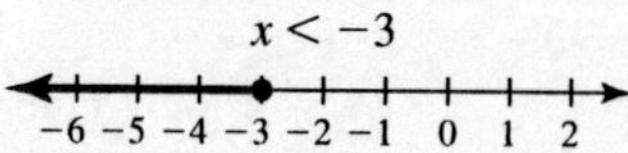

b.

$$\begin{aligned} 3(x+2) &\geq 2x+5 \\ 3x+6 &\geq 2x+5 \\ 3x+6-6 &\geq 2x+5-6 \\ 3x &\geq 2x-1 \\ 3x-2x &\geq 2x-2x-1 \\ x &\geq -1 \end{aligned}$$

Any number greater than or equal to −1 is a solution.

$x \geq -1$

−4 −3 −2 −1 0 1 2 3 4

4. a.

$$\begin{aligned} 4x+3 &\leq 2x+5 \\ 4x+3-3 &\leq 2x+5-3 \\ 4x &\leq 2x+2 \\ 4x-2x &\leq 2x-2x+2 \\ 2x &\leq 2 \\ \frac{2x}{2} &\leq \frac{2}{2} \\ x &\leq 1 \end{aligned}$$

Any number less than or equal to 1 is a solution.

$x \leq 1$

−4 −3 −2 −1 0 1 2 3 4

b.

$$\begin{aligned} 5(x-1) &> 3x+1 \\ 5x-5 &> 3x+1 \\ 5x-5+5 &> 3x+1+5 \\ 5x &> 3x+6 \\ 5x-3x &> 3x-3x+6 \\ 2x &> 6 \\ \frac{2x}{2} &> \frac{6}{2} \\ x &> 3 \end{aligned}$$

Any number greater than 3 is a solution.

$x > 3$

−2 −1 0 1 2 3 4 5 6

5. a.

$$\begin{aligned} -4x &< 20 \\ \frac{-4x}{-4} &> \frac{20}{-4} \\ x &> -5 \end{aligned}$$

Any number greater than −5 is a solution.

b.

$$\begin{aligned} \frac{-x}{3} &> 2 \\ -3\left(\frac{-x}{3}\right) &< -3 \cdot 2 \\ x &< -6 \end{aligned}$$

Any number less than −6 is a solution.

c.

$$\begin{aligned} 2(x-1) &\leq 4x+1 \\ 2x-2 &\leq 4x+1 \\ 2x-2+2 &\leq 4x+1+2 \\ 2x &\leq 4x+3 \\ 2x-4x &\leq 4x-4x+3 \\ -2x &\leq 3 \\ \frac{-2x}{-2} &\geq \frac{3}{-2} \\ x &\geq -\frac{3}{2} \end{aligned}$$

Any number greater than or equal to $-\frac{3}{2}$ is a solution.

6.

$$\begin{aligned} \frac{-x}{3}+\frac{x}{4} &< \frac{x-8}{4} \\ 12 \cdot \left(\frac{-x}{3}\right)+12 \cdot \frac{x}{4} &< 12 \cdot \frac{x-8}{4} \\ -4x+3x &< 3(x-8) \\ -x &< 3x-24 \\ -x-3x &< 3x-3x-24 \\ -4x &< -24 \\ \frac{-4x}{-4} &> \frac{-24}{-4} \\ x &> 6 \end{aligned}$$

Any number greater than 6 is a solution.

7. a. $2 < x$ and $x < 4$ can be written as $2 < x < 4$. The solution consists of all numbers between 2 and 4.

−3 −2 −1 0 1 2 3 4 5

b. $3 \ge -x$
$-1 \cdot 3 \le -1 \cdot (-x)$
$-3 \le x$

Now $-3 \le x$ and $x \le -1$ can be written as $-3 \le x \le -1$. The solution consists of all numbers between –3 and –1, inclusive.

−4 −3 −2 −1 0 1 2 3 4

c. $x + 2 \le 6$
$x + 2 - 2 \le 6 - 2$
$x \le 4$

$-3x < 6$
$\frac{-3x}{-3} > \frac{6}{-3}$
$x > -2$

Rearranging, we have $-2 < x \le 4$. The solution consists of all numbers between –2 and 4 and the number 4.

−3 −2 −1 0 1 2 3 4 5

8. $P = 20 - 2x < 10$
$20 - 2x < 10$
$20 - 20 - 2x < 10 - 20$
$-2x < -10$
$\frac{-2x}{-2} > \frac{-10}{-2}$
$x > 5$

This means that more than 5 years after 1997, or in 2003, the percent of smokers in this group is less than 10%.

Exercises 2.7

1. Since 8 is to the left of 9, $8 < 9$.

3. Since –4 is to the right of –9, $-4 > -9$.

5. Since $\frac{1}{4}$ is to the left of $\frac{1}{3}$, $\frac{1}{4} < \frac{1}{3}$.

7. Since $-\frac{2}{3}$ is to the right of –1, $-\frac{2}{3} > -1$.

9. Since $-3\frac{1}{4}$ is to the left of –3, $-3\frac{1}{4} < -3$.

11. $2x + 6 \le 8$
$2x + 6 - 6 \le 8 - 6$
$2x \le 2$
$\frac{2x}{2} \le \frac{2}{2}$
$x \le 1$

Any number less than or equal to 1 is a solution.

0 1

13. $-3y - 4 \ge -10$
$-3y - 4 + 4 \ge -10 + 4$
$-3y \ge -6$
$\frac{-3y}{-3} \le \frac{-6}{-3}$
$y \le 2$

Any number less than or equal to 2 is a solution.

0 2

15. $-5x + 1 < -14$
$-5x + 1 - 1 < -14 - 1$
$-5x < -15$
$\frac{5x}{-5} > \frac{-15}{-5}$
$x > 3$

Any number greater than 3 is a solution.

0 3

17. $3a + 4 \le a + 10$
$3a + 4 - 4 \le a + 10 - 4$
$3a \le a + 6$
$3a - a \le a - a + 6$
$2a \le 6$
$\frac{2a}{2} \le \frac{6}{2}$
$a \le 3$

Any number less than or equal to 3 is a solution.

0 3

19. $5z - 12 \ge 6z - 8$
$5z - 12 + 12 \ge 6z - 8 + 12$
$5z \ge 6z + 4$
$5z - 6z \ge 6z - 6z + 4$
$-z \ge 4$
$z \le -4$

Any number less than or equal to –4 is a

solution.

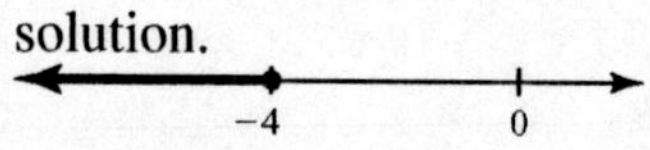

21.
$$10-3x \le 7-6x$$
$$10-10-3x \le 7-10-6x$$
$$-3x \le -3-6x$$
$$-3x+6x \le -3-6x+6x$$
$$3x \le -3$$
$$\frac{3x}{3} \le \frac{-3}{3}$$
$$x \le -1$$

Any number less than or equal to -1 is a solution.

23.

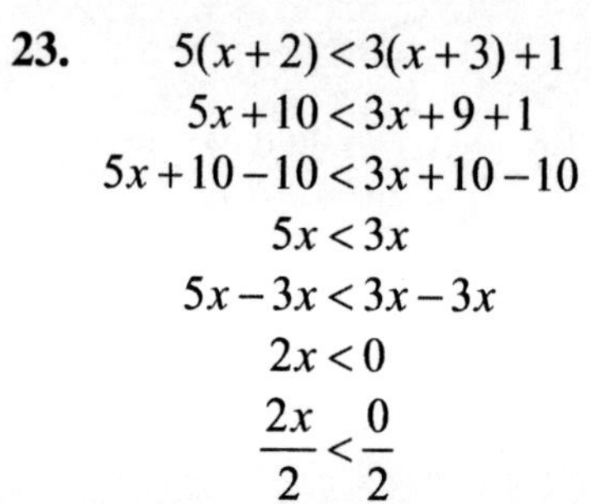

$$5(x+2) < 3(x+3)+1$$
$$5x+10 < 3x+9+1$$
$$5x+10-10 < 3x+10-10$$
$$5x < 3x$$
$$5x-3x < 3x-3x$$
$$2x < 0$$
$$\frac{2x}{2} < \frac{0}{2}$$
$$x < 0$$

Any number less than 0 is a solution.

25.
$$-2x+\frac{1}{4} \ge 2x+\frac{4}{5}$$
$$20(-2x)+20\left(\frac{1}{4}\right) \ge 20(2x)+20\left(\frac{4}{5}\right)$$
$$-40x+5 \ge 40x+16$$
$$-40x+5-5 \ge 40x+16-5$$
$$-40x \ge 40x+11$$
$$-40x-40x \ge 40x-40x+11$$
$$-80x \ge 11$$
$$\frac{-80x}{-80} \le \frac{11}{-80}$$
$$x \le -\frac{11}{80}$$

Any number less than or equal to $-\frac{11}{80}$ is a solution.

27.
$$\frac{x}{5}-\frac{x}{4} \le 1$$
$$20 \cdot \frac{x}{5} - 20 \cdot \frac{x}{4} \le 20 \cdot 1$$
$$4x-5x \le 20$$
$$-x \le 20$$
$$x \ge -20$$

Any number greater than or equal to -20 is a solution.

29.
$$\frac{7x+2}{6}+\frac{1}{2} \ge \frac{3}{4}x$$
$$12 \cdot \frac{7x+2}{6} + 12 \cdot \frac{1}{2} \ge 12 \cdot \frac{3}{4}x$$
$$2(7x+2)+6 \ge 3 \cdot 3x$$
$$14x+4+6 \ge 9x$$
$$14x-14x+10 \ge 9x-14x$$
$$10 \ge -5x$$
$$\frac{10}{-5} \le \frac{-5x}{-5}$$
$$-2 \le x$$
$$x \ge -2$$

Any number greater than or equal to -2 is a solution.

31. $x < 3$ and $-x < -2$ (or $x > 2$) can be rewritten as $2 < x < 3$. The solution consists of all numbers between 2 and 3.

33.
$$x+1 < 4 \quad \text{and} \quad -x < -1$$
$$x+1-1 < 4-1 \qquad -1 \cdot (-x) > -1 \cdot (-1)$$
$$x < 3 \qquad x > 1$$

Now $x < 3$ and $x > 1$ can be rewritten as $1 < x < 3$. The solution consists of all numbers between 1 and 3.

35.
$$x-2 < 3 \quad \text{and} \quad 2 > -x$$
$$x-2+2 < 3+2 \qquad -1 \cdot 2 < -1 \cdot (-x)$$
$$x < 5 \qquad -2 < x$$

Now $x < 5$ and $-2 < x$ can be rewritten as $-2 < x < 5$. The solution consists of all the numbers between -2 and 5.

37. $x+2<3$ and $-4<x+1$

$x<1 \qquad -5<x$

Now $x<1$ and $-5<x$ can be rewritten as $-5<x<1$. The solution consists of all numbers between -5 and 1.

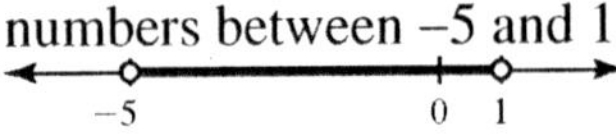

39. $x-1>2$ and $x+7<12$

$x>3 \qquad x<5$

Now $x>3$ and $x<5$ can be rewritten as $3<x<5$. The solution consists of all numbers between 3 and 5.

41. $20°\text{F} < t < 40°\text{F}$

43. $\$12{,}000 < s < \$13{,}000$

45. $2 \le e \le 7$

47. $\$3.50 < c < \$4.00$

49. $a < 41$ ft

51. $N = 720 + 5t > 800$

$$720+5t>800$$
$$720-720+5t>800-720$$
$$5t>80$$
$$\frac{5t}{5}>\frac{80}{5}$$
$$t>16$$

$1980 + 16 = 1996$

After 1996, you would expect that the NCAA would have more than 800 teams.

53. $C = 127 + 17t > 300$

$$127+17t>300$$
$$127-127+17t>300-127$$
$$17t>173$$
$$\frac{17t}{17}>\frac{173}{17}$$
$$t>10.2 \text{ or } 11$$

$1980 + 11 = 1991$

After 1991, you would expect the average hospital room rate to surpass the \$300 per day mark.

55. $C = 3.91 - 0.13t < 3$

$$3.91-0.13t<3$$
$$3.91-3.91-0.13t<3-3.91$$
$$-0.13t<-0.91$$
$$\frac{-0.13t}{-0.13}>\frac{-0.91}{-0.13}$$
$$t>7$$

$1989 + 7 = 1996$

After 1996, or in 1997, you would expect the per capita consumption to fall below 3 pounds per year.

57. $-5 \cdot 8 = -40$

59. $-5(-8) = 40$

61. $\dfrac{-18}{9} = -2$

63. $\dfrac{-35}{-5} = 7$

65. $J = 5$ ft $= 60$ in.

67. $F = S - 3$

69. $S = 6$ ft 5 in. $= 77$ in.

71. $F = S - 3 = 77 - 3 = 74$

$B > F$

$B > 74$ in.

or $B > 6$ ft 2 in.

73. You have to change the direction of the inequality when multiplying or dividing by a negative number.

75. Answers may vary.

77. $3 \ge -x$ and $x \le -1$

$$-1\cdot 3 \le -1\cdot(-x)$$
$$-3 \le x$$

Now $-3 \le x$ and $x \le -1$ can be rewritten as $-3 \le x \le -1$. The solution consists of all numbers between -3 and -1, inclusive.

79.
$$\frac{-x}{3}+\frac{x}{4}<\frac{x-4}{4}$$
$$12\cdot\left(\frac{-x}{3}\right)+12\cdot\frac{x}{4}<12\cdot\frac{x-4}{4}$$
$$-4x+3x<3(x-4)$$
$$-x<3x-12$$
$$-x-3x<3x-3x-12$$
$$-4x<-12$$
$$\frac{-4x}{-4}>\frac{-12}{-4}$$
$$x>3$$
Any number greater than 3 is a solution.

81.
$$3(x-1)>x+3$$
$$3x-3>x+3$$
$$3x-3+3>x+3+3$$
$$3x>x+6$$
$$3x-x>x-x+6$$
$$2x>6$$
$$\frac{2x}{2}>\frac{6}{2}$$
$$x>3$$
Any number greater than 3 is a solution.

83.
$$3(x+1)\geq 2x+5$$
$$3x+3\geq 2x+5$$
$$3x+3-3\geq 2x+5-3$$
$$3x\geq 2x+2$$
$$3x-2x\geq 2x-2x+2$$
$$x\geq 2$$
Any number greater than or equal to 2 is a solution.

85. $x<1$
Any number less than 1 is a solution.

87. Since −2 is to the left of −1, $-2<-1$.

89. Since $\frac{1}{3}$ is to the right of −3, $\frac{1}{3}>-3$.

Review Exercises

1. a. If $x = 5$, then $7 = 14 - x$ becomes $7 = 14 - 5$, which is a *false* statement. Hence, 5 is not a solution of the equation.

b. If $x = 4$, then $13 = 17 - x$ becomes $13 = 17 - 4$ which is a *true* statement. Thus 4 is a solution of the equation.

c. If $x = -2$, then $8 = 6 - x$ becomes $8 = 6 - (-2)$, which is a *true* statement. Thus −2 is a solution of the equation.

2. a.
$$x-\frac{1}{3}=\frac{1}{3}$$
$$x-\frac{1}{3}+\frac{1}{3}=\frac{1}{3}+\frac{1}{3}$$
$$x=\frac{2}{3}$$
The solution is $\frac{2}{3}$.

b.
$$x-\frac{5}{7}=\frac{2}{7}$$
$$x-\frac{5}{7}+\frac{5}{7}=\frac{2}{7}+\frac{5}{7}$$
$$x=\frac{7}{7}$$
$$x=1$$
The solution is 1.

c.
$$x-\frac{5}{9}=\frac{1}{9}$$
$$x-\frac{5}{9}+\frac{5}{9}=\frac{1}{9}+\frac{5}{9}$$
$$x=\frac{6}{9}$$
$$x=\frac{2}{3}$$
The solution is $\frac{2}{3}$.

3. a.
$$\begin{aligned}-3x+\frac{5}{9}+4x-\frac{2}{9}&=\frac{5}{9}\\ x+\frac{3}{9}&=\frac{5}{9}\\ x+\frac{3}{9}-\frac{3}{9}&=\frac{5}{9}-\frac{3}{9}\\ x&=\frac{2}{9}\end{aligned}$$
The solution is $\frac{2}{9}$.

b.
$$\begin{aligned}-2x+\frac{4}{7}+3x-\frac{2}{7}&=\frac{6}{7}\\ x+\frac{2}{7}&=\frac{6}{7}\\ x+\frac{2}{7}-\frac{2}{7}&=\frac{6}{7}-\frac{2}{7}\\ x&=\frac{4}{7}\end{aligned}$$
The solution is $\frac{4}{7}$.

c.
$$\begin{aligned}-4x+\frac{5}{6}+5x-\frac{1}{6}&=\frac{5}{6}\\ x+\frac{4}{6}&=\frac{5}{6}\\ x+\frac{4}{6}-\frac{4}{6}&=\frac{5}{6}-\frac{4}{6}\\ x&=\frac{1}{6}\end{aligned}$$
The solution is $\frac{1}{6}$.

4. a.
$$\begin{aligned}3&=4(x-1)+2-3x\\ 3&=4x-4+2-3x\\ 3&=x-2\\ 3+2&=x-2+2\\ 5&=x \text{ or } x=5\end{aligned}$$
The solution is 5.

b.
$$\begin{aligned}4&=5(x-1)+9-4x\\ 4&=5x-5+9-4x\\ 4&=x+4\\ 4-4&=x+4-4\\ 0&=x \text{ or } x=0\end{aligned}$$
The solution is 0.

c.
$$\begin{aligned}5&=6(x-1)+8-5x\\ 5&=6x-6+8-5x\\ 5&=x+2\\ 5-2&=x+2-2\\ 3&=x \text{ or } x=3\end{aligned}$$
The solution is 3.

5. a.
$$\begin{aligned}6+3(x+1)&=2+3x\\ 6+3x+3&=2+3x\\ 9+3x&=2+3x\\ 9+3x-3x&=2+3x-3x\\ 9&=2\end{aligned}$$
Since this statement is *false*, the equation has no solution.

b.
$$\begin{aligned}-2+4(x-1)&=-7-4x\\ -2+4x-4&=-7-4x\\ -6+4x&=-7-4x\\ -6+6+4x&=-7+6-4x\\ 4x&=-1-4x\\ 4x+4x&=-1-4x+4x\\ 8x&=-1\\ \frac{8x}{8}&=\frac{-1}{8}\\ x&=-\frac{1}{8}\end{aligned}$$
The solution is $-\frac{1}{8}$.

c.
$$\begin{aligned}-1-2(x+1)&=3-2x\\ -1-2x-2&=3-2x\\ -3-2x&=3-2x\\ -3-2x+2x&=3-2x+2x\\ -3&=3\end{aligned}$$
Since this statement is *false*, the equation has no solution.

6. a.
$$\begin{aligned}5+2(x+1)&=2x+7\\ 5+2x+2&=2x+7\\ 2x+7&=2x+7\end{aligned}$$
Since both sides are *identical*, the equation is an identity. The solution is all real numbers.

b. $-2+3(x-1)=-5+3x$
$-2+3x-3=-5+3x$
$-5+3x=-5+3x$
Since both sides are *identical*, the equation is an identity. The solution is all real numbers.

c. $-3-4(x-1)=1-4x$
$-3-4x+4=1-4x$
$1-4x=1-4x$
Since both sides are *identical*, the equation is an identity. The solution is all real numbers.

7. a. $\frac{1}{5}x=-3$
$5\cdot\frac{1}{5}x=5\cdot(-3)$
$x=-15$
The solution is −15.

b. $\frac{1}{7}x=-2$
$7\cdot\frac{1}{7}x=7\cdot(-2)$
$x=-14$
The solution is −14.

c. $5x=-10$
$\frac{5x}{5}=\frac{-10}{5}$
$x=-2$
The solution is −2.

8. a. $-\frac{3}{4}x=-9$
$-\frac{4}{3}\cdot\left(-\frac{3}{4}x\right)=-\frac{4}{3}\cdot(-9)$
$x=12$
The solution is 12.

b. $-\frac{3}{5}x=-9$
$-\frac{5}{3}\cdot\left(-\frac{3}{5}x\right)=-\frac{5}{3}\cdot(-9)$
$x=15$
The solution is 15.

c. $-\frac{2}{3}x=-6$
$-\frac{3}{2}\cdot\left(-\frac{2}{3}x\right)=-\frac{3}{2}\cdot(-6)$
$x=9$
The solution is 9.

9. a. $\frac{x}{3}+\frac{2x}{4}=5$
$12\cdot\frac{x}{3}+12\cdot\frac{2x}{4}=12\cdot5$
$4x+6x=60$
$10x=60$
$\frac{10x}{10}=\frac{60}{10}$
$x=6$
The solution is 6.

b. $\frac{x}{4}+\frac{3x}{2}=6$
$4\cdot\frac{x}{4}+4\cdot\frac{3x}{2}=4\cdot6$
$x+6x=24$
$7x=24$
$\frac{7x}{7}=\frac{24}{7}$
The solution is $\frac{24}{7}$.

c. $\frac{x}{5}+\frac{3x}{10}=10$
$10\cdot\frac{x}{5}+10\cdot\frac{3x}{10}=10\cdot10$
$2x+3x=100$
$5x=100$
$\frac{5x}{5}=\frac{100}{5}$
$x=20$
The solution is 20.

10. a. $\frac{x}{3}-\frac{x}{4}=1$
$12\cdot\frac{x}{3}-12\cdot\frac{x}{4}=12\cdot1$
$4x-3x=12$
$x=12$
The solution is 12.

b.
$$\frac{x}{2}-\frac{x}{7}=10$$
$$14\cdot\frac{x}{2}-14\cdot\frac{x}{7}=14\cdot 10$$
$$7x-2x=140$$
$$5x=140$$
$$\frac{5x}{5}=\frac{140}{5}$$
$$x=28$$
The solution is 28.

c.
$$\frac{x}{4}-\frac{x}{5}=2$$
$$20\cdot\frac{x}{4}-20\cdot\frac{x}{5}=20\cdot 2$$
$$5x-4x=40$$
$$x=40$$
The solution is 40.

11. a.
$$\frac{x-1}{4}-\frac{x+1}{6}=1$$
$$12\cdot\frac{x-1}{4}-12\cdot\frac{x+1}{6}=12\cdot 1$$
$$3(x-1)-2(x+1)=12$$
$$3x-3-2x-2=12$$
$$x-5=12$$
$$x-5+5=12+5$$
$$x=17$$
The solution is 17.

b.
$$\frac{x-1}{6}-\frac{x+1}{8}=0$$
$$24\cdot\frac{x-1}{6}-24\cdot\frac{x+1}{8}=24\cdot 0$$
$$4(x-1)-3(x+1)=0$$
$$4x-4-3x-3=0$$
$$x-7=0$$
$$x-7+7=0+7$$
$$x=7$$
The solution is 7.

c.
$$\frac{x-1}{8}-\frac{x+1}{10}=0$$
$$40\cdot\frac{x-1}{8}-40\cdot\frac{x+1}{10}=40\cdot 0$$
$$5(x-1)-4(x+1)=0$$
$$5x-5-4x-4=0$$
$$x-9=0$$
$$x-9+9=0+9$$
$$x=9$$
The solution is 9.

12. a. What percent of 30 is 6?
$$x\cdot 30=6$$
$$\frac{x\cdot 30}{30}=\frac{6}{30}$$
$$x=\frac{1}{5}=\frac{20}{100}$$
Thus, 6 is 20% of 30.

b. What percent of 40 is 4?
$$x\cdot 40=4$$
$$\frac{x\cdot 40}{40}=\frac{4}{40}$$
$$x=\frac{1}{10}=\frac{10}{100}$$
Thus, 4 is 10% of 40.

c. What percent of 50 is 10?
$$x\cdot 50=10$$
$$\frac{x\cdot 50}{50}=\frac{10}{50}$$
$$x=\frac{1}{5}=\frac{20}{100}$$
Thus, 10 is 20% of 50.

13. a. 20 is 40% of what number?
$$20=\frac{40}{100}\cdot n$$
$$20=\frac{2}{5}\cdot n$$
$$\frac{5}{2}\cdot 20=\frac{5}{2}\cdot\frac{2}{5}\cdot n$$
$$50=n$$
Thus 20 is 40% of 50.

b. 30 is 90% of what number?

$$30=\frac{90}{100}\cdot n$$
$$30=\frac{9}{10}\cdot n$$
$$\frac{10}{9}\cdot 30=\frac{10}{9}\cdot\frac{9}{10}\cdot n$$
$$\frac{100}{3}=n$$
$$n=33\frac{1}{3}$$

Thus, 30 is 90% of $33\frac{1}{3}$.

c. 25 is 75% of what number?

$$25=\frac{75}{100}\cdot n$$
$$25=\frac{3}{4}\cdot n$$
$$\frac{4}{3}\cdot 25=\frac{4}{3}\cdot\frac{3}{4}\cdot n$$
$$\frac{100}{3}=n$$
$$n=33\frac{1}{3}$$

Thus, 25 is 75% of $33\frac{1}{3}$.

14. a.

$$\frac{1}{5}-\frac{x}{4}=\frac{19(x+4)}{20}$$
$$20\cdot\frac{1}{5}-20\cdot\frac{x}{4}=20\cdot\frac{19(x+4)}{20}$$
$$4-5x=19(x+4)$$
$$4-5x=19x+76$$
$$4-4-5x=19x+76-4$$
$$-5x=19x+72$$
$$-5x-19x=19x-19x+72$$
$$-24x=72$$
$$\frac{-24x}{-24}=\frac{72}{-24}$$
$$x=-3$$

The solution is –3.

b.

$$\frac{1}{5}-\frac{x}{4}=\frac{6(x+5)}{5}$$
$$20\cdot\frac{1}{5}-20\cdot\frac{x}{4}=20\cdot\frac{6(x+5)}{5}$$
$$4-5x=24(x+5)$$
$$4-5x=24x+120$$
$$4-4-5x=24x+120-4$$
$$-5x=24x+116$$
$$-5x-24x=24x-24x+116$$
$$-29x=116$$
$$\frac{-29x}{-29}=\frac{116}{-29}$$
$$x=-4$$

The solution is –4.

c.

$$\frac{1}{5}-\frac{x}{4}=\frac{29(x+6)}{20}$$
$$20\cdot\frac{1}{5}-20\cdot\frac{x}{4}=20\cdot\frac{29(x+6)}{20}$$
$$4-5x=29(x+6)$$
$$4-5x=29x+174$$
$$4-4-5x=29x+174-4$$
$$-5x=29x+170$$
$$-5x-29x=29x-29x+170$$
$$-34x=170$$
$$\frac{-34x}{-34}=\frac{170}{-34}$$
$$x=-5$$

The solution is –5.

15. a.

$$A=\frac{1}{2}bh$$
$$2\cdot A=2\cdot\frac{1}{2}bh$$
$$2A=bh$$
$$\frac{2A}{b}=\frac{bh}{b}$$
$$\frac{2A}{b}=h \text{ or } h=\frac{2A}{b}$$

b.

$$C=2\pi r$$
$$\frac{C}{2\pi}=\frac{2\pi r}{2\pi}$$
$$\frac{C}{2\pi}=r \text{ or } r=\frac{C}{2\pi}$$

c.
$$V = \frac{bh}{3}$$
$$3 \cdot V = 3 \cdot \frac{bh}{3}$$
$$3V = bh$$
$$\frac{3V}{h} = \frac{bh}{h}$$
$$\frac{3V}{h} = b \text{ or } b = \frac{3V}{h}$$

16. a. Let n = first number.
Then $n + 20$ = other number.
$$n + (n + 20) = 84$$
$$n + n + 20 = 84$$
$$2n + 20 = 84$$
$$2n + 20 - 20 = 84 - 20$$
$$2n = 64$$
$$\frac{2n}{2} = \frac{64}{2}$$
$$n = 32$$
$$n + 20 = 52$$
The numbers are 32 and 52.

b. Let n = first number.
Then $n + 19$ = other number.
$$n + (n + 19) = 47$$
$$n + n + 19 = 47$$
$$2n + 19 = 47$$
$$2n + 19 - 19 = 47 - 19$$
$$2n = 28$$
$$\frac{2n}{2} = \frac{28}{2}$$
$$n = 14$$
$$n + 19 = 33$$
The numbers are 14 and 33.

c. Let n = first number.
Then $n + 23$ = second number.
$$n + (n + 23) = 81$$
$$n + n + 23 = 81$$
$$2n + 23 = 81$$
$$2n + 23 - 23 = 81 - 23$$
$$2n = 58$$
$$\frac{2n}{2} = \frac{58}{2}$$
$$n = 29$$
$$n + 23 = 52$$
The numbers are 29 and 52.

17. a. Let c = calories in chicken breast.
Then $c + 22$ = calories in pie.
$$c + (c + 22) = 578$$
$$2c + 22 = 578$$
$$2c + 22 - 22 = 578 - 22$$
$$2c = 556$$
$$\frac{2c}{2} = \frac{556}{2}$$
$$c = 278$$
$$c + 22 = 300$$
The chicken breast has 278 calories and the pie has 300 calories.

b. Let c = calories in chicken breast.
Then $c + 38$ = calories in pie.
$$c + (c + 38) = 620$$
$$2c + 38 = 620$$
$$2c + 38 - 38 = 620 - 38$$
$$2c = 582$$
$$\frac{2c}{2} = \frac{582}{2}$$
$$c = 291$$
$$c + 38 = 329$$
The chicken breast has 291 calories and the pie has 329 calories.

c. Let c = calories in chicken breast.
Then $c + 42$ = calories in pie.

$$
\begin{aligned}
c+(c+42)&=650\\
2c+42&=650\\
2c+42-42&=650-42\\
2c&=608\\
\frac{2c}{2}&=\frac{608}{2}\\
c&=304\\
c+42&=346
\end{aligned}
$$

The chicken breast has 304 calories and the pie has 346 calories.

18. Let m = measure of angle.
$90 - m$ = complement and
$180 - m$ = supplement.

a.

$$
\begin{aligned}
180-m&=3(90-m)-20\\
180-m&=270-3m-20\\
180-m&=250-3m\\
180-180-m&=250-180-3m\\
-m&=70-3m\\
-m+3m&=70-3m+3m\\
2m&=70\\
\frac{2m}{2}&=\frac{70}{2}\\
m&=35
\end{aligned}
$$

The measure of the angle is 35°.

b.

$$
\begin{aligned}
180-m&=3(90-m)-30\\
180-m&=270-3m-30\\
180-m&=240-3m\\
180-180-m&=240-180-3m\\
-m&=60-3m\\
-m+3m&=60-3m+3m\\
2m&=60\\
\frac{2m}{2}&=\frac{60}{2}\\
m&=30
\end{aligned}
$$

The measure of the angle is 30°.

c.

$$\begin{aligned}180-m&=3(90-m)-40\\180-m&=270-3m-40\\180-m&=230-3m\\180-180-m&=230-180-3m\\-m&=50-3m\\-m+3m&=50-3m+3m\\2m&=50\\\frac{2m}{2}&=\frac{50}{2}\\m&=25\end{aligned}$$

The measure of the angle is 25°.

19. a.

	R ×	T =	D
first car	40	$T+1$	$40(T+1)$
second car	50	T	$50T$

$$\begin{aligned}40(T+1)&=50T\\40T+40&=50T\\40&=10T\\\frac{40}{10}&=\frac{10T}{10}\\4&=T\end{aligned}$$

It takes the second car 4 hours to overtake the first car.

b.

	R ×	T =	D
first car	30	$T+1$	$30(T+1)$
second car	50	T	$50T$

$$\begin{aligned}30(T+1)&=50T\\30T+30&=50T\\30&=20T\\\frac{30}{20}&=\frac{20T}{20}\\\frac{3}{2}&=T \text{ or } T=1\frac{1}{2}\end{aligned}$$

It takes the second car $1\frac{1}{2}$ hours to overtake the first car.

c.

	R ×	T =	D
first car	40	$T+1$	$40(T+1)$
second car	60	T	$60T$

$$40(T+1)=60T$$
$$40T+40=60T$$
$$40=20T$$
$$\frac{40}{20}=\frac{20T}{20}$$
$$2=T$$

It takes the second car 2 hours to overtake the first car.

20. a.

	price per pound	×	pounds	=	price
product	1.50		x		$1.5x$
another product	3.00		15		3(15)
mixture	2.40		$15+x$		$2.4(15+x)$

$$1.5x+3(15)=2.4(15+x)$$
$$1.5x+45=36+2.4x$$
$$1.5x+9=2.4x$$
$$9=0.9x$$
$$10=x$$

10 pounds should be mixed.

b.

	price per pound	×	pounds	=	price
product	2.00		x		$2x$
another product	3.00		15		3(15)
mixture	2.50		$15+x$		$2.5(15+x)$

$$2x+3(15)=2.5(15+x)$$
$$2x+45=37.5+2.5x$$
$$2x+7.5=2.5x$$
$$7.5=0.5x$$
$$15=x$$

15 pounds should be mixed.

c.

	price per pound	×	pounds	=	price
product	6.00		x		$6x$
another product	2.00		15		2(15)
mixture	4.50		$15+x$		$4.5(15+x)$

$$6x+2(15)=4.5(15+x)$$
$$6x+30=67.5+4.5x$$
$$6x=37.5+4.5x$$
$$1.5x=37.5$$
$$x=25$$

25 pounds should be mixed.

21. a.

	P	$\times$ r =	I
5%	x	0.05	$0.05x$
6%	$30{,}000 - x$	0.06	$0.06(30{,}000 - x)$

$$\begin{aligned} 0.05x + 0.06(30,000 - x) &= 1600 \\ 0.05x + 1800 - 0.06x &= 1600 \\ 1800 - 0.01x &= 1600 \\ -0.01x &= -200 \\ x &= 20,000 \\ 30,000 - x &= 10,000 \end{aligned}$$

\$20,000 is invested at 5% and \$10,000 is invested at 6%.

b.

	P	$\times$ r =	I
7%	x	0.07	$0.07x$
9%	$30{,}000 - x$	0.09	$0.09(30{,}000 - x)$

$$\begin{aligned} 0.07x + 0.09(30,000 - x) &= 2300 \\ 0.07x + 2700 - 0.09x &= 2300 \\ 2700 - 0.02x &= 2300 \\ -0.02x &= -400 \\ x &= 20,000 \\ 30,000 - x &= 10,000 \end{aligned}$$

\$20,000 is invested at 7% and \$10,000 is invested at 9%.

c.

	P	$\times$ r =	I
6%	x	0.06	$0.06x$
10%	$30{,}000 - x$	0.10	$0.10(30{,}000 - x)$

$$\begin{aligned} 0.06x + 0.10(30,000 - x) &= 2000 \\ 100 \cdot 0.06x + 100 \cdot 0.10(30,000 - x) &= 100 \cdot 2000 \\ 6x + 10(30,000 - x) &= 200,000 \\ 6x + 300,000 - 10x &= 200,000 \\ 300,000 - 4x &= 200,000 \\ -4x &= -100,000 \\ x &= 25,000 \\ 30,000 - x &= 5000 \end{aligned}$$

\$25,000 is invested at 6% and \$5000 is invested at 10%.

22. a. $C = 3.05m + 3$

$C - 3 = 3.05m$

$\frac{C-3}{3.05} = m$ or $m = \frac{C-3}{3.05}$

Let $C = 27.40$.

$m = \frac{27.40 - 3}{3.05} = 8$

The call lasted 8 minutes.

b. $C = 3.15m + 3$

$C - 3 = 3.15m$

$\frac{C-3}{3.15} = m$ or $m = \frac{C-3}{3.15}$

Let $C = 34.50$.

$m = \frac{34.50 - 3}{3.15} = 10$

The call lasted 10 minutes.

c. $C = 3.25m + 2$

$C - 2 = 3.25m$

$\frac{C-2}{3.25} = m$ or $m = \frac{C-2}{3.25}$

Let $C = 21.50$.

$m = \frac{21.50 - 2}{3.25} = 6$

The call lasted 6 minutes.

23. a. The angles are supplementary.

$(3x - 20) + 2x = 180$

$5x - 20 = 180$

$5x = 200$

$x = 40$

$3(40) - 20 = 100$ and $2(40) = 80$

The angles measure 100° and 80°.

b. Vertical angles must be equal.

$7x - 10 = 3x + 30$

$7x = 3x + 40$

$4x = 40$

$x = 10$

$7(10) - 10 = 60$ and $3(10) + 30 = 60$

The angles each measure 60°.

c. The angles are complementary.

$(5x + 15) + (2x + 5) = 90$

$7x + 20 = 90$

$7x = 70$

$x = 10$

$5(10) + 15 = 65$ and $2(10) + 5 = 25$

The angles measure 65° and 25°.

24. a. Since −8 is to the left of −7, $-8 < -7$.

b. Since $\frac{1}{2}$ is to the right of −3, $\frac{1}{2} > -3$.

c. Since 4 is to the left of $4\frac{1}{3}$, $4 < 4\frac{1}{3}$.

25. a. $4x - 2 < 2(x + 2)$

$4x - 2 < 2x + 4$

$4x - 2 + 2 < 2x + 4 + 2$

$4x < 2x + 6$

$4x - 2x < 2x - 2x + 6$

$2x < 6$

$\frac{2x}{2} < \frac{6}{2}$

$x < 3$

Any number less than 3 is a solution.

0 3

b. $5x - 4 < 2(x + 1)$

$5x - 4 < 2x + 2$

$5x - 4 + 4 < 2x + 2 + 4$

$5x < 2x + 6$

$5x - 2x < 2x - 2x + 6$

$3x < 6$

$\frac{3x}{3} < \frac{6}{3}$

$x < 2$

Any number less than 2 is a solution.

0 2

c.
$$\begin{aligned}7x-1&<3(x+1)\\7x-1&<3x+3\\7x-1+1&<3x+3+1\\7x&<3x+4\\7x-3x&<3x-3x+4\\4x&<4\\\frac{4x}{4}&<\frac{4}{4}\\x&<1\end{aligned}$$

Any number less than 1 is a solution.

26. a.
$$\begin{aligned}6(x-1)&\geq 4x+2\\6x-6&\geq 4x+2\\6x-6+6&\geq 4x+2+6\\6x&\geq 4x+8\\6x-4x&\geq 4x-4x+8\\2x&\geq 8\\\frac{2x}{2}&\geq\frac{8}{2}\\x&\geq 4\end{aligned}$$

Any number greater than or equal to 4 is a solution.

b.
$$\begin{aligned}5(x-1)&\geq 2x+1\\5x-5&\geq 2x+1\\5x-5+5&\geq 2x+1+5\\5x&\geq 2x+6\\5x-2x&\geq 2x-2x+6\\3x&\geq 6\\\frac{3x}{3}&\geq\frac{6}{3}\\x&\geq 2\end{aligned}$$

Any number greater than or equal to 2 is a solution.

c.
$$\begin{aligned}4(x-2)&\geq 2x+2\\4x-8&\geq 2x+2\\4x-8+8&\geq 2x+2+8\\4x&\geq 2x+10\\4x-2x&\geq 2x-2x+10\\2x&\geq 10\\\frac{2x}{2}&\geq\frac{10}{2}\\x&\geq 5\end{aligned}$$

Any number greater than or equal to 5 is a solution.

27. a.
$$\begin{aligned}-\frac{x}{3}+\frac{x}{6}&\leq\frac{x-1}{6}\\6\cdot\left(-\frac{x}{3}\right)+6\cdot\frac{x}{6}&\leq 6\cdot\frac{x-1}{6}\\-2x+x&\leq x-1\\-x&\leq x-1\\-x-x&\leq x-x-1\\-2x&\leq -1\\\frac{-2x}{-2}&\geq\frac{-1}{-2}\\x&\geq\frac{1}{2}\end{aligned}$$

Any number greater than or equal to $\frac{1}{2}$ is a solution.

b.
$$\begin{aligned}-\frac{x}{4}+\frac{x}{7}&\leq\frac{x-1}{7}\\28\cdot\left(-\frac{x}{4}\right)+28\cdot\frac{x}{7}&\leq 28\cdot\frac{x-1}{7}\\-7x+4x&\leq 4(x-1)\\-3x&\leq 4x-4\\-3x-4x&\leq 4x-4x-4\\-7x&\leq -4\\\frac{-7x}{-7}&\geq\frac{-4}{-7}\\x&\geq\frac{4}{7}\end{aligned}$$

Any number greater than or equal to $\frac{4}{7}$ is a solution.

c.

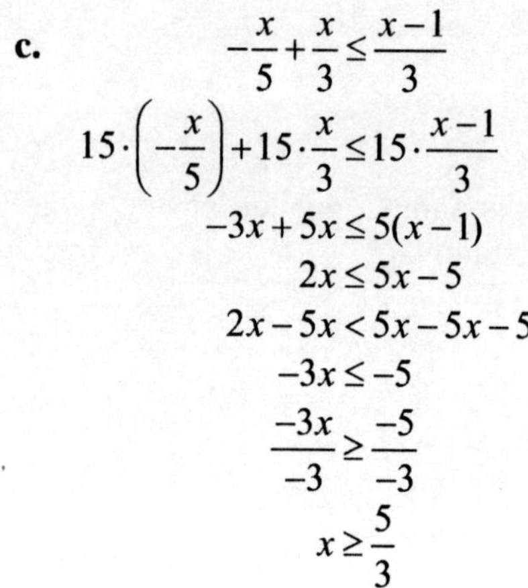

$$-\frac{x}{5}+\frac{x}{3}\le\frac{x-1}{3}$$
$$15\cdot\left(-\frac{x}{5}\right)+15\cdot\frac{x}{3}\le15\cdot\frac{x-1}{3}$$
$$-3x+5x\le5(x-1)$$
$$2x\le5x-5$$
$$2x-5x<5x-5x-5$$
$$-3x\le-5$$
$$\frac{-3x}{-3}\ge\frac{-5}{-3}$$
$$x\ge\frac{5}{3}$$

Any number greater than or equal to $\frac{5}{3}$ is a solution.

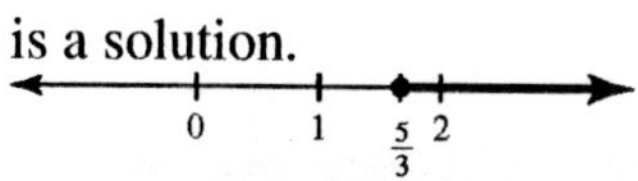

28. **a.** $x+2\le4$ and $-2x<6$

$x\le2$ $\quad \frac{-2x}{-2}>\frac{6}{-2}$

$x>-3$

Now $x\le2$ and $x>-3$ can be rewritten as $-3<x\le2$. The solution consists of all numbers between -3 and 2, including 2.

b. $x+3\le5$ and $-3x<9$

$x\le2$ $\quad \frac{-3x}{-3}>\frac{9}{-3}$

$x>-3$

Now $x\le2$ and $x>-3$ can be rewritten as $-3<x\le2$. The solution consists of all numbers between -3 and 2, including 2.

c. $x+1\le2$ and $-4x<8$

$x\le1$ $\quad \frac{-4x}{-4}>\frac{8}{-4}$

$x>-2$

Now $x\le1$ and $x>-2$ can be rewritten as $-2<x\le1$. The solution consists of all numbers between -2 and 1, including 1.

Cumulative Review Chapters 1–2

1. The additive inverse of -7 is 7.

2. $\left|-9\frac{9}{10}\right|=9\frac{9}{10}$

3. The LCD is 63.

$-\frac{2}{7}=-\frac{18}{63}$ and $-\frac{2}{9}=-\frac{14}{63}$

$-\frac{2}{7}+\left(-\frac{2}{9}\right)=-\frac{18}{63}+\left(-\frac{14}{63}\right)=-\frac{32}{63}$

4. $-0.7-(-8.9)=-0.7+8.9=8.2$

5. $(-2.4)(3.6)=-8.64$

6. $-(2^4)=-(2\cdot2\cdot2\cdot2)=-(16)=-16$

7. $-\frac{7}{8}\div\left(-\frac{5}{24}\right)=-\frac{7}{8}\cdot\left(-\frac{24}{5}\right)=\frac{168}{40}=\frac{21}{5}$

8. $y\div5\cdot x-z=60\div5\cdot6-3$
$=12\cdot6-3$
$=72-3$
$=69$

9. In $9\cdot(8\cdot5)=9\cdot(5\cdot8)$, we changed the *order* of multiplication. The commutative property of multiplication was used.

10. $6(5x+7)=6(5x)+6(7)$
$=30x+42$

11. $-5cd-(-6cd)=-5cd+6cd$
$=(-5+6)cd$
$=1cd$
$=cd$

12. $2x-2(x+4)-3(x+1)$
$=2x-2x-8-3x-3$
$=(2x-2x-3x)+(-8-3)$
$=-3x+(-11)$
$=-3x-11$

13. The quotient of $(a-4b)$ and c is written as $\frac{a-4b}{c}$.

14. If $x=4$, then $11=15-x$ becomes $11=15-4$ which is a *true* statement. Thus 4 is a solution of the equation.

15.
$$5=4(x-3)+4-3x$$
$$5=4x-12+4-3x$$
$$5=x-8$$
$$5+8=x-8+8$$
$$13=x \text{ or } x=13$$
The solution is 13.

16.
$$-\frac{7}{3}x=-21$$
$$-\frac{3}{7}\cdot\left(-\frac{7}{3}x\right)=-\frac{3}{7}\cdot(-21)$$
$$x=3\cdot3$$
$$x=9$$
The solution is 9.

17.
$$\frac{x}{3}-\frac{x}{5}=2$$
$$15\cdot\frac{x}{3}-15\cdot\frac{x}{5}=15\cdot2$$
$$5x-3x=30$$
$$2x=30$$
$$\frac{2x}{2}=\frac{30}{2}$$
$$x=15$$
The solution is 15.

18.
$$4-\frac{x}{4}=\frac{2(x+1)}{9}$$
$$36\cdot4-36\cdot\frac{x}{4}=36\cdot\frac{2(x+1)}{9}$$
$$144-9x=4\cdot2(x+1)$$
$$144-9x=8(x+1)$$
$$144-9x=8x+8$$
$$144-144-9x=8x+8-144$$
$$-9x=8x-136$$
$$-9x-8x=8x-8x-136$$
$$-17x=-136$$
$$\frac{-17x}{-17}=\frac{-136}{-17}$$
$$x=8$$
The solution is 8.

19.
$$S=6a^2b$$
$$\frac{S}{6a^2}=\frac{6a^2b}{6a^2}$$
$$\frac{S}{6a^2}=b \text{ or } b=\frac{S}{6a^2}$$

20. Let n = first number.
Then $n+35$ = second number.
$$n+(n+35)=155$$
$$2n+35=155$$
$$2n+35-35=155-35$$
$$2n=120$$
$$\frac{2n}{2}=\frac{120}{2}$$
$$n=60$$
$$n+35=95$$
The numbers are 60 and 95.

21. Let b = annual return from bonds.
Then $b+105$ = annual return from stocks.
$$b+(b+105)=595$$
$$2b+105=595$$
$$2b+105-105=595-105$$
$$2b=490$$
$$\frac{2b}{2}=\frac{490}{2}$$
$$b=245$$
$$b+105=350$$
The bonds return \$245 and the stocks return \$350.

22.

	R ×	T =	D
Train A	40	$T+6$	$40(T+6)$
Train B	50	T	$50T$

$$\begin{aligned} 40(T+6) &= 50T \\ 40T+240 &= 50T \\ 240 &= 10T \\ 24 &= T \end{aligned}$$

It takes 24 hours for train B to catch train A.

23.

	P ×	r =	I
bonds	x	0.12	$0.12x$
certificates	$5000-x$	0.14	$0.14(5000-x)$

$$\begin{aligned} 0.12x+0.14(5000-x) &= 660 \\ 0.12x+700-0.14x &= 660 \\ 700-0.02x &= 660 \\ -0.02x &= -40 \\ x &= 2000 \\ 5000-x &= 3000 \end{aligned}$$

Arlene invested \$2000 in bonds and \$3000 in certificates of deposit.

24.

$$\begin{aligned} -\frac{x}{6}+\frac{x}{5} &\le \frac{x-5}{5} \\ 30\cdot\left(-\frac{x}{6}\right)+30\cdot\frac{x}{5} &\le 30\cdot\frac{x-5}{5} \\ -5x+6x &\le 6(x-5) \\ x &\le 6x-30 \\ x-6x &\le 6x-6x-30 \\ -5x &\le -30 \\ \frac{-5x}{-5} &\ge \frac{-30}{-5} \\ x &\ge 6 \end{aligned}$$

Any number greater than or equal to 6 is a solution.

0 6

Chapter 3 Graphs of Linear Equations

3.1 Graphs and Applications

Problems 3.1

1. **a.** Start at the origin. Go 2 units right and 4 units up.

   b. Start at the origin. Go 4 units right and 2 units down.

   c. Start at the origin. Go 3 units left and 2 units up.

   d. Start at the origin. Go 4 units left and 4 units down.

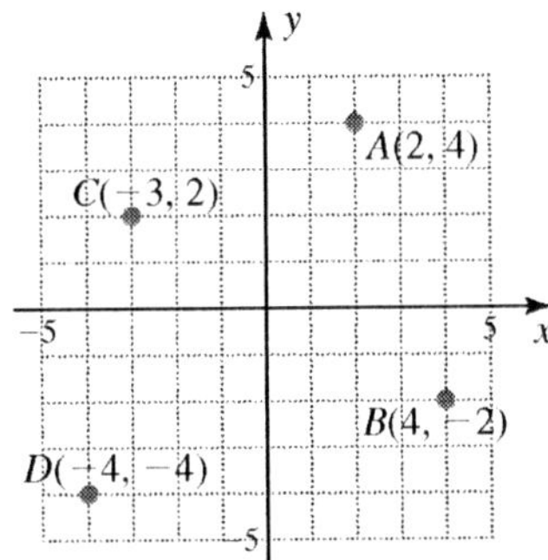

2.

Point	Start at the origin, move:	Coordinates
A	4 units left, 2 units up	(−4, 2)
B	4 units right, 1 unit up	(4, 1)
C	2 units left, 0 units up	(−2, 0)
D	0 units right, 4 units up	(0, 4)
E	3 units right, 2 units down	(3, −2)
F	1 unit left, 4 units down	(−1, −4)

3. **a.** Five human years is equivalent to 40 dog years.

   b. Start at (0, 0) and move right to 11 on the horizontal axis. Then go up until you reach the graph. The y-coordinate is 70. Thus, 11 years old in human years is equivalent to 70 years old in dog years.

   c. We need to find how many human years are represented by 21 dog years. At the point for which y is 21 on the graph, x is about 2. Thus, the equivalent drinking age for dogs is 2 human years.

4. **a.** To find the payment corresponding to 48 months, move right on the horizontal axis to the category labeled 48 months and then vertically to the end of the bar. According to the vertical scale, the monthly payment will be $30.

   b. If you pay the minimum $25 payment for 60 months, you would pay $1500. If you pay $30 for 48 months, you would pay $1440 and would save $60.

5.

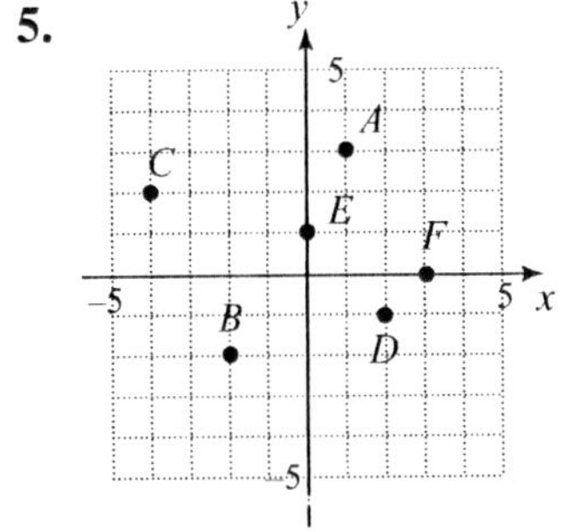

   a. The point $A(1, 3)$ is in the first quadrant.

   b. The point $B(-2, -2)$ is in the third quadrant.

   c. The point $C(-4, 2)$ is in the second quadrant.

   d. The point $D(2, -1)$ is in the fourth quadrant.

   e. The point $E(0, 1)$ is on the y-axis (no quadrant).

   f. The point $F(3, 0)$ is on the x-axis (no quadrant).

6. a. (16, 80) means that if a cat's actual age is 16, its equivalent human age is 80 years.

b. The ordered pair whose first coordinate is 12 is (12, 64), so $h = 64$.

c. The ordered pair whose second coordinate is 72 is (14, 72), so $c = 14$.

d. Extending the pattern of the ordered pairs, the next two ordered pairs after (31, 140) are (32, 144) and (33, 148). The cat's actual age is 33 years.

Exercises 3.1

1. a. Start at the origin. Go 1 unit right and 2 units up.

b. Start at the origin. go 2 units left and 3 units up.

c. Start at the origin. Go 3 units left and 1 unit up.

d. Start at the origin. Go 4 units left and 1 unit down.

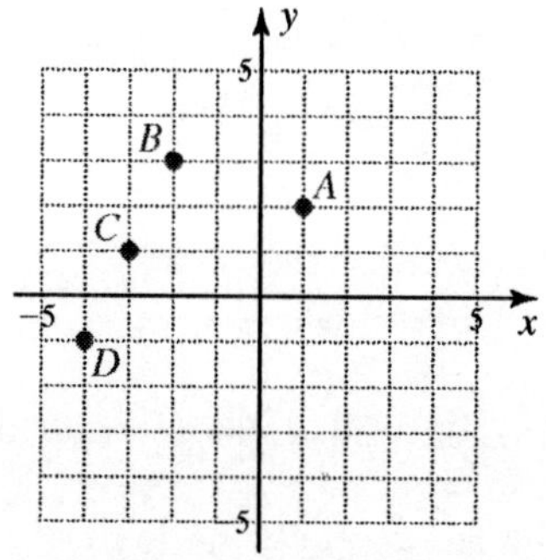

3. a. Start at the origin. Go 0 units right and 2 units up.

b. Start at the origin. Go 3 units left and 0 units up.

c. Start at the origin. Go $3\frac{1}{2}$ units right and 0 units up.

d. Start at the origin. Go 0 units right and $1\frac{1}{4}$ units down.

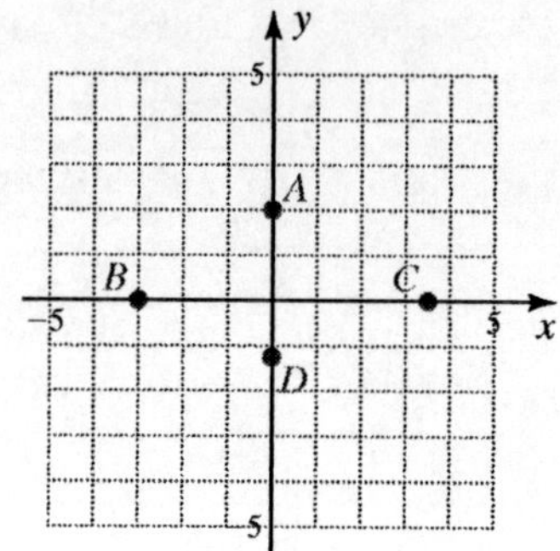

5. a. Start at the origin. Go 0 units right and 40 units up.

b. Start at the origin. Go 35 units left and 0 units up.

c. Start at the origin. Go 40 units left and 15 units down.

d. Start at the origin. Go 0 units right and 25 units down.

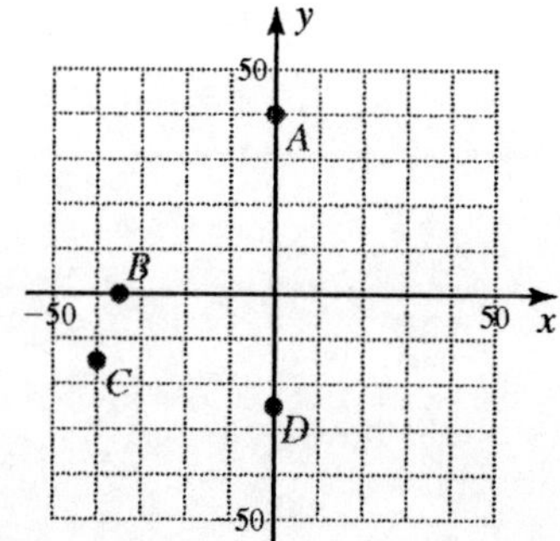

7.

Point	Start at the origin, move:	Coordinates	Quadrant
A	0 units right, $2\frac{1}{2}$ units up	$\left(0, 2\frac{1}{2}\right)$	y-axis
B	3 units left, 1 unit up	$(-3, 1)$	II
C	2 units left, 2 units down	$(-2, -2)$	III
D	0 units right, $3\frac{1}{2}$ units down	$\left(0, -3\frac{1}{2}\right)$	y-axis
E	$1\frac{1}{2}$ units right, 2 units down	$\left(1\frac{1}{2}, -2\right)$	IV

9.

Point	Start at the origin, move:	Coordinates	Quadrant
A	10 units right, 10 units up	$(10, 10)$	I
B	30 units left, 20 units up	$(-30, 20)$	II
C	25 units left, 20 units down	$(-25, -20)$	III
D	0 units right, 40 units down	$(0, -40)$	y-axis
E	15 units right, 15 units down	$(15, -15)$	IV

11. Where the age is 20 years, the lower edge of the blue region corresponds to a pulse rate of 140 beats per minute. The ordered pair is (20, 140).

13. Where the age is 45 years, the upper edge of the blue region corresponds to a pulse rate of 150 beats per minute. The ordered pair is (45, 150).

15. The point on the green graph above the year 1960 is labeled 2.7. In 1960, 2.7 pounds of waste were generated per person each day.

17. Subtract the 1960 amount from the 1999 amount: $4.6 - 2.7 = 1.9$. There were 1.9 more pounds generated per person in 1999 than in 1960.

19. Since the amount increased by 0.1 pound between 1990 and 1999, we might expect a similar increase over another time period of the same length. So, we expect $4.6 + 0.1$ or 4.7 pounds of waste per person in 2008.

21. The point on the blue graph above the year 1999 is labeled 229.9. In 1999, 229.9 million tons of waste were generated.

23. Subtract the 1990 amount from the 1999 amount: $229.9 - 205.2 = 24.7$. There were 24.7 million more tons of waste generated in 1999 than in 1990.

25. The point on the red curve above 6 months corresponds to \$950. If you are paying \$25 a month, the balance after 6 months is \$950.

27. The point on the red curve above 18 months corresponds to \$800. If you are paying \$25 a month, the balance after 18 months is \$800.

29. The point on the graph for the new closing time above 12 A.M. corresponds to about 70 people.

31. The greatest vertical distance between the two graphs occurs above 3 A.M.

33. There were about 80 border crossers at 5 A.M. with the old closing time and 25 with the new closing time. The difference is 80 – 25 or 55 people.

35. The number at the top of the green bar is 195; 195 were before 3 A.M. The number at the top of the pink bar is 288; 288 of those were after 3 A.M.

37. The number at the top of the green bar is 162; 162 were before 3 A.M. The number at the top of the pink bar is 13; 13 of those were after 3 A.M.

39. **a.** The tallest bar corresponds to auto batteries.

b. The shortest bar corresponds to tires.

41. The number at the top of the bar for tires is 26.5; 26.5% of tires were recycled. Of the 70 million tons, 1 million were tires; $\frac{1}{70} \approx 0.0143$, so about 1.43% of the 70 million tons were tires.

43. The top of the bar for a down payment of \$0 corresponds to a monthly payment of \$500.

45. We subtract the monthly payment for the \$2500 down payment from the monthly payment for the \$1500 down payment: 463 – 438 = 25. The difference is \$25.

47.

Dog's age	WEIGHT (POUNDS)			
	15–30	30–49	50–74	75–100
1	12	12	14	16
2	19	21	23	23
3	25	25	26	29
4	32	32	35	35
5	36	36	37	39
6	40	42	43	45
7	44	45	46	48
8	48	50	52	52
9	52	53	54	56
10	56	56	58	61
11	60	60	62	65
12	62	63	66	70
13	66	67	70	75
14	70	70	75	80
15	72	76	80	85

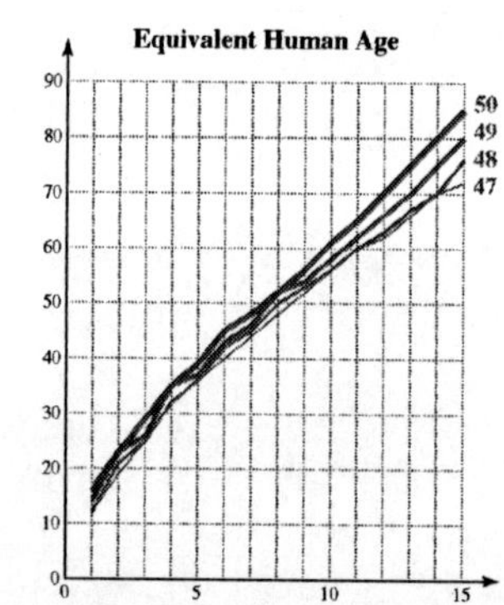

49.

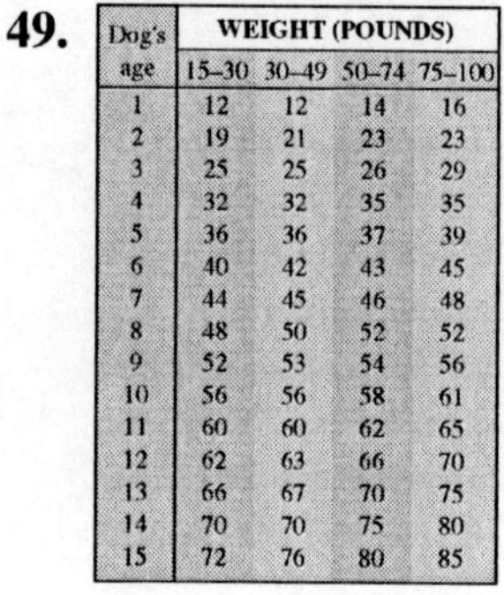

Dog's age	WEIGHT (POUNDS)			
	15–30	30–49	50–74	75–100
1	12	12	14	16
2	19	21	23	23
3	25	25	26	29
4	32	32	35	35
5	36	36	37	39
6	40	42	43	45
7	44	45	46	48
8	48	50	52	52
9	52	53	54	56
10	56	56	58	61
11	60	60	62	65
12	62	63	66	70
13	66	67	70	75
14	70	70	75	80
15	72	76	80	85

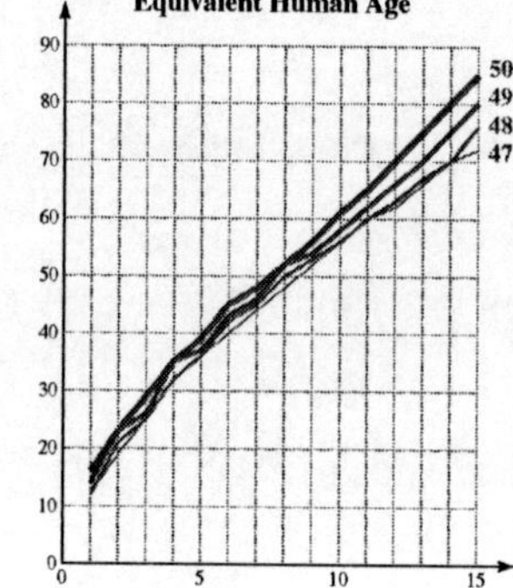

51. The last line of the table corresponds to a dog's age of 15 years. The largest value in this row, 85, occurs in the column for 75–100 pounds.

53. The less a dog weighs, the younger it appears, or the more a dog weighs, the older it appears.

55. No two human ages are the same in the rows for dog ages of 6, 7, 9, 12, 13, and 15.

57.
$$\begin{aligned} 3x+y&=9 \\ 3(0)+y&=9 \\ 0+y&=9 \\ y&=9 \end{aligned}$$

59.
$$\begin{aligned} 3x+y&=6 \\ 3(1)+y&=6 \\ 3+y&=6 \\ 3+y-3&=6-3 \\ y&=3 \end{aligned}$$

61. The highest temperature in the safe zone when the humidity is 50% is 86°F.

63. If the temperature is 100°F, the humidity must be less than 10% so that it is safe to exercise.

65. When the humidity is 60%, the danger zone starts at a temperature of about 98%. Since the temperature is 86°F, the temperature can rise 12°F before you get to the danger zone.

67. For a point (a, b) in the second quadrant, a is negative and b is positive. Since a and b have opposite signs in the second quadrant, it is not possible that $a = b$.

69. If the ordered pairs (a, b) and (c, d) have the same point as their graphs, then $a = c$ and $b = d$.

71. Start at the origin. Go 4 units right and 2 units up. See graph below.

73. Start at the origin. Go 2 units left and 1 unit up. See graph below.

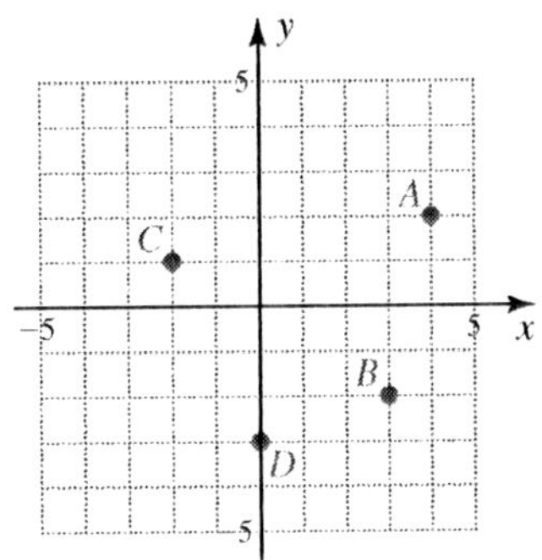

75. Above 1, place a point at a height of 25.
Above 2, place a point at a height of 23.
Above 3, place a point at a height of 22.
Above 4, place a point at a height of 20.

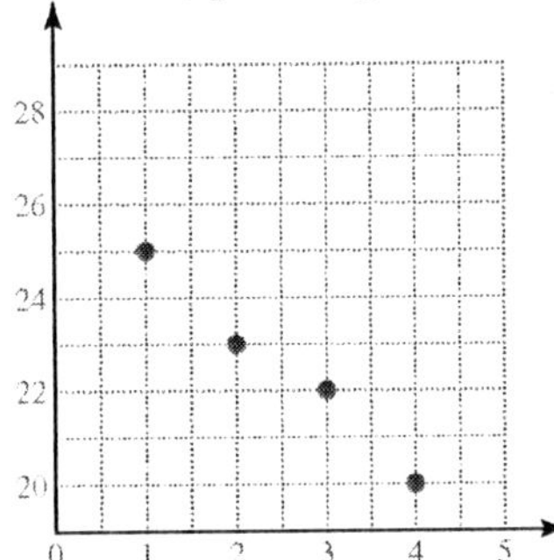

77. The point on the graph at a height of 70 corresponds to 11 human years.

79.

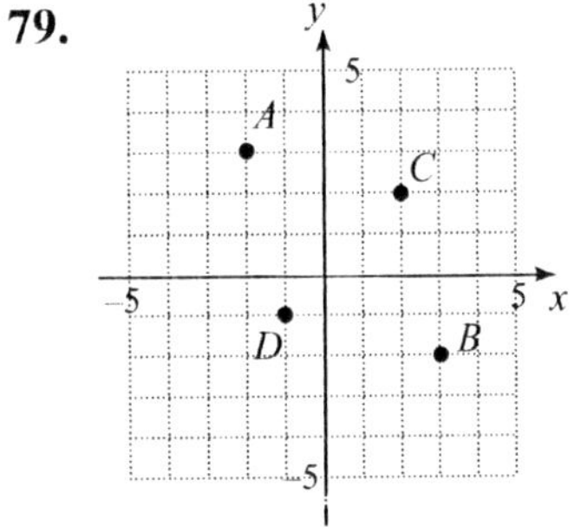

a. The point $A(-2, 3)$ is in the second quadrant.

b. The point $B(3, -2)$ is in the fourth quadrant.

c. The point $C(2, 2)$ is in the first quadrant.

d. The point $A(-1, -1)$ is in the third quadrant.

3.2 Graphing Linear Equations in Two Variables

Problems 3.2

1. a.
$$3x+2y=10$$
$$3(2)+2(2)=10$$
$$6+4=10$$
yes

b.
$$3x+2y=10$$
$$3(-3)+2(4)=10$$
$$-9+8=10$$
no

c.
$$3x+2y=10$$
$$3(-4)+2(11)=10$$
$$-12+22=10$$
yes

2. a. $y=3x+4$
$7=3x+4$
$3=3x$
$1=x$
$(1, 7)$

b. $y = 3x + 4$
$y = 3(-2) + 4$
$y = -6 + 4$
$y = -2$
$(-2, -2)$

3. a. The cholesterol level at the end of 4 weeks corresponds to the point (3, 202) on the graph. Thus, the cholesterol level at the end of 4 weeks is about 202.

b. From the equation $C = -3w + 215$, the cholesterol level at the end of 4 weeks ($w = 4$) is given by
$C = -3(4) + 215 = 203$.

4. The graph is a straight line. We find two points and a check point. Let $x = 0$.
$2x+y=4$
$2\cdot 0+y=4$
$y=4$
Thus, (0, 4) is on the graph.
Let $y = 0$.
$2x+y=4$
$2x+0=4$
$2x=4$
$x=2$
Thus, (2, 0) is on the graph.
Let $x = 1$.
$2x+y=4$
$2\cdot 1+y=4$
$2+y=4$
$y=2$
Thus, (1, 2) is on the graph.

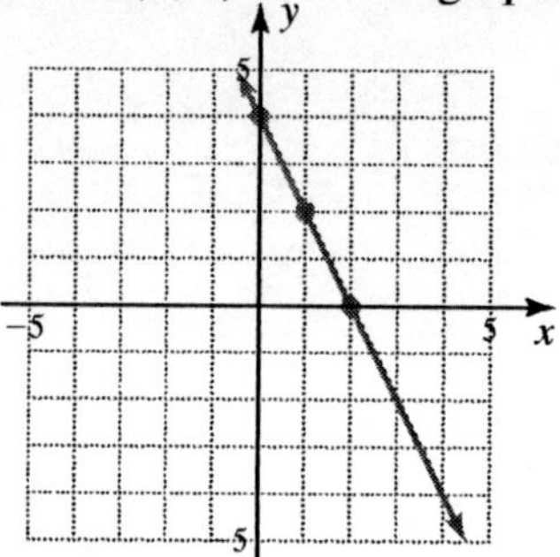

5. The graph is a straight line. We find two points and a check point. Let $x = 0$.
$y = -3x + 6$
$y = -3 \cdot 0 + 6 = 6$
Thus, (0, 6) is on the graph. Let $y = 0$.
$y=-3x+6$
$0=-3x+6$
$3x=6$
$x=2$
Thus, (2, 0) is on the graph. Let $x = 1$.
$y = -3x + 6$
$y = -3 \cdot 1 + 6$
$y = -3 + 6 = 3$
Thus, (1, 3) is on the graph.

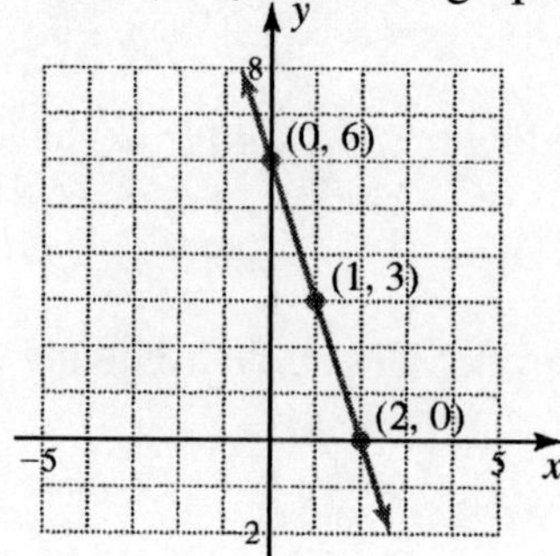

6. For $t = -5$, $W = 1.2(-5) - 20 = -26$. Graph $(-5, -26)$. For $t = 0$, $W = 1.2(0) - 20 = -20$. Graph $(0, -20)$. For $t = 5$, $W = 1.2(5) - 20 = -14$. Graph $(5, -14)$.

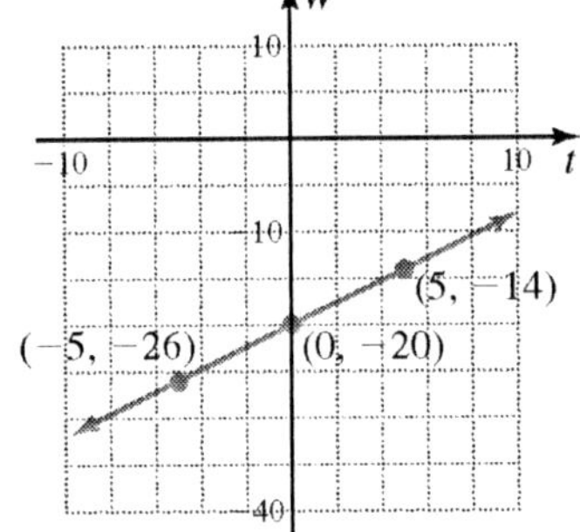

Exercises 3.2

1. $x + 2y = 7$
$3 + 2 \cdot 2 = 7$
$3 + 4 = 7$
yes

3. $2x - 5y = -5$
$2 \cdot 5 - 5 \cdot 3 = -5$
$10 - 15 = -5$
yes

5. $-5x = 2y + 4$
$-5 \cdot 2 = 2 \cdot 3 + 4$
$-10 = 6 + 4$
no

7. $2x - y = 6$
$2 \cdot 3 - y = 6$
$6 - y = 6$
$-y = 0$
$y = 0$
The missing coordinate is 0.

9. $3x + 2y = -2$
$3x + 2 \cdot 2 = -2$
$3x + 4 = -2$
$3x = -6$
$x = -2$
The missing coordinate is −2.

11. $3x - y = 3$
$3 \cdot 0 - y = 3$
$-y = 3$
$y = -3$
The missing coordinate is −3.

13. $2x - y = 6$
$2x - 0 = 6$
$2x = 6$
$x = 3$

15. $-2x + y = 8$
$-2(-3) + y = 8$
$6 + y = 8$
$y = 2$
The missing coordinate is 2.

17. Let $x = 0$.
$2x + y = 4$
$2 \cdot 0 + y = 4$
$y = 4$
Thus, (0, 4) is on the graph. Let $y = 0$.
$2x + 0 = 4$
$2x = 4$
$x = 2$
Thus, (2, 0) is on the graph.

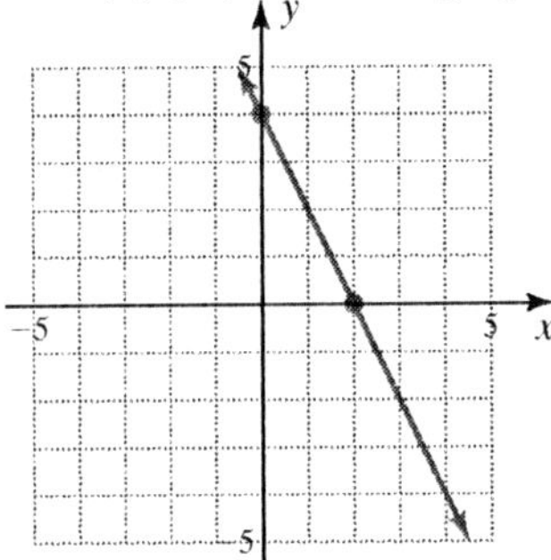

19. Let $x = 0$.
$-2x - 5y = -10$
$-2 \cdot 0 - 5y = -10$
$-5y = -10$
$y = 2$
Thus, (0, 2) is on the graph. Let $y = 0$.
$-2x - 5y = -10$
$-2x - 5 \cdot 0 = -10$
$-2x = -10$
$x = 5$
Thus, (5, 0) is on the graph.

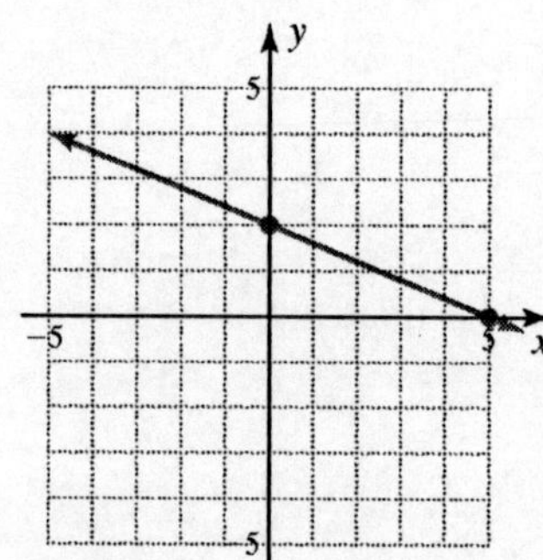

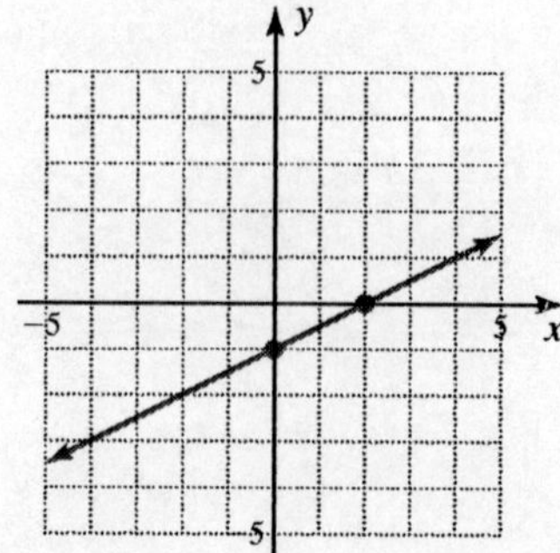

21. Let $x = 0$.

$y+3=3x$
$y+3=3\cdot 0$
$y+3=0$
$y=-3$

Thus, (0, –3) is on the graph. Let $y = 0$.

$y+3=3x$
$0+3=3x$
$3=3x$
$1=x$

Thus, (1, 0) is on the graph.

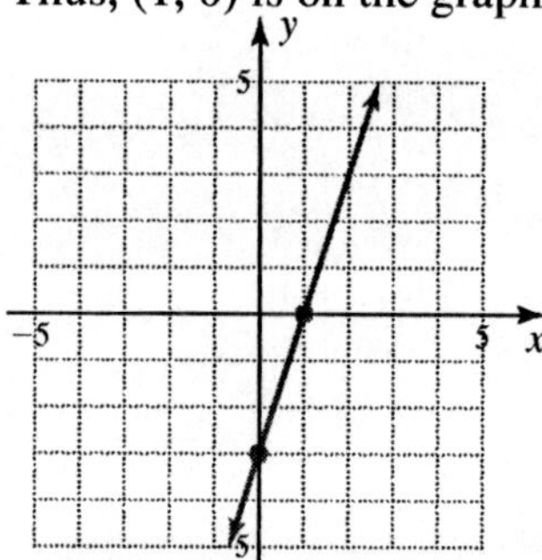

23. Let $x = 0$.

$6=3x-6y$
$6=3\cdot 0-6y$
$6=-6y$
$-1=y$

Thus, (0, –1) is on the graph. Let $y = 0$.

$6=3x-6y$
$6=3x-6\cdot 0$
$6=3x$
$2=x$

Thus, (2, 0) is on the graph.

25. Let $x = 0$.

$-3y=4x+12$
$-3y=4\cdot 0+12$
$-3y=12$
$y=-4$

Thus, (0, –4) is on the graph. Let $y = 0$.

$-3y=4x+12$
$-3\cdot 0=4x+12$
$0=4x+12$
$-12=4x$
$-3=x$

Thus, (–3, 0) is on the graph.

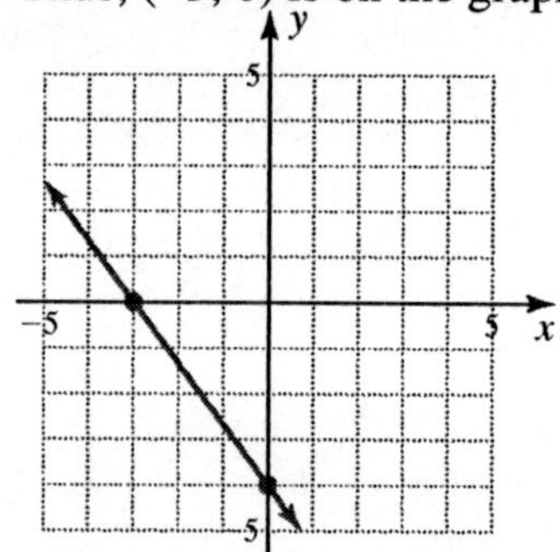

27. Let $x = 0$.

$-2y=-x+4$
$-2y=-0+4$
$-2y=4$
$y=-2$

Thus, (0, –2) is on the graph. Let $y = 0$.

$-2y=-x+4$
$-2\cdot 0=-x+4$
$0=-x+4$
$x=4$

Thus, (4, 0) is on the graph.

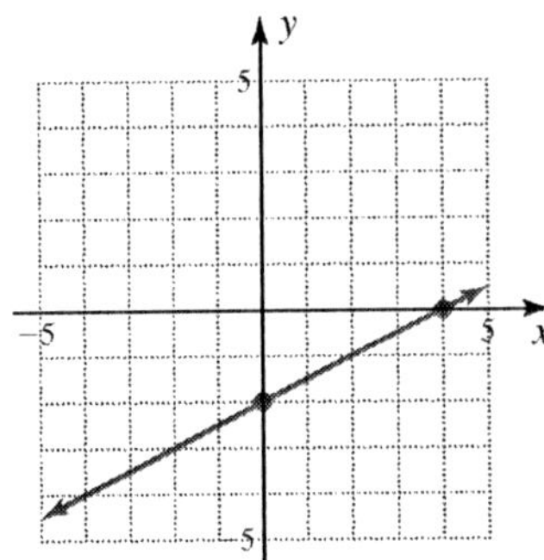

29. Let $x = 0$.

$-3y = -6x + 3$
$-3y = -6 \cdot 0 + 3$
$-3y = 3$
$y = -1$

Thus, $(0, -1)$ is on the graph. Let $y = 0$.

$-3y = -6x + 3$
$-3 \cdot 0 = -6x + 3$
$0 = -6x + 3$
$6x = 3$
$x = \frac{1}{2}$

Thus, $\left(\frac{1}{2}, 0\right)$ is on the graph.

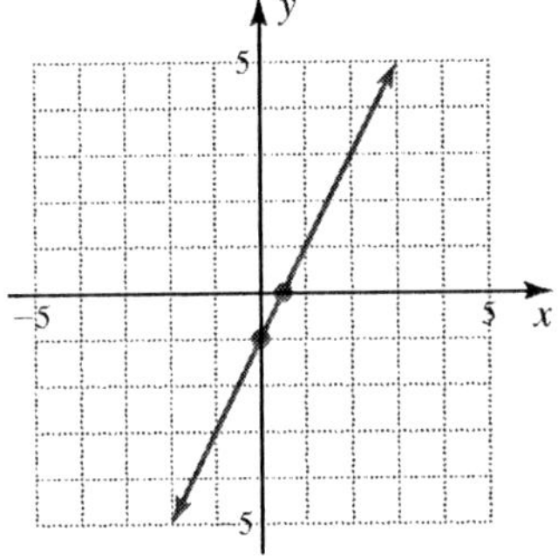

31. Let $x = 0$.

$y = 2x + 4$
$y = 2 \cdot 0 + 4$
$y = 4$

Thus, $(0, 4)$ is on the graph. Let $y = 0$.

$y = 2x + 4$
$0 = 2x + 4$
$-2x = 4$
$x = -2$

Thus, $(-2, 0)$ is on the graph.

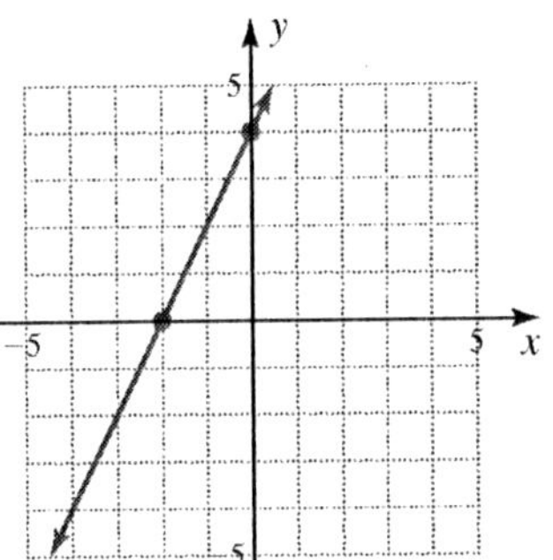

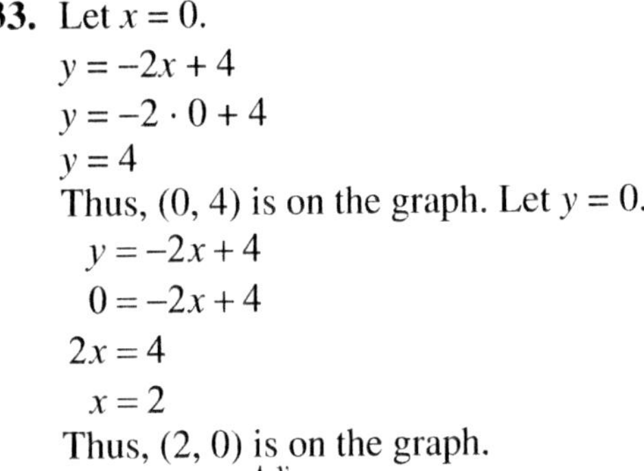

33. Let $x = 0$.

$y = -2x + 4$
$y = -2 \cdot 0 + 4$
$y = 4$

Thus, $(0, 4)$ is on the graph. Let $y = 0$.

$y = -2x + 4$
$0 = -2x + 4$
$2x = 4$
$x = 2$

Thus, $(2, 0)$ is on the graph.

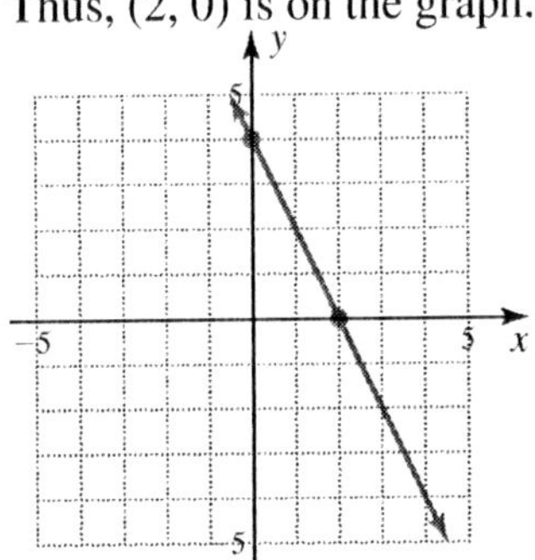

35. Let $x = 0$.

$y = -3x - 6$
$y = -3 \cdot 0 - 6$
$y = -6$

Thus, $(0, -6)$ is on the graph. Let $y = 0$.

$y = -3x - 6$
$0 = -3x - 6$
$3x = -6$
$x = -2$

Thus, $(-2, 0)$ is on the graph.

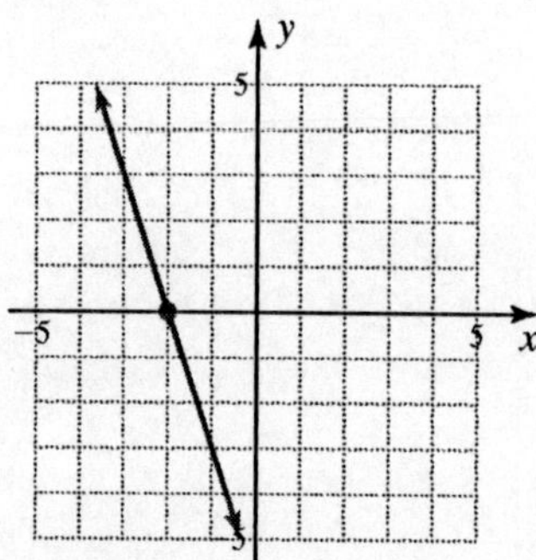

37. Let $x = 0$.

$$y = \frac{1}{2}x - 2$$
$$y = \frac{1}{2} \cdot 0 - 2$$
$$y = -2$$

Thus, (0, −2) is on the graph. Let $y = 0$.

$$y = \frac{1}{2}x - 2$$
$$0 = \frac{1}{2}x - 2$$
$$2 = \frac{1}{2}x$$
$$4 = x$$

Thus, (4, 0) is on the graph.

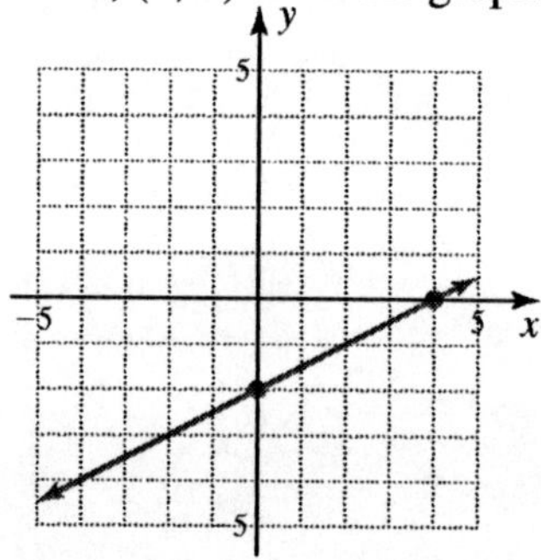

39. Let $x = 0$.

$$y = -\frac{1}{2}x - 2$$
$$y = -\frac{1}{2} \cdot 0 - 2$$
$$y = -2$$

Thus, (0, −2) is on the graph. Let $y = 0$.

$$y = -\frac{1}{2}x - 2$$
$$0 = -\frac{1}{2}x - 2$$
$$2 = -\frac{1}{2}x$$
$$-4 = x$$

Thus, (−4, 0) is on the graph.

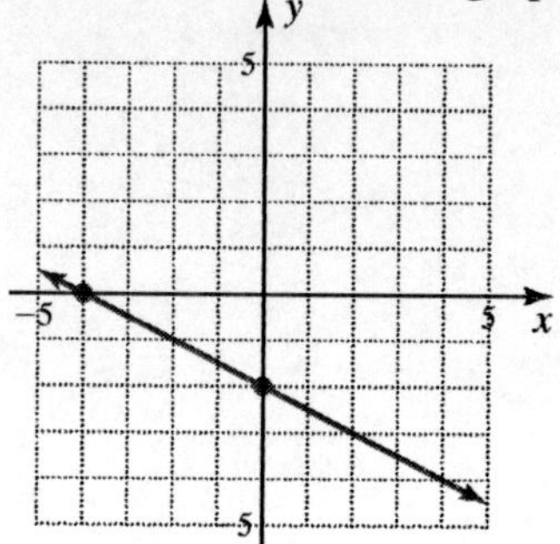

41. **a.** $W = 1.1 \cdot 0 - 9 = -9$

b. $W = 1.1 \cdot 10 - 9 = 2$

c.

43. **a.** For 1996, $t = 0$.
$C = 200 \cdot 0 + 7000 = \$7000$

b. For 2001, $t = 5$.
$C = 200 \cdot 5 + 7000 = \$8000$

c. For 2006, $t = 10$.
$C = 200 \cdot 10 + 7000 = \$9000$

d.

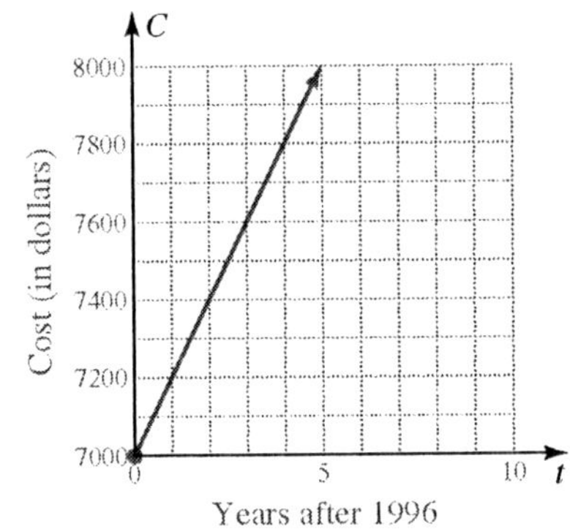

45. $2x+4=0$
$2x=-4$
$x=-2$

47. $0=0-0.1t$
$0=-0.1t$
$0=t$

49. $H=2.894(18)+27.811\approx 80$
Since 80" is 6'8", the bone could belong to a 6'8" basketball player.

51. $H=3.343(14)+31.978\approx 79$
Since 79" is 6'7", the bone could not belong to a 7' female.
$84=3.343r+31.978$
$52.022=3.343r$
$16\approx r$
Sandy Allen's radius should be 16 inches long.

53. $H=3.271(10)+33.829\approx 66.5$
The bone could belong to a man 66.5" tall.

55. Answers may vary.

57. Answers may vary.

59. a. $3x+2y=10$
$3\cdot 1+2\cdot 2=10$
$3+4=10$
no

b. $3x+2y=10$
$3\cdot 2+2\cdot 1=10$
$6+2=10$
no

61. Let $x=0$.
$y=-3x+6$
$y=-3\cdot 0+6$
$y=6$
Thus, (0, 6) is on the graph. Let $y=0$.
$y=-3x+6$
$0=-3x+6$
$3x=6$
$x=2$
Thus, (2, 0) is on the graph.

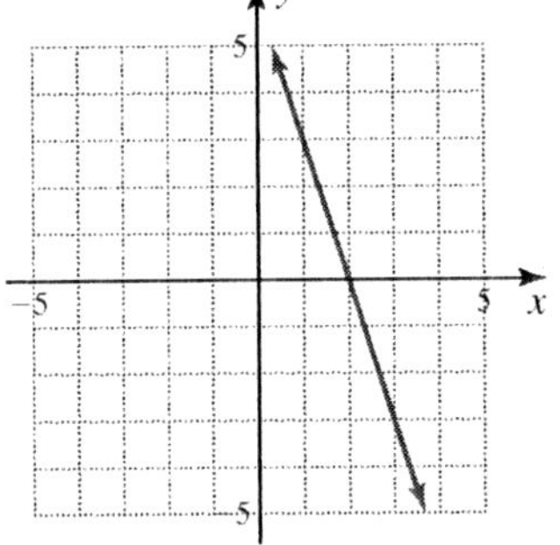

63. a. $W=1.3\cdot 0-18=-18$

b. $W=1.3\cdot 10-18=-5$

c.

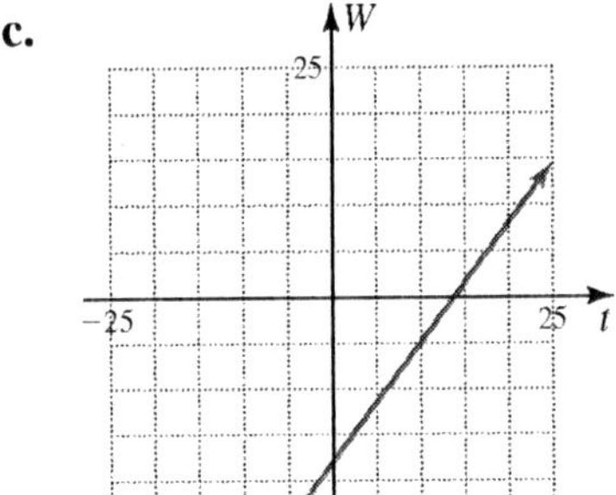

3.3 Graphing Lines Using Intercepts

Problems 3.3

1. $2x+5y=10$
Let $x=0$.
$2(0)+5y=10$
$5y=10$
$y=2$
(0, 2) is the y-intercept.

Let $y = 0$.
$2x + 5(0) = 10$
$2x = 10$
$x = 5$
(5, 0) is the x-intercept.

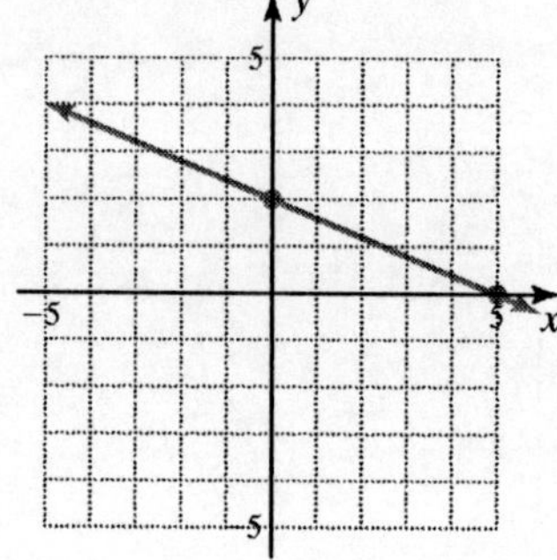

2. $3x - 2y - 6 = 0$
Let $y = 0$.
$3x - 2(0) - 6 = 0$
$3x - 6 = 0$
$3x = 6$
$x = 2$
(2, 0) is the x-intercept. Let $x = 0$.
$3(0) - 2y - 6 = 0$
$-2y - 6 = 0$
$-2y = 6$
$y = -3$
(0, −3) is the y-intercept.

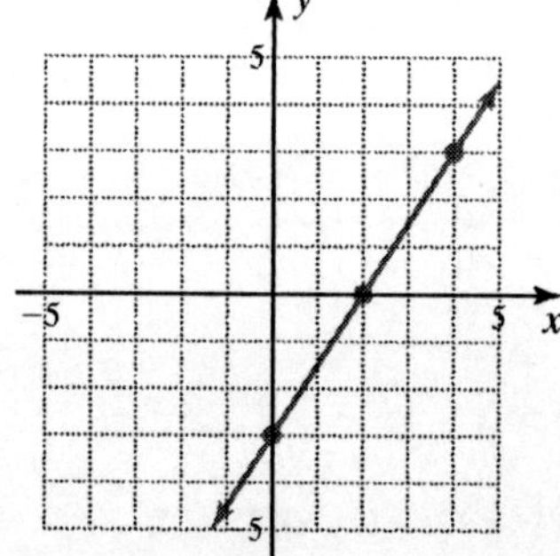

3. $3x + y = 0$ is of the form $Ax + By = 0$, so the line goes through the origin. Let $y = 3$ to get another point.
$3x + 3 = 0$
$3x = -3$
$x = -1$
(−1, 3) is on the graph.

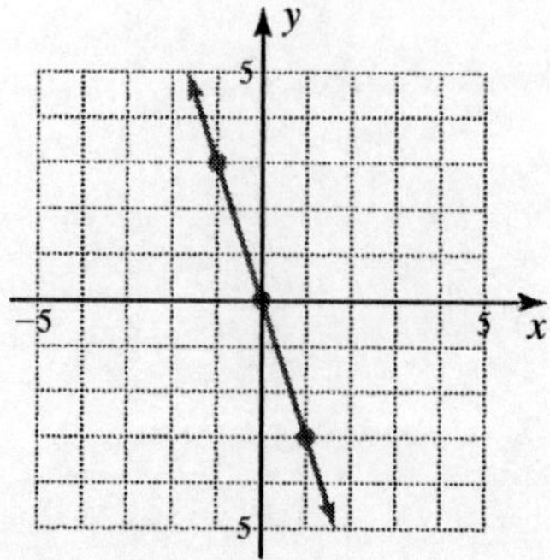

4. a. $3x - 3 = 0$
$3x = 3$
$x = 1$
This is a vertical line crossing the x-axis at 1.

b. $4 + 2y = 0$
$2y = -4$
$y = -2$
This is a horizontal line crossing the y-axis at −2.

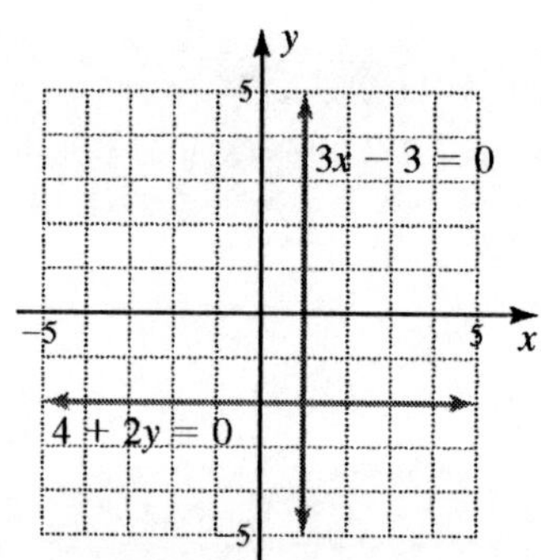

5. a. $d = 9 - 0.2t$
Let $t = 0$.
$d = 9 - 0.2(0)$
$d = 9$
(0, 9) is the d-intercept. Let $d = 0$.
$0 = 9 - 0.2t$
$0.2t = 9$
$t = 45$
(45, 0) is the t-intercept.

b. $d = 8 - 0.2t$
Let $t = 0$.
$d = 8 - 0.2(0)$
$d = 8$
(0, 8) is the d-intercept. Let $d = 0$.

$0 = 8 - 0.2t$
$0.2t = 8$
$t = 40$
(40, 0) is the t-intercept.

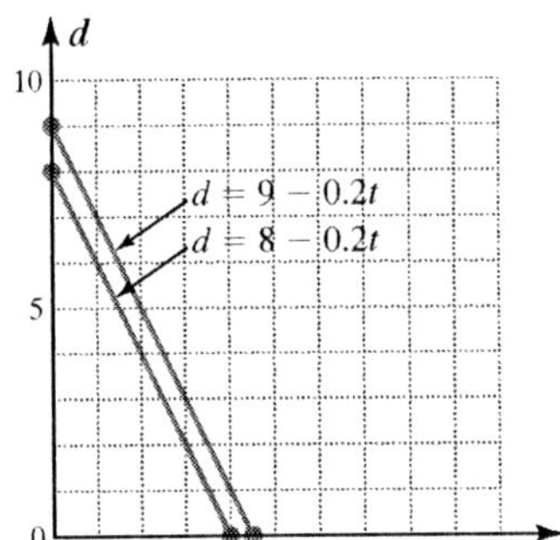

Exercises 3.3

1. $x + 2y = 4$
Let $x = 0$.
$0 + 2y = 4$
$2y = 4$
$y = 2$
(0, 2) is the y-intercept. Let $y = 0$.
$x + 2(0) = 4$
$x = 4$
(4, 0) is the x-intercept.

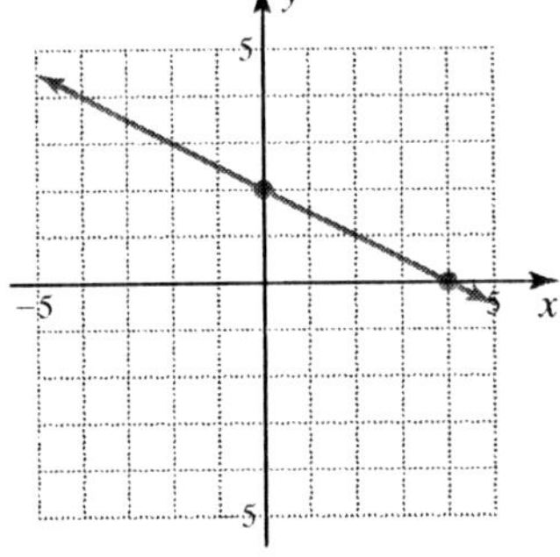

3. $-5x - 2y = -10$
Let $x = 0$.
$-5(0) - 2y = -10$
$-2y = -10$
$y = 5$
(0, 5) is the y-intercept. Let $y = 0$.
$-5x - 2(0) = -10$
$-5x = -10$
$x = 2$
(2, 0) is the x-intercept.

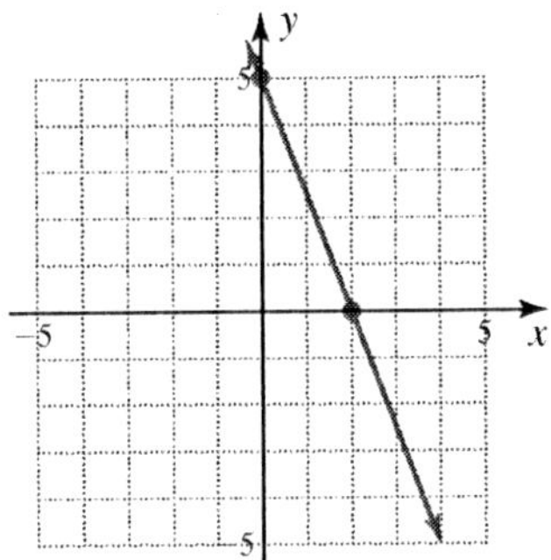

5. $y - 3x - 3 = 0$
Let $y = 0$.
$0 - 3x - 3 = 0$
$-3x = 3$
$x = -1$
(−1, 0) is the x-intercept. Let $x = 0$.
$y - 3(0) - 3 = 0$
$y - 3 = 0$
$y = 3$
(0, 3) is the y-intercept.

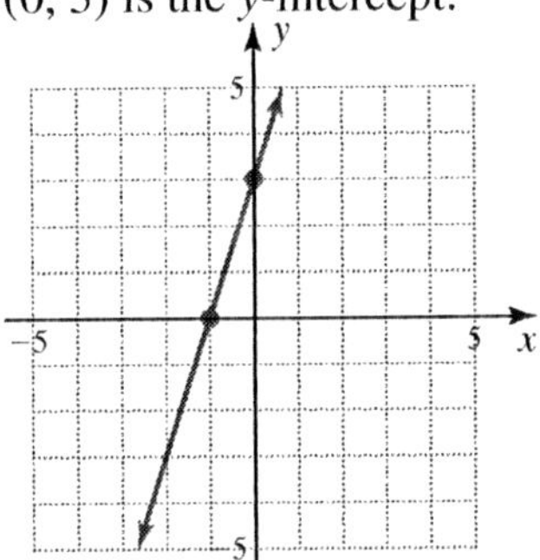

7. $6 = 6x - 3y$
Let $x = 0$.
$6 = 6(0) - 3y$
$6 = -3y$
$-2 = y$
(0, −2) is the y-intercept. Let $y = 0$.
$6 = 6x - 3(0)$
$6 = 6x$
$1 = x$
(1, 0) is the x-intercept.

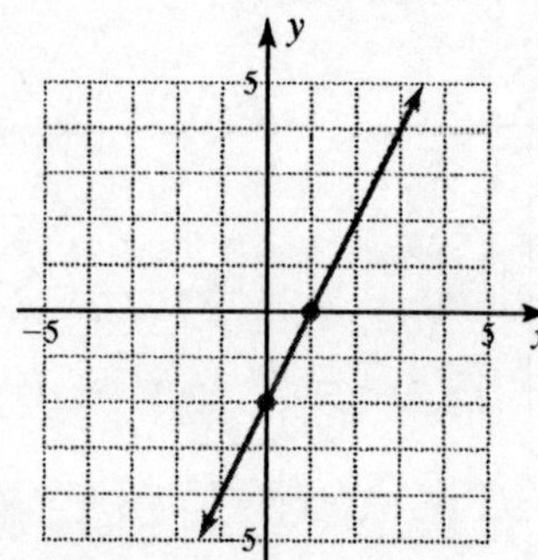

9. $3x + 4y + 12 = 0$
Let $x = 0$.

$$\begin{aligned} 3(0) + 4y + 12 &= 0 \\ 4y + 12 &= 0 \\ 4y &= -12 \\ y &= -3 \end{aligned}$$

$(0, -3)$ is the y-intercept. Let $y = 0$.

$$\begin{aligned} 3x + 4(0) + 12 &= 0 \\ 3x + 12 &= 0 \\ 3x &= -12 \\ x &= -4 \end{aligned}$$

$(-4, 0)$ is the x-intercept.

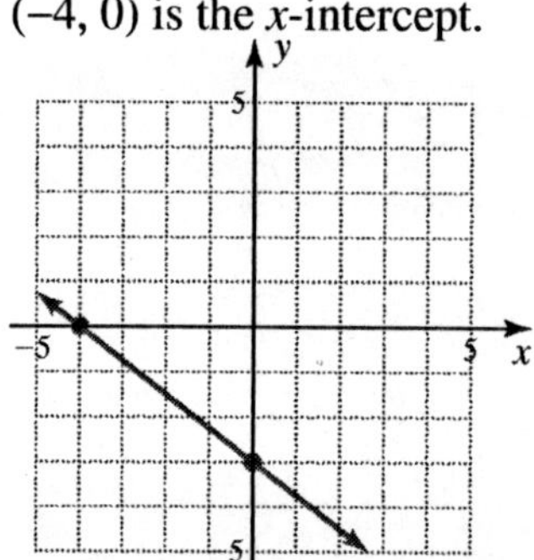

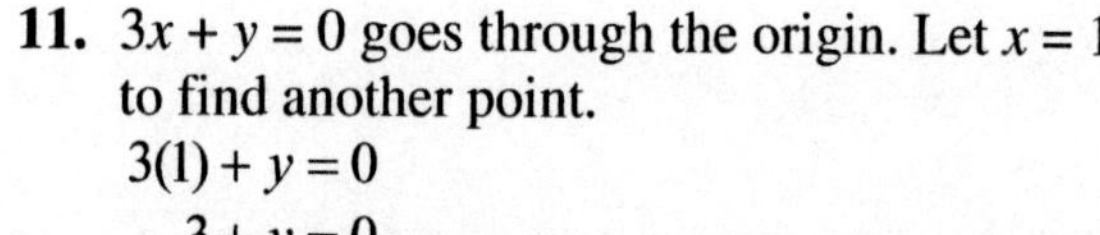

11. $3x + y = 0$ goes through the origin. Let $x = 1$ to find another point.

$$\begin{aligned} 3(1) + y &= 0 \\ 3 + y &= 0 \\ y &= -3 \end{aligned}$$

$(1, -3)$ is on the graph.

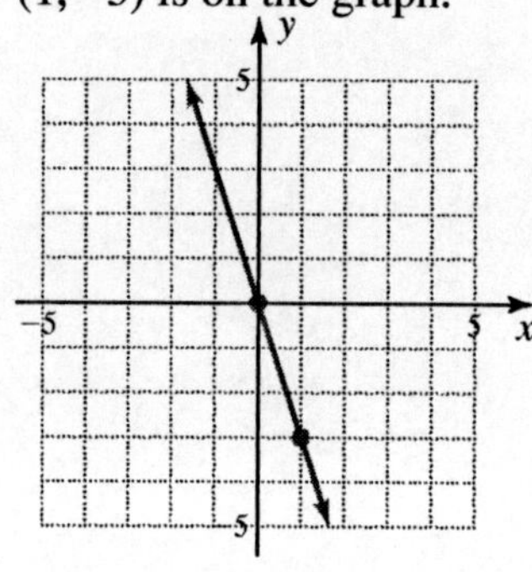

13. $2x + 3y = 0$ goes through the origin. Let $x = 3$ to find another point.

$$\begin{aligned} 2(3) + 3y &= 0 \\ 6 + 3y &= 0 \\ 3y &= -6 \\ y &= -2 \end{aligned}$$

$(3, -2)$ is on the graph.

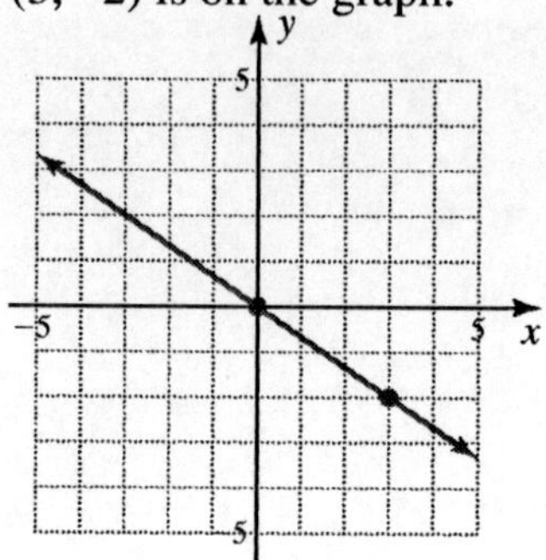

15. $-2x + y = 0$ goes through the origin. Let $x = 1$ to find another point.

$$\begin{aligned} -2(1) + y &= 0 \\ -2 + y &= 0 \\ y &= 2 \end{aligned}$$

$(1, 2)$ is on the graph.

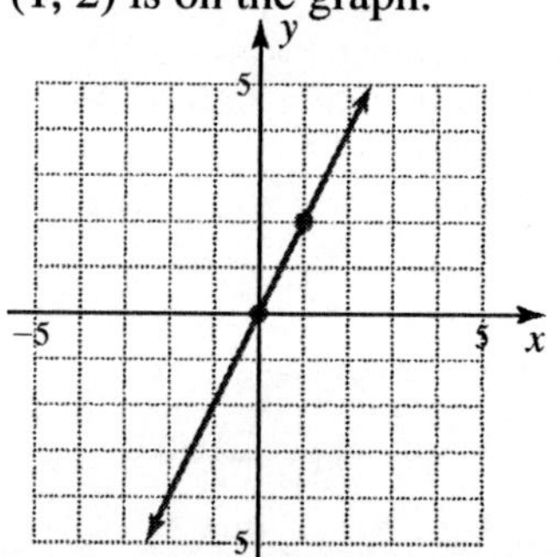

17. $2x - 3y = 0$ goes through the origin. Let $x = 3$ to find another point.

$$\begin{aligned} 2(3) - 3y &= 0 \\ 6 - 3y &= 0 \\ -3y &= -6 \\ y &= 2 \end{aligned}$$

$(3, 2)$ is on the graph.

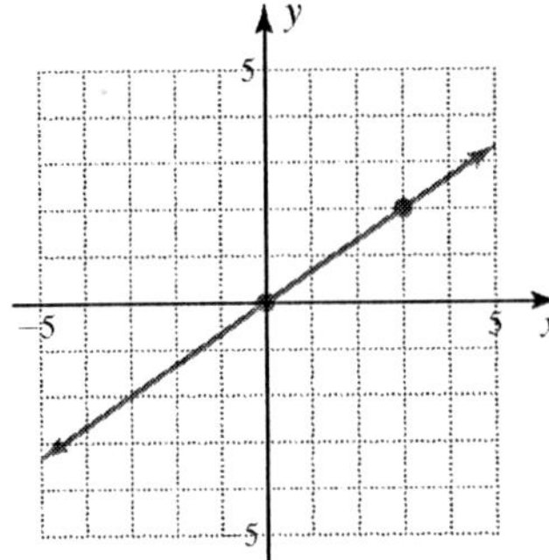

19. $-3x = -2y$ goes through the origin. Let $x = 2$ to find another point.

$$-3(2) = -2y$$
$$-6 = -2y$$
$$3 = y$$

(2, 3) is on the graph.

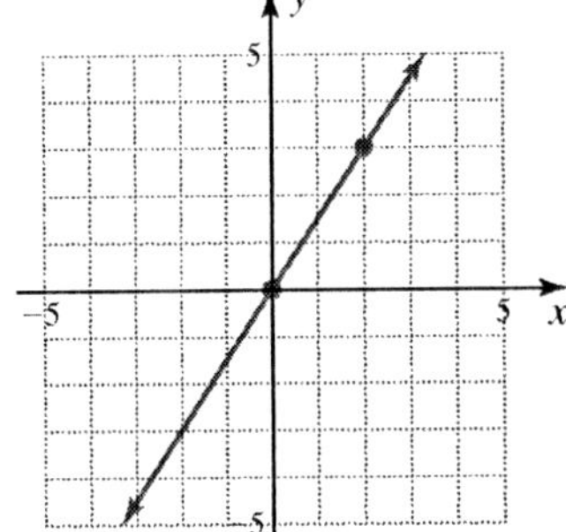

21. $y = -4$ is a horizontal line crossing the y-axis at -4.

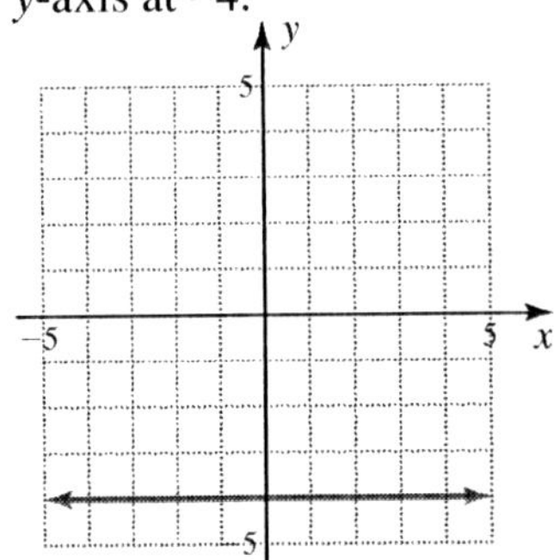

23. $2y + 6 = 0$

$$2y = -6$$
$$y = -3$$

This is a horizontal line crossing the y-axis at -3.

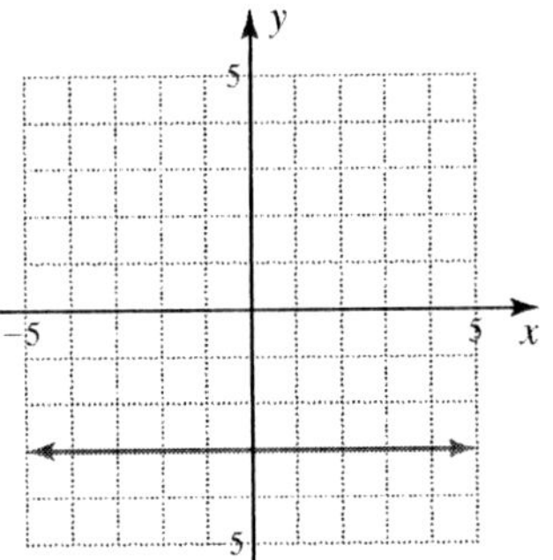

25. $x = -\frac{5}{2}$ is a vertical line crossing the x-axis at $-\frac{5}{2}$.

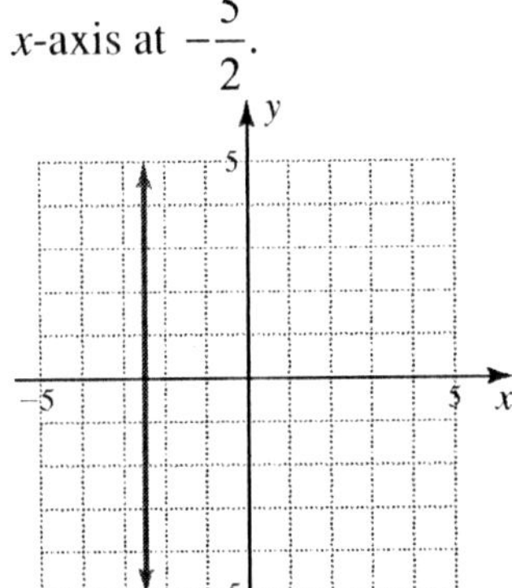

27. $2x + 4 = 0$

$$2x = -4$$
$$x = -2$$

This is a vertical line crossing the x-axis at -2.

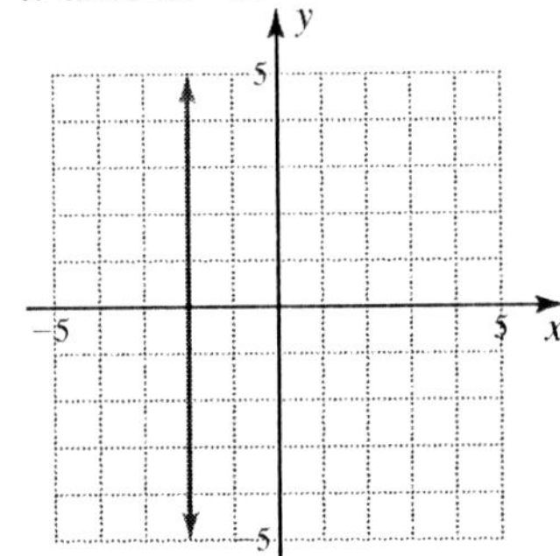

29. $2x - 9 = 0$

$$2x = 9$$
$$x = \frac{9}{2}$$

This is a vertical line crossing the x-axis at $\frac{9}{2} = 4\frac{1}{2}$.

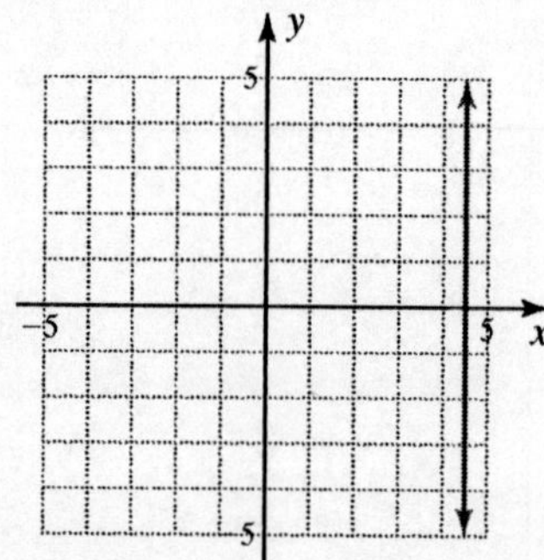

31. $d = 5.6 - 0.08t$

a. Let $t = 0$.
$d = 5.6 - 0.08(0)$
$d = 5.6$
The d-intercept is 5.6.

b. Let $d = 0$.
$0 = 5.6 - 0.08t$
$0.08t = 5.6$
$t = 70$
The t-intercept is 70.

c.

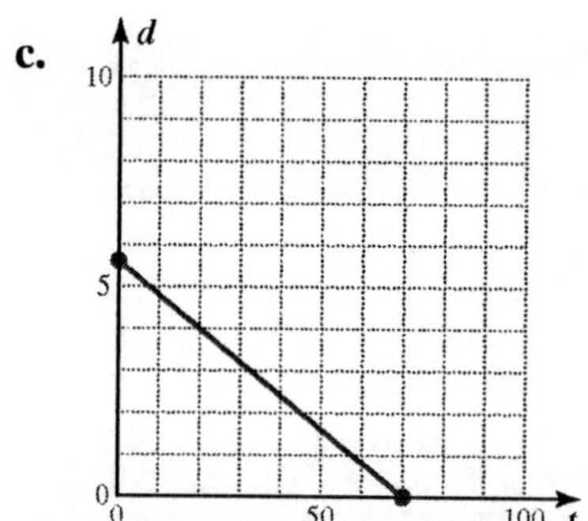

33. $g = 140 + t$

a. 1950 corresponds to $t = 0$.
$g = 140 + 0$
$g = 140$
The daily fat intake per person was 140 g in 1950.

b. 1990 corresponds to $t = 40$.
$g = 140 + 40$
$g = 180$
The daily fat intake per person was 180 g in 1990.

c. 2000 corresponds to $t = 50$.
$g = 140 + 50$
$g = 190$
The daily fat intake per person is predicted to be 190 g in 2000.

d.

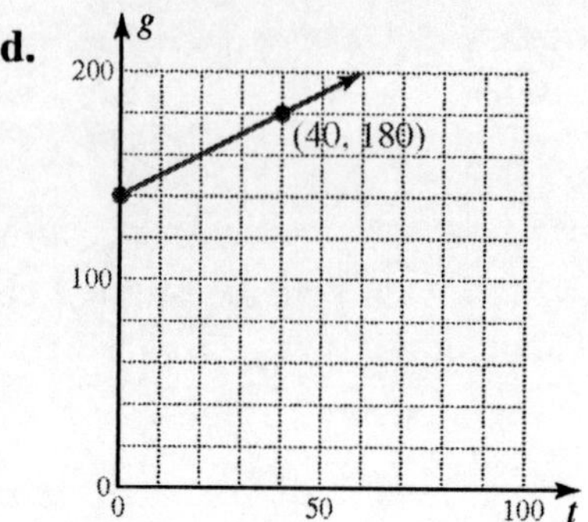

35. $3 - (-6) = 3 + 6 = 9$

37. $-6 - 3 = -9$

39. $C = -3w + 215$
Initially $w = 0$.
$C = -3(0) + 215$
$C = 215$
The initial cholesterol level was 215.

41.

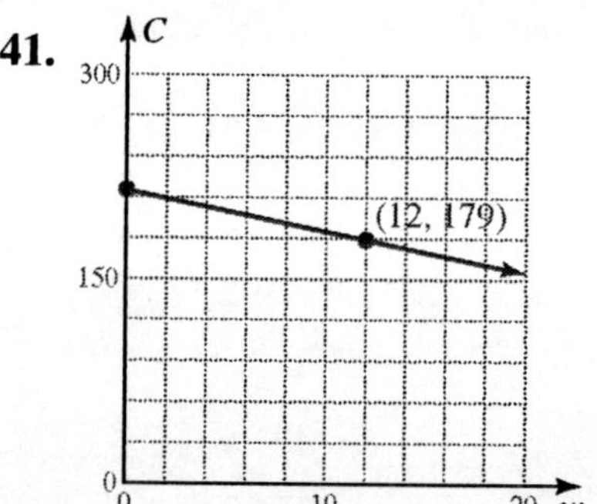

43. If $A = 0$ in $Ax + By = C$ with B and C not zero, the equation becomes $By = C$ or $y = \frac{C}{B}$. The graph of $y = \frac{C}{B}$ is a horizontal line.

45. If $C = 0$ in $Ax + By = C$ with A and B not zero, the equation has the form $Ax + By = 0$. The graph is a line that passes through the origin.

47. Two points are needed to graph a line.

49. $-2x + y = 4$
Let $x = 0$.
$-2(0) + y = 4$
$y = 4$
(0, 4) is the y-intercept. Let $y = 0$.
$-2x + 0 = 4$
$-2x = 4$
$x = -2$
(−2, 0) is the x-intercept.

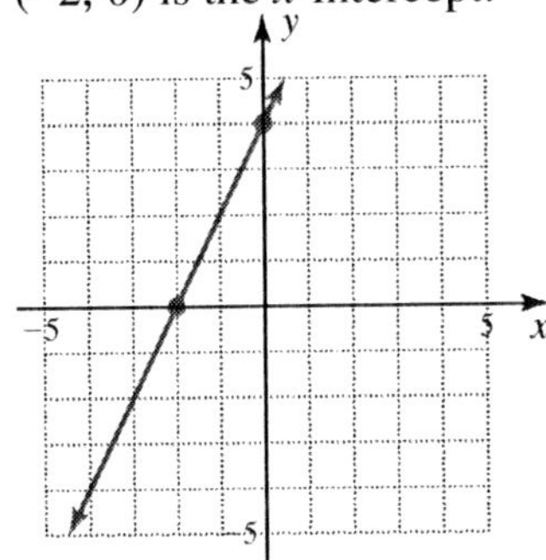

51. $4 + 2y = 0$
$2y = -4$
$y = -2$
This is a horizontal line that crosses the y-axis at −2.

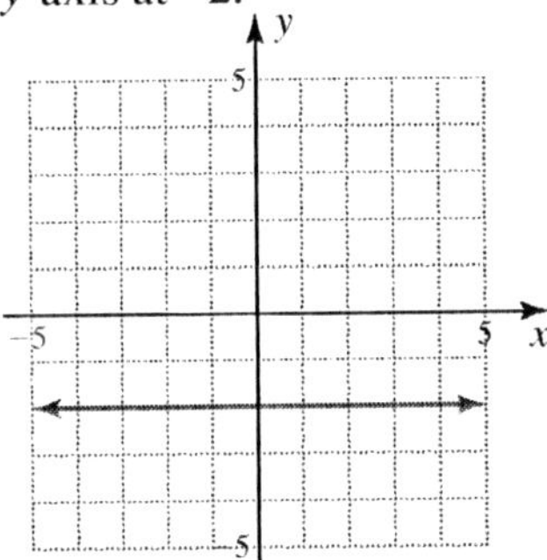

53. $-x + 4y = 0$ goes through the origin. Let $y = 1$ to find another point.
$-x + 4(1) = 0$
$-x + 4 = 0$
$-x = -4$
$x = 4$
(4, 1) is on the graph.

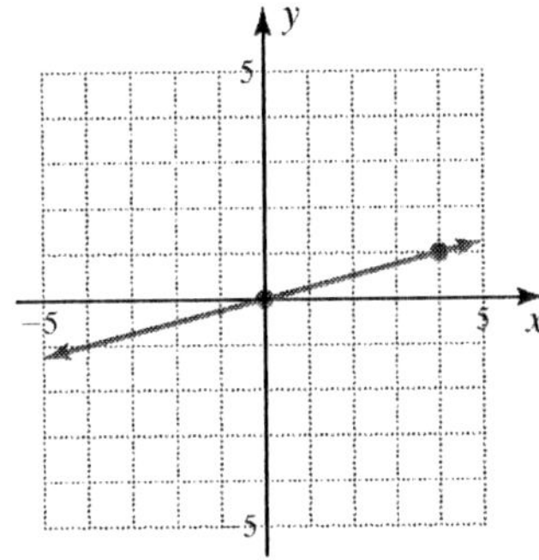

55. a. $C = A + 30$
Let $A = 0$.
$C = 0 + 30$
$C = 30$
The C-intercept is 30.
Let $C = 0$.
$0 = A + 30$
$-30 = A$
The A-intercept is −30.

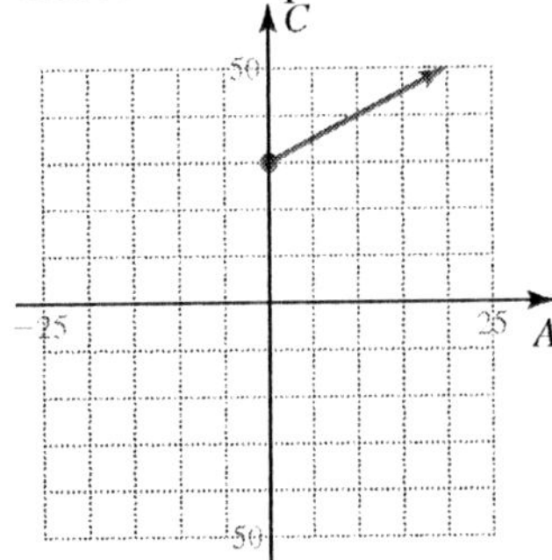

b. Since C and A represent dress sizes, which cannot be negative, the graph should be in Quadrant I.

3.4 The Slope of a Line

Problems 3.4

1. Choose $(x_1, y_1) = (0, -6)$ and $(x_2, y_2) = (2, 2)$.
$$m = \frac{2-(-6)}{2-0} = \frac{8}{2} = 4$$

2. Let $(x_1, y_1) = (2, -4)$ and $(x_2, y_2) = (-3, 2)$.
$$m = \frac{2-(-4)}{-3-2} = \frac{6}{-5} = -\frac{6}{5}$$

3. a. Let $(x_1, y_1) = (4, 1)$ and $(x_2, y_2) = (-3, 1)$.

$$m = \frac{1-1}{-3-4} = \frac{0}{-7} = 0$$

This is a horizontal line.

b. Let $(x_1, y_1) = (-2, 4)$ and $(x_2, y_2) = (-2, 1)$.

$$m = \frac{1-4}{-2-(-2)} = \frac{-3}{0} \text{ which is undefined.}$$

This is a vertical line.

4. a.
$$3x + 2y = 6$$
$$2y = -3x + 6$$
$$y = -\frac{3}{2}x + \frac{6}{2}$$
$$y = -\frac{3}{2}x + 3$$

Since the coefficient of x is $-\frac{3}{2}$, the slope is $-\frac{3}{2}$.

b.
$$2x - 3y = 9$$
$$-3y = -2x + 9$$
$$y = \frac{-2}{-3}x + \frac{9}{-3}$$
$$y = \frac{2}{3}x - 3$$

Since the coefficient of x is $\frac{2}{3}$, the slope is $\frac{2}{3}$.

5. a.
$$x - 2y = 3$$
$$-2y = -x + 3$$
$$y = \frac{1}{2}x - \frac{3}{2}$$

The slope is $\frac{1}{2}$.

$$2x - 4y = 8$$
$$-4y = -2x + 8$$
$$y = \frac{1}{2}x - 2$$

The slope is $\frac{1}{2}$.

Since the slopes are equal and the y-intercepts are different, the lines are parallel.

b.
$$3x + y = 6$$
$$y = -3x + 6$$

The slope is -3.

$$x + y = 2$$
$$y = -x + 2$$

The slope is -1.

The slopes are not equal and their product is not -1, so the lines are neither parallel nor perpendicular.

c.
$$3x + y = 6$$
$$y = -3x + 6$$

The slope is -3.

$$x - 3y = 5$$
$$-3y = -x + 5$$
$$y = \frac{1}{3}x - \frac{5}{3}$$

The slope is $\frac{1}{3}$.

Since $-3\left(\frac{1}{3}\right) = -1$, the lines are perpendicular.

6. a. The slope of $C = -3w + 215$ is -3.

b. The slope -3 represents the weekly *decrease* in cholesterol level. Thus, the cholesterol level is dropping by 3 points per week.

Exercises 3.4

1. $(x_1, y_1) = (1, 2)$; $(x_2, y_2) = (3, 4)$

$$m = \frac{4-2}{3-1} = \frac{2}{2} = 1$$

3. $(x_1, y_1) = (0, 5)$; $(x_2, y_2) = (5, 0)$

$$m = \frac{0-5}{5-0} = \frac{-5}{5} = -1$$

5. $(x_1, y_1) = (-1, -3); (x_2, y_2) = (7, -4)$

$$m = \frac{-4-(-3)}{7-(-1)} = \frac{-1}{8} = -\frac{1}{8}$$

7. $(x_1, y_1) = (0, 0); (x_2, y_2) = (12, 3)$

$$m = \frac{3-0}{12-0} = \frac{3}{12} = \frac{1}{4}$$

9. $(x_1, y_1) = (3, 5); (x_2, y_2) = (-2, 5)$

$$m = \frac{5-5}{-2-3} = \frac{0}{-5} = 0$$

11. $(x_1, y_1) = (4, 7); (x_2, y_2) = (-5, 7)$

$$m = \frac{7-7}{-5-4} = \frac{0}{-9} = 0$$

13. $(x_1, y_1) = \left(-\frac{1}{2}, -\frac{1}{3}\right); (x_2, y_2) = \left(2, -\frac{1}{3}\right)$

$$m = \frac{-\frac{1}{3}-\left(-\frac{1}{3}\right)}{2-\left(-\frac{1}{2}\right)} = \frac{0}{\frac{5}{2}} = 0$$

15. $y = 3x + 7$
The slope is $m = 3$.

17. $-3y = 2x - 4$

$$y = -\frac{2}{3}x + \frac{4}{3}$$

The slope is $m = -\frac{2}{3}$.

19. $x + 3y = 6$

$$3y = -x + 6$$
$$y = -\frac{1}{3}x + 2$$

The slope is $m = -\frac{1}{3}$.

21. $-2x + 5y = 5$

$$5y = 2x + 5$$
$$y = \frac{2}{5}x + 1$$

The slope is $m = \frac{2}{5}$.

23. $y = 6$
$y = 0x + 6$
The slope is $m = 0$.

25. $2x - 4 = 0$

$$2x = 4$$
$$x = 2$$

This is a vertical line. The slope is undefined.

27. $y = 2x + 5$
The slope is 2.

$$4x - 2y = 7$$
$$-2y = -4x + 7$$
$$y = 2x - \frac{7}{2}$$

The slope is 2.
The slopes are equal and the y-intercepts are different, so the lines are parallel.

29. $2x + 5y = 8$

$$5y = -2x + 8$$
$$y = -\frac{2}{5}x + \frac{8}{5}$$

The slope is $-\frac{2}{5}$.

$$5x - 2y = -9$$
$$-2y = -5x - 9$$
$$y = \frac{5}{2}x + \frac{9}{2}$$

The slope is $\frac{5}{2}$.

The product of the slopes is $\left(-\frac{2}{5}\right)\left(\frac{5}{2}\right) = -1$, so the lines are perpendicular.

31. $x + 7y = 7$

$$7y = -x + 7$$
$$y = -\frac{1}{7}x + 1$$

The slope is $-\frac{1}{7}$.

$$2x + 14y = 21$$
$$14y = -2x + 21$$
$$y = -\frac{1}{7}x + \frac{3}{2}$$

The slope is $-\frac{1}{7}$.
The slopes are equal and the y-intercepts are different, so the lines are parallel.

33. $2x + y = 7$
$y = -2x + 7$
The slope is -2.
$-2x - y = 9$
$-y = 2x + 9$
$y = -2x - 9$
The slope is -2.
The slopes are equal and the y-intercepts are different, so the lines are parallel.

35. $2y - 4 = 0$
$2y = 4$
$y = 2$
$y = 0x + 2$
The slope is 0.
$3y - 6 = 0$
$3y = 6$
$y = 2$
$y = 0x + 2$
The slope is 0.
The slopes are equal, but the y-intercepts are the same. Thus, the lines are neither parallel nor perpendicular, since they coincide.

37. $3x = 7$
$x = \frac{7}{3}$
This is a vertical line.
$2y = 7$
$y = \frac{7}{2}$
This is a horizontal line.
Since one line is vertical and the other is horizontal, the lines are perpendicular.

39. $S = 5 - 0.1t$

a. $S = -0.1t + 5$
The slope is $m = -0.1$.

b. Since the slope is negative, the average hospital stay is decreasing.

c. The slope represents the rate at which the average hospital stay is changing.

41. $y = 0.15t + 80$

a. The slope is $m = 0.15$.

b. Since the slope is positive, the average life span of American women is increasing.

c. The slope of 0.15 represents the annual increase in the average life span of American women.

43. $v = 128 - 32t$

a. $v = -32t + 128$
The slope is $m = -32$.

b. Since the slope is negative, the velocity of the ball is decreasing.

c. The slope of -32 represents the decrease in the velocity of the ball each second.

45. $f = 165 + 0.4t$

a. $f = 0.4t + 165$
The slope is $m = 0.4$.

b. Since the slope is positive, the consumption of fat by Americans is increasing.

c. The slope of 0.4 represents the annual increase in the amount of fat consumed daily by the average American.

47. $2[x - (-4)] = 2[x + 4] = 2x + 8$

49. $-2[x - (-1)] = -2[x + 1] = -2x - 2$

51. A line with positive slope rises as it goes from left to right.

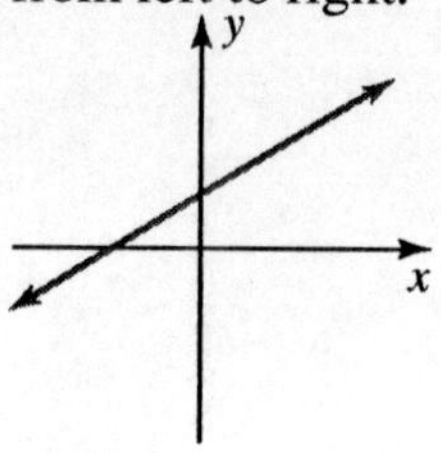

53. A line whose slope is undefined is a vertical line.

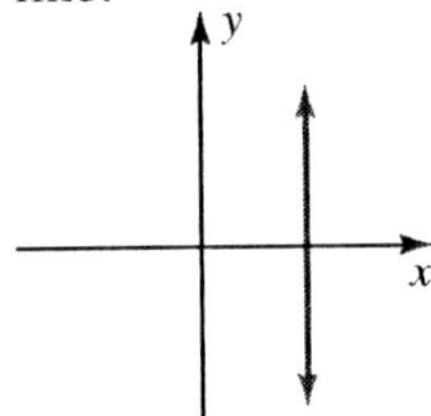

55. The slope of a line measures the change in the y-values for a 1-unit change in the x-values. If a line is horizontal, there is no change in the y-values, so the slope is 0.

57. $(x_1, y_1) = (2, -3); (x_2, y_2) = (4, -5)$

$$m = \frac{-5-(-3)}{4-2} = \frac{-2}{2} = -1$$

59. $(x_1, y_1) = (-4, 2); (x_2, y_2) = (-4, 5)$

$$m = \frac{5-2}{-4-(-4)} = \frac{3}{0} \text{ which is undefined.}$$

61.
$$\begin{aligned} 3x + 2y &= 6 \\ 2y &= -3x + 6 \\ y &= -\frac{3}{2}x + 3 \end{aligned}$$

The slope is $m = -\frac{3}{2}$.

63.
$$\begin{aligned} -x + 3y &= -6 \\ 3y &= x - 6 \\ y &= \frac{1}{3}x - 2 \end{aligned}$$

The slope is $\frac{1}{3}$.

$$\begin{aligned} 2x - 6y &= -7 \\ -6y &= -2x - 7 \\ y &= \frac{1}{3}x + \frac{7}{6} \end{aligned}$$

The slope is $\frac{1}{3}$.

The slopes are equal and the y-intercepts are different, so the lines are parallel.

65.
$$\begin{aligned} 3x - 2y &= 6 \\ -2y &= -3x + 6 \\ y &= \frac{3}{2}x - 3 \end{aligned}$$

The slope is $\frac{3}{2}$.

$$\begin{aligned} -2x - 3y &= 6 \\ -3y &= 2x + 6 \\ y &= -\frac{2}{3}x - 2 \end{aligned}$$

The slope is $-\frac{2}{3}$.

The product of the slopes is $\frac{3}{2}\left(-\frac{2}{3}\right) = -1$,

so the lines are perpendicular.

Review Exercises

1. a. Start at the origin. Go 1 unit left and 2 units up.

b. Start at the origin. Go 2 units left and 1 unit up.

c. Start at the origin. Go 3 units left and 3 units up.

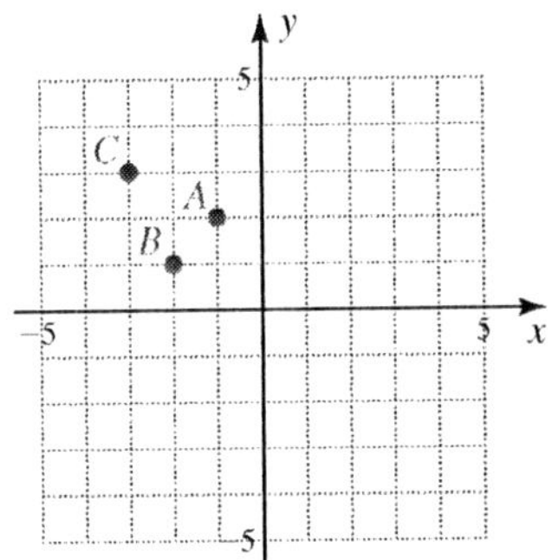

2.

Point	Start at the origin, move:	Coordinates
A	1 unit right, 1 unit up	(1, 1)
B	1 unit left, 2 units down	(−1, −2)
C	3 units right, 3 units down	(3, −3)

3. a. The point (10, −10) means that when the air temperature is 5°F, for a wind speed of 10 mi/hr, the wind chill temperature is −10°F by the new wind chill formula.

b. To find the wind chill temperature corresponding to a wind speed of 40 mi/hr, move right on the horizontal axis to the line labeled 40. Then go up until you reach the graph for the new wind chill formula. The new wind chill temperature is approximately −22°F.

c. To find the wind chill temperature corresponding to a wind speed of 40 mi/hr, move right on the horizontal axis to the line labeled 40. Then go up until you reach the graph for the old wind chill formula. The old wind chill temperature is approximately −45°F.

4. a. To find the payment corresponding to 60 months, move right on the horizontal axis to the category labeled 60 months and then vertically to the end of the bar. According to the vertical scale, the monthly payment will be $10.

b. To find the payment corresponding to 48 months, move right on the horizontal axis to the category labeled 48 months and then vertically to the end of the bar. According to the vertical scale, the monthly payment will be $12.

c. To find the payment corresponding to 24 months, move right on the horizontal axis to the category labeled 24 months and then vertically to the end of the bar. According to the vertical scale, the monthly payment will be $20.

5. a. Point A is in the second quadrant.

b. Point B is in the third quadrant.

c. Point C is in the fourth quadrant.

6. a.

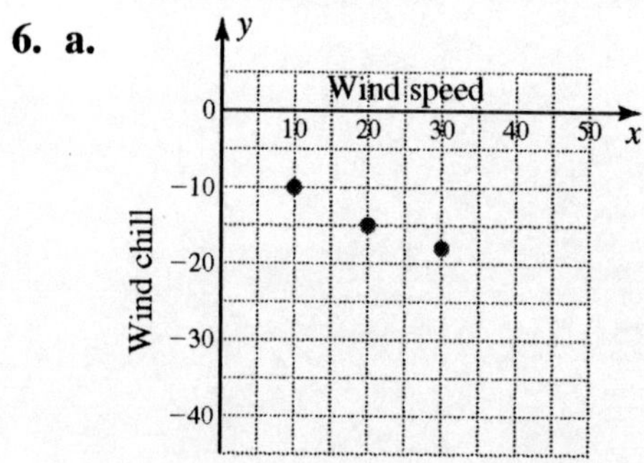

b. The point (10, −10) means that for a wind speed of 10 mi/hr, the wind chill temperature is −10°F.

c. Since a wind speed of 20 mi/hr corresponds to a wind chill temperature of −15°F, the number s in (s, −15) is 20.

7. **a.** $x - 2y = -3$
$1 - 2(-2) = -3$
$1 + 4 = -3$
no

b. $x - 2y = -3$
$2 - 2(-1) = -3$
$2 + 2 = -3$
no

c. $x - 2y = -3$
$-1 - 2(1) = -3$
$-1 - 2 = -3$
yes

8. **a.** $2x - y = 4$
$2x - 2 = 4$
$2x = 6$
$x = 3$

b. $2x - y = 4$
$2x - 4 = 4$
$2x = 8$
$x = 4$

c. $2x - y = 4$
$2x - 0 = 4$
$2x = 4$
$x = 2$

9. **a.** Let $x = 0$.
$x + y = 4$
$0 + y = 4$
$y = 4$
Thus (0, 4) is on the graph.
Let $y = 0$.
$x + 0 = 4$
$x = 4$
Thus (4, 0) is on the graph.

b. Let $x = 0$.
$x + y = 2$
$0 + y = 2$
$y = 2$
Thus (0, 2) is on the graph. Let $y = 0$.
$x + 0 = 2$
$x = 2$
Thus (2, 0) is on the graph.

c. Let $x = 0$.
$x + 2y = 2$
$0 + 2y = 2$
$2y = 2$
$y = 1$
Thus (0, 1) is on the graph. Let $y = 0$.
$x + 2(0) = 2$
$x = 2$
Thus (2, 0) is on the graph.

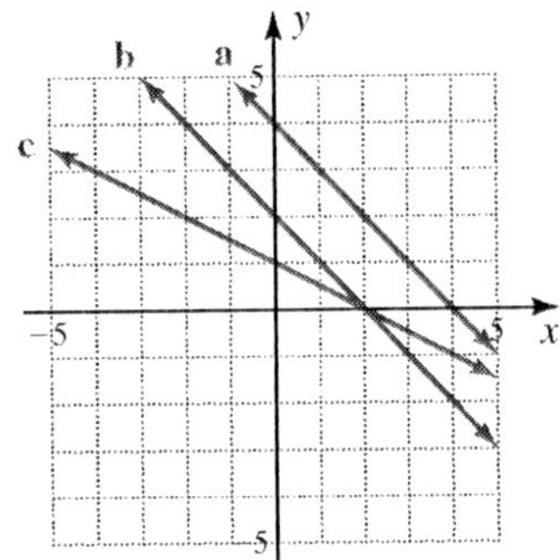

10. **a.** Let $x = 0$.
$y = \frac{3}{2}x + 3$
$y = \frac{3}{2}(0) + 3$
$y = 3$
Thus (0, 3) is on the graph. Let $x = -2$.
$y = \frac{3}{2}(-2) + 3$
$y = -3 + 3$
$y = 0$
Thus (−2, 0) is on the graph.

b. Let $x = 0$.
$y = -\frac{3}{2}x + 3$
$y = -\frac{3}{2}(0) + 3$
$y = 3$
Thus (0, 3) is on the graph. Let $x = 2$.
$y = -\frac{3}{2}(2) + 3$
$y = -3 + 3$
$y = 0$
Thus (2, 0) is on the graph.

c. Let $x = 0$.

$y = \frac{3}{4}x + 4$

$y = \frac{3}{4}(0) + 4$

$y = 4$

Thus (0, 4) is on the graph. Let $x = -4$.

$y = \frac{3}{4}(-4) + 4$

$y = -3 + 4$

$y = 1$

Thus (−4, 1) is on the graph.

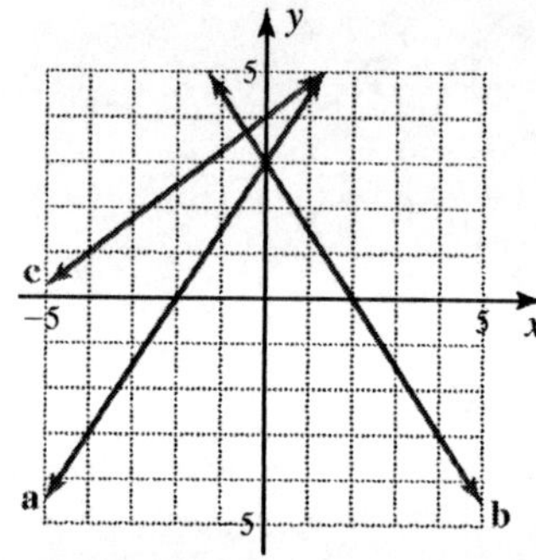

11. a. Let $t = 0$.

$g = 30 - 0.2t$

$g = 30 - 0.2(0)$

$g = 30$

Thus (0, 30) is on the graph. Let $t = 25$.

$g = 30 - 0.2(25)$

$g = 30 - 5$

$g = 25$

Thus (25, 25) is on the graph.

b. Let $t = 0$.

$g = 20 - 0.2t$

$g = 20 - 0.2(0)$

$g = 20$

Thus (0, 20) is on the graph. Let $t = 25$.

$g = 20 - 0.2(25)$

$g = 20 - 5$

$g = 15$

Thus (25, 15) is on the graph.

c. Let $t = 0$.

$g = 10 - 0.2t$

$g = 10 - 0.2(0)$

$g = 10$

Thus (0, 10) is on the graph. Let $t = 25$.

$g = 10 - 0.2(25)$

$g = 10 - 5$

$g = 5$

Thus (25, 5) is on the graph.

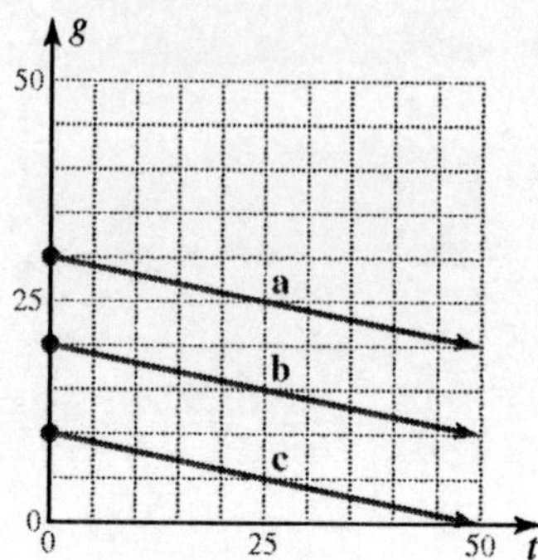

12. a. Let $x = 0$.

$2x - 3y - 12 = 0$

$2(0) - 3y - 12 = 0$

$-3y - 12 = 0$

$-3y = 12$

$y = -4$

The y-intercept is (0, −4). Let $y = 0$.

$2x - 3(0) - 12 = 0$

$2x - 12 = 0$

$2x = 12$

$x = 6$

The x-intercept is (6, 0). Let $x = 3$.

$2(3) - 3y - 12 = 0$

$6 - 3y - 12 = 0$

$-3y - 6 = 0$

$-3y = 6$

$y = -2$

(3, −2) is also on the graph.

b. Let $x = 0$.

$3x - 2y - 12 = 0$

$3(0) - 2y - 12 = 0$

$-2y - 12 = 0$

$-2y = 12$

$y = -6$

The y-intercept is (0, −6). Let $y = 0$.

$3x - 2(0) - 12 = 0$

$3x - 12 = 0$

$3x = 12$

$x = 4$

The x-intercept is (4, 0). Let $y = -3$.

$$3x - 2(-3) - 12 = 0$$
$$3x + 6 - 12 = 0$$
$$3x - 6 = 0$$
$$3x = 6$$
$$x = 2$$

(2, –3) is also on the graph.

c. Let $x = 0$.

$$2x + 3y + 12 = 0$$
$$2(0) + 3y + 12 = 0$$
$$3y + 12 = 0$$
$$3y = -12$$
$$y = -4$$

The y-intercept is (0, –4). Let $y = 0$.

$$2x + 3(0) + 12 = 0$$
$$2x + 12 = 0$$
$$2x = -12$$
$$x = -6$$

The x-intercept is (–6, 0). Let $x = -3$.

$$2(-3) + 3y + 12 = 0$$
$$-6 + 3y + 12 = 0$$
$$3y + 6 = 0$$
$$3y = -6$$
$$y = -2$$

(–3, –2) is also on the graph.

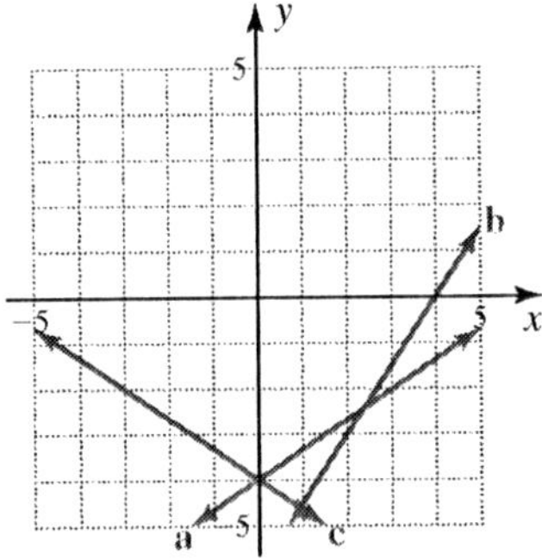

13. a. $3x + y = 0$ goes through the origin. Let $x = 1$ to find another point.

$$3(1) + y = 0$$
$$3 + y = 0$$
$$y = -3$$

(1, –3) is also on the graph.

b. $-2x + 3y = 0$ goes through the origin. Let $x = 3$ to find another point.

$$-2(3) + 3y = 0$$
$$-6 + 3y = 0$$
$$3y = 6$$
$$y = 2$$

(3, 2) is also on the graph.

c. $-3x + 2y = 0$ goes through the origin. Let $x = 2$ to find another point.

$$-3(2) + 2y = 0$$
$$-6 + 2y = 0$$
$$2y = 6$$
$$y = 3$$

(2, 3) is also on the graph.

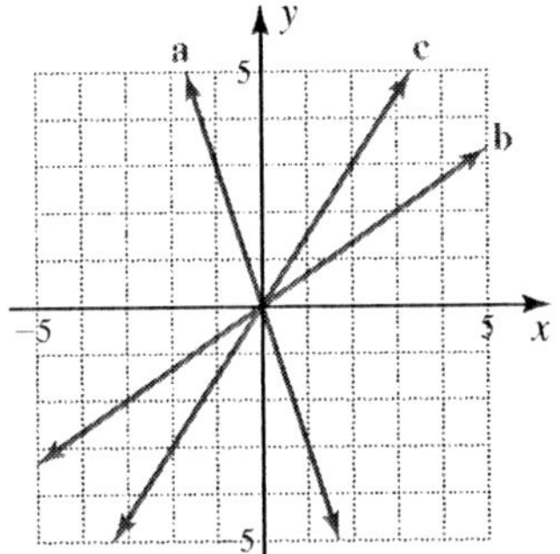

14. a.

$$2x - 6 = 0$$
$$2x = 6$$
$$x = 3$$

This is a vertical line that crosses the x-axis at 3.

b.

$$2x - 2 = 0$$
$$2x = 2$$
$$x = 1$$

This is a vertical line that crosses the x-axis at 1.

c.

$$2x - 4 = 0$$
$$2x = 4$$
$$x = 2$$

This is a vertical line that crosses the x-axis at 2.

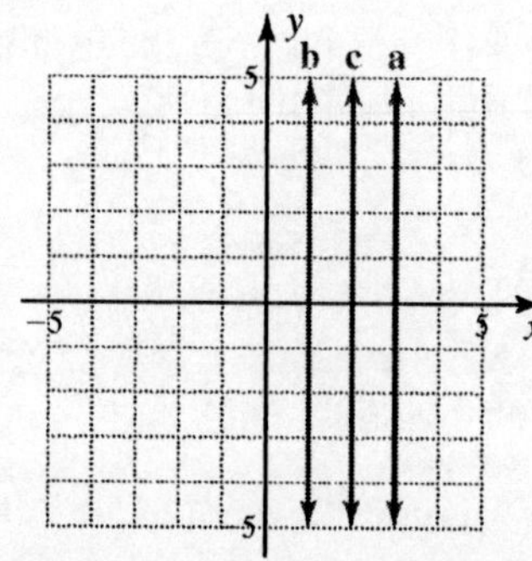

15. a. $y = -1$ is a horizontal line that crosses the y-axis at -1.

b. $y = -3$ is a horizontal line that crosses the y-axis at -3.

c. $y = -4$ is a horizontal line that crosses the y-axis at -4.

16. a. Let $(x_1, y_1) = (1, -4)$ and $(x_2, y_2) = (2, -3)$.

$$m = \frac{-3-(-4)}{2-1} = \frac{1}{1} = 1$$

b. Let $(x_1, y_1) = (5, -2)$ and $(x_2, y_2) = (8, 5)$.

$$m = \frac{5-(-2)}{8-5} = \frac{7}{3}$$

c. Let $(x_1, y_1) = (3, -4)$ and $(x_2, y_2) = (4, -8)$.

$$m = \frac{-8-(-4)}{4-3} = \frac{-4}{1} = -4$$

17. a. Let $(x_1, y_1) = (-5, 2)$ and $(x_2, y_2) = (-5, 4)$.

$$m = \frac{4-2}{-5-(-5)} = \frac{2}{0} \text{ which is undefined.}$$

b. Let $(x_1, y_1) = (-3, 5)$ and $(x_2, y_2) = (3, 5)$.

$$m = \frac{5-5}{3-(-3)} = \frac{0}{6} = 0$$

c. Let $(x_1, y_1) = (-2, -1)$ and $(x_2, y_2) = (-2, -5)$.

$$m = \frac{-5-(-1)}{-2-(-2)} = \frac{-4}{0} \text{ which is undefined.}$$

18. a.

$$\begin{aligned} 3x + 2y &= 6 \\ 2y &= -3x + 6 \\ y &= -\frac{3}{2}x + 3 \end{aligned}$$

The slope is $m = -\frac{3}{2}$.

b.

$$\begin{aligned} x + 4y &= 4 \\ 4y &= -x + 4 \\ y &= -\frac{1}{4}x + 1 \end{aligned}$$

The slope is $m = -\frac{1}{4}$.

c.

$$\begin{aligned} -2x + 3y &= 6 \\ 3y &= 2x + 6 \\ y &= \frac{2}{3}x + 2 \end{aligned}$$

The slope is $m = \frac{2}{3}$.

19. a.

$$\begin{aligned} 2x + 3y &= 6 \\ 3y &= -2x + 6 \\ y &= -\frac{2}{3}x + 2 \end{aligned}$$

The slope is $-\frac{2}{3}$.

$$\begin{aligned} 6x &= 6 - 4y \\ 4y + 6x &= 6 \\ 4y &= -6x + 6 \\ y &= -\frac{3}{2}x + \frac{3}{2} \end{aligned}$$

The slope is $-\frac{3}{2}$.

The slopes are different and their product is $\left(-\frac{2}{3}\right)\left(-\frac{3}{2}\right) = 1 \neq -1$, so the lines are neither parallel nor perpendicular.

b. $3x+2y=4$

$$2y=-3x+4$$
$$y=-\frac{3}{2}x+2$$

The slope is $-\frac{3}{2}$.

$$-2x+3y=4$$
$$3y=2x+4$$
$$y=\frac{2}{3}x+\frac{4}{3}$$

The product of the slopes is $\left(-\frac{3}{2}\right)\left(\frac{2}{3}\right)=-1$, so the lines are perpendicular.

c. $2x+3y=6$

$$3y=-2x+6$$
$$y=-\frac{2}{3}x+2$$

The slope is $-\frac{2}{3}$.

$$-2x-3y=6$$
$$-3y=2x+6$$
$$y=-\frac{2}{3}x-2$$

The slope is $-\frac{2}{3}$.

The slopes are equal and the y-intercepts are different, so the lines are parallel.

20. $N = 0.6t + 15$

a. The slope is $m = 0.6$.

b. The slope represents the change in the number of theaters per year. Since the slope is positive, the number of theaters is increasing.

c. 0.6 thousand theaters, or $0.6(1000) = 600$ theaters were added each year.

Cumulative Review Chapters 1–3

1. The additive inverse of -1 is $-(-1) = 1$.

2. $\left|-3\frac{1}{7}\right|=-\left(-3\frac{1}{7}\right)=3\frac{1}{7}$

3. The LCM of 6 and 8 is 24.

$$-\frac{1}{6}+\left(-\frac{3}{8}\right)=-\frac{1\cdot 4}{6\cdot 4}+\left(-\frac{3\cdot 3}{8\cdot 3}\right)$$
$$=-\frac{4}{24}+\left(-\frac{9}{24}\right)$$
$$=-\frac{4}{24}-\frac{9}{24}$$
$$=\frac{-4-9}{24}$$
$$=\frac{-13}{24}$$
$$=-\frac{13}{24}$$

4. $9.7 - (-3.3) = 9.7 + 3.3 = 13.0$

5. $(-2.6)(7.6) = -19.76$

6. $(-6)^4=(-6)(-6)(-6)(-6)=1296$

7. $-\frac{6}{7}\div\left(-\frac{1}{14}\right)=-\frac{6}{7}\cdot\left(-\frac{14}{1}\right)$

$$=-\frac{6}{\cancel{7}_1}\cdot\left(-\frac{\cancel{14}^2}{1}\right)$$
$$=\frac{6\cdot 2}{1\cdot 1}$$
$$=12$$

8. $y\div 2\cdot x-z=8\div 2\cdot 2-3=4\cdot 2-3=8-3=5$

9. Since the order of the addends inside the parentheses changed, the statement illustrates the commutative law of addition.

10. $3(6x - 7) = 3(6x) + 3(-7) = 18x - 21$

11. $-4cd^2-(-5cd^2)=-4cd^2+5cd^2$
$=(-4+5)cd^2$
$=1cd^2$
$=cd^2$

12. $3x-3(x+4)-(x+2)=3x-3x-12-x-2$
$=3x-3x-x-12-2$
$=-x-14$

13. The quotient of $(m + n)$ and p is $\dfrac{m+n}{p}$.

14. $1 = 15 - x$
$1 = 15 - (-14)$
$1 = 15 + 14$
No, −14 does not satisfy the equation.

15. $1=5(x-2)+5-4x$
$1=5x-10+5-4x$
$1=5x-4x-10+5$
$1=x-5$
$6=x$

16. $-\dfrac{8}{3}x=-24$

$x=\left(-\dfrac{3}{8}\right)(-24)$ (with 8 cancelled to 1 and −24 cancelled to −3)

$x=(-3)(-3)$
$x=9$

17. The LCM of 6 and 9 is 18.
$\dfrac{x}{6}-\dfrac{x}{9}=3$
$18\left(\dfrac{x}{6}-\dfrac{x}{9}\right)=18(3)$
$18\left(\dfrac{x}{6}\right)-18\left(\dfrac{x}{9}\right)=54$
$3x-2x=54$
$x=54$

18. The LCM of 5 and 11 is 55.
$4-\dfrac{x}{5}=\dfrac{2(x+1)}{11}$

$55\left(4-\dfrac{x}{5}\right)=55\left[\dfrac{2(x+1)}{11}\right]$ (with 55 cancelled to 5 and 11 cancelled to 1)

$55(4)-55\left(\dfrac{x}{5}\right)=5[2(x+1)]$ (with 55 cancelled to 11 and 5 cancelled to 1)

$220-11x=5[2x+2]$
$220-11x=10x+10$
$210-11x=10x$
$210=21x$
$10=x$

19. $S=7c^2d$
$\dfrac{S}{7c^2}=\dfrac{7c^2d}{7c^2}$
$\dfrac{S}{7c^2}=d$

20. Let x be one of the numbers. Then the other number is 40 more than x, or $x + 40$. The sum of x and $x + 40$ is 170.
$x+x+40=170$
$2x+40=170$
$2x=130$
$x=65$
$x + 40 = 65 + 40 = 105$
The numbers are 65 and 105.

21. Let x be the amount Dave receives from the bonds. Then he receives \$245 more, or $x + 245$ from the stocks. The total annual return is $x + x + 245$, which is 625.
$x+x+245=625$
$2x+245=625$
$2x=380$
$x=190$
$x + 245 = 190 + 245 = 435$
Dave has \$190 invested in bonds and \$435 invested in stocks.

22. Let x be the length of time that train B has been traveling. In x hours, train B travels $40x$ miles. Since train A left 2 hours before train B, the length of time that train A has been traveling is $x + 2$. In $(x + 2)$ hours, train A travels $20(x + 2)$ miles. Both trains have traveled the same distance when train B catches up to train A, thus $40x = 20(x + 2)$.

$$40x = 20(x+2)$$
$$40x = 20x + 40$$
$$20x = 40$$
$$x = 2$$

It takes train B 2 hours to catch up to train A.

23. Let x be the amount invested at 10%. Then the amount invested at 11% is $25{,}000 - x$. $0.10x$ is the interest on the 10% investment, while $0.11(25{,}000 - x)$ is the interest on the 11% investment. The total interest received is \$2630.

$$0.10x + 0.11(25{,}000 - x) = 2630$$
$$0.10x + 2750 - 0.11x = 2630$$
$$2750 - 0.01x = 2630$$
$$-0.01x = -120$$
$$x = 12{,}000$$

$25{,}000 - x = 25{,}000 - 12{,}000 = 13{,}000$

Martin has \$12,000 invested in bonds yielding 10% and \$13,000 invested in certificates of deposit yielding 11%.

24.
$$-\frac{x}{7} + \frac{x}{6} \le \frac{x-6}{6}$$
$$42\left(-\frac{x}{7} + \frac{x}{6}\right) \le 42\left(\frac{x-6}{6}\right)$$
$$-6x + 7x \le 7(x-6)$$
$$x \le 7x - 42$$
$$42 + x \le 7x$$
$$42 \le 6x$$
$$7 \le x$$

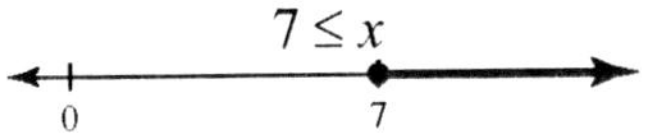

25. Start at the origin. Go 3 units left and 4 units up.

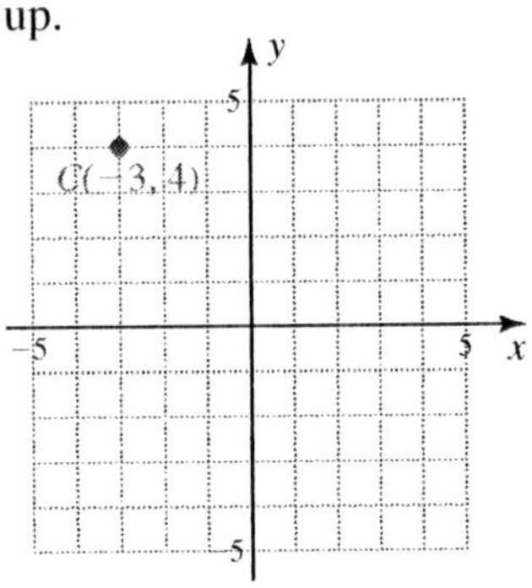

26. Start at the origin. Move 2 units right and 4 units down. The coordinates are (2, −4).

27.
$$4x + y = -21$$
$$4(-5) + 1 = -21$$
$$-20 + 1 = -21$$

No, (−5, 1) is not a solution of $4x + y = -21$.

28.
$$3x - y = -8$$
$$3x - 2 = -8$$
$$3x = -6$$
$$x = -2$$

29. Let $y = 0$.
$$2x + y = 8$$
$$2x + 0 = 8$$
$$2x = 8$$
$$x = 4$$

(4, 0) is on the graph. Let $x = 2$.
$$2(2) + y = 8$$
$$4 + y = 8$$
$$y = 4$$

(2, 4) is on the graph.

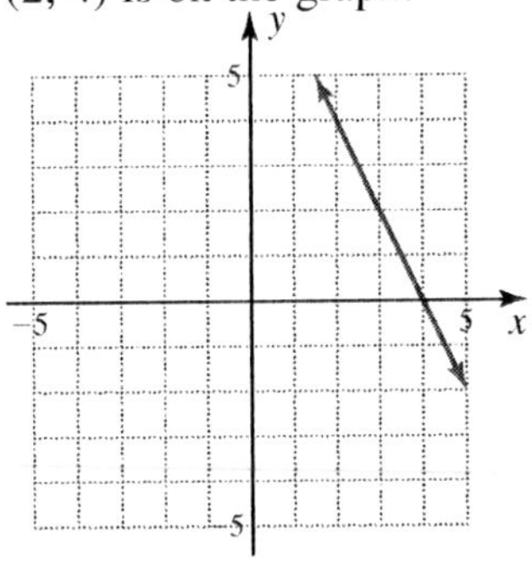

30. $4x-8=0$

$4x=8$

$x=2$

This is a vertical line that crosses the x-axis at 2.

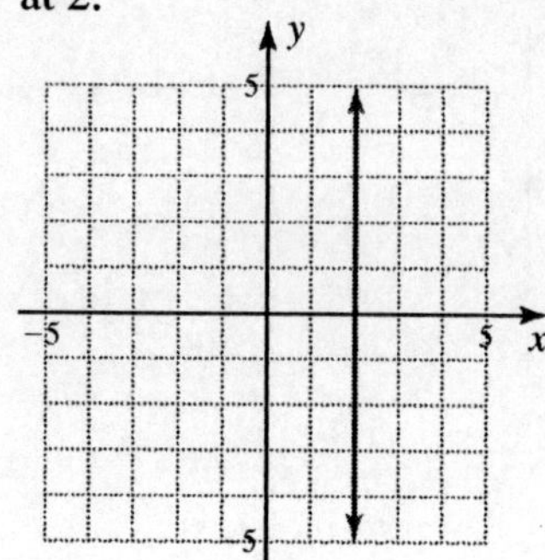

31. Let $(x_1, y_1)=(-5, 3)$ and $(x_2, y_2)=(-6, -6)$.

$$m=\frac{-6-3}{-6-(-5)}=\frac{-9}{-1}=9$$

32. $4x-2y=18$

$-2y=-4x+18$

$y=2x-9$

The slope is $m = 2$.

33. **(1)** $-5y=4x+7$

$$y=-\frac{4}{5}x-\frac{7}{5}$$

The slope is $-\frac{4}{5}$.

(2) $-15y+12x=7$

$-15y=-12x+7$

$$y=\frac{4}{5}x-\frac{7}{15}$$

The slope is $\frac{4}{5}$.

(3) $12x+15y=7$

$15y=-12x+7$

$$y=-\frac{4}{5}x+\frac{7}{15}$$

The slope is $-\frac{4}{5}$.

Since lines (1) and (3) have the same slope and different y-intercepts, they are parallel.

Chapter 4 Exponents and Polynomials

4.1 The Product, Quotient, and Power Rules for Exponents

Problems 4.1

1. a. $(3x^5)(4x^7) = (3\cdot 4)(x^5\cdot x^7)$
$= 12x^{5+7}$
$= 12x^{12}$

b. $(4ab^3c^2)(3ab^3)(2bc)$
$= (4\cdot 3\cdot 2)(a^1\cdot a^1)(b^3\cdot b^3\cdot b^1)(c^2\cdot c^1)$
$= 24a^{1+1}b^{3+3+1}c^{2+1}$
$= 24a^2b^7c^3$

2. a. $(2a^3bc)(-5ac^5)$
$= (2)(-5)(a^3\cdot a^1)(b^1)(c^1\cdot c^5)$
$= -10a^{3+1}b^1c^{1+5}$
$= -10a^4bc^6$

b. $(-3x^3yz^4)(-2xyz^4)$
$= [(-3)(-2)](x^3\cdot x^1)(y^1\cdot y^1)(z^4\cdot z^4)$
$= 6x^{3+1}y^{1+1}z^{4+4}$
$= 6x^4y^2z^8$

3. $\dfrac{25a^3b^7}{-5ab^3} = \dfrac{25}{-5}\cdot\dfrac{a^3}{a}\cdot\dfrac{b^7}{b^3}$
$= -5\cdot a^{3-1}\cdot b^{7-3}$
$= -5a^2b^4$

4. a. $(3^2)^4 = 3^{2\cdot 4} = 3^8$

b. $(a^3)^5 = a^{3\cdot 5} = a^{15}$

c. $(b^5)^3 = b^{5\cdot 3} = b^{15}$

5. a. $(2a^3b^2)^4 = 2^4(a^3)^4(b^2)^4$
$= 16a^{12}b^8$

b. $(-3a^3b^2)^3 = (-3)^3(a^3)^3(b^2)^3$
$= -27a^9b^6$

6. a. $\left(\dfrac{2}{3}\right)^3 = \dfrac{2^3}{3^3} = \dfrac{8}{27}$

b. $\left(\dfrac{3a^2}{b^4}\right)^3 = \dfrac{3^3\cdot(a^2)^3}{(b^4)^3}$
$= \dfrac{27a^{2\cdot 3}}{b^{4\cdot 3}}$
$= \dfrac{27a^6}{b^{12}}$

7. a. $(2a^3)^4(-3b^2)^3 = (2)^4(a^3)^4(-3)^3(b^2)^3$
$= 16a^{3\cdot 4}(-27)b^{2\cdot 3}$
$= 16\cdot(-27)a^{12}b^6$
$= -432a^{12}b^6$

b. $\left(\dfrac{2}{3}\right)^3\cdot 3^5 = \dfrac{2^3}{3^3}\cdot 3^5$
$= \dfrac{2^3}{3^3}\cdot\dfrac{3^5}{1}$
$= \dfrac{2^3\cdot 3^5}{3^3}$
$= 2^3\cdot 3^2$
$= 8\cdot 9$
$= 72$

8. a. Let $r = 1$ and $h = 4$.
$V = \dfrac{1}{3}\pi r^2h = \dfrac{1}{3}\pi(1)^2(4) = \dfrac{4\pi}{3}$

The volume is $\dfrac{4\pi}{3}$ in^3 or approximately 4 in^3.

b. Let $r = 1$.
$S = \dfrac{2}{3}\pi r^3 = \dfrac{2}{3}\pi(1)^3 = \dfrac{2\pi}{3}$

The volume of the mound is $\dfrac{2\pi}{3}$ in^3 or approximately 2 in^3.

c. The total amount of ice cream is $\frac{4\pi}{3}+\frac{2\pi}{3}=\frac{6\pi}{3}=2\pi\text{ in}^3$ or about 6 in^3.

Exercises 4.1

1. $(4x)(6x^2)=(4\cdot 6)(x^1\cdot x^2)=24x^{1+2}=24x^3$

3. $(5ab^2)(6a^3b)=(5\cdot 6)(a^1\cdot a^3)(b^2\cdot b^1)=30a^{1+3}b^{2+1}=30a^4b^3$

5. $(-xy^2)(-3x^2y)=[(-1)\cdot(-3)](x^1\cdot x^2)(y^2\cdot y^1)=3x^{1+2}y^{2+1}=3x^3y^3$

7. $b^3\left(\frac{-b^2c}{5}\right)=-\frac{(b^3\cdot b^2)\cdot c}{5}=-\frac{b^{3+2}\cdot c}{5}=-\frac{b^5c}{5}$

9. $\left(\frac{-5xy^2z}{2}\right)\left(\frac{-3x^2yz^5}{5}\right)=\frac{(-5)(-3)(x^1\cdot x^2)(y^2\cdot y^1)\cdot(z^1\cdot z^5)}{(2)(5)}=\frac{3x^{1+2}y^{2+1}z^{1+5}}{2}=\frac{3x^3y^3z^6}{2}$

11. $(-2xyz)(3x^2yz^3)(5x^2yz^4)=(-2)(3)(5)(x^1\cdot x^2\cdot x^2)(y^1\cdot y^1\cdot y^1)(z^1\cdot z^3\cdot z^4)$
$=-30x^{1+2+2}y^{1+1+1}z^{1+3+4}$
$=-30x^5y^3z^8$

13. $(a^2c^3)(-3b^2c)(-5a^2b)=(-3)(-5)(a^2\cdot a^2)(b^2\cdot b^1)(c^3\cdot c^1)=15a^{2+2}b^{2+1}c^{3+1}=15a^4b^3c^4$

15. $(-2abc)(-3a^2b^2c^2)(-4c)(-b^2c)=(-2)(-3)(-4)(-1)(a^1\cdot a^2)(b^1\cdot b^2\cdot b^2)(c^1\cdot c^2\cdot c^1\cdot c^1)$
$=24a^{1+2}b^{1+2+2}c^{1+2+1+1}$
$=24a^3b^5c^5$

17. $\frac{x^7}{x^3}=x^{7-3}=x^4$

19. $\frac{-8a^4}{16a^2}=\frac{-8}{16}\cdot a^{4-2}=-\frac{a^2}{2}$

21. $\frac{12x^5y^3}{6x^2y}=\frac{12}{6}\cdot\frac{x^5}{x^2}\cdot\frac{y^3}{y}=2x^{5-2}y^{3-1}=2x^3y^2$

23. $\frac{-6x^6y^3}{12x^3y}=\frac{-6}{12}\cdot\frac{x^6}{x^3}\cdot\frac{y^3}{y}=-\frac{1}{2}\cdot x^{6-3}\cdot y^{3-1}=-\frac{x^3y^2}{2}$

25. $\frac{-14a^8y^6}{-21a^5y^2}=\frac{-14}{-21}\cdot\frac{a^8}{a^5}\cdot\frac{y^6}{y^2}=\frac{2}{3}a^{8-5}y^{6-2}=\frac{2a^3y^4}{3}$

27. $\dfrac{-27a^2b^8c^3}{-36ab^5c^2} = \dfrac{-27}{-36} \cdot \dfrac{a^2}{a} \cdot \dfrac{b^8}{b^5} \cdot \dfrac{c^3}{c^2}$
$= \dfrac{3}{4}a^{2-1}b^{8-5}c^{3-2}$
$= \dfrac{3ab^3c}{4}$

29. $\dfrac{3a^3 \cdot a^5}{2a^4} = \dfrac{3a^{3+5}}{2a^4} = \dfrac{3a^8}{2a^4} = \dfrac{3a^{8-4}}{2} = \dfrac{3a^4}{2}$

31. $\dfrac{(2x^2y^3)(-3x^5y)}{6xy^3} = \dfrac{2(-3)}{6} \cdot \dfrac{x^2 \cdot x^5}{x} \cdot \dfrac{y^3 \cdot y}{y^3}$
$= -1 \cdot x^{2+5-1}y^{3+1-3}$
$= -x^6y$

33. $\dfrac{(-x^2y)(x^3y^2)}{x^3y} = \dfrac{-x^2 \cdot x^3}{x^3} \cdot \dfrac{y \cdot y^2}{y}$
$= -x^{2+3-3}y^{1+2-1}$
$= -x^2y^2$

35. $(2^2)^3 = 2^{2\cdot3} = 2^6 = 64$

37. $(3^2)^1 = 3^{2\cdot1} = 3^2 = 9$

39. $(x^3)^3 = x^{3\cdot3} = x^9$

41. $(y^3)^2 = y^{3\cdot2} = y^6$

43. $(-a^2)^3 = -a^{2\cdot3} = -a^6$

45. $(2x^3y^2)^3 = 2^3(x^3)^3(y^2)^3 = 8x^9y^6$

47. $(2x^2y^3)^2 = 2^2(x^2)^2(y^3)^2 = 4x^4y^6$

49. $(-3x^3y^2)^3 = (-3)^3(x^3)^3(y^2)^3 = -27x^9y^6$

51. $(-3x^6y^3)^2 = (-3)^2(x^6)^2(y^3)^2 = 9x^{12}y^6$

53. $(-2x^4y^4)^3 = (-2)^3(x^4)^3(y^4)^3 = -8x^{12}y^{12}$

55. $\left(\dfrac{2}{3}\right)^4 = \dfrac{2^4}{3^4} = \dfrac{16}{81}$

57. $\left(\dfrac{3x^2}{2y^3}\right)^3 = \dfrac{3^3(x^2)^3}{2^3(y^3)^3} = \dfrac{27x^{2\cdot3}}{8y^{3\cdot3}} = \dfrac{27x^6}{8y^9}$

59. $\left(\dfrac{-2x^2}{3y^3}\right)^4 = \dfrac{(-2)^4(x^2)^4}{3^4(y^3)^4} = \dfrac{16x^{2\cdot4}}{81y^{3\cdot4}} = \dfrac{16x^8}{81y^{12}}$

61. $(2x^3)^2(3y^3) = 2^2 \cdot 3x^{3\cdot2}y^3 = 12x^6y^3$

63. $(-3a)^2(-4b)^3 = (-3)^2 \cdot a^2 \cdot (-4)^3 \cdot b^3$
$= 9 \cdot (-64) \cdot a^2 \cdot b^3$
$= -576a^2b^3$

65. $-(4a^2)^2(-3b^3)^2 = -(4^2)(a^2)^2 \cdot (-3)^2 \cdot (b^3)^2$
$= -16 \cdot a^{2\cdot2} \cdot 9 \cdot b^{3\cdot2}$
$= -144a^4b^6$

67. $\left(\dfrac{2}{3}\right)^5 \cdot 3^6 = \dfrac{2^5}{3^5} \cdot 3^6$
$= \dfrac{2^5}{3^5} \cdot \dfrac{3^6}{1}$
$= \dfrac{2^5 \cdot 3^6}{3^5}$
$= 2^5 \cdot 3^1$
$= 32 \cdot 3$
$= 96$

69. $\left(\dfrac{x}{y}\right)^5 \cdot y^7 = \dfrac{x^5}{y^5} \cdot y^7$
$= \dfrac{x^5}{y^5} \cdot \dfrac{y^7}{1}$
$= \dfrac{x^5 \cdot y^7}{y^5}$
$= x^5y^2$

71. Let $P = 1000$, $r = 0.08$, and $n = 3$.
$$A = P(1+r)^n$$
$$= 1000(1+0.08)^3$$
$$= 1000(1.08)^3$$
$$= 1000(1.259712)$$
$$= 1259.712$$
You will have \$1259.71.

73. Let $P = 1000$, $r = 0.10$, and $n = 4$.
$$A = P(1+r)^n$$
$$= 1000(1+0.10)^4$$
$$= 1000(1.1)^4$$
$$= 1000(1.4641)$$
$$= 1464.1$$
You will have \$1464.10.

75. $V = lwh$
$$V = x \cdot x \cdot \frac{2}{3}x$$
$$V = \frac{2}{3} \cdot (x \cdot x \cdot x)$$
$$V = \frac{2}{3} \cdot (x^{1+1+1})$$
$$V = \frac{2}{3}x^3$$

77. $\dfrac{144 \text{ in}^3}{1 \text{ in}^3} = 144$
144 sandwiches can be made.

79. $\dfrac{1.08}{144} = 0.0075$
The cost of pickles in the sandwich is \$0.0075 or 0.75¢.

81. $V = lwh$
$$V = x \cdot 2x \cdot \frac{1}{2}x$$
$$V = \left(2 \cdot \frac{1}{2}\right)(x \cdot x \cdot x)$$
$$V = 1 \cdot x^{1+1+1}$$
$$V = x^3$$

83. $S = \dfrac{2}{3}\pi r^3$
$$S = \frac{2}{3}(3)(2)^3$$
$$S = \frac{2}{3}(3)(8)$$
$$S = 2(8)$$
$$S = 16$$
The volume of one serving is 16 in^3.

85. \$1.99 – \$0.24 = \$1.75
13.5(\$1.75) = \$23.625
The restaurant makes about \$23.63 per container.

87. $V = \dfrac{3}{4} \cdot \dfrac{3}{4}x^3 \text{ in}^3$
$$V = \frac{9}{16}x^3 \text{ in}^3$$

89. $\dfrac{562.5}{15} = 37.5$
There are 37.5 servings of beans in a pot of beans.

91. $8.1 \cdot 10 = 81$

93. $9.142 \cdot 10^3 = 9142$

95. **a.** $(11{,}111)^2 = 123{,}454{,}321$

b. $(111{,}111)^2 = 12{,}345{,}654{,}321$

97. **a.** $1+3+5+7+9 = 5^2 = 25$

b. $1+3+5+7+9+11+13 = 7^2 = 49$

99. The bases are different in $x^2 \cdot y^3$.

101. Answers may vary.

103. $(5a^3)(6a^8) = (5 \cdot 6)(a^3 \cdot a^8)$
$$= 30a^{3+8}$$
$$= 30a^{11}$$

105. $(-3x^3yz)(4xz^4)$
$=(-3\cdot 4)(x^3\cdot x^1)(y^1)(z^1\cdot z^4)$
$=-12x^{3+1}yz^{1+4}$
$=-12x^4yz^5$

107. $\dfrac{15x^4y^7}{-3x^2y}=\dfrac{15}{-3}\cdot x^{4-2}y^{7-1}=-5x^2y^6$

109. $\dfrac{-3a^2b^9c^2}{-12ab^2c}=\dfrac{-3}{-12}\cdot a^{2-1}\cdot b^{9-2}\cdot c^{2-1}=\dfrac{ab^7c}{4}$

111. $(y^5)^4=y^{5\cdot 4}=y^{20}$

113. $\left(\dfrac{3y^2}{x^5}\right)^3=\dfrac{3^3(y^2)^3}{(x^5)^3}=\dfrac{27y^{2\cdot 3}}{x^{5\cdot 3}}=\dfrac{27y^6}{x^{15}}$

115. $\left(\dfrac{1}{2^5}\right)\cdot 2^7=\dfrac{1}{2^5}\cdot\dfrac{2^7}{1}=\dfrac{2^7}{2^5}=2^{7-5}=2^2=4$

4.2 Integer Exponents

Problems 4.2

1. a. $2^{-3}=\dfrac{1}{2^3}=\dfrac{1}{2\cdot 2\cdot 2}=\dfrac{1}{8}$

b. $3^{-3}=\dfrac{1}{3^3}=\dfrac{1}{3\cdot 3\cdot 3}=\dfrac{1}{27}$

c. $\left(\dfrac{1}{2}\right)^{-3}=2^3=8$

d. $\left(\dfrac{1}{a}\right)^{-4}=a^4$

2. a. $\dfrac{2^{-4}}{3^{-3}}=\dfrac{3^3}{2^4}=\dfrac{27}{16}$

b. $\dfrac{x^{-9}}{y^{-4}}=\dfrac{y^4}{x^9}$

c. $\dfrac{a^5}{b^{-8}}=a^5b^8$

3. a. $\dfrac{1}{7^6}=7^{-6}$

b. $\dfrac{1}{8^5}=8^{-5}$

c. $\dfrac{1}{a^9}=a^{-9}$

d. $\dfrac{7}{a^6}=7\cdot\dfrac{1}{a^6}=7a^{-6}$

4. a. $3^{-4}\cdot 3^6=3^{-4+6}=3^2=9$

b. $2^4\cdot 2^{-6}=2^{4+(-6)}=2^{-2}=\dfrac{1}{2^2}=\dfrac{1}{4}$

c. $b^{-3}\cdot b^{-5}=b^{-3+(-5)}=b^{-8}=\dfrac{1}{b^8}$

d. $x^{-7}\cdot x^7=x^{-7+7}=x^0=1$

5. a. $\dfrac{3^5}{3^{-3}}=3^{5-(-3)}=3^{5+3}=3^8$

b. $\dfrac{a}{a^{-6}}=a^{1-(-6)}=a^{1+6}=a^7$

c. $\dfrac{b^{-5}}{b^{-5}}=b^{-5-(-5)}=b^{-5+5}=b^0=1$

d. $\dfrac{a^{-4}}{a^6}=a^{-4-6}=a^{-10}=\dfrac{1}{a^{10}}$

6. $\left(\dfrac{3a^5b^4}{4a^6b^{-3}}\right)^{-3} = \left(\dfrac{3a^{5-6}b^{4-(-3)}}{4}\right)^{-3}$

$= \left(\dfrac{3a^{-1}b^7}{4}\right)^{-3}$

$= \dfrac{3^{-3}a^3b^{-21}}{4^{-3}}$

$= \dfrac{4^3a^3}{3^3b^{21}}$

$= \dfrac{64a^3}{27b^{21}}$

7. a. The value in 3 years will be
$10{,}000(1-0.10)^3 = 10{,}000(0.90)^3 = \$7290.$

b. The value of the car 2 years ago was
$10{,}000(1-0.10)^{-2} = 10{,}000(0.90)^{-2} = \$12{,}345.68.$

Exercises 4.2

1. $4^{-2} = \dfrac{1}{4^2} = \dfrac{1}{16}$

3. $5^{-3} = \dfrac{1}{5^3} = \dfrac{1}{125}$

5. $\left(\dfrac{1}{8}\right)^{-2} = 8^2 = 64$

7. $\left(\dfrac{1}{x}\right)^{-1} = x^7$

9. $\dfrac{4^{-2}}{3^{-3}} = \dfrac{3^3}{4^2} = \dfrac{27}{16}$

11. $\dfrac{5^{-2}}{3^{-4}} = \dfrac{3^4}{5^2} = \dfrac{81}{25}$

13. $\dfrac{a^{-5}}{b^{-6}} = \dfrac{b^6}{a^5}$

15. $\dfrac{x^{-9}}{y^{-9}} = \dfrac{y^9}{x^9}$

17. $\dfrac{1}{2^3} = 2^{-3}$

19. $\dfrac{1}{y^5} = y^{-5}$

21. $\dfrac{1}{q^5} = q^{-5}$

23. $3^5 \cdot 3^{-4} = 3^{5+(-4)} = 3^1 = 3$

25. $2^{-5} \cdot 2^7 = 2^{-5+7} = 2^2 = 4$

27. $4^{-6} \cdot 4^4 = 4^{-6+4} = 4^{-2} = \dfrac{1}{4^2} = \dfrac{1}{16}$

29. $6^{-1} \cdot 6^{-2} = 6^{-1+(-2)} = 6^{-3} = \dfrac{1}{6^3} = \dfrac{1}{216}$

31. $2^{-4} \cdot 2^{-2} = 2^{-4+(-2)} = 2^{-6} = \dfrac{1}{2^6} = \dfrac{1}{64}$

33. $x^6 \cdot x^{-4} = x^{6+(-4)} = x^2$

35. $y^{-3} \cdot y^5 = y^{-3+5} = y^2$

37. $a^3 \cdot a^{-8} = a^{3+(-8)} = a^{-5} = \dfrac{1}{a^5}$

39. $x^{-5} \cdot x^3 = x^{-5+3} = x^{-2} = \dfrac{1}{x^2}$

41. $x \cdot x^{-3} = x^{1+(-3)} = x^{-2} = \dfrac{1}{x^2}$

43. $a^{-2} \cdot a^{-3} = a^{-2+(-3)} = a^{-5} = \dfrac{1}{a^5}$

45. $b^{-3} \cdot b^3 = b^{-3+3} = b^0 = 1$

47. $\frac{3^4}{3^{-1}} = 3^{4-(-1)} = 3^{4+1} = 3^5 = 243$

49. $\frac{4^{-1}}{4^2} = 4^{-1-2} = 4^{-3} = \frac{1}{4^3} = \frac{1}{64}$

51. $\frac{y}{y^3} = y^{1-3} = y^{-2} = \frac{1}{y^2}$

53. $\frac{x}{x^{-2}} = x^{1-(-2)} = x^{1+2} = x^3$

55. $\frac{x^{-3}}{x^{-1}} = x^{-3-(-1)} = x^{-3+1} = x^{-2} = \frac{1}{x^2}$

57. $\frac{x^{-3}}{x^4} = x^{-3-4} = x^{-7} = \frac{1}{x^7}$

59. $\frac{x^{-2}}{x^{-5}} = x^{-2-(-5)} = x^{-2+5} = x^3$

61. $\left(\frac{a}{b^3}\right)^2 = \frac{a^2}{b^{3\cdot 2}} = \frac{a^2}{b^6}$

63. $\left(\frac{-3a}{2b^2}\right)^{-3} = \frac{(-3)^{-3}a^{-3}}{2^{-3}b^{-6}} = \frac{2^3b^6}{(-3)^3a^3} = -\frac{8b^6}{27a^3}$

65. $\left(\frac{a^{-4}}{b^2}\right)^{-2} = \frac{a^8}{b^{-4}} = a^8b^4$

67. $\left(\frac{x^5}{y^{-2}}\right)^{-3} = \frac{x^{-15}}{y^6} = \frac{1}{x^{15}y^6}$

69. $\left(\frac{x^{-4}y^3}{x^5y^5}\right)^{-3} = (x^{-4-5}y^{3-5})^{-3}$
$= (x^{-9}y^{-2})^{-3}$
$= x^{27}y^6$

71. $8.39 \times 10^2 = 839$

73. $8.16 \times 10^{-2} = 0.0816$

75. In 2 years, the population would be
$(1.02)^2P = (1.02)^2(6{,}000{,}000{,}000)$
$= 6{,}242{,}400{,}000$
or 6.242 billion.

77. The population 10 years ago (in 1990) would have been
$(1.02)^{-10}P = (1.02)^{-10}(6{,}000{,}000{,}000)$
$\approx 4{,}922{,}000{,}000$
or 4.922 billion.

79. **a.** Let $P = 400$, $r = 0.03$, and $n = 1$.
$C = P(1+r)^n$
$= 400(1+0.03)^1$
$= 400(1.03)^1$
$= 412$
Next year they would cost $412.

b. Let $P = 400$, $r = 0.03$, and $n = 5$.
$C = 400(1+0.03)^5$
$= 400(1.03)^5$
≈ 463.71
In 5 years, they would cost $463.71.

81. $12 \cdot 5.14 = 61.68$
It should cost $61.68.

83. Let $P = 170{,}000$, $r = 0.02$, and $n = 3$.
$C = P(1+r)^n$
$= 170{,}000(1+0.02)^3$
$= 170{,}000(1.02)^3$
$= 180{,}405.36$
In 3 years, the average price would be $180,405.36.

85. Let $P = 100$, $r = 0.015$, and $n = 4$.
$C = P(1+r)^n$
$= 100(1+0.015)^4$
$= 100(1.015)^4$
≈ 106.14
You would owe $106.14 at the end of 4 months.

87. a. $\frac{60}{40}=1.5, \frac{90}{60}=1.5$

The percent rate of growth is 50% day.

b. Let $P = 40$, $r = 0.50$ and $n = n$.

$A = 40(1+0.50)^n$

c. From Monday to Friday, 4 days have passed. Let $n = 4$.

$$\begin{aligned} A &= 40(1+0.50)^n \\ &= 40(1+0.50)^4 \\ &= 40(1.5)^4 \\ &\approx 203 \end{aligned}$$

Expect 203 ants on Friday.

d. Let $n = 14$.

$A = 40(1+0.50)^{14} = 40(1.5)^{14} \approx 11{,}677$

Expect 11,677 ants in two weeks.

e. Answers may vary.

89. Answers may vary.

91. No; answers may vary. Sample answer: Let $x = 2$ and $y = 3$.

$$\begin{aligned} x^{-2}+y^{-2} &= 2^{-2}+3^{-2} \\ &= \frac{1}{2^2}+\frac{1}{3^2} \\ &= \frac{1}{4}+\frac{1}{9} \\ &= \frac{9}{36}+\frac{4}{36} \\ &= \frac{13}{36} \end{aligned}$$

$(x+y)^{-2} = (2+3)^{-2} = 5^{-2} = \frac{1}{5^2} = \frac{1}{25}$

93. $\frac{7^{-2}}{6^{-2}} = \frac{6^2}{7^2} = \frac{36}{49}$

95. $\left(\frac{1}{9}\right)^{-2} = 9^2 = 81$

97. $\frac{1}{7^6} = 7^{-6}$

99. $5^6 \cdot 5^{-4} = 5^{6+(-4)} = 5^2 = 25$

101. $\frac{z^{-5}}{z^{-7}} = z^{-5-(-7)} = z^{-5+7} = z^2$

103.
$$\begin{aligned} \left(\frac{2xy^3}{3x^3y^{-3}}\right)^{-4} &= \left(\frac{2x^{1-3}y^{3-(-3)}}{3}\right)^{-4} \\ &= \left(\frac{2x^{-2}y^6}{3}\right)^{-4} \\ &= \frac{2^{-4}x^8y^{-24}}{3^{-4}} \\ &= \frac{3^4x^8}{2^4y^{24}} \\ &= \frac{81x^8}{16y^{24}} \end{aligned}$$

105. a. Let $P = 5000$, $r = -0.10$, and $n = 3$.

$$\begin{aligned} C &= P(1+r)^n \\ &= 5000(1-0.10)^3 \\ &= 5000(0.90)^3 \\ &= 3645 \end{aligned}$$

The car will be valued at \$3645 in 3 years.

b. Let $P = 5000$, $r = -0.10$, and $n = -2$.

$$\begin{aligned} C &= P(1+r)^n \\ &= 5000(1-0.10)^{-2} \\ &= 5000(0.90)^{-2} \\ &\approx 6172.84 \end{aligned}$$

The car was valued at \$6172.84 two years ago.

4.3 Application of Exponents: Scientific Notation

Problems 4.3

1. $239{,}000 = 2.39 \times 10^5$

$0.012 = 1.2 \times 10^{-2}$

2. Move the decimal point 6 places right.

$10.1 \times 10^6 = 10{,}100{,}000$

Move the decimal point 2 places left.

$5.25 \times 10^{-2} = 0.0525$

3. a.
$$\begin{aligned}(6 \times 10^4) \times (5.2 \times 10^5) &= (6 \times 5.2) \times (10^4 \times 10^5)\\ &= (31.2) \times (10^{4+5})\\ &= (3.12 \times 10) \times 10^9\\ &= 3.12 \times 10^{1+9}\\ &= 3.12 \times 10^{10}\end{aligned}$$

b.
$$\begin{aligned}(3.1 \times 10^3) \times (5 \times 10^{-6}) &= (3.1 \times 5) \times (10^3 \times 10^{-6})\\ &= (15.5) \times (10^{3+(-6)})\\ &= (1.55 \times 10) \times (10^{-3})\\ &= 1.55 \times 10^{1-3}\\ &= 1.55 \times 10^{-2}\end{aligned}$$

4.
$$\begin{aligned}(1.23 \times 10^{-3}) \div (4.1 \times 10^{-4}) &= (1.23 \div 4.1) \times (10^{-3} \div 10^{-4})\\ &= 0.3 \times (10^{-3-(-4)})\\ &= (3 \times 10^{-1}) \times (10^{-3+4})\\ &= 3 \times (10^{-1} \times 10^1)\\ &= 3 \times 10^{-1+1}\\ &= 3 \times 10^0\\ &= 3\end{aligned}$$

5. $\dfrac{3.3\times10^4}{7.5\times10^{-1}}=(3.3\div7.5)\times(10^4\div10^{-1})$

$$\begin{aligned}&=0.44\times(10^{4-(-1)})\\&=(4.4\times10^{-1})\times10^{4+1}\\&=4.4\times(10^{-1}\times10^5)\\&=4.4\times10^{-1+5}\\&=4.4\times10^4\\&=44{,}000\end{aligned}$$

The population density is $44{,}000/\text{mi}^2$.

6. a. Pennies: $1{,}025{,}740{,}000=1.02574\times10^9$

Nickels: $66{,}183{,}600=6.61836\times10^7$

Dimes: $233{,}530{,}000=2.3353\times10^8$

Quarters: $466{,}850{,}000=4.6685\times10^8$

Half-dollars: $15{,}355{,}000=1.5355\times10^7$

b.

$$\begin{array}{rcrcr} & & & & 1.02574\times10^9\\ 6.61836\times10^7 & = & (0.0661836\times10^2)\times10^7 & = & 0.0661836\times10^9\\ 2.3353\times10^8 & = & (0.23353\times10)\times10^8 & = & 0.23353\times10^9\\ 4.6685\times10^8 & = & (0.46685\times10)\times10^8 & = & 0.46685\times10^9\\ 1.5355\times10^7 & = & 0.015355\times10^2\times10^7 & = & \underline{0.015355\times10^9}\\ & & & & 1.8076586\times10^9\end{array}$$

There was \$1.8076586 billion minted.

Exercises 4.3

1. $54{,}000{,}000=5.4\times10^7$

3. $287{,}190{,}520=2.8719052\times10^8$

5. $1{,}900{,}000{,}000=1.9\times10^9$

7. $0.00024=2.4\times10^{-4}$

9. $0.000000002=2\times10^{-9}$

11. Move the decimal point 2 places right.
$1.53\times10^2=153$

13. Move the decimal point 6 places right.
$8\times10^6 = 8{,}000{,}000$

15. Move the decimal point 9 places right.
$6.85\times10^9 = 6{,}850{,}000{,}000$

17. Move the decimal point 1 place left.
$2.3\times10^{-1} = 0.23$

19. Move the decimal point 4 places left.
$2.5\times10^{-4} = 0.00025$

21. $(3\times10^4)\times(5\times10^5) = (3\times5)\times(10^4\times10^5)$
$= 15\times10^{4+5}$
$= (1.5\times10)\times10^9$
$= 1.5\times10^{1+9}$
$= 1.5\times10^{10}$

23. $(6\times10^{-3})\times(5.1\times10^6)$
$= (6\times5.1)\times(10^{-3}\times10^6)$
$= 30.6\times10^{-3+6}$
$= (3.06\times10)\times10^3$
$= 3.06\times10^{1+3}$
$= 3.06\times10^4$

25. $(4\times10^{-2})\times(3.1\times10^{-3})$
$= (4\times3.1)\times(10^{-2}\times10^{-3})$
$= 12.4\times10^{-2+(-3)}$
$= (1.24\times10)\times10^{-5}$
$= 1.24\times10^{1+(-5)}$
$= 1.24\times10^{-4}$

27. $\dfrac{4.2\times10^5}{2.1\times10^2} = (4.2\div2.1)\times(10^5\div10^2)$
$= 2\times10^{5-2}$
$= 2\times10^3$

29. $\dfrac{2.2\times10^4}{8.8\times10^6} = (2.2\div8.8)\times(10^4\div10^6)$
$= 0.25\times10^{4-6}$
$= (2.5\times10^{-1})\times10^{-2}$
$= 2.5\times10^{-1+(-2)}$
$= 2.5\times10^{-3}$

31. a. $(8\times10^1)\times(2.9\times10^8)$
$= (8\times2.9)\times(10^1\times10^8)$
$= 23.2\times10^{1+8}$
$= (2.32\times10)\times10^9$
$= 2.32\times10^{1+9}$
$= 2.32\times10^{10}$

b. Move the decimal point 10 places to the right.
$2.32\times10^{10} = 23{,}200{,}000{,}000$

33. $\dfrac{(1.485\times10^8)\times(2\times10^3)}{(2.9\times10^8)\times(3.6\times10^2)}$
$= \dfrac{(1.485\times2)\times(10^8\times10^3)}{(2.9\times3.6)\times(10^8\times10^2)}$
$= \dfrac{2.97\times10^{8+3}}{10.44\times10^{8+2}}$
$= \dfrac{2.97\times10^{11}}{(1.044\times10)\times10^{10}}$
$= \dfrac{2.97\times10^{11}}{1.044\times10^{11}}$
$= (2.97\div1.044)\times(10^{11}\div10^{11})$
$= 2.844827586\times10^{11-11}$
$= 2.844827586\times10^0$
$= 2.844827586\times1$
$= 2.844827586$

35. $\dfrac{4.7\times10^9}{235} = \dfrac{4.7\times10^9}{2.35\times10^2}$
$= (4.7\div2.35)\times(10^9\div10^2)$
$= 2\times10^{9-2}$
$= 2\times10^7$

37. Now:

$$\frac{730\text{ million}}{365}=\frac{7.3\times10^8}{3.65\times10^2}$$
$$=(7.3\div3.65)\times(10^8\div10^2)$$
$$=2\times10^{8-2}$$
$$=2\times10^6$$
$$=2{,}000{,}000$$

Later:

$$\frac{4.38\text{ billion}}{365}=\frac{4.38\times10^9}{3.65\times10^2}$$
$$=(4.38\div3.65)\times(10^9\div10^2)$$
$$=1.2\times10^{9-2}$$
$$=1.2\times10^7$$
$$=12{,}000{,}000$$

Travelocity is sending 2,000,000 now and projects to send 12,000,000 later.

39. $$\frac{(36\text{ million})\times365}{288\text{ million}}$$
$$=\frac{(3.6\times10^7)\times(3.65\times10^2)}{2.88\times10^8}$$
$$=\frac{(3.6\times3.65)\times(10^7\times10^2)}{2.88\times10^8}$$
$$=\frac{13.14\times10^9}{2.88\times10^8}$$
$$=(13.14\div2.88)\times(10^9\div10^8)$$
$$=4.5625\times10^{9-8}$$
$$=4.5625\times10$$
$$=45.625$$

Each person will send 45.625 e-mails per year.

41. $-8(3)^2+80=-8(9)+80$
$=-72+80$
$=8$

43. $-4\cdot8\div2+20=-32\div2+20=-16+20=4$

45. $299{,}792{,}458=2.99792458\times10^8$

47. $(2.06\times10^5)\times(1.5\times10^8)$
$=(2.06\times1.5)\times(10^5\times10^8)$
$=3.09\times10^{5+8}$
$=3.09\times10^{13}$

49. $$\frac{3.09\times10^{13}}{9.46\times10^{12}}=(3.09\div9.46)\times(10^{13}\div10^{12})$$
$$=0.327\times10^{13-12}$$
$$=(3.27\times10^{-1})\times10^1$$
$$=3.27\times10^0$$
$$=3.27\times1$$
$$=3.27$$

51. Answers may vary.

53. The distance is 289,000 miles or 2.89×10^5 miles. The mass is 0.12456 or 1.2456×10^{-1} that of the earth.

55. $(2.52\times10^{-2})\div(4.2\times10^{-3})$
$=(2.52\div4.2)\times(10^{-2}\div10^{-3})$
$=0.6\times10^{-2-(-3)}$
$=(6\times10^{-1})\times10^1$
$=6\times10^0$
$=6\times1$
$=6$

57. $(6\times10^4)\times(2.2\times10^3)$
$=(6\times2.2)\times(10^4\times10^3)$
$=13.2\times10^{4+3}$
$=(1.32\times10^1)\times10^7$
$=1.32\times10^{1+7}$
$=1.32\times10^8$
$=132{,}000{,}000$

4.4 Polynomials: An Introduction

Problems 4.4

1. **a.** –5 has only one term; it is a monomial.

   b. $-3+4y+6y^2$ has three terms; it is a trinomial.

   c. $8x - 3$ has two terms; it is a binomial.

   d. $8(x + 9) - 3(x - 1)$ has two terms; it is a binomial.

2. **a.** The degree of 9 is, by convention, 0.

   b. The highest exponent of the variable z in the polynomial $-5z^2+2z-8$ is 2; thus the degree of the polynomial is 2.

   c. 0 is the zero polynomial; it does not have a degree.

   d. Since $y = y^1$, $-8y + 1$ can be written as $-8y^1+1$, making the degree of the polynomial 1.

3. **a.** $-4x^2+3x^3-8+2x=3x^3-4x^2+2x-8$

   b. $-3y+y^2-1=y^2-3y-1$

4. Let $t = 2$.

$$P(t)=-16t^2+90$$
$$\begin{aligned}P(2)&=-16(2)^2+90\\&=-16(4)+90\\&=-64+90\\&=26\end{aligned}$$

5. Let $x = 3$.

$$R(x)=5x^2-3x+9$$
$$\begin{aligned}R(3)&=5(3)^2-3(3)+9\\&=5(9)-3(3)+9\\&=45-9+9\\&=36+9\\&=45\end{aligned}$$

6. **a.** Let $t = 1961 - 1960 = 1$.

$$G(t)=-0.001t^3+0.06t^2+2.6t+88.6$$
$$\begin{aligned}&G(1)\\&=-0.001(1)^3+0.06(1)^2+2.6(1)+88.6\\&=-0.001+0.06+2.6+88.6\\&=91.259\end{aligned}$$

   In 1961, 91.259 million tons were generated.

   b. Let $t = 2000 - 1960 = 40$.

$$\begin{aligned}&G(40)\\&=-0.001(40)^3+0.06(40)^2+2.6(40)+88.6\\&=-64+96+104+88.6\\&=224.6\end{aligned}$$

   In 2000, 224.6 million tons were generated.

7. **a.** Locate $x = 3$ on the x-axis. Move vertically until you reach the blue line and then left to the y-axis. The y-value at that point is 0.082. This means that the BAL of a male 3 hours after consuming 3 ounces of alcohol is about 0.082.

   b. This time, locate 3 on the x-axis and move vertically until you reach the red line. Moving left to the y-axis yields a value of 0.09. This means that the BAL of a female 3 hours after drinking is 0.09.

   c. Let $x = 3$.

$$\begin{aligned}y&=-0.0226x+0.1509\\&=-0.0226(3)+0.1509\\&=0.0831\end{aligned}$$

   d. Let $x = 3$.

$$\begin{aligned}y&=-0.0257x+0.1663\\&=-0.0257(3)+0.1663\\&=0.0892\end{aligned}$$

Exercises 4.4

1. $-5x + 7$ has two terms, so it is a binomial (B). Since $-5x=-5x^1$, the degree is 1.

3. $7x$ has only one term, so it is a monomial (M). Since $7x = 7x^1$, the degree of the polynomial is 1.

5. $-2x + 7x^2 + 9$ has three terms, so it is a trinomial (T). Since the highest exponent of the variable x is 2, the degree of the polynomial is 2.

7. 18 has only one term, so it is a monomial (M). The degree of 18 is, by convention, 0.

9. $9x^3 - 2x$ has two terms, so it is a binomial (B). Since the highest exponent of the variable x is 3, the degree of the polynomial is 3.

11. $-3x + 8x^3 = 8x^3 - 3x$
The degree of the polynomial is 3.

13. $4x - 7 + 8x^2 = 8x^2 + 4x - 7$
The degree of the polynomial is 2.

15. $5x + x^2 = x^2 + 5x$
The degree of the polynomial is 2.

17. $3 + x^3 - x^2 = x^3 - x^2 + 3$
The degree of the polynomial is 3.

19. $4x^5 + 2x^2 - 3x^3 = 4x^5 - 3x^3 + 2x^2$
The degree of the polynomial is 5.

21. a. Let $x = 2$.
$3x - 2 = 3(2) - 2 = 6 - 2 = 4$

b. Let $x = -2$.
$3x - 2 = 3(-2) - 2 = -6 - 2 = -8$

23. a. Let $x = 2$.
$2x^2 - 1 = 2(2)^2 - 1 = 2(4) - 1 = 8 - 1 = 7$

b. Let $x = -2$.
$2x^2 - 1 = 2(-2)^2 - 1 = 2(4) - 1 = 8 - 1 = 7$

25. $P(x) = 3x^2 - x - 1$

a.
$$\begin{aligned} P(2) &= 3(2)^2 - 2 - 1 \\ &= 3(4) - 2 - 1 \\ &= 12 - 2 - 1 \\ &= 9 \end{aligned}$$

b.
$$\begin{aligned} P(-2) &= 3(-2)^2 - (-2) - 1 \\ &= 3(4) + 2 - 1 \\ &= 12 + 2 - 1 \\ &= 13 \end{aligned}$$

27. $R(x) = 3x - 1 + x^2$

a. $R(2) = 3(2) - 1 + 2^2 = 6 - 1 + 4 = 9$

b.
$$\begin{aligned} R(-2) &= 3(-2) - 1 + (-2)^2 \\ &= -6 - 1 + 4 \\ &= -3 \end{aligned}$$

29. $T(y) = -3 + y + y^2$

a. $T(2) = -3 + 2 + 2^2 = -3 + 2 + 4 = 3$

b.
$$\begin{aligned} T(-2) &= -3 + (-2) + (-2)^2 \\ &= -3 - 2 + 4 \\ &= -1 \end{aligned}$$

31. a. Let $k = 150$.
$-16t^2 + k = -16t^2 + 150$
The height after t seconds is $(-16t^2 + 150)$ feet.

b. Let $t = 1$.
$$\begin{aligned} -16t^2 + 150 &= -16(1)^2 + 150 \\ &= -16 + 150 \\ &= 134 \end{aligned}$$
The height after 1 second is 134 feet.

c. Let $t = 2$.
$$\begin{aligned} -16t^2 + 150 &= -16(2)^2 + 150 \\ &= -16(4) + 150 \\ &= -64 + 150 \\ &= 86 \end{aligned}$$
The height after 2 seconds is 86 feet.

33. a. Let $k = 200$.

$-4.9t^2 + k = -4.9t^2 + 200$

The height after t seconds is $(-4.9t^2 + 200)$ meters.

b. Let $t = 1$.

$$\begin{aligned} -4.9t^2 + 200 &= -4.9(1)^2 + 200 \\ &= -4.9 + 200 \\ &= 195.1 \end{aligned}$$

The height after 1 second is 195.1 meters.

c. Let $t = 2$.

$$\begin{aligned} -4.9t^2 + 200 &= -4.9(2)^2 + 200 \\ &= -4.9(4) + 200 \\ &= -19.6 + 200 \\ &= 180.4 \end{aligned}$$

the height after 2 seconds is 180.4 meters.

35. $R(t) = 1.76t^2 - 17.24t + 251$

a. Let $t = 0$.

$$\begin{aligned} R(0) &= 1.76(0)^2 - 17.24(0) + 251 \\ &= 0 - 0 + 251 \\ &= 251 \end{aligned}$$

The number of robberies in 1980 was 251 per 100,000.

b. Let $t = 2000 - 1980 = 20$.

$$\begin{aligned} R(20) &= 1.76(20)^2 - 17.24(20) + 251 \\ &= 704 - 344.8 + 251 \\ &\approx 610 \end{aligned}$$

Let $t = 2010 - 1980 = 30$.

$$\begin{aligned} R(30) &= 1.76(30)^2 - 17.24(30) + 251 \\ &= 1584 - 517.2 + 251 \\ &\approx 1318 \end{aligned}$$

The number of robberies is predicted to be about 610 per 100,000 in 2000 and about 1318 per 100,000 in 2010.

37. $R(t) = 0.07t^3 - 1.15t^2 + 4t + 30$

a. Let $t = 0$.

$$\begin{aligned} R(0) &= 0.07(0)^3 - 1.15(0)^2 + 4(0) + 30 \\ &= 30 \end{aligned}$$

The percent in 1992 was 30%.

b. Let $t = 2002 - 1992 = 10$.

$$\begin{aligned} &R(10) \\ &= 0.07(10)^3 - 1.15(10)^2 + 4(10) + 30 \\ &= 25 \end{aligned}$$

The percent in 2002 was 25%.

c. Since $25\% < 30\%$, rock music sales were decreasing.

d. $9.02(0.30) \approx 2.71$

In 1992, \$2.71 billion was spent on rock music.

e. $13.0(0.25) = 3.25$

In 2002, \$3.25 billion was spent on rock music.

39. $L(m) = -4m^2 + 57m - 175$

a. Let $m = 7$ for July.

$L(7) = -4(7)^2 + 57(7) - 175 = 28$

The record low for July was 28°.

b. $L(1) = -4(1)^2 + 57(1) - 175 = -122$

The record low for January would be −122°. Since 122° below 0 is unreasonable, $m = 1$ is not one of the choices.

41. $N(t) = 0.5t^2 + 4t + 2.1$

a. Locate year 2003 on the x-axis. Move vertically until you reach the line and then left to the y-axis. The y-value at that point is 34. Therefore, the number of users in 2003 is 34 million.

b. Let $t = 2003 - 1998 = 5$.

$N(5) = 0.5(5)^2 + 4(5) + 2.1 = 34.6$

This result is close to the number found on the graph.

43. $-3ab + (-4ab) = -7ab$

45. $-3x^2y + 8x^2y - 2x^2y = 3x^2y$

47. $5xy^2 - (-3xy^2) = 5xy^2 + 3xy^2 = 8xy^2$

49. Let $t = t$ and $v_0 = -10$.
$-32t + v_0 = -32t - 10$
The velocity is $(-32t - 10)$ feet per second.

51. **a.** Let $t = 1$ and $v_0 = -2$.
$-9.8t + v_0 = -9.8(1) - 2 = -11.8$
The velocity after 1 second is -11.8 meters per second.

b. Let $t = 2$ and $v_0 = -2$.
$-9.8t + v_0 = -9.8(2) - 2 = -21.6$
The velocity after 2 seconds in -21.6 meters per second.

53. Answers may vary.

55. $x^{-2} + x + 3$ is not a polynomial because there is a negative exponent on x.

57. The degree of 7^4 is 0 because $7^4 = 2401$ and it can be written as $2401x^0$.

59.
$$\begin{aligned} 2x^2 - 3x + 10 &= 2(2)^2 - 3(2) + 10 \\ &= 2(4) - 3(2) + 10 \\ &= 8 - 6 + 10 \\ &= 12 \end{aligned}$$

61.
$$\begin{aligned} -16t^2 + 118 &= -16(2.5)^2 + 118 \\ &= -16(6.25) + 118 \\ &= -100 + 118 \\ &= 18 \end{aligned}$$

63. Since the highest exponent of the variable x is 8, the degree is 8.

65. 0 is the zero polynomial and has no degree.

67. $-8 + 5x^2 - 3x = 5x^2 - 3x - 8$

69. $-5y$ has only one term, so it is a monomial.

71. $2x^3 - x^2 + x - 1$ has four terms, so it is a polynomial.

73. **a.** Locate $x = 2$ on the x-axis. Move vertically until you reach the red line. Moving left to the y-axis yields a value of 0.115. This means that the BAL of a female after 2 hours in 0.115.

b. Locate $x = 2$ on the x-axis. Move vertically until you reach the blue line. Moving left to the y-axis yields a value of 0.105. This means that the BAL of a male after 2 hours is 0.105.

c. Let $x = 2$.
$$\begin{aligned} y &= -0.0257x + 0.1663 \\ &= -0.0257(2) + 0.1663 \\ &= 0.1149 \end{aligned}$$
This is very close to 0.115.

d. Let $x = 2$.
$$\begin{aligned} y &= -0.0226x + 0.1509 \\ &= -0.0226(2) + 0.1509 \\ &= 0.1057 \end{aligned}$$
This is very close to 0.105.

4.5 Addition and Subtraction of Polynomials

Problems 4.5

1.
$$\begin{aligned} &(5x + 8x^2 - 3) + (-3x^2 + 8 - 5x) \\ &= (8x^2 + 5x - 3) + (-3x^2 - 5x + 8) \\ &= (8x^2 - 3x^2) + (5x - 5x) + (-3 + 8) \\ &= 5x^2 + 0 + 5 \\ &= 5x^2 + 5 \end{aligned}$$

2.
$$\begin{array}{rr} & 8y^2 - 5y - 3 \\ (+) & -5y^2 + 6y - 4 \\ \hline & 3y^2 + y - 7 \end{array}$$

3.
$$\begin{array}{r} 6y^2-4y \\ (-)\ 8y^2+5y-4 \\ \hline \end{array} \rightarrow \begin{array}{r} 6y^2-4y \\ (+)\ -8y^2-5y+4 \\ \hline -2y^2-9y+4 \end{array}$$

4.
$$\begin{array}{r} y^3-4y^2+3y-6 \\ -6y^2+3y-9 \\ (+)\ 6y^3\qquad\quad -5y+8 \\ \hline 7y^3-10y^2+y-7 \end{array}$$

5. Area of A + Area of B + Area of C + Area of D $= 6y+3y+3y+(4y)^2 = 12y+16y^2$ or $6y^2+12y$

6. Let $t = 2000 - 1985 = 15$.
$6.6t^2 + 71t + 975 = 6.6(15)^2 + 71(15) + 975 = 3525$
In 2000, \$3525 was spent on hospital care and doctors' services.

7. $y = -0.0257x + 0.1663 - 0.015x = -0.0407x + 0.1663$

Exercises 4.5

1. $(5x^2+2x+5)+(7x^2+3x+1)=(5x^2+7x^2)+(2x+3x)+(5+1)=12x^2+5x+6$

3.
$$\begin{aligned}(-3x+5x^2-1)+(-7+2x-7x^2)&=(5x^2-3x-1)+(-7x^2+2x-7)\\&=(5x^2-7x^2)+(-3x+2x)+(-1-7)\\&=-2x^2+(-x)+(-8)\\&=-2x^2-x-8\end{aligned}$$

5.
$$\begin{aligned}(2x+5x^2-2)+(-3+5x-8x^2)&=(5x^2+2x-2)+(-8x^2+5x-3)\\&=(5x^2-8x^2)+(2x+5x)+(-2-3)\\&=-3x^2+7x+(-5)\\&=-3x^2+7x-5\end{aligned}$$

7.
$$\begin{aligned}(-2+5x)+(-3-x^2-5x)&=(5x-2)+(-x^2-5x-3)\\&=-x^2+(5x-5x)+(-2-3)\\&=-x^2+0+(-5)\\&=-x^2-5\end{aligned}$$

9. $(x^3-2x+3)+(-2x^2+x-5)=(x^3)+(-2x^2)+(-2x+x)+(3-5)=x^3-2x^2-x-2$

11.
$$\begin{aligned}(-6x^3+2x^4-x)+(2x^2+5+2x-2x^3)&=(2x^4-6x^3-x)+(-2x^3+2x^2+2x+5)\\&=(2x^4)+(-6x^3-2x^3)+(2x^2)+(-x+2x)+5\\&=2x^4-8x^3+2x^2+x+5\end{aligned}$$

13. $\left(\frac{1}{3}-\frac{2}{5}x^2+\frac{3}{4}x\right)+\left(\frac{1}{4}x-\frac{1}{5}x^2+\frac{2}{3}\right)=\left(-\frac{2}{5}x^2+\frac{3}{4}x+\frac{1}{3}\right)+\left(-\frac{1}{5}x^2+\frac{1}{4}x+\frac{2}{3}\right)$

$$=\left(-\frac{2}{5}x^2-\frac{1}{5}x^2\right)+\left(\frac{3}{4}x+\frac{1}{4}x\right)+\left(\frac{1}{3}+\frac{2}{3}\right)$$

$$=-\frac{3}{5}x^2+x+1$$

15. $(0.2x-0.3+0.5x^2)+\left(-\frac{1}{10}+\frac{1}{10}x-\frac{1}{10}x^2\right)=(0.5x^2+0.2x-0.3)+\left(-\frac{1}{10}x^2+\frac{1}{10}x-\frac{1}{10}\right)$

$$=\left(0.5x^2-\frac{1}{10}x^2\right)+\left(0.2x+\frac{1}{10}x\right)+\left(-0.3-\frac{1}{10}\right)$$

$$=(0.5x^2-0.1x^2)+(0.2x+0.1x)+(-0.3-0.1)$$

$$=0.4x^2+0.3x-0.4$$

17.
$$\begin{array}{r} -3x^2+2x-4 \\ (+)\ x^2-4x+7 \\ \hline -2x^2-2x+3 \end{array}$$

19.
$$\begin{array}{r} 3x^4 \qquad -3x+4 \\ (+) \qquad x^3-2x-5 \\ \hline 3x^4+x^3-5x-1 \end{array}$$

21.
$$\begin{array}{r} -5x^4 \qquad -5x^2+3 \\ (+) \qquad 5x^3+3x^2-5 \\ \hline -5x^4+5x^3-2x^2-2 \end{array}$$

23.
$$\begin{array}{r} 5x^3 \ -x^2 \qquad -3 \\ 5x+9 \\ (+) \quad -3x^2 \qquad -7 \\ \hline 5x^3-4x^2+5x-1 \end{array}$$

25.
$$\begin{array}{r} -\frac{2}{7}x^3+\frac{1}{6}x^2 \qquad +2 \\ \frac{1}{7}x^3 \qquad +5x-3 \\ (+) \qquad -\frac{5}{6}x^2 \qquad +1 \\ \hline -\frac{1}{7}x^3-\frac{2}{3}x^2+5x \end{array}$$

27.

$$\begin{array}{rrrr} & -\frac{1}{7}x^3 & & +2 \\ & & -\frac{1}{9}x^2 & -x-3 \\ (+) & -\frac{2}{7}x^3 & +\frac{2}{9}x^2 & +2x-5 \\ \hline & -\frac{3}{7}x^3 & +\frac{1}{9}x^2 & +x-6 \end{array}$$

29.

$$\begin{array}{rrrrr} & -6x^3 & +2x^2 & & +1 \\ -x^4 & +3x^3 & -5x^2 & +3x & \\ & -x^3 & & -7x & +2 \\ (+)\ -3x^4 & & & +3x & -1 \\ \hline -4x^4 & -4x^3 & -3x^2 & -x & +2 \end{array}$$

31. $(7x^2+2)-(3x^2-5)=7x^2+2-3x^2+5=(7x^2-3x^2)+(2+5)=4x^2+7$

33. $(3x^2-2x-1)-(4x^2+2x+5)=3x^2-2x-1-4x^2-2x-5$
$=(3x^2-4x^2)+(-2x-2x)+(-1-5)$
$=-x^2-4x-6$

35. $(-1+7x^2-2x)-(5x+3x^2-7)=-1+7x^2-2x-5x-3x^2+7$
$=(7x^2-3x^2)+(-2x-5x)+(-1+7)$
$=4x^2-7x+6$

37. $(5x^2-2x+5)-(3x^3-x^2+5)=5x^2-2x+5-3x^3+x^2-5$
$=(-3x^3)+(5x^2+x^2)+(-2x)+(5-5)$
$=-3x^3+6x^2+(-2x)+0$
$=-3x^3+6x^2-2x$

39. $(6x^3-2x^2-3x+1)-(-x^3-x^2-5x+7)=6x^3-2x^2-3x+1+x^3+x^2+5x-7$
$=(6x^3+x^3)+(-2x^2+x^2)+(-3x+5x)+(1-7)$
$=7x^3-x^2+2x-6$

41.

$$\begin{array}{r} 6x^2-3x+5 \\ (-)\ 3x^2+4x-2 \\ \hline \end{array} \rightarrow \begin{array}{r} 6x^2-3x+5 \\ (+)\ -3x^2-4x+2 \\ \hline 3x^2-7x+7 \end{array}$$

43.

$$\begin{array}{r} 3x^2-2x-1 \\ (+)\ 3x^2-2x-1 \\ \hline \end{array} \rightarrow \begin{array}{r} 3x^2-2x-1 \\ (+)\ -3x^2+2x+1 \\ \hline 0 \end{array}$$

45. $$\begin{array}{r} 4x^3 \quad -2x+5 \\ (-)\ 3x^2+5x-1 \\ \hline \end{array} \rightarrow \begin{array}{r} 4x^3 \quad\quad -2x+5 \\ (+)\quad -3x^2-5x+1 \\ \hline 4x^3-3x^2-7x+6 \end{array}$$

47. $$\begin{array}{r} 3x^3 \quad\quad -2 \\ (-)\quad 2x^2-x+6 \\ \hline \end{array} \rightarrow \begin{array}{r} 3x^3 \quad\quad -2 \\ (+)\quad -2x^2+x-6 \\ \hline 3x^3-2x^2+x-8 \end{array}$$

49. $$\begin{array}{r} -5x^3 \quad +x-2 \\ (-)\quad 5x^2-3x+7 \\ \hline \end{array} \rightarrow \begin{array}{r} -5x^3 \quad\quad +x-2 \\ (+)\quad -5x^2+3x-7 \\ \hline -5x^3-5x^2+4x-9 \end{array}$$

51. Area of A + Area of B + Area of C + Area of D $= x^2+2x+4x+x^2 = 2x^2+6x$

53. Area of A + Area of B + Area of C + Area of D $= x^2+x^2+4x+x^2 = 3x^2+4x$

55. Area of A + Area of B + Area of C + Area of D $= (3x)^2+2x(5)+(3x)^2+(3x)^2$
$= 9x^2+10x+9x^2+9x^2$
$= 27x^2+10x$

57. a. $C(t)+T(t) = (0.05t^3-0.7t^2-t+36)+(-0.15t^2+1.6t+9)$
$P(t) = 0.05t^3+(-0.7t^2-0.15t^2)+(-t+1.6t)+(36+9)$
$P(t) = 0.05t^3-0.85t^2+0.6t+45$

b. Let t = 1995 – 1985 = 10.
$P(10) = 0.05(10)^3-0.85(10)^2+0.6(10)+45 = 16$
Let t = 2005 – 1985 = 20.
$P(20) = 0.05(20)^3-0.85(20)^2+0.6(20)+45 = 117$
In 1995, 16 pounds per person was consumed and in 2005, 117 pounds per person is predicted to be consumed.

59. a. $T(t)+B(t)+R(t) = (45t^2+110t+3356)+(27.5t+680)+(32t^2+200t+4730)$
$= (45t^2+32t^2)+(110t+27.5t+200t)+(3356+680+4730)$
$= 77t^2+337.5t+8766$

b. Let $t = 0$.
$77t^2+337.5t+8766 = 77(0)^2+337.5(0)+8766 = 8766$
The cost in 2000 was \$8766.

c. Let t = 2005 – 2000 = 5.
$77(5)^2+337.5(5)+8766 = 12{,}378.5$
The cost in 2005 is predicted to be \$12,378.50.

61. $T(t)+B(t)+R(t)-S(t)$
$=(45t^2+110t+3356)+(27.5t+680)+(32t^2+200t+4730)-(562.5t^2+312.5t+2625)$
$=45t^2+110t+3356+27.5t+680+32t^2+200t+4730-562.5t^2-312.5t-2625$
$=(45t^2+32t^2-562.5t^2)+(110t+27.5t+200t-312.5t)+(3356+680+4730-2625)$
$=-485.5t^2+25t+6141$

63. $(-2x^4)\cdot(3x^5)=(-2\cdot 3)x^{4+5}=-6x^9$

65. $6(y-4)=6\cdot y+6\cdot(-4)=6y-24$

67. $R=S-C$
$R=4x-(3x+50)$
$R=4x-3x-50$
$R=x-50$

69. $R=S-C$
$R=9x-(7x)$
$R=2x$

71. $R=2x-100$
$-40=2x-100$
$60=2x$
$30=x$
There were 30 items manufactured.

73. Answers may vary.

75. Answers may vary.

77. $(-5x+8x^2-3)+(-3x^2+4+8x)=(8x^2-5x-3)+(-3x^2+8x+4)$
$=(8x^2-3x^2)+(-5x+8x)+(-3+4)$
$=5x^2+3x+1$

79. $(10+7x^2+5x^3)-(9+x^3-3x^2)=10+7x^2+5x^3-9-x^3+3x^2$
$=(5x^3-x^3)+(7x^2+3x^2)+(10-9)$
$=4x^3+10x^2+1$

81. Area of A + Area of B + Area of C + Area of D $=3x+(2x)^2+3x+2x=4x^2+8x$

4.6 Multiplication of Polynomials

Problems 4.6

1. $(-4y^3)(5y^4) = (-4 \cdot 5)(y^3 \cdot y^4) = -20y^{3+4} = -20y^7$

2. a. $4(a - 3b) = 4a - 4 \cdot 3b = 4a - 12b$

   b. $(a^2 + 3a)4a^5 = a^2 \cdot 4a^5 + 3a \cdot 4a^5 = 4 \cdot a^2 \cdot a^5 + 3 \cdot 4 \cdot a \cdot a^5 = 4a^7 + 12a^6$

3. a. $(a+4)(a-3) = a \cdot a - 3a + 4a - 4 \cdot 3 = a^2 + a - 12$

   b. $(a-5)(a+4) = a \cdot a + 4a - 5a - 5 \cdot 4 = a^2 - a - 20$

4. a. $(3a+5)(2a-3) = (3a)(2a) + (3a)(-3) + 5(2a) + 5(-3) = 6a^2 - 9a + 10a - 15 = 6a^2 + a - 15$

   b. $(2a-3)(4a-1) = (2a)(4a) + (2a)(-1) + (-3)(4a) + (-3)(-1)$
   $= 8a^2 - 2a - 12a + 3$
   $= 8a^2 - 14a + 3$

5. a. $(4a+3b)(3a+5b) = (4a)(3a) + (4a)(5b) + (3b)(3a) + (3b)(5b)$
   $= 12a^2 + 20ab + 9ab + 15b^2$
   $= 12a^2 + 29ab + 15b^2$

   b. $(2a-b)(3a-4b) = (2a)(3a) + (2a)(-4b) + (-b)(3a) + (-b)(-4b)$
   $= 6a^2 - 8ab - 3ab + 4b^2$
   $= 6a^2 - 11ab + 4b^2$

6. $(y-2L)(y-3L) = y \cdot y + (y)(-3L) + (-2L)(y) + (-2L)(-3L)$
   $= y^2 - 3yL - 2yL + 6L^2$
   $= y^2 - 5yL + 6L^2$

7. $C(t) \times D(t) = (200 + 15t)(10 - 0.1t)$
   $= 200 \cdot 10 + 200(-0.1t) + 15t \cdot 10 + 15t(-0.1t)$
   $= 2000 - 20t + 150t - 1.5t^2$
   $= -1.5t^2 + 130t + 2000$

Exercises 4.6

1. $(5x^3)(9x^2) = (5 \cdot 9)(x^3 \cdot x^2) = 45x^{3+2} = 45x^5$

3. $(-2x)(5x^2) = (-2 \cdot 5)(x \cdot x^2) = -10x^{1+2} = -10x^3$

5. $(-2y^2)(-3y) = -2 \cdot (-3)(y^2 \cdot y) = 6y^{2+1} = 6y^3$

7. $3(x+y)=3\cdot x+3\cdot y=3x+3y$

9. $5(2x-y)=5\cdot 2x-5\cdot y=10x-5y$

11. $-4x(2x-3)=-4x(2x)-4x\cdot(-3)$
$=-8x^2+12x$

13. $(x^2+4x)x^3=x^2\cdot x^3+4x\cdot x^3=x^5+4x^4$

15. $(x-x^2)4x=x\cdot 4x-x^2\cdot 4x$
$=4\cdot x\cdot x-4\cdot x^2\cdot x$
$=4x^2-4x^3$

17. $(x+y)3x=x\cdot 3x+y\cdot 3x$
$=3\cdot x\cdot x+3\cdot x\cdot y$
$=3x^2+3xy$

19. $(2x-3y)(-4y^2)$
$=2x\cdot(-4y^2)-3y\cdot(-4y^2)$
$=2\cdot(-4)\cdot x\cdot y^2-3\cdot(-4)\cdot y\cdot y^2$
$=-8xy^2+12y^3$

21. $(x+1)(x+2)=x\cdot x+x\cdot 2+1\cdot x+1\cdot 2$
$=x^2+2x+x+2$
$=x^2+3x+2$

23. $(y+4)(y-9)$
$=y\cdot y+y\cdot(-9)+4\cdot y+4\cdot(-9)$
$=y^2-9y+4y-36$
$=y^2-5y-36$

25. $(x-7)(x+2)$
$=x\cdot x+x\cdot 2+(-7)\cdot x+(-7)(2)$
$=x^2+2x-7x-14$
$=x^2-5x-14$

27. $(x-3)(x-9)$
$=x\cdot x+x\cdot(-9)+(-3)\cdot x+(-3)(-9)$
$=x^2-9x-3x+27$
$=x^2-12x+27$

29. $(y-3)(y-3)$
$=y\cdot y+y\cdot(-3)+(-3)\cdot y+(-3)(-3)$
$=y^2-3y-3y+9$
$=y^2-6y+9$

31. $(2x+1)(3x+2)=2x\cdot 3x+2x\cdot 2+1\cdot 3x+1\cdot 2$
$=6x^2+4x+3x+2$
$=6x^2+7x+2$

33. $(3y+5)(2y-3)$
$=3y\cdot 2y+3y\cdot(-3)+5\cdot 2y+5\cdot(-3)$
$=6y^2-9y+10y-15$
$=6y^2+y-15$

35. $(5z-1)(2z+9)=5z\cdot 2z+5z\cdot 9-1\cdot 2z-1\cdot 9$
$=10z^2+45z-2z-9$
$=10z^2+43z-9$

37. $(2x-4)(3x-11)$
$=2x\cdot 3x+2x\cdot(-11)-4\cdot 3x-4\cdot(-11)$
$=6x^2-22x-12x+44$
$=6x^2-34x+44$

39. $(4z+1)(4z+1)=4z\cdot 4z+4z\cdot 1+1\cdot 4z+1\cdot 1$
$=16z^2+4z+4z+1$
$=16z^2+8z+1$

41. $(3x+y)(2x+3y)$
$=3x\cdot 2x+3x\cdot 3y+y\cdot 2x+y\cdot 3y$
$=6x^2+9xy+2xy+3y^2$
$=6x^2+11xy+3y^2$

43. $(2x+3y)(x-y)$
$=2x\cdot x+2x\cdot(-y)+3y\cdot x+3y\cdot(-y)$
$=2x^2-2xy+3xy-3y^2$
$=2x^2+xy-3y^2$

45. $(5z-y)(2z+3y)$
$=5z\cdot 2z+5z\cdot 3y-y\cdot 2z-y\cdot 3y$
$=10z^2+15yz-2yz-3y^2$
$=10z^2+13yz-3y^2$

47. $(3x-2z)(4x-z) = 3x\cdot 4x + 3x\cdot(-z) - 2z\cdot 4x - 2z\cdot(-z)$
$= 12x^2 - 3xz - 8xz + 2z^2$
$= 12x^2 - 11xz + 2z^2$

49. $(2x-3y)(2x-3y) = 2x\cdot 2x + 2x\cdot(-3y) + (-3y)\cdot(2x) + (-3y)\cdot(-3y)$
$= 4x^2 - 6xy - 6xy + 9y^2$
$= 4x^2 - 12xy + 9y^2$

51. $(3+4x)(2+3x) = 3\cdot 2 + 3\cdot 3x + 4x\cdot 2 + 4x\cdot 3x = 6 + 9x + 8x + 12x^2 = 6 + 17x + 12x^2$

53. $(2-3x)(3+x) = 2\cdot 3 + 2\cdot x + (-3x)\cdot 3 + (-3x)\cdot x = 6 + 2x - 9x - 3x^2 = 6 - 7x - 3x^2$

55. $(2-5x)(4+2x) = 2\cdot 4 + 2\cdot 2x - 5x\cdot 4 - 5x\cdot 2x = 8 + 4x - 20x - 10x^2 = 8 - 16x - 10x^2$

57. $A = LW = (x+5)(x+2) = x\cdot x + x\cdot 2 + 5\cdot x + 5\cdot 2 = x^2 + 2x + 5x + 10 = x^2 + 7x + 10$
The area is $(x^2 + 7x + 10)$ square units.

59. $16t(6-t) = 16t\cdot 6 + 16t\cdot(-t) = 16\cdot 6\cdot t - 16\cdot t\cdot t = 96t - 16t^2$
The height reached is $(96t - 16t^2)$ feet.

61. $(V_2 - V_1)(CP + PR) = V_2\cdot CP + V_2\cdot PR - V_1\cdot CP - V_1\cdot PR = V_2CP + V_2PR - V_1CP - V_1PR$

63. $A = LW = 48\cdot 25 = 1200$
The new area is 1200 square feet.

65. $800 + 160 + 40 + 200 = 1200\ \text{ft}^2$

67. **a.** $S_1 = x\cdot 20 = 20x\ \text{ft}^2$

b. $S_2 = x\cdot y = xy\ \text{ft}^2$

c. $S_3 = 40\cdot y = 40y\ \text{ft}^2$

69. $S_1 + S_2 + S_3 + 800 = 20x + xy + 40y + 800 = 800 + 40y + 20x + xy$
Yes; they are the same.

71. $(3x)^2 = 3^2x^2 = 9x^2$

73. $(-3A)^2 = (-3)^2A^2 = 9A^2$

75. $P = np - E$
$P = (-3p + 60)p - (-5p + 100)$
$P = -3p \cdot p + 60 \cdot p + 5p - 100$
$P = -3p^2 + 65p - 100$

77. Let $p = 2$.
$P = -3p^2 + 65p - 100$
$= -3(2)^2 + 65(2) - 100$
$= -12 + 130 - 100$
$= 18$
The profit was \$18.

79. Yes; answers may vary.

81. No; consider the results in Problem 82.

83. $(-7x^4)(5x^2) = -7 \cdot 5 \cdot x^4 \cdot x^2$
$= -35x^{4+2}$
$= -35x^6$

85. $(x + 7)(x - 3) = x \cdot x + x \cdot (-3) + 7 \cdot x + 7 \cdot (-3)$
$= x^2 - 3x + 7x - 21$
$= x^2 + 4x - 21$

87. $(3x + 4)(3x - 1)$
$= 3x \cdot 3x + 3x \cdot (-1) + 4 \cdot 3x + 4 \cdot (-1)$
$= 9x^2 - 3x + 12x - 4$
$= 9x^2 + 9x - 4$

89. $(5x - 2y)(2x - 3y)$
$= 5x \cdot 2x + 5x(-3y) - 2y \cdot 2x - 2y \cdot (-3y)$
$= 10x^2 - 15xy - 4xy + 6y^2$
$= 10x^2 - 19xy + 6y^2$

91. $6(x - 3y) = 6 \cdot x + 6 \cdot (-3y) = 6x - 18y$

4.7 Special Products of Polynomials

Problems 4.7

1. a. $(y + 6)^2 = y^2 + 2 \cdot 6 \cdot y + 6^2$
$= y^2 + 12y + 36$

b. $(2y + 1)^2 = (2y)^2 + 2 \cdot 1 \cdot 2y + 1^2$
$= 4y^2 + 4y + 1$

c. $(2x + 3y)^2 = (2x)^2 + 2(3y)(2x) + (3y)^2$
$= 4x^2 + 12xy + 9y^2$

2. a. $(y - 4)^2 = y^2 - 2 \cdot 4 \cdot y + 4^2$
$= y^2 - 8y + 16$

b. $(2y - 3)^2 = (2y)^2 - 2 \cdot 3 \cdot 2y + 3^2$
$= 4y^2 - 12y + 9$

c. $(3x - 2y)^2 = (3x)^2 - 2 \cdot 2y \cdot 3x + (2y)^2$
$= 9x^2 - 12xy + 4y^2$

3. a. $(y + 9)(y - 9) = y^2 - 9^2 = y^2 - 81$

b. $(3x + y)(3x - y) = (3x)^2 - y^2 = 9x^2 - y^2$

c. $(3x - 2y)(3x + 2y) = (3x)^2 - (2y)^2$
$= 9x^2 - 4y^2$

4. $(y - 2)(y^2 - y - 3)$
$= y(y^2 - y - 3) - 2(y^2 - y - 3)$
$= y^3 - y^2 - 3y - 2y^2 + 2y + 6$
$= y^3 + (-y^2 - 2y^2) + (-3y + 2y) + 6$
$= y^3 - 3y^2 - y + 6$

5. $2y(y + 2)(y + 3) = 2y(y^2 + 3y + 2y + 6)$
$= 2y(y^2 + 5y + 6)$
$= 2y^3 + 10y^2 + 12y$

6. $(y+2)^3 = (y+2)(y+2)(y+2)$
$= (y+2)(y^2+4y+4)$
$= y(y^2+4y+4)+2(y^2+4y+4)$
$= y^3+4y^2+4y+2y^2+8y+8$
$= y^3+6y^2+12y+8$

7. $\left(3y^2-\frac{1}{3}\right)^2 = (3y^2)^2 - 2\cdot\frac{1}{3}\cdot(3y^2)+\left(\frac{1}{3}\right)^2$
$= 9y^4-2y^2+\frac{1}{9}$

8. $(3y^2+2)(3y^2-2) = (3y^2)^2-(2)^2$
$= 9y^4-4$

9. $A = \pi r^2$
$= \pi(x-2)^2$
$= \pi(x^2-4x+4)$
$= \pi x^2 - 4\pi x + 4\pi$

Exercises 4.7

1. $(x+1)^2 = x^2+2\cdot 1\cdot x+1^2 = x^2+2x+1$

3. $(2x+1)^2 = (2x)^2+2\cdot 1\cdot 2x+1^2$
$= 4x^2+4x+1$

5. $(3x+2y)^2 = (3x)^2+2\cdot 2y\cdot 3x+(2y)^2$
$= 9x^2+12xy+4y^2$

7. $(x-1)^2 = x^2-2\cdot 1\cdot x+1^2 = x^2-2x+1$

9. $(2x-1)^2 = (2x)^2-2\cdot 1\cdot 2x+1^2$
$= 4x^2-4x+1$

11. $(3x-y)^2 = (3x)^2-2\cdot y\cdot 3x+y^2$
$= 9x^2-6xy+y^2$

13. $(6x-5y)^2 = (6x)^2-2\cdot 5y\cdot 6x+(5y)^2$
$= 36x^2-60xy+25y^2$

15. $(2x-7y)^2 = (2x)^2-2\cdot 7y\cdot 2x+(7y)^2$
$= 4x^2-28xy+49y^2$

17. $(x+2)(x-2) = x^2-2^2 = x^2-4$

19. $(x+4)(x-4) = x^2-4^2 = x^2-16$

21. $(3x+2y)(3x-2y) = (3x)^2-(2y)^2$
$= 9x^2-4y^2$

23. $(x-6)(x+6) = x^2-6^2 = x^2-36$

25. $(x-12)(x+12) = x^2-12^2 = x^2-144$

27. $(3x-y)(3x+y) = (3x)^2-y^2 = 9x^2-y^2$

29. $(2x-7y)(2x+7y) = (2x)^2-(7y)^2$
$= 4x^2-49y^2$

31. $(x^2+2)(x^2+5) = (x^2)^2+5x^2+2x^2+2\cdot 5$
$= x^4+7x^2+10$

33. $(x^2+y)^2 = (x^2)^2+2\cdot y\cdot x^2+y^2$
$= x^4+2x^2y+y^2$

35. $(3x^2-2y^2)^2$
$= (3x^2)^2+2\cdot(-2y^2)\cdot(3x^2)+(2y^2)^2$
$= 9x^4-12x^2y^2+4y^4$

37. $(x^2-2y^2)(x^2+2y^2) = (x^2)^2-(2y^2)^2$
$= x^4-4y^4$

39. $(2x+4y^2)(2x-4y^2) = (2x)^2-(4y^2)^2$
$= 4x^2-16y^4$

41. $(x+3)(x^2+x+5)$
$= x(x^2+x+5)+3(x^2+x+5)$
$= x^3+x^2+5x+3x^2+3x+15$
$= x^3+4x^2+8x+15$

43. $(x+4)(x^2-x+3)=x(x^2-x+3)+4(x^2-x+3)=x^3-x^2+3x+4x^2-4x+12=x^3+3x^2-x+12$

45. $(x+3)(x^2-x-2)=x(x^2-x-2)+3(x^2-x-2)=x^3-x^2-2x+3x^2-3x-6=x^3+2x^2-5x-6$

47. $$\begin{aligned}(x-2)(x^2+2x+4)&=x(x^2+2x+4)-2(x^2+2x+4)\\&=x^3+2x^2+4x-2x^2-4x-8\\&=x^3+0x^2+0x-8\\&=x^3-8\end{aligned}$$

49. $$\begin{aligned}-(x-1)(x^2-x+2)&=-[x(x^2-x+2)-1(x^2-x+2)]\\&=-(x^3-x^2+2x-x^2+x-2)\\&=-(x^3-2x^2+3x-2)\\&=-x^3+2x^2-3x+2\end{aligned}$$

51. $$\begin{aligned}-(x-4)(x^2-4x-1)&=-[x(x^2-4x-1)-4(x^2-4x-1)]\\&=-(x^3-4x^2-x-4x^2+16x+4)\\&=-(x^3-8x^2+15x+4)\\&=-x^3+8x^2-15x-4\end{aligned}$$

53. $2x(x+1)(x+2)=2x(x^2+2x+x+2)=2x(x^2+3x+2)=2x^3+6x^2+4x$

55. $3x(x-1)(x+2)=3x(x^2+2x-x-2)=3x(x^2+x-2)=3x^3+3x^2-6x$

57. $4x(x-1)(x-2)=4x(x^2-2x-x+2)=4x(x^2-3x+2)=4x^3-12x^2+8x$

59. $5x(x+1)(x-5)=5x(x^2-5x+x-5)=5x(x^2-4x-5)=5x^3-20x^2-25x$

61. $$\begin{aligned}(x+5)^3&=(x+5)(x+5)(x+5)\\&=(x+5)(x^2+10x+25)\\&=x(x^2+10x+25)+5(x^2+10x+25)\\&=x^3+10x^2+25x+5x^2+50x+125\\&=x^3+15x^2+75x+125\end{aligned}$$

63. $$\begin{aligned}(2x+3)^3&=(2x+3)(2x+3)(2x+3)\\&=(2x+3)(4x^2+12x+9)\\&=2x(4x^2+12x+9)+3(4x^2+12x+9)\\&=8x^3+24x^2+18x+12x^2+36x+27\\&=8x^3+36x^2+54x+27\end{aligned}$$

65. $$\begin{aligned}(2x+3y)^3 &= (2x+3y)(2x+3y)(2x+3y)\\ &= (2x+3y)(4x^2+12xy+9y^2)\\ &= 2x(4x^2+12xy+9y^2)+3y(4x^2+12xy+9y^2)\\ &= 8x^3+24x^2y+18xy^2+12x^2y+36xy^2+27y^3\\ &= 8x^3+36x^2y+54xy^2+27y^3\end{aligned}$$

67. $(4t^2+3)^2 = (4t^2)^2+2\cdot 3\cdot 4t^2+3^2 = 16t^4+24t^2+9$

69. $(4t^2+3u)^2 = (4t^2)^2+2\cdot 3u\cdot 4t^2+(3u)^2 = 16t^4+24t^2u+9u^2$

71. $\left(3t^2-\frac{1}{3}\right)^2 = (3t^2)^2-2\cdot\frac{1}{3}\cdot 3t^2+\left(\frac{1}{3}\right)^2 = 9t^4-2t^2+\frac{1}{9}$

73. $\left(3t^2-\frac{1}{3}u\right)^2 = (3t^2)^2-2\cdot\frac{1}{3}u\cdot 3t^2+\left(\frac{1}{3}u\right)^2 = 9t^4-2t^2u+\frac{1}{9}u^2$

75. $(3x^2+5)(3x^2-5) = (3x^2)^2-5^2 = 9x^4-25$

77. $(3x^2+5y^2)(3x^2-5y^2) = (3x^2)^2-(5y^2)^2 = 9x^4-25y^4$

79. $(4x^3-5y^3)(4x^3+5y^3) = (4x^3)^2-(5y^3)^2 = 16x^6-25y^6$

81. $$\begin{aligned}V &= \frac{1}{3}\pi r^2h\\ &= \frac{1}{3}\pi(x+2)^2(x+4)\\ &= \frac{1}{3}\pi(x+2)(x+2)(x+4)\\ &= \frac{1}{3}\pi(x+2)(x^2+6x+8)\\ &= \frac{1}{3}\pi[x(x^2+6x+8)+2(x^2+6x+8)]\\ &= \frac{1}{3}\pi(x^3+6x^2+8x+2x^2+12x+16)\\ &= \frac{1}{3}\pi(x^3+8x^2+20x+16)\\ &= \frac{1}{3}\pi x^3+\frac{8}{3}\pi x^2+\frac{20}{3}\pi x+\frac{16}{3}\pi\end{aligned}$$

83. $V = \frac{4}{3}\pi r^3$
$= \frac{4}{3}\pi(x+1)^3$
$= \frac{4}{3}\pi(x+1)(x+1)(x+1)$
$= \frac{4}{3}\pi(x+1)(x^2+2x+1)$
$= \frac{4}{3}\pi[x\cdot(x^2+2x+1)+1\cdot(x^2+2x+1)]$
$= \frac{4}{3}\pi(x^3+2x^2+x+x^2+2x+1)$
$= \frac{4}{3}\pi(x^3+3x^2+3x+1)$
$= \frac{4}{3}\pi x^3 + 4\pi x^2 + 4\pi x + \frac{4}{3}\pi$

85. $V = \pi r^2 h$
$= \pi(x+1)^2(x+2)$
$= \pi(x+1)(x+1)(x+2)$
$= \pi(x+1)(x^2+3x+2)$
$= \pi[x(x^2+3x+2)+1\cdot(x^2+3x+2)]$
$= \pi(x^3+3x^2+2x+x^2+3x+2)$
$= \pi(x^3+4x^2+5x+2)$
$= \pi x^3 + 4\pi x^2 + 5\pi x + 2\pi$

87. $I_D = 0.004\left(1+\frac{V}{2}\right)^2$
$= 0.004\left[1+2\cdot\frac{V}{2}+\left(\frac{V}{2}\right)^2\right]$
$= 0.004\left(1+V+\frac{V^2}{4}\right)$
$= 0.004 + 0.004V + 0.001V^2$

89. $(T_1^2+T_2^2)(T_1^2-T_2^2) = (T_1^2)^2-(T_2^2)^2$
$= T_1^4 - T_2^4$

91. $K(t_n-t_a)^2 = K(t_n^2 - 2\cdot t_n t_a + t_a^2)$
$= Kt_n^2 - 2Kt_n t_a + Kt_a^2$

93. $\frac{-5x^2}{10x^2} = \frac{-5}{10}x^{2-2} = -\frac{1}{2}x^0 = -\frac{1}{2}$

95. a. $(x+y)^2 = (1+2)^2 = 3^2 = 9$

b. $x^2+y^2 = 1^2+2^2 = 1+4 = 5$

c. No, because $9 \neq 5$.

97. a. $lw = x\cdot x = x^2$

b. $lw = y\cdot x = xy$

c. $lw = y\cdot y = y^2$

d. $lw = y\cdot x = xy$

99. They are equal.

101. Answers may vary.

103. The exception is when the binomials are the sum and difference of the same two terms. Then you get a binomial.

105. $(x+7)^2 = x^2 + 2\cdot 7\cdot x + 7^2 = x^2 + 14x + 49$

107. $(2x+5y)^2 = (2x)^2 + 2(5y)(2x) + (5y)^2$
$= 4x^2 + 20xy + 25y^2$

109. $(x^2+2y)^2 = (x^2)^2 + 2\cdot 2y\cdot x^2 + (2y)^2$
$= x^4 + 4x^2y + 4y^2$

111. $(3x-2y)(3x+2y) = (3x)^2 - (2y)^2$
$= 9x^2 - 4y^2$

113. $(3x+4)(5x-6)$
$= 3x\cdot 5x + 3x\cdot(-6) + 4\cdot 5x + 4\cdot(-6)$
$= 15x^2 - 18x + 20x - 24 = 15x^2 + 2x - 24$

115. $$\begin{aligned}(7x-2y)(3x-4y) &= 7x\cdot 3x+7x\cdot(-4y)+(-2y)\cdot(3x)+(-2y)(-4y)\\ &= 21x^2-28xy-6xy+8y^2\\ &= 21x^2-34xy+8y^2\end{aligned}$$

117. $(x+2)(x^2+x+1)=x(x^2+x+1)+2(x^2+x+1)=x^3+x^2+x+2x^2+2x+2=x^3+3x^2+3x+2$

119. $(x^2-2)(x^2+x+1)=x^2(x^2+x+1)-2(x^2+x+1)=x^4+x^3+x^2-2x^2-2x-2=x^4+x^3-x^2-2x-2$

121. $$\begin{aligned}(x+2y)^3 &= (x+2y)(x+2y)(x+2y)\\ &= (x+2y)(x^2+4xy+4y^2)\\ &= x(x^2+4xy+4y^2)+2y(x^2+4xy+4y^2)\\ &= x^3+4x^2y+4xy^2+2x^2y+8xy^2+8y^3\\ &= x^3+6x^2y+12xy^2+8y^3\end{aligned}$$

123. $3x(x^2+1)(x^2-1)=3x[(x^2)^2-1^2]=3x(x^4-1)=3x^5-3x$

4.8 Division of Polynomials

Problems 4.8

1. a. $$\frac{27y^4-18y^3}{9y^2}=\frac{27y^4}{9y^2}-\frac{18y^3}{9y^2}=3y^2-2y$$

b. $$\begin{aligned}\frac{8y^3-4y^2+12y}{4y^2} &= \frac{8y^3}{4y^2}-\frac{4y^2}{4y^2}+\frac{12y}{4y^2}\\ &= 2y-1+\frac{3}{y}\end{aligned}$$

2. $$\begin{array}{r|l} & \quad y+3 \\ y-2 & \quad y^2+y-7 \\ & (-)\underline{y^2-2y} \\ & \qquad 3y-7 \\ & \quad (+)\underline{3y-6} \\ & \qquad\quad -1 \end{array}$$

Thus $(y^2+y-7)\div(y-2)=y+3+\dfrac{-1}{y-2}$.

3.
$$\begin{array}{r}
y^2-2y-1\\
y+3\overline{)\ y^3+y^2-7y-3}\\
(-)\underline{y^3+3y^2}\\
-2y^2-7y\\
(-)\underline{-2y^2-6y}\\
-y-3\\
(-)\underline{-y-3}\\
0
\end{array}$$

Thus
$(y^3+y^2-7y-3)\div(3+y)=y^2-2y-1.$

4.
$$\begin{array}{r}
y^2+y\\
y^2-2\overline{)\ y^4+y^3-2y^2+y+1}\\
(-)\underline{y^4\quad -2y^2}\\
y^3\quad +y\\
(-)\underline{y^3\quad -2y}\\
3y+1
\end{array}$$

Thus $(y^4+y^3-2y^2+y+1)\div(y^2-2)$
$=y^2+y+\dfrac{3y+1}{y^2-2}.$

Exercises 4.8

1. $\dfrac{3x+9y}{3}=\dfrac{3x}{3}+\dfrac{9y}{3}=x+3y$

3. $\dfrac{10x-5y}{5}=\dfrac{10x}{5}-\dfrac{5y}{5}=2x-y$

5. $\dfrac{8y^3-32y^2+16y}{-4y^2}=\dfrac{8y^3}{-4y^2}+\dfrac{-32y^2}{-4y^2}+\dfrac{16y}{-4y^2}$
$=-2y+8-\dfrac{4}{y}$

7. $\dfrac{10x^2+8x}{x}=\dfrac{10x^2}{x}+\dfrac{8x}{x}=10x+8$

9. $\dfrac{15x^3-10x^2}{5x^2}=\dfrac{15x^3}{5x^2}-\dfrac{10x^2}{5x^2}=3x-2$

11.
$$\begin{array}{r}
x+2\\
x+3\overline{)\ x^2+5x+6}\\
(-)\underline{x^2+3x}\\
2x+6\\
(-)\underline{2x+6}\\
0
\end{array}$$

Thus, $(x^2+5x+6)\div(x+3)=x+2.$

13.
$$\begin{array}{r}
y-2\\
y+5\overline{)\ y^2+3y-11}\\
(-)\underline{y^2+5y}\\
-2y-11\\
(-)\underline{-2y-10}\\
-1
\end{array}$$

Thus $(y^2+3y-11)\div(y+5)=(y-2)$ R -1
or $y-2+\dfrac{-1}{y+5}.$

15.
$$\begin{array}{r}
x+6\\
x-4\overline{)\ x^2+2x-24}\\
(-)\underline{x^2-4x}\\
6x-24\\
(-)\underline{6x-24}\\
0
\end{array}$$

Thus $(2x+x^2-24)\div(x-4)=x+6.$

17.
$$\begin{array}{r}
3x-4\\
x+2\overline{)\ 3x^2+2x-8}\\
(-)\underline{3x^2+6x}\\
-4x-8\\
(-)\underline{-4x-8}\\
0
\end{array}$$

Thus $(-8+2x+3x^2)\div(2+x)=3x-4.$

19.
$$\begin{array}{r}
2y-5 \\
y+7\overline{\smash{)}\;2y^2+9y-36} \\
(-)\underline{2y^2+14y} \\
-5y-36 \\
(-)\underline{-5y-35} \\
-1
\end{array}$$

Thus
$(2y^2+9y-36)\div(7+y)=(2y-5)$ R -1 or
$2y-5+\dfrac{-1}{y+7}$.

21.
$$\begin{array}{r}
x^2-x-1 \\
2x+2\overline{\smash{)}\;2x^3+0x^2-4x-2} \\
(-)\underline{2x^3+2x^2} \\
-2x^2-4x \\
(-)\underline{-2x^2-2x} \\
-2x-2 \\
(-)\underline{-2x-2} \\
0
\end{array}$$

Thus $(2x^3-4x-2)\div(2x+2)=x^2-x-1$.

23.
$$\begin{array}{r}
y^2-y-1 \\
y^2+y+1\overline{\smash{)}\;y^4+0y^3-y^2-2y-1} \\
(-)\underline{y^4+y^3+y^2} \\
-y^3-2y^2-2y \\
(-)\underline{-y^3-y^2-y} \\
-y^2-y-1 \\
(-)\underline{-y^2-y-1} \\
0
\end{array}$$

Thus
$(y^4-y^2-2y-1)\div(y^2+y+1)=y^2-y-1$.

25.
$$\begin{array}{r}
4x^2+3x+7 \\
2x-3\overline{\smash{)}\;8x^3-6x^2+5x-9} \\
(-)\underline{8x^3-12x^2} \\
6x^2+5x \\
(-)\underline{6x^2-9x} \\
14x--9 \\
(-)\underline{14x-21} \\
12
\end{array}$$

Thus $(8x^3-6x^2+5x-9)\div(2x-3)$
$=(4x^2+3x+7)$ R 12
or $4x+3x+7+\dfrac{12}{2x-3}$.

27.
$$\begin{array}{r}
x^2+2x+4 \\
x-2\overline{\smash{)}\;x^3+0x^2+0x-8} \\
(-)\underline{x^3-2x^2} \\
2x^2+0x \\
(1)\underline{2x^2-4x} \\
4x-8 \\
(-)\underline{4x-8} \\
0
\end{array}$$

Thus $(x^3-8)\div(x-2)=x^2+2x+4$.

29.
$$\begin{array}{r}
4y^2+8y+16 \\
2y-4\overline{\smash{)}\;8y^3+0y^2+0y-64} \\
(-)\underline{8y^3-16y^2} \\
16y^2+0y \\
(-)\underline{16y^2-32y} \\
32y-64 \\
(-)\underline{32y-64} \\
0
\end{array}$$

Thus $(8y^3-64)\div(2y-4)=4y^2+8y+16$.

31.

$$
\begin{array}{r}
x^2+x+1 \\
x^2-x-1\overline{)\ x^4+0x^3-x^2-2x+2} \\
(-)\underline{x^4-x^3-x^2} \qquad\qquad \\
x^3 \qquad -2x \qquad \\
(-)\underline{x^3-x^2-x} \qquad \\
x^2-x+2 \\
(-)\underline{x^2-x-1} \\
3
\end{array}
$$

Thus $(x^4-x^2-2x+2)\div(x^2-x-1)$
$=(x^2+x+1)$ R 3

or $x^2+x+1+\dfrac{3}{x^2-x-1}$.

33.

$$
\begin{array}{r}
x^3+x^2-x+1 \\
x^2-2x+3\overline{)\ x^5-x^4+0x^3+6x^2-5x+3} \\
(-)\underline{x^5-2x^4+3x^3} \qquad\qquad\qquad \\
x^4-3x^3+6x^2 \qquad\qquad \\
(-)\underline{x^4-2x^3+3x^2} \qquad\qquad \\
-x^3+3x^2-5x \qquad \\
(-)\underline{-x^3+2x^2-3x} \qquad \\
x^2-2x+3 \\
(-)\underline{x^2-2x+3} \\
0
\end{array}
$$

Thus $(x^5-x^4+6x^2-5x+3)\div(x^2-2x+3)=x^3+x^2-x+1$.

35.

$$
\begin{array}{r}
m^2-8 \\
m^2-3\overline{)\ m^4-11m^2+34} \\
(-)\underline{m^4-3m^2} \qquad \\
-8m^2+34 \\
(-)\underline{-8m^2+24} \\
10
\end{array}
$$

Thus $(m^4-11m^2+34)\div(m^2-3)=(m^2-8)$ R 10 or $m^2-8+\dfrac{10}{m^2-3}$.

37. $x-y\overline{\big)\,x^3+0x^2y+0xy^2-y^3}$ with quotient x^2+xy+y^2

$$\begin{array}{r} x^3+0x^2y+0xy^2-y^3 \\ (-)\underline{x^3-x^2y} \\ x^2y+0xy^2 \\ (-)\underline{x^2y-xy^2} \\ xy^2-y^3 \\ (-)\underline{xy^2-y^3} \\ 0 \end{array}$$

Thus $\dfrac{x^3-y^3}{x-y}=x^2+xy+y^2.$

39. $x-2\overline{\big)\,x^3+0x^2+0x+8}$ with quotient x^2+2x+4

$$\begin{array}{r} x^3+0x^2+0x+8 \\ (-)\underline{x^3-2x^2} \\ 2x^2+0x \\ (-)\underline{2x^2-4x} \\ 4x+8 \\ (-)\underline{4x-8} \\ 16 \end{array}$$

Thus $\dfrac{x^3+8}{x-2}=(x^2+2x+4)$ R 16 or

$x^2+2x+4+\dfrac{16}{x-2}.$

41. $20=2\cdot2\cdot5$
$18=2\cdot3\cdot3$
$\text{LCM}=2\cdot2\cdot3\cdot3\cdot5=180$

43. $40=2\cdot2\cdot2\cdot5$
$12=2\cdot2\cdot3$
$\text{LCM}=2\cdot2\cdot2\cdot3\cdot5=120$

45. $20=2\cdot2\cdot5$
$30=2\cdot3\cdot5$
$18=2\cdot3\cdot3$
$\text{LCM}=2\cdot2\cdot3\cdot3\cdot5=180$

47. $\overline{C(x)}=\dfrac{C(x)}{x}=\dfrac{3x^2+5x}{x}=\dfrac{3x^2}{x}+\dfrac{5x}{x}=3x+5$

49. $\overline{P(x)}=\dfrac{P(x)}{x}$
$=\dfrac{50x+x^2-7000}{x}$
$=\dfrac{50x}{x}+\dfrac{x^2}{x}-\dfrac{7000}{x}$
$=50+x-\dfrac{7000}{x}$

51. Answers may vary.

53. Answers may vary.

55. $x-1\overline{\big)\,2x^3+0x^2+x-3}$ with quotient $2x^2+2x+3$

$$\begin{array}{r} 2x^3+0x^2+x-3 \\ (-)\underline{2x^3-2x^2} \\ 2x^2+x \\ (-)\underline{2x^2-2x} \\ 3x-3 \\ (-)\underline{3x-3} \\ 0 \end{array}$$

Thus $(2x^3+x-3)\div(x-1)=2x^2+2x+3.$

57. $\dfrac{24x^4-18x^3}{6x^2}=\dfrac{24x^4}{6x^2}-\dfrac{18x^3}{6x^2}=4x^2-3x$

59. $\dfrac{-6y^3+12y^2+3}{-3y^2}=\dfrac{-6y^3}{-3y^2}+\dfrac{12y^2}{-3y^2}+\dfrac{3}{-3y^2}$
$=2y-4-\dfrac{1}{y^2}$

Review Exercises

1. a. (i) $(3a^2b)(-5ab^3)$
$=(3)(-5)(a^2\cdot a^1)(b^1\cdot b^3)$
$=-15a^{2+1}b^{1+3}$
$=-15a^3b^4$

(ii) $(4a^2b)(-6ab^4)$
$=(4)(-6)(a^2\cdot a^1)(b^1\cdot b^4)$
$=-24a^{2+1}b^{1+4}$
$=-24a^3b^5$

(iii) $(5a^2b)(-7ab^3)$
$= (5)(-7)(a^2 \cdot b^1)(b^1 \cdot b^3)$
$= -35a^{2+1}b^{1+3}$
$= -35a^3b^4$

b. **(i)** $(-2xy^2z)(-3x^2yz^4)$
$= (-2)(-3)(x^1 \cdot x^2)(y^2 \cdot y^1)(z^1 \cdot z^4)$
$= 6x^{1+2}y^{2+1}z^{1+4}$
$= 6x^3y^3z^5$

(ii) $(-3x^2yz^2)(-4xy^3z)$
$= [(-3)(-4)](x^2 \cdot x^1)(y^1 \cdot y^3)(z^2 \cdot z^1)$
$= 12x^{2+1}y^{1+3}z^{2+1}$
$= 12x^3y^4z^3$

(iii) $(-4xyz)(-5xy^2z^3)$
$= [(-4)(-5)](x^1 \cdot x^1)(y^1 \cdot y^2)(z^1 \cdot z^3)$
$= 20x^{1+1}y^{1+2}z^{1+3}$
$= 20x^2y^3z^4$

2. a. **(i)** $\dfrac{16x^6y^8}{-8xy^4} = \dfrac{16}{-8} \cdot \dfrac{x^6}{x} \cdot \dfrac{y^8}{y^4}$
$= -2x^{6-1}y^{8-4}$
$= -2x^5y^4$

(ii) $\dfrac{24x^7y^6}{-4xy^3} = \dfrac{24}{-4} \cdot \dfrac{x^7}{x} \cdot \dfrac{y^6}{y^3}$
$= -6x^{7-1}y^{6-3}$
$= -6x^6y^3$

(iii) $\dfrac{-18x^8y^7}{9xy^4} = \dfrac{-18}{9} \cdot \dfrac{x^8}{x} \cdot \dfrac{y^7}{y^4}$
$= -2x^{8-1}y^{7-4}$
$= -2x^7y^3$

b. **(i)** $\dfrac{-8x^9y^7}{-16x^4y} = \dfrac{-8}{-16} \cdot \dfrac{x^9}{x^4} \cdot \dfrac{y^7}{y}$
$= \dfrac{1}{2}x^{9-4}y^{7-1}$
$= \dfrac{x^5y^6}{2}$

(ii) $\dfrac{-5x^7y^8}{-10x^6y} = \dfrac{-5}{-10} \cdot \dfrac{x^7}{x^6} \cdot \dfrac{y^8}{y}$
$= \dfrac{1}{2}x^{7-6}y^{8-1}$
$= \dfrac{xy^7}{2}$

(iii) $\dfrac{-3x^9y^7}{-9x^8y} = \dfrac{-3}{-9} \cdot \dfrac{x^9}{x^8} \cdot \dfrac{y^7}{y}$
$= \dfrac{1}{3}x^{9-8}y^{7-1}$
$= \dfrac{xy^6}{3}$

3. a. $(2^2)^3 = 2^{2\cdot3} = 2^6 = 64$

b. $(2^2)^2 = 2^{2\cdot2} = 2^4 = 16$

c. $(3^2)^2 = 3^{2\cdot2} = 3^4 = 81$

4. a. $(y^3)^2 = y^{3\cdot2} = y^6$

b. $(x^2)^3 = x^{2\cdot3} = x^6$

c. $(a^4)^5 = a^{4\cdot5} = a^{20}$

5. a. $(4xy^3)^2 = 4^2x^2(y^3)^2 = 16x^2y^6$

b. $(2x^2y)^3 = 2^3(x^2)^3y^3 = 8x^6y^3$

c. $(3x^2y^2)^3 = 3^3(x^2)^3(y^2)^3 = 27x^6y^6$

6. a. $(-2xy^3)^3 = (-2)^3x^3(y^3)^3 = -8x^3y^9$

b. $(-3x^2y^3)^2 = (-3)^2(x^2)^2(y^3)^2 = 9x^4y^6$

c. $(-2x^2y^2)^3 = (-2)^3(x^2)^3(y^2)^3 = -8x^6y^6$

7. a. $\left(\dfrac{2y^2}{x^4}\right)^2 = \dfrac{2^2(y^2)^2}{(x^4)^2} = \dfrac{4y^{2\cdot 2}}{x^{4\cdot 2}} = \dfrac{4y^4}{x^8}$

b. $\left(\dfrac{3x}{y^3}\right)^3 = \dfrac{3^3x^3}{(y^3)^3} = \dfrac{27x^3}{y^{3\cdot 3}} = \dfrac{27x^3}{y^9}$

c. $\left(\dfrac{2x^2}{y^4}\right)^4 = \dfrac{2^4(x^2)^4}{(y^4)^4} = \dfrac{16x^{2\cdot 4}}{y^{4\cdot 4}} = \dfrac{16x^8}{y^{16}}$

8. a. $(2x^4)^3(-2y^2)^2 = (2)^3(x^4)^3(-2)^2(y^2)^2$
$= 8x^{4\cdot 3}(4)y^{2\cdot 2}$
$= 8\cdot 4x^{12}y^4$
$= 32x^{12}y^4$

b. $(3x^2)^2(-2y^3)^3 = 3^2(x^2)^2(-2)^3(y^3)^3$
$= 9x^{2\cdot 2}(-8)y^{3\cdot 3}$
$= 9(-8)x^4y^9$
$= -72x^4y^9$

c. $(4x^3)^2(-2y^4)^4 = 4^2(x^3)^2(-2)^4(y^4)^4$
$= 16x^{3\cdot 2}16y^{4\cdot 4}$
$= 16\cdot 16x^6y^{16}$
$= 256x^6y^{16}$

9. a. $2^{-3} = \dfrac{1}{2^3} = \dfrac{1}{8}$

b. $3^{-4} = \dfrac{1}{3^4} = \dfrac{1}{81}$

c. $5^{-2} = \dfrac{1}{5^2} = \dfrac{1}{25}$

10. a. $\left(\dfrac{1}{x}\right)^{-4} = x^4$

b. $\left(\dfrac{1}{y}\right)^{-3} = y^3$

c. $\left(\dfrac{1}{z}\right)^{-5} = z^5$

11. a. $\dfrac{1}{x^5} = x^{-5}$

b. $\dfrac{1}{y^7} = y^{-7}$

c. $\dfrac{1}{z^8} = z^{-8}$

12. a. **(i)** $2^8\cdot 2^{-5} = 2^{8+(-5)} = 2^3 = 8$

(ii) $2^6\cdot 2^{-4} = 2^{6+(-4)} = 2^2 = 4$

(iii) $3^6\cdot 3^{-3} = 3^{6+(-3)} = 3^3 = 27$

b. **(i)** $y^{-3}\cdot y^{-5} = y^{-3+(-5)} = y^{-8} = \dfrac{1}{y^8}$

(ii) $y^{-2}\cdot y^{-3} = y^{-2+(-3)} = y^{-5} = \dfrac{1}{y^5}$

(iii) $y^{-4}\cdot y^{-2} = y^{-4+(-2)} = y^{-6} = \dfrac{1}{y^6}$

13. a. **(i)** $\dfrac{x}{x^5} = x^{1-5} = x^{-4} = \dfrac{1}{x^4}$

(ii) $\dfrac{x}{x^7} = x^{1-7} = x^{-6} = \dfrac{1}{x^6}$

(iii) $\dfrac{x}{x^9} = x^{1-9} = x^{-8} = \dfrac{1}{x^8}$

b. **(i)** $\dfrac{a^{-2}}{a^{-2}} = a^{-2-(-2)} = a^{-2+2} = a^0 = 1$

(ii) $\dfrac{a^{-4}}{a^{-4}} = a^{-4-(-4)} = a^{-4+4} = a^0 = 1$

(iii) $\dfrac{a^{-10}}{a^{-10}} = a^{-10-(-10)}$
$= a^{-10+10}$
$= a^0$
$= 1$

c. **(i)** $\dfrac{x^{-2}}{x^{-3}} = x^{-2-(-3)} = x^{-2+3} = x^1 = x$

(ii) $\dfrac{x^{-5}}{x^{-8}} = x^{-5-(-8)} = x^{-5+8} = x^3$

(iii) $\dfrac{x^{-7}}{x^{-9}} = x^{-7-(-9)} = x^{-7+9} = x^2$

14. a. $\left(\dfrac{2x^3y^4}{3x^4y^{-3}}\right)^{-2} = \left(\dfrac{2x^{3-4}y^{4-(-3)}}{3}\right)^{-2}$
$= \left(\dfrac{2x^{-1}y^7}{3}\right)^{-2}$
$= \dfrac{2^{-2}x^2y^{-14}}{3^{-2}}$
$= \dfrac{3^2x^2}{2^2y^{14}}$
$= \dfrac{9x^2}{4y^{14}}$

b. $\left(\dfrac{3x^5y^{-3}}{2x^7y^4}\right)^{-3} = \left(\dfrac{3x^{5-7}y^{-3-4}}{2}\right)^{-3}$
$= \left(\dfrac{3x^{-2}y^{-7}}{2}\right)^{-3}$
$= \dfrac{3^{-3}x^6y^{21}}{2^{-3}}$
$= \dfrac{2^3x^6y^{21}}{3^3}$
$= \dfrac{8x^6y^{21}}{27}$

c. $\left(\dfrac{3x^{-5}y^{-4}}{2x^{-6}y^{-8}}\right)^{-2} = \left(\dfrac{3x^{-5-(-6)}y^{-4-(-8)}}{2}\right)^{-2}$
$= \left(\dfrac{3xy^4}{2}\right)^{-2}$
$= \dfrac{3^{-2}x^{-2}y^{-8}}{2^{-2}}$
$= \dfrac{2^2}{3^2x^2y^8}$
$= \dfrac{4}{9x^2y^8}$

15. a. **(i)** $44{,}000{,}000 = 4.4\times10^7$

(ii) $4{,}500{,}000 = 4.5\times10^6$

(iii) $460{,}000 = 4.6\times10^5$

b. **(i)** $0.0014 = 1.4\times10^{-3}$

(ii) $0.00015 = 1.5\times10^{-4}$

(iii) $0.000016 = 1.6\times10^{-5}$

16. a. **(i)** $(2\times10^2)\times(1.1\times10^3)$
$= (2\times1.1)\times(10^2\times10^3)$
$= 2.2\times10^{2+3}$
$= 2.2\times10^5$

(ii) $(3\times10^2)\times(3.1\times10^4)$
$= (3\times3.1)\times(10^2\times10^4)$
$= 9.3\times10^{2+4}$
$= 9.3\times10^6$

(iii) $(4\times10^2)\times(3.1\times10^5)$
$= (4\times3.1)\times(10^2\times10^5)$
$= 12.4\times10^{2+5}$
$= (1.24\times10)\times10^7$
$= 1.24\times10^{1+7}$
$= 1.24\times10^8$

b. (i) $\dfrac{1.15\times10^{-3}}{2.3\times10^{-4}}$
$=(1.15\div2.3)\times(10^{-3}\div10^{-4})$
$=0.5\times10^{-3-(-4)}$
$=(5\times10^{-1})\times10^{1}$
$=5\times10^{-1+1}$
$=5\times10^{0}$
$=5\times1$
$=5$

(ii) $\dfrac{1.38\times10^{-3}}{2.3\times10^{-4}}$
$=(1.38\div2.3)\times(10^{-3}\div10^{-4})$
$=0.6\times10^{-3-(-4)}$
$=(6\times10^{-1})\times10^{1}$
$=6\times10^{-1+1}$
$=6\times10^{0}$
$=6\times1$
$=6$

(iii) $\dfrac{1.61\times10^{-3}}{2.3\times10^{-4}}$
$=(1.61\div2.3)\times(10^{-3}\div10^{-4})$
$=0.7\times10^{-3-(-4)}$
$=(7\times10^{-1})\times10^{1}$
$=7\times10^{-1+1}$
$=7\times10^{0}$
$=7\times1$
$=7$

17. a. $9x^2-9+7x$ has three terms, so it is a trinomial (T).

b. $7x^2$ has only one term, so it is a monomial (M).

c. $3x-1$ has two terms, so it is a binomial (B).

18. a. The highest exponent of the variable x is 4; thus the degree of the polynomial is 4.

b. The highest exponent of the variable x is 2; thus the degree of the polynomial is 2.

c. The highest exponent of the variable x is 2; thus the degree of the polynomial is 2.

19. a. $4x^2-8x+9x^4=9x^4+4x^2-8x$

b. $-3x+4x^2-3=4x^2-3x-3$

c. $8+3x-4x^2=-4x^2+3x+8$

20. a. Let $t=1$.
$-16t^2+300=-16(1)^2+300$
$=-16(1)+300$
$=-16+300$
$=284$

b. Let $t=3$.
$-16t^2+300=-16(3)^2+300$
$=-16(9)+300$
$=-144+300$
$=156$

c. Let $t=5$.
$-16t^2+300=-16(5)^2+300$
$=-16(25)+300$
$=-400+300$
$=-100$

21. a. $(-5x+7x^2-3)+(-2x^2-7+4x)$
$=(7x^2-5x-3)+(-2x^2+4x-7)$
$=(7x^2-2x^2)+(-5x+4x)+[-3+(-7)]$
$=5x^2-x-10$

b. $(-3x^2+8x-1)+(3+7x-2x^2)$
$=(-3x^2+8x-1)+(-2x^2+7x+3)$
$=[-3x^2+(-2x^2)]+(8x+7x)+(-1+3)$
$=-5x^2+15x+2$

c.
$$\begin{array}{r} 3x^2-5x-4 \\ (+)\ \underline{6x^2-2x+5} \\ 9x^2-7x+1 \end{array}$$

22. **a.** $(6x^2-4x)-(3x-4+7x^2)$
$=6x^2-4x-3x+4-7x^2$
$=(6x^2-7x^2)+(-4x-3x)+4$
$=-x^2-7x+4$

b. $(9x^2-2x)-(5x-3+2x^2)$
$=9x^2-2x-5x+3-2x^2$
$=(9x^2-2x^2)+(-2x-5x)+3$
$=7x^2-7x+3$

c. $(2x-5)-(6-2x+5x^2)$
$=2x-5-6+2x-5x^2$
$=-5x^2+(2x+2x)+(-5-6)$
$=-5x^2+4x-11$

23. **a.** $(-6x^2)(3x^5)=(-6\cdot 3)(x^2\cdot x^5)$
$=-18x^{2+5}$
$=-18x^7$

b. $(-8x^3)(5x^6)=(-8\cdot 5)(x^3\cdot x^6)$
$=-40x^{3+6}$
$=-40x^9$

c. $(-9x^4)(3x^7)=(-9\cdot 3)(x^4\cdot x^7)$
$=-27x^{4+7}$
$=-27x^{11}$

24. **a.** $-2x^2(x+2y)=-2x^2\cdot x-2x^2\cdot 2y$
$=-2x^3-4x^2y$

b. $-3x^3(2x+3y)=-3x^3\cdot 2x-3x^3\cdot 3y$
$=-3\cdot 2x^3\cdot x-3\cdot 3x^3y$
$=-6x^4-9x^3y$

c. $-4x^3(5x+7y)=-4x^3\cdot 5x-4x^3\cdot 7y$
$=-4\cdot 5x^3\cdot x-4\cdot 7x^3\cdot y$
$=-20x^4-28x^3y$

25. **a.** $(x+6)(x+9)=x\cdot x+9x+6x+6\cdot 9$
$=x^2+15x+54$

b. $(x+2)(x+3)=x\cdot x+3x+2x+2\cdot 3$
$=x^2+5x+6$

c. $(x+7)(x+9)=x\cdot x+9x+7x+7\cdot 9$
$=x^2+16x+63$

26. **a.** $(x+7)(x-3)=x\cdot x-3x+7x+7\cdot(-3)$
$=x^2+4x-21$

b. $(x+6)(x-2)=x\cdot x-2x+6x+6\cdot(-2)$
$=x^2+4x-12$

c. $(x+5)(x-1)=x\cdot x-x+5x+5\cdot(-1)$
$=x^2+4x-5$

27. **a.** $(x+3)(x-7)=x\cdot x-7x+3x+3\cdot(-7)$
$=x^2-4x-21$

b. $(x+2)(x-6)=x\cdot x-6x+2x+2\cdot(-6)$
$=x^2-4x-12$

c. $(x+1)(x-5)=x\cdot x-5x+x+1\cdot(-5)$
$=x^2-4x-5$

28. a. $(3x-2y)(2x-3y) = (3x)(2x)+(3x)(-3y)+(-2y)(2x)+(-2y)(-3y)$
$= 6x^2 - 9xy - 4xy + 6y^2$
$= 6x^2 - 13xy + 6y^2$

b. $(5x-3y)(4x-3y) = (5x)(4x)+(5x)(-3y)+(-3y)(4x)+(-3y)(-3y)$
$= 20x^2 - 15xy - 12xy + 9y^2$
$= 20x^2 - 27xy + 9y^2$

c. $(4x-3y)(2x-5y) = (4x)(2x)+(4x)(-5y)+(-3y)(2x)+(-3y)(-5y)$
$= 8x^2 - 20xy - 6xy + 15y^2$
$= 8x^2 - 26xy + 15y^2$

29. a. $(2x+3y)^2 = (2x)^2 + 2(3y)(2x) + (3y)^2 = 4x^2 + 12xy + 9y^2$

b. $(3x+4y)^2 = (3x)^2 + 2(4y)(3x) + (4y)^2 = 9x^2 + 24xy + 16y^2$

c. $(4x+5y)^2 = (4x)^2 + 2(4x)(5y) + (5y)^2 = 16x^2 + 40xy + 25y^2$

30. a. $(2x-3y)^2 = (2x)^2 - 2(3y)(2x) + (3y)^2 = 4x^2 - 12xy + 9y^2$

b. $(3x-2y)^2 = (3x)^2 - 2(2y)(3x) + (2y)^2 = 9x^2 - 12xy + 4y^2$

c. $(5x-2y)^2 = (5x)^2 - 2(2y)(5x) + (2y)^2 = 25x^2 - 20xy + 4y^2$

31. a. $(3x-5y)(3x+5y) = (3x)^2 - (5y)^2 = 9x^2 - 25y^2$

b. $(3x-2y)(3x+2y) = (3x)^2 - (2y)^2 = 9x^2 - 4y^2$

c. $(3x-4y)(3x+4y) = (3x)^2 - (4y)^2 = 9x^2 - 16y^2$

32. a. $(x+1)(x^2+3x+2) = x(x^2+3x+2) + 1(x^2+3x+2)$
$= x^3 + 3x^2 + 2x + x^2 + 3x + 2$
$= x^3 + (3x^2 + x^2) + (2x + 3x) + 2$
$= x^3 + 4x^2 + 5x + 2$

b. $(x+2)(x^2+3x+2) = x(x^2+3x+2) + 2(x^2+3x+2)$
$= x^3 + 3x^2 + 2x + 2x^2 + 6x + 4$
$= x^3 + (3x^2 + 2x^2) + (2x + 6x) + 4$
$= x^3 + 5x^2 + 8x + 4$

c. $(x+3)(x^2+3x+2)$
$= x(x^2+3x+2)+3(x^2+3x+2)$
$= x^3+3x^2+2x+3x^2+9x+6$
$= x^3+(3x^2+3x^2)+(2x+9x)+6$
$= x^3+6x^2+11x+6$

33. a. $3x(x+1)(x+2) = 3x(x^2+2x+x+2)$
$= 3x(x^2+3x+2)$
$= 3x^3+9x^2+6x$

b. $4x(x+1)(x+2) = 4x(x^2+2x+x+2)$
$= 4x(x^2+3x+2)$
$= 4x^3+12x^2+8x$

c. $5x(x+1)(x+2) = 5x(x^2+2x+x+2)$
$= 5x(x^2+3x+2)$
$= 5x^3+15x^2+10x$

34. a. $(x+2)^3$
$= (x+2)(x+2)(x+2)$
$= (x+2)(x^2+4x+4)$
$= x(x^2+4x+4)+2(x^2+4x+4)$
$= x^3+4x^2+4x+2x^2+8x+8$
$= x^3+6x^2+12x+8$

b. $(x+3)^3$
$= (x+3)(x+3)(x+3)$
$= (x+3)(x^2+6x+9)$
$= x(x^2+6x+9)+3(x^2+6x+9)$
$= x^3+6x^2+9x+3x^2+18x+27$
$= x^3+9x^2+27x+27$

c. $(x+4)^3$
$= (x+4)(x+4)(x+4)$
$= (x+4)(x^2+8x+16)$
$= x(x^2+8x+16)+4(x^2+8x+16)$
$= x^3+8x^2+16x+4x^2+32x+64$
$= x^3+12x^2+48x+64$

35. a. $\left(5x^2-\frac{1}{2}\right)^2$
$= (5x^2)^2 - 2\cdot\frac{1}{2}\cdot(5x^2)+\left(\frac{1}{2}\right)^2$
$= 25x^4-5x^2+\frac{1}{4}$

b. $\left(7x^2-\frac{1}{2}\right)^2$
$= (7x^2)^2 - 2\cdot\frac{1}{2}\cdot(7x^2)+\left(\frac{1}{2}\right)^2$
$= 49x^4-7x^2+\frac{1}{4}$

c. $\left(9x^2-\frac{1}{2}\right)^2$
$= (9x^2)^2 - 2\cdot\frac{1}{2}\cdot(9x^2)+\left(\frac{1}{2}\right)^2$
$= 81x^4-9x^2+\frac{1}{4}$

36. a. $(3x^2+2)(3x^2-2) = (3x^2)^2-2^2$
$= 9x^4-4$

b. $(3x^2+4)(3x^2-4) = (3x^2)^2-4^2$
$= 9x^4-16$

c. $(2x^2+5)(2x^2-5) = (2x^2)^2-5^2$
$= 4x^4-25$

37. a. $\frac{18x^3-9x^2}{9x} = \frac{18x^3}{9x}-\frac{9x^2}{9x} = 2x^2-x$

b. $\frac{20x^3-10x^2}{5x} = \frac{20x^3}{5x}-\frac{10x^2}{5x} = 4x^2-2x$

c. $\frac{24x^3-12x^2}{6x} = \frac{24x^3}{6x}-\frac{12x^2}{6x} = 4x^2-2x$

38. a.

$$\begin{array}{r} x+6 \\ x-2\overline{)\;x^2+4x-12} \\ (-)\underline{x^2-2x} \\ 6x-12 \\ (-)\underline{6x-12} \\ 0 \end{array}$$

Thus $(x^2+4x-12)\div(x-2)=x+6$.

b.

$$\begin{array}{r} x+7 \\ x-3\overline{)\;x^2+4x-21} \\ (-)\underline{x^2-3x} \\ 7x-21 \\ (-)\underline{7x-21} \\ 0 \end{array}$$

Thus $(x^2+4x-21)\div(x-3)=x+7$.

c.

$$\begin{array}{r} x+8 \\ x-4\overline{)\;x^2+4x-32} \\ (-)\underline{x^2-4x} \\ 8x-32 \\ (-)\underline{8x-32} \\ 0 \end{array}$$

Thus $(x^2+4x-32)\div(x-4)=x+8$.

39. a.

$$\begin{array}{r} 4x^2-4x-4 \\ 2x+2\overline{)\;8x^3+0x^2-16x-8} \\ (-)\underline{8x^3+8x^2} \\ -8x^2-16x \\ (-)\underline{-8x^2-8x} \\ -8x-8 \\ (-)\underline{-8x-8} \\ 0 \end{array}$$

Thus

$(8x^3-16x-8)\div(2+2x)=4x^2-4x-4$.

b.

$$\begin{array}{r} 6x^2-6x-6 \\ 2x+2\overline{)\;12x^3+0x^2-24x-12} \\ (-)\underline{12x^3+12x^2} \\ -12x^2-24x \\ (-)\underline{-12x^2-12x} \\ -12x-12 \\ (-)\underline{-12x-12} \\ 0 \end{array}$$

Thus $(12x^3-24x-12)\div(2+2x)$
$=6x^2-6x-6$.

c.

$$\begin{array}{r} 2x^2-2x-2 \\ 2x+2\overline{)\;4x^3+0x^2-8x-4} \\ (-)\underline{4x^3+4x^2} \\ -4x^2-8x \\ (-)\underline{-4x^2-4x} \\ -4x-4 \\ (-)\underline{-4x-4} \\ 0 \end{array}$$

Thus

$(4x^3-8x-4)\div(2+2x)=2x^2-2x-2$.

40. a.

$$\begin{array}{r} 2x^2+6x-2 \\ x-3\overline{)\;2x^3+0x^2-20x+8} \\ (-)\underline{2x^3-6x^2} \\ 6x^2-20x \\ (-)\underline{6x^2-18x} \\ -2x+8 \\ (-)\underline{-2x+6} \\ 2 \end{array}$$

Thus $(2x^3-20x+8)\div(x-3)$
$=(2x^2+6x-2)$ R 2

or $2x^2+6x-2+\dfrac{2}{x-3}$.

b. $$\begin{array}{r|l} & 2x^2+6x-3 \\ \hline x-3 & 2x^3+0x^2-21x+12 \\ & (-)\underline{2x^3-6x^2} \\ & 6x^2-21x \\ & (-)\underline{6x^2-18x} \\ & -3x+12 \\ & (-)\underline{-3x+9} \\ & 3 \end{array}$$

Thus $(2x^3-21x+12)\div(x-3)$
$=(2x^2+6x-3)$ R 3
or $2x^2+6x-3+\dfrac{3}{x-3}$.

c. $$\begin{array}{r|l} & 3x^2+3x-1 \\ \hline x-1 & 3x^3+0x^2-4x+5 \\ & (-)\underline{3x^3-3x^2} \\ & 3x^2-4x \\ & (-)\underline{3x^2-3x} \\ & -x+5 \\ & (-)\underline{-x+1} \\ & 4 \end{array}$$

Thus $(3x^3-4x+5)\div(x-1)$
$=(3x^2+3x-1)$ R 4
or $3x^2+3x-1+\dfrac{4}{x-1}$.

41. a. $$\begin{array}{r|l} & x^2+x \\ \hline x^2-4 & x^4+x^3-4x^2+0x+1 \\ & (-)\underline{x^4\quad -4x^2} \\ & x^3\quad +0x \\ & (-)\underline{x^3\quad -4x} \\ & 4x+1 \end{array}$$

Thus $(x^4+x^3-4x^2+1)\div(x^2-4)$
$=(x^2+x)$ R $(4x+1)$
or $x^2+x+\dfrac{4x+1}{x^2-4}$.

b. $$\begin{array}{r|l} & x^2+x \\ \hline x^2-5 & x^4+x^3-5x^2+0x+1 \\ & (-)\underline{x^4\quad -5x^2} \\ & x^3\quad +0x \\ & (-)\underline{x^3\quad -5x} \\ & 5x+1 \end{array}$$

Thus $(x^4+x^3-5x^2+1)\div(x^2-5)$
$=(x^2+x)$ R $(5x+1)$
or $x^2+x+\dfrac{5x+1}{x^2-5}$.

c. $$\begin{array}{r|l} & x^2+x \\ \hline x^2-6 & x^4+x^3-6x^2+0x+1 \\ & (-)\underline{x^4\quad -6x^2} \\ & x^3\quad +0x \\ & (-)\underline{x^3\quad -6x} \\ & 6x+1 \end{array}$$

Thus $(x^4+x^3-6x^2+1)\div(x^2-6)$
$=(x^2+x)$ R $(6x+1)$
or $x^2+x+\dfrac{6x+1}{x^2-6}$.

Cumulative Review Chapters 1–4

1. The additive inverse of –7 is 7.

2. $\left|-9\dfrac{9}{10}\right|=9\dfrac{9}{10}$

3. The LCD is 18.
$-\dfrac{1}{9}=-\dfrac{2}{18}$ and $-\dfrac{1}{6}=-\dfrac{3}{18}$
$-\dfrac{1}{9}+\left(-\dfrac{1}{6}\right)=\left(-\dfrac{2}{18}\right)+\left(-\dfrac{3}{18}\right)=-\dfrac{5}{18}$

4. 7.2 – (–6.4) = 7.2 + 6.4 = 13.6

5. (–2.4)(2.6) = –6.24

6. $-(2^4)=-(2\cdot2\cdot2\cdot2)=-(16)=-16$

7. $-\frac{3}{4}\div\left(-\frac{7}{8}\right)=-\frac{3}{4}\cdot\left(-\frac{8}{7}\right)=\frac{24}{28}=\frac{6}{7}$

8. $y\div 5\cdot x-z=50\div 5\cdot 5-3$
$=10\cdot 5-3$
$=50-3$
$=47$

9. In 6 · (5 · 2) = (6 · 5) · 2, we changed the *grouping* of the numbers. The associative law of multiplication was used.

10. $3(2x-8)=3(2x)+3(-8)=6x-24$

11. $-8xy^3-(-xy^3)=-8xy^3+xy^3$
$=(-8+1)xy^3$
$=-7xy^3$

12. $5x+(x+4)-2(x-3)$
$=5x+x+4-2x+6$
$=(5x+x-2x)+(4+6)$
$=4x+10$

13. The quotient of $(m+3n)$ and p is written as $\frac{m+3n}{p}$.

14. If $x=-3$, then $12=15-x$ becomes $12=15-(-3)$, which is a *false* statement. Hence, -3 is not a solution of the equation.

15.
$$5=5(x-3)+5-4x$$
$$5=5x-15+5-4x$$
$$5=x-10$$
$$5+10=x-10+10$$
$$15=x \text{ or } x=15$$
The solution is 15.

16.
$$-\frac{8}{7}x=-56$$
$$-\frac{7}{8}\cdot\left(-\frac{8}{7}x\right)=-\frac{7}{8}\cdot(-56)$$
$$x=-7\cdot(-7)$$
$$x=49$$
The solution is 49.

17.
$$\frac{x}{4}-\frac{x}{9}=5$$
$$36\cdot\frac{x}{4}-36\cdot\frac{x}{9}=36\cdot 5$$
$$9x-4x=180$$
$$5x=180$$
$$\frac{5x}{5}=\frac{180}{5}$$
$$x=36$$
The solution is 36.

18.
$$5-\frac{x}{3}=\frac{3(x+1)}{7}$$
$$21\cdot 5-21\cdot\frac{x}{3}=21\cdot\frac{3(x+1)}{7}$$
$$105-7x=3\cdot 3(x+1)$$
$$105-7x=9(x+1)$$
$$105-7x=9x+9$$
$$105-105-7x=9x+9-105$$
$$-7x=9x-96$$
$$-7x-9x=9x-9x-96$$
$$-16x=-96$$
$$\frac{-16x}{-16}=\frac{-96}{-16}$$
$$x=6$$
The solution is 6.

19.
$$S=6a^2b$$
$$\frac{S}{6a^2}=\frac{6a^2b}{6a^2}$$
$$\frac{S}{6a^2}=b \text{ or } b=\frac{S}{6a^2}$$

20. Let n = first number.
Then $n+20$ = second number.
$$n+(n+20)=100$$
$$2n+20=100$$
$$2n+20-20=100-20$$
$$2n=80$$
$$\frac{2n}{2}=\frac{80}{2}$$
$$n=40$$
$$n+20=60$$
The numbers are 40 and 60.

21. Let b = amount in bonds.
Then 245 + b = amount in stocks.

$$\begin{aligned} b+(245+b) &= 615 \\ 2b+245 &= 615 \\ 2b &= 370 \\ b &= 185 \\ 245+b &= 430 \end{aligned}$$

Dave has \$185 from bonds and \$430 from stocks.

22.

	R	×	T	=	D
Train A	30		$T+6$		$30(T+6)$
Train B	40		T		$40T$

$$\begin{aligned} 30(T+6) &= 40T \\ 30T+180 &= 40T \\ 180 &= 10T \\ 18 &= T \end{aligned}$$

It takes train B 18 hours to catch train A.

23.

	P	×	r	=	T
bonds	x		0.08		$0.08x$
certificates	$17{,}000-x$		0.11		$0.11(17{,}000-x)$

$$\begin{aligned} 0.08x+0.11(17{,}000-x) &= 1630 \\ 0.08x+1870-0.11x &= 1630 \\ 1870-0.03x &= 1630 \\ -0.03x &= -240 \\ x &= 8000 \\ 17{,}000-x &= 9000 \end{aligned}$$

Susan has invested \$8000 in bonds and \$9000 in certificates of deposit.

24.

$$\begin{aligned} -\frac{x}{2}+\frac{x}{9} &\ge \frac{x-9}{9} \\ 18\cdot\left(-\frac{x}{2}\right)+18\cdot\frac{x}{9} &\ge 18\cdot\frac{x-9}{9} \\ -9x+2x &\ge 2(x-9) \\ -7x &\ge 2x-18 \\ -7x-2x &\ge 2x-2x-18 \\ -9x &\ge -18 \\ \frac{-9x}{-9} &\le \frac{-18}{-9} \\ x &\le 2 \end{aligned}$$

Any number less than or equal to 2 is a solution.

0 2

25. Start at the origin. To reach the point (4, –4), go 4 units to the right and 4 units down.

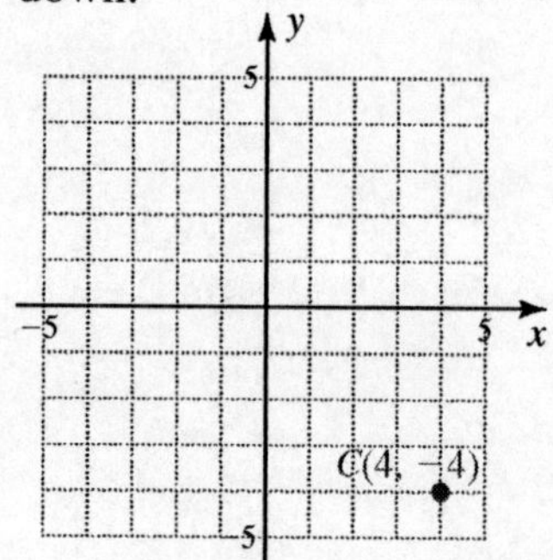

26. Point *A* is 3 units to the right of the origin and 2 units below the horizontal axis. The ordered pair corresponding to *A* is (3, –2).

27. In the ordered pair (–5, 1), $x = -5$ and $y = 1$. Substituting these into $5x + 3y = -22$, we get $5(-5) + 3(1) = -25 + 3 = -22$, which is true. Yes, (–5, 1) is a solution.

28. In the ordered pair $(x, -2)$, $y = -2$. Substituting –2 for *y* in the given equation, we have

$$5x-4y=13$$
$$5x+8=13$$
$$5x=5$$
$$x=1$$

Thus, $x = 1$ and the ordered pair is (1, –2).

29. Graph two points and join them.

For $x = 0$, $5x+y=5$

$$5(0)+y=5$$
$$y=5$$

Thus (0, 5) is on the graph.

For $y = 0$, $5x+y=5$

$$5x+0=5$$
$$5x=5$$
$$x=1$$

Thus (1, 0) is on the graph.

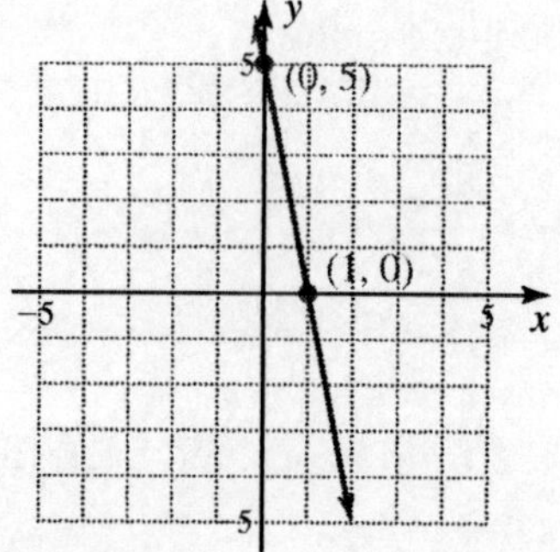

30. Solve for *y*.

$$3y-15=0$$
$$3y=15$$
$$y=5$$

The graph of $y = 5$ is a horizontal line crossing the *y*-axis at 5.

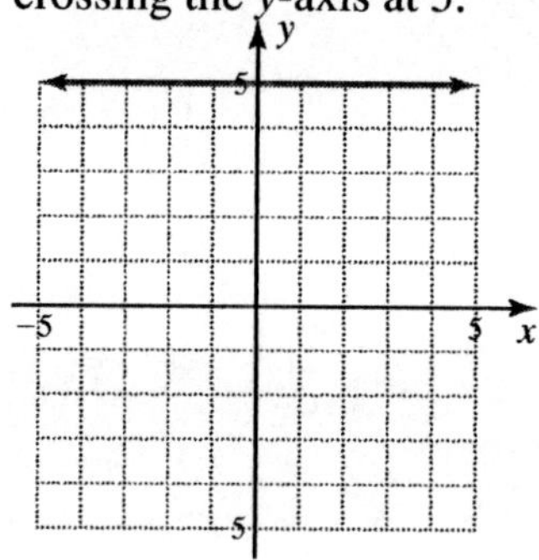

31. Let $(x_1, y_1)=(-7, -1)$ and $(x_2, y_2)=(-2, 6)$.

$$m=\frac{y_2-y_1}{x_2-x_1}=\frac{6-(-1)}{-2-(-7)}=\frac{6+1}{-2+7}=\frac{7}{5}$$

32. Solve for *y*.

$$8x-4y=14$$
$$-4y=-8x+14$$
$$y=2x-\frac{7}{2}$$

Since the coefficient of *x* is 2, the slope is 2.

33. Find the slope of each line by solving for *y*.

(1) $18y+24x=8$

$$18y=-24x+8$$
$$y=-\frac{4}{3}x+\frac{4}{9}$$

The slope is $-\frac{4}{3}$.

(2) $3y = 4x + 8$

$y = \frac{4}{3}x + \frac{8}{3}$

The slope is $\frac{4}{3}$.

(3) $24x - 18y = 8$

$-18y = -24x + 8$

$y = \frac{4}{3}x - \frac{4}{9}$

The slope is $\frac{4}{3}$.

Since the slopes of (2) and (3) are equal and the y-intercepts are different, the lines are parallel.

34. $(-3x^4y)(-6x^3y^2)$
$= (-3)(-6)\cdot(x^4\cdot x^3)\cdot(y^1\cdot y^2)$
$= 18x^{4+3}y^{1+2}$
$= 18x^7y^3$

35. $\frac{25x^2y^5}{-5xy^7} = \frac{25}{-5}\cdot\frac{x^2}{x}\cdot\frac{y^5}{y^7}$
$= -5x^{2-1}y^{5-7}$
$= -5x^1y^{-2}$
$= -\frac{5x}{y^2}$

36. $\frac{x^{-6}}{x^{-9}} = x^{-6-(-9)} = x^{-6+9} = x^3$

37. $x^8\cdot x^{-3} = x^{8+(-3)} = x^5$

38. $(4x^4y^{-4})^3 = 4^3(x^4)^3(y^{-4})^3$
$= 64x^{4\cdot 3}y^{-4\cdot 3}$
$= 64x^{12}y^{-12}$
$= \frac{64x^{12}}{y^{12}}$

39. $8{,}000{,}000 = 8.0\times 10^6$

40. $(26.04\times 10^{-5})\div(6.2\times 10^3)$
$= (26.04\div 6.2)\times(10^{-5}\div 10^3)$
$= 4.2\times 10^{-5-3}$
$= 4.2\times 10^{-8}$

41. Since $2x^2$ has only one term, it is a monomial.

42. The highest exponent of the variable x is 2; thus the degree of the polynomial is 2.

43. $-3x^2 - x - 3x^3 + 4 = -3x^3 - 3x^2 - x + 4$

44. Let $x = 2$.
$3x^3 + 2x^2 = 3(2)^3 + 2(2)^2$
$= 3(8) + 2(4)$
$= 24 + 8$
$= 32$

45. $(x + 6x^3 - 2) + (-2x^3 - 1 + 8x)$
$= (6x^3 + x - 2) + (-2x^3 + 8x - 1)$
$= (6x^3 - 2x^3) + (x + 8x) + (-2 - 1)$
$= 4x^3 + 9x - 3$

46. $-3x(3x^2 + 4y) = (-3x)\cdot(3x^2) + (-3x)\cdot(4y)$
$= -3\cdot 3\cdot x\cdot x^2 + (-3)\cdot 4\cdot x\cdot y$
$= -9x^3 - 12xy$

47. $(3x + 2y)^2 = (3x)^2 + 2(2y)(3x) + (2y)^2$
$= 9x^2 + 12xy + 4y^2$

48. $(4x - 3y)(4x + 3y) = (4x)^2 - (3y)^2$
$= 16x^2 - 9y^2$

49. $\left(5x^2 - \frac{1}{2}\right)^2 = (5x^2)^2 - 2\cdot\frac{1}{2}\cdot(5x^2) + \left(\frac{1}{2}\right)^2$
$= 25x^4 - 5x^2 + \frac{1}{4}$

50. $(4x^2 + 9)(4x^2 - 9) = (4x^2)^2 - 9^2$
$= 16x^4 - 81$

51.
$$
\begin{array}{r}
3x^2-5x+4 \\
x-5\overline{\big)\ 3x^3-20x^2+29x-17} \\
(-)\underline{3x^3-15x^2} \qquad\qquad \\
-5x^2+29x \qquad \\
(-)\underline{-5x^2+25x} \qquad \\
4x-17 \\
(-)\underline{4x-20} \\
3
\end{array}
$$

Thus $(3x^3-20x^2+29x-17)\div(x-5)$
$=(3x^2-5x+4)$ R 3
or $3x^2-5x+4+\dfrac{3}{x-5}$.

Chapter 5 Factoring

5.1 Common Factors and Graphing

Problems 5.1

1. a. $30 = 2 \cdot 3 \cdot 5$
$45 = 3^2 \cdot 5$
The common factors are 3 and 5, so the GCF is $3 \cdot 5 = 15$.

b. $45 = 3^2 \cdot 5$
$60 = 2^2 \cdot 3 \cdot 5$
$108 = 2^2 \cdot 3^3$
The only common factor is 3, so the GCF is 3.

c. $20 = 2^2 \cdot 5$
$13 = 13$
$18 = 2 \cdot 3^2$
There are no primes common to all three numbers, so the GCF is 1.

2. a. $12x^8 = 2^2 \cdot 3x^8$
$9x^5 = 3^2 \cdot x^5$
$-30x^8 = -1 \cdot 2 \cdot 3 \cdot 5x^8$
The common factors are 3 and x^5, so the GCF is $3x^5$.

b. x^4y^3, y^4x^6, x^3y^8, y^9
There are no factors containing x in y^9, so the GCF is y^3.

3. a. $6a + 18 = 6 \cdot a + 6 \cdot 3 = 6(a + 3)$

b. $-9y + 27 = -9 \cdot y - 9(-3) = -9(y - 3)$

c. $15a^2 - 45a^3 = 15a^2 \cdot 1 - 15a^2 \cdot 3a$
$= 15a^2(1 - 3a)$

4. a. $5x^3 + 15x^2 + 20x$
$= 5x \cdot x^2 + 5x \cdot 3x + 5x \cdot 4$
$= 5x(x^2 + 3x + 4)$

b. $15x^7 - 20x^6 + 10x^5 + 25x^3$
$= 5x^3 \cdot 3x^4 - 5x^3 \cdot 4x^3 + 5x^3 \cdot 2x^2 + 5x^3 \cdot 5$
$= 5x^3(3x^4 - 4x^3 + 2x^2 + 5)$

c. $3a^3 + 9a^4 + 12a^5$
$= 3a^3 \cdot 1 + 3a^3 \cdot 3a + 3a^3 \cdot 4a^2$
$= 3a^3(1 + 3a + 4a^2)$

5. $\frac{4}{5}a^2 - \frac{1}{5}a + \frac{2}{5} = \frac{1}{5} \cdot 4a^2 - \frac{1}{5} \cdot a + \frac{1}{5} \cdot 2$
$= \frac{1}{5}(4a^2 - a + 2)$

6. a. $2a^3 + 6a^2 + 5a + 15$
$= (2a^3 + 6a^2) + (5a + 15)$
$= 2a^2(a + 3) + 5(a + 3)$
$= (a + 3)(2a^2 + 5)$

b. $6a^3 - 2a^2 - 3a + 1$
$= (6a^3 - 2a^2) + (-3a + 1)$
$= 2a^2(3a - 1) - 1(3a - 1)$
$= (3a - 1)(2a^2 - 1)$

7. a. $3a^3 - 9a^2 - a + 3$
$= (3a^3 - 9a^2) + (-a + 3)$
$= 3a^2(a - 3) - 1(a - 3)$
$= (a - 3)(3a^2 - 1)$

b. $6a^4 - 4a^2 + 9a^2 - 6$
$= (6a^4 - 4a^2) + (9a^2 - 6)$
$= 2a^2(3a^2 - 2) + 3(3a^2 - 2)$
$= (3a^2 - 2)(2a^2 + 3)$

8. $6a^2 + 3ab - 4ab - 2b^2$
$= 3a(2a + b) - 2b(2a + b)$
$= (2a + b)(3a - 2b)$

Exercises 5.1

1. $20 = 2^2 \cdot 5$

$24 = 2^3 \cdot 3$

The common factor is 2^2, so the GCF is $2 \cdot 2 = 4$.

3. $16 = 2^4$

$48 = 2^4 \cdot 3$

$88 = 2^3 \cdot 11$

The common factor is 2^3, so the GCF is $2 \cdot 2 \cdot 2 = 8$.

5. $8 = 2^3$

$19 = 19$

$12 = 2^2 \cdot 3$

There are no primes common to all three numbers, so the GCF is 1.

7. a^3, a^8

The smallest exponent is 3, so the GCF is a^3.

9. x^3, x^6, x^{10}

The smallest exponent is 3, so the GCF is x^3.

11. $5y^6 = 5y^6 \cdot 1$

$10y^7 = 5y^6 \cdot 2y$

The GCF is $5y^6$.

13. $8x^3 = 2x^3 \cdot 4$

$6x^7 = 2x^3 \cdot 3x^4$

$10x^9 = 2x^3 \cdot 5x^6$

The GCF is $2x^3$.

15. $9b^2c = 3bc \cdot 3b$

$12bc^2 = 3bc \cdot 4c$

$15b^2c^2 = 3bc \cdot 5bc$

The GCF is $3bc$.

17. $9y^2 = 3 \cdot 3y^2$

$6x^3 = 3 \cdot 2x^3$

$3x^3y = 3 \cdot x^3y$

The GCF is 3.

19. $18a^4b^3z^4 = 9ab^3z \cdot 2a^3z^3$

$27a^5b^3z^4 = 9ab^3z \cdot 3a^4z^3$

$81ab^3z = 9ab^3z \cdot 9$

The GCF is $9ab^3z$.

21. $3x + 15 = 3 \cdot x + 3 \cdot 5 = 3(x + 5)$

23. $9y - 18 = 9 \cdot y - 9 \cdot 2 = 9(y - 2)$

25. $-5y + 20 = -5 \cdot y - 5 \cdot (-4) = -5(y - 4)$

27. $-3x - 27 = -3 \cdot x - 3 \cdot 9 = -3(x + 9)$

29. $4x^2 + 32x = 4x \cdot x + 4x \cdot 8 = 4x(x + 8)$

31. $6x - 42x^2 = 6x \cdot 1 - 6x \cdot 7x = 6x(1 - 7x)$

33. $-5x^2 - 25x^4 = -5x^2 \cdot 1 - 5x^2 \cdot 5x^2$
$= -5x^2(1 + 5x^2)$

35. $3x^3 + 6x^2 + 9x = 3x \cdot x^2 + 3x \cdot 2x + 3x \cdot 3$
$= 3x(x^2 + 2x + 3)$

37. $9y^3 - 18y^2 + 27y$
$= 9y \cdot y^2 + 9y(-2y) + 9y \cdot 3$
$= 9y(y^2 - 2y + 3)$

39. $6x^6 + 12x^5 - 18x^4 + 30x^2$
$= 6x^2 \cdot x^4 + 6x^2 \cdot 2x^3 + 6x^2(-3x^2) + 6x^2 \cdot 5$
$= 6x^2(x^4 + 2x^3 - 3x^2 + 5)$

41. $8y^8 + 16y^5 - 24y^4 + 8y^3$
$= 8y^3 \cdot y^5 + 8y^3 \cdot 2y^2 + 8y^3(-3y) + 8y^3 \cdot 1$
$= 8y^3(y^5 + 2y^2 - 3y + 1)$

43. $\frac{4}{7}x^3+\frac{3}{7}x^2-\frac{9}{7}x+\frac{3}{7}$
$=\frac{1}{7}\cdot 4x^3+\frac{1}{7}\cdot 3x^2+\frac{1}{7}(-9x)+\frac{1}{7}\cdot 3$
$=\frac{1}{7}(4x^3+3x^2-9x+3)$

45. $\frac{7}{8}y^9+\frac{3}{8}y^6-\frac{5}{8}y^4+\frac{5}{8}y^2=\frac{1}{8}y^2\cdot 7y^7+\frac{1}{8}y^2\cdot 3y^4+\frac{1}{8}y^2(-5y^2)+\frac{1}{8}y^2\cdot 5$
$=\frac{1}{8}y^2(7y^7+3y^4-5y^2+5)$

47. $3(x+4)-y(x+4)=(x+4)(3-y)$

49. $x(y-2)-(y-2)=(y-2)(x-1)$

51. $c(t+s)-(t+s)=(t+s)(c-1)$

53. $4x^3+4x^4-12x^5=4x^3\cdot 1+4x^3\cdot x+4x^3(-3x^2)=4x^3(1+x-3x^2)$

55. $6y^7-12y^9-6y^{11}=6y^7\cdot 1+6y^7(-2y^2)+6y^7(-y^4)=6y^7(1-2y^2-y^4)$

57. $x^3+2x^2+x+2=x^2(x+2)+1(x+2)=(x+2)(x^2+1)$

59. $y^3-3y^2+y-3=y^2(y-3)+1(y-3)=(y-3)(y^2+1)$

61. $4x^3+6x^2+2x+3=2x^2(2x+3)+1(2x+3)=(2x+3)(2x^2+1)$

63. $6x^3-2x^2+3x-1=2x^2(3x-1)+1(3x-1)=(3x-1)(2x^2+1)$

65. $4y^3+8y^2+y+2=4y^2(y+2)+1(y+2)=(y+2)(4y^2+1)$

67. $2a^3+3a^2+2a+3=a^2(2a+3)+1(2a+3)=(2a+3)(a^2+1)$

69. $3x^4+12x^2+x^2+4=3x^2(x^2+4)+1(x^2+4)=(x^2+4)(3x^2+1)$

71. $6y^4+9y^2+2y^2+3=3y^2(2y^2+3)+1(2y^2+3)=(2y^2+3)(3y^2+1)$

73. $4y^4+12y^2+y^2+3=4y^2(y^2+3)+1(y^2+3)=(y^2+3)(4y^2+1)$

75. $3a^4-6a^2-2a^2+4=3a^2(a^2-2)-2(a^2-2)=(a^2-2)(3a^2-2)$

77. $6a-5b+12ad-10bd=1(6a-5b)+2d(6a-5b)=(6a-5b)(1+2d)$

79. $x^2 - y - 3x^2z + 3yz$
$= 1(x^2 - y) - 3z(x^2 - y)$
$= (x^2 - y)(1 - 3z)$

81. $\alpha Lt_2 - \alpha Lt_1 = \alpha L \cdot t_2 - \alpha L \cdot t_1$
$= \alpha L(t_2 - t_1)$

83. $R^2 - R - R + 1 = R(R-1) - 1(R-1)$
$= (R-1)(R-1)$

85. $2\pi rh + 2\pi r^2 = 2\pi r \cdot h + 2\pi r \cdot r = 2\pi r(h + r)$

87. $(x+5)(x+3) = x(x+3) + 5(x+3)$
$= x^2 + 3x + 5x + 15$
$= x^2 + 8x + 15$

89. $(x-1)(x-4) = x(x-4) - 1(x-4)$
$= x^2 - 4x - x + 4$
$= x^2 - 5x + 4$

91. $(x+1)(2x-3) = x(2x-3) + 1(2x-3)$
$= 2x^2 - 3x + 2x - 3$
$= 2x^2 - x - 3$

93. $-wl + wz = -w \cdot l - w(-z) = -w(l - z)$

95. $a^2 + 2as = a \cdot a + a \cdot 2s = a(a + 2s)$

97. $-16t^2 + 80t + 240$
$= -16 \cdot t^2 - 16(-5t) - 16(-15)$
$= -16(t^2 - 5t - 15)$

99. Sample answer: First, find the GCF of all terms of the polynomial. Then, factor out the GCF from each term using the distributive property. The GCF of $\frac{1}{2}x$ and $\frac{1}{2}$ is $\frac{1}{2}$, so $\frac{1}{2}x + \frac{1}{2} = \frac{1}{2} \cdot x + \frac{1}{2} \cdot 1 = \frac{1}{2}(x+1)$.

101. $6x^4 + 2x^2 - 9x^2 - 3$
$= 2x^2(3x^2 + 1) - 3(3x^2 + 1)$
$= (3x^2 + 1)(2x^2 - 3)$

103. $6x^3 - 9x^2 - 2x + 3 = 3x^2(2x-3) - 1(2x-3)$
$= (2x-3)(3x^2 - 1)$

105. $3x^6 - 6x^5 + 12x^4 + 27x^2$
$= 3x^2 \cdot x^4 + 3x^2(-2x^3) + 3x^2 \cdot 4x^2 + 3x^2 \cdot 9$
$= 3x^2(x^4 - 2x^3 + 4x^2 + 9)$

107. $12x^2 + 6xy - 10xy - 5y^2$
$= 6x(2x + y) - 5y(2x + y)$
$= (2x + y)(6x - 5y)$

109. $-3y + 21 = -3 \cdot y - 3(-7) = -3(y - 7)$

111. $5x(x + b) + 6y(x + b) = (x + b)(5x + 6y)$

113. $14 = 2 \cdot 7$
$24 = 2^3 \cdot 3$
The common factor is 2, so the GCF is 2.

5.2 Factoring $x^2 + bx + c$

Problems 5.2

1. $8 + 9x + x^2 = x^2 + 9x + 8$
$8 + 1 = 9$ and $8 \cdot 1 = 8$, so
$x^2 + 9x + 8 = (x+8)(x+1)$.

2. $x^2 - 8x + 7$
$-7 + (-1) = -8$ and $-7(-1) = 7$, so
$x^2 - 8x + 7 = (x-7)(x-1)$.

3. $x^2 - 2x - 8$
$-4 + 2 = -2$ and $-4 \cdot 2 = -8$, so
$x^2 - 2x - 8 = (x-4)(x+2)$.

4. a. $y^2 + 4y - 14$
There are no factors of -14 that sum to 4, so the trinomial is prime.

b. $y^2 + 3y - 40$
$-5 + 8 = 3$ and $-5 \cdot 8 = -40$, so
$y^2 + 3y - 40 = (y+8)(y-5)$.

5. $x^2 + 6ax + 8a^2$

$4a + 2a = 6a$ and $4a \cdot 2a = 8a^2$, so

$x^2 + 6ax + 8a^2 = (x + 4a)(x + 2a)$.

6. a. $y^5 - 4y^4 + 4y^3$

The GCF is y^3, so

$y^5 - 4y^4 + 4y^3 = y^3(y^2 - 4y + 4)$.

$-2 + (-2) = -4$ and $(-2)^2 = 4$, so

$y^3(y^2 - 4y + 4) = y^3(y - 2)^2$.

b. $b^2y^6 - b^2y^5 - 20b^2y^4$

The GCF is b^2y^4, so

$b^2y^6 - b^2y^5 - 20b^2y^4$

$= b^2y^4(y^2 - y - 20)$.

$-5 + 4 = -1$ and $-5 \cdot 4 = -20$, so

$b^2y^4(y^2 - y - 20) = b^2y^4(y + 4)(y - 5)$.

Exercises 5.2

1. $2 + 4 = 6$ and $2 \cdot 4 = 8$, so

$y^2 + 6y + 8 = (y + 2)(y + 4)$.

3. $2 + 5 = 7$ and $2 \cdot 5 = 10$, so

$x^2 + 7x + 10 = (x + 2)(x + 5)$.

5. $-2 + 5 = 3$ and $-2 \cdot 5 = -10$, so

$y^2 + 3y - 10 = (y + 5)(y - 2)$.

7. $-2 + 7 = 5$ and $-2 \cdot 7 = -14$, so

$x^2 + 5x - 14 = (x + 7)(x - 2)$.

9. $-7 + 1 = -6$ and $-7 \cdot 1 = -7$, so

$x^2 - 6x - 7 = (x + 1)(x - 7)$.

11. $-7 + 2 = -5$ and $-7 \cdot 2 = -14$, so

$y^2 - 5y - 14 = (y + 2)(y - 7)$.

13. $-2 + (-1) = -3$ and $-2(-1) = 2$, so

$y^2 - 3y + 2 = (y - 2)(y - 1)$.

15. $-4 + (-1) = -5$ and $-4(-1) = 4$, so

$x^2 - 5x + 4 = (x - 4)(x - 1)$.

17. There are no factors of 4 that sum to 3, so $x^2 + 3x + 4$ is prime and, therefore, not factorable.

19. There are no factors of -7 that sum to -7, so $-7 - 7x + x^2 = x^2 - 7x - 7$ is prime, and, therefore, not factorable.

21. $a + 2a = 3a$ and $a \cdot 2a = 2a^2$, so

$x^2 + 3ax + 2a^2 = (x + a)(x + 2a)$.

23. $3b + 3b = 6b$ and $(3b)^2 = 9b^2$, so

$z^2 + 6bz + 9b^2 = (z + 3b)(z + 3b)$.

25. $-3a + 4a = a$ and $-3a \cdot 4a = -12a^2$, so

$r^2 + ar - 12a^2 = (r + 4a)(r - 3a)$.

27. $-a + 10a = 9a$ and $-a \cdot 10a = -10a^2$, so

$x^2 + 9ax - 10a^2 = (x + 10a)(x - a)$.

29. $-b^2 - 2by + 3y^2 = -(b^2 + 2by - 3y^2)$

$-y + 3y = 2y$ and $-y \cdot 3y = -3y^2$, so

$-b^2 - 2by + 3y^2 = -(b - y)(b + 3y)$.

31. $-2a + a = -a$ and $-2a \cdot a = -2a^2$, so

$m^2 - am - 2a^2 = (m - 2a)(m + a)$.

33. The GCF is $2t$, so

$2t^3 + 10t^2 + 8t = 2t(t^2 + 5t + 4)$.

$1 + 4 = 5$ and $1 \cdot 4 = 4$, so

$2t(t^2 + 5t + 4) = 2t(t + 1)(t + 4)$.

35. The GCF is ax^2, so

$a^2x^3 + 3a^2x^2 + 2ax^2 = ax^2(ax + 3a + 2)$.

37. The GCF is b^3x^5, so
$b^3x^7 + b^3x^6 - 12b^3x^5 = b^3x^5(x^2 + x - 12)$.
$-3 + 4 = 1$ and $-3 \cdot 4 = -12$, so
$b^3x^5(x^2 + x - 12) = b^3x^5(x-3)(x+4)$.

39. The GCF is $2c^5z^4$, so
$2c^5z^6 + 4c^5z^5 - 30c^5z^4$
$= 2c^5z^4(z^2 + 2z - 15)$.
$-3 + 5 = 2$ and $-3 \cdot 5 = -15$, so
$2c^5z^4(z^2 + 2z - 15) = 2c^5z^4(z-3)(z+5)$.

41. a. $V_0 = 5$ and $h = 10$, so
$-5t^2 + V_0t + h = -5t^2 + 5t + 10$.

b. $-5t^2 + 5t + 10 = -5(t^2 - t - 2)$
$-2 + 1 = -1$ and $-2 \cdot 1 = -2$, so
$-5(t^2 - t - 2) = -5(t-2)(t+1)$.

c. $-5(t-2)(t+1) = 0$ when $t - 2 = 0$ or $t + 1 = 0$. The product is 0 when $t = 2$ or $t = -1$.

d. It takes $t = 2$ seconds for the object to return to the ground.

43. a. $-16t^2 + 32t + 240 = -16(t^2 - 2t - 15)$
$-5 + 3 = -2$ and $-5 \cdot 3 = -15$, so
$-16(t^2 - 2t - 15) = -16(t-5)(t+3)$.

b. $-16(t-5)(t+3) = 0$ when $t - 5 = 0$ or $t + 3 = 0$. The product is 0 when $t = 5$ or $t = -3$.

c. It takes $t = 5$ seconds for the rock to return to the ground.

45. $(x+4)^2 = (x+4)(x+4)$
$= x^2 + 4x + 4x + 16$
$= x^2 + 8x + 16$

47. $(x-3)^2 = (x-3)(x-3)$
$= x^2 - 3x - 3x + 9$
$= x^2 - 6x + 9$

49. $(3x+2y)^2 = (3x+2y)(3x+2y)$
$= 9x^2 + 6xy + 6xy + 4y^2$
$= 9x^2 + 12xy + 4y^2$

51. $(5x-2y)^2 = (5x-2y)(5x-2y)$
$= 25x^2 - 10xy - 10xy + 4y^2$
$= 25x^2 - 20xy + 4y^2$

53. $(3x+4y)(3x-4y)$
$= 9x^2 - 12xy + 12xy - 16y^2$
$= 9x^2 - 16y^2$

55. Sample answer: Factor out the GCF of all three terms.

57. Sample answer: Assuming the polynomial is $x^2 - 5x + 6$, both are correct.
$(2-x)(3-x) = [(-1)(x-2)][(-1)(x-3)]$
$= (-1)^2(x-2)(x-3)$
$= (x-2)(x-3)$

59. Sample answer: *A* and *B* are positive.

61. Sample answer: *A* and *B* have opposite signs.

63. $3y^3 - 6y^2 - y + 2 = 3y^2(y-2) - 1(y-2)$
$= (y-2)(3y^2 - 1)$

65. $2y^3 - 6y^2 - y + 3 = 2y^2(y-3) - 1(y-3)$
$= (y-3)(2y^2 - 1)$

67. $4y + 3y = 7y$ and $4y \cdot 3y = 12y^2$, so
$x^2 + 7xy + 12y^2 = (x+4y)(x+3y)$.

69. $-2y + (-5y) = -7y$ and $-2y(-5y) = 10y^2$, so
$x^2 - 7xy + 10y^2 = (x-2y)(x-5y)$.

71. There are no factors of -25 that sum to -10, so $z^2 - 10z - 25$ is prime and, therefore, not factorable.

73. $2y^3 + 6y^5 + 10y^7 = 2y^3(1 + 3y^2 + 5y^4)$

5.3 Factoring $ax^2 + bx + c, a \neq 1$

Problems 5.3

1. a. $4y^2 + 3y + 2$

$ac = 4 \cdot 2 = 8$

The factors of 8 are 4 and 2, and 8 and 1. Since neither pair sums to 3, the polynomial is not factorable.

b. $3y^2 + 5y + 2$

$ac = 3 \cdot 2 = 6$

The factors of 6 are 1 and 6, and 2 and 3. Since $b = 2 + 3 = 5$, the polynomial is factorable.

2. a. $9x^2 - 2 - 3x = 9x^2 - 3x - 2$

$ac = 9(-2) = -18$, $-18 = -6 \cdot 3$, and $b = -6 + 3 = -3$. So,

$$\begin{aligned} 9x^2 - 3x - 2 &= (9x^2 - 6x) + (3x - 2) \\ &= 3x(3x - 2) + 1(3x - 2) \\ &= (3x - 2)(3x + 1) \end{aligned}$$

b. $3 - 4x + 2x^2 = 2x^2 - 4x + 3$

$ac = 2 \cdot 3 = 6$

It's impossible to find two numbers with product 6 and sum -4. The polynomial is not factorable (prime).

3. $4x^2 - 13x + 3$

$ac = 4 \cdot 3 = 12$, $12 = -12(-1)$, and $b = -12 + (-1) = -13$.

$$\begin{aligned} 4x^2 - 13x + 3 &= 4x^2 - 12x - x + 3 \\ &= 4x(x - 3) - 1(x - 3) \\ &= (x - 3)(4x - 1) \end{aligned}$$

4. $4x^2 - 4xy - 3y^2$

$ac = 4(-3y^2) = -12y^2, -12y^2 = -6y \cdot 2y,$

and $b = -6y + 2y = -4y$.

$$\begin{aligned} 4x^2 - 4xy - 3y^2 &= 4x^2 - 6xy + 2xy - 3y^2 \\ &= 2x(2x - 3y) + y(2x - 3y) \\ &= (2x - 3y)(2x + y) \end{aligned}$$

5. $$\begin{aligned} 3x^2 + 5x + 2 &= 3x^2 + 3x + 2x + 2 \\ &= (3x + 2)(x + 1) \end{aligned}$$

6. $$\begin{aligned} 18x^2 + 3x - 6 &= 3(6x^2 + x - 2) \\ &= 3(6x^2 + 4x - 3x - 2) \\ &= 3(2x - 1)(3x + 2) \end{aligned}$$

7. $$\begin{aligned} 6x^2 - 5xy - 6y^2 &= 6x^2 - 9xy + 4xy - 6y^2 \\ &= (3x + 2y)(2x - 3y) \end{aligned}$$

Exercises 5.3

1. $2 \cdot 3 = 6$, $6 = 2 \cdot 3$, and $2 + 3 = 5$, so

$$\begin{aligned} 2x^2 + 5x + 3 &= 2x^2 + 2x + 3x + 3 \\ &= 2x(x + 1) + 3(x + 1) \\ &= (x + 1)(2x + 3). \end{aligned}$$

3. $6 \cdot 3 = 18$, $18 = 9 \cdot 2$, and $9 + 2 = 11$, so

$$\begin{aligned} 6x^2 + 11x + 3 &= 6x^2 + 2x + 9x + 3 \\ &= 2x(3x + 1) + 3(3x + 1) \\ &= (3x + 1)(2x + 3). \end{aligned}$$

5. $6 \cdot 4 = 24$, $24 = 8 \cdot 3$, and $8 + 3 = 11$, so

$$\begin{aligned} 6x^2 + 11x + 4 &= 6x^2 + 8x + 3x + 4 \\ &= 2x(3x + 4) + 1(3x + 4) \\ &= (3x + 4)(2x + 1). \end{aligned}$$

7. $2(-2) = -4$, $-4 = 4(-1)$, and $4 + (-1) = 3$, so

$$\begin{aligned} 2x^2 + 3x - 2 &= 2x^2 + 4x - x - 2 \\ &= 2x(x + 2) - 1(x + 2) \\ &= (2x - 1)(x + 2). \end{aligned}$$

9. $3(-12) = -36$, $-36 = -2 \cdot 18$, and $-2 + 18 = 16$, so

$$\begin{aligned} 3x^2 + 16x - 12 &= 3x^2 - 2x + 18x - 12 \\ &= x(3x - 2) + 6(3x - 2) \\ &= (3x - 2)(x + 6). \end{aligned}$$

11. $4 \cdot 6 = 24$, $24 = -8(-3)$, and $-8 + (-3) = -11$, so

$$\begin{aligned} 4y^2 - 11y + 6 &= 4y^2 - 8y - 3y + 6 \\ &= 4y(y - 2) - 3(y - 2) \\ &= (y - 2)(4y - 3). \end{aligned}$$

13. $4y^2 - 8y + 6 = 2(2y^2 - 4y + 3)$
$2 \cdot 3 = 6$
It's impossible to find two numbers with product 6 and sum –4, so the polynomial cannot be factored further.

15. $6y^2 - 10y - 4 = 2(3y^2 - 5y - 2)$
$3(-2) = -6$, $-6 = -6 \cdot 1$, and $-6 + 1 = -5$, so
$$\begin{aligned}2(3y^2 - 5y - 2) &= 2(3y^2 - 6y + y - 2)\\ &= 2[3y(y-2) + 1(y-2)]\\ &= 2(y-2)(3y+1).\end{aligned}$$

17. $12(-6) = -72$, $-72 = -9 \cdot 8$, and $-9 + 8 = -1$, so
$$\begin{aligned}12y^2 - y - 6 &= 12y^2 + 8y - 9y - 6\\ &= 4y(3y+2) - 3(3y+2)\\ &= (3y+2)(4y-3).\end{aligned}$$

19. $18y^2 - 21y - 9 = 3(6y^2 - 7y - 3)$
$6(-3) = -18$, $-18 = -9 \cdot 2$, and $-9 + 2 = -7$, so
$$\begin{aligned}3(6y^2 - 7y - 3) &= 3(6y^2 + 2y - 9y - 3)\\ &= 3[2y(3y+1) - 3(3y+1)]\\ &= 3(3y+1)(2y-3).\end{aligned}$$

21. $3x^2 + 2 + 7x = 3x^2 + 7x + 2$
$3 \cdot 2 = 6$, $6 = 6 \cdot 1$, and $6 + 1 = 7$, so
$$\begin{aligned}3x^2 + 7x + 2 &= 3x^2 + 6x + x + 2\\ &= 3x(x+2) + 1(x+2)\\ &= (x+2)(3x+1).\end{aligned}$$

23. $5x^2 + 2 + 11x = 5x^2 + 11x + 2$
$5 \cdot 2 = 10$, $10 = 10 \cdot 1$, and $10 + 1 = 11$, so
$$\begin{aligned}5x^2 + 11x + 2 &= 5x^2 + 10x + x + 2\\ &= 5x(x+2) + 1(x+2)\\ &= (x+2)(5x+1).\end{aligned}$$

25. $6x^2 - 5 + 15x = 6x^2 + 15x - 5$
$6(-5) = -30$
It's impossible to find two numbers with product –30 and sum 15. The polynomial is not factorable (prime).

27. $3x^2 - 2 - 5x = 3x^2 - 5x - 2$
$3(-2) = -6$, $-6 = -6 \cdot 1$, and $-6 + 1 = -5$, so
$$\begin{aligned}3x^2 - 5x - 2 &= 3x^2 - 6x + x - 2\\ &= 3x(x-2) + 1(x-2)\\ &= (x-2)(3x+1).\end{aligned}$$

29. $15x^2 - 2 + x = 15x^2 + x - 2$
$15(-2) = -30$, $-30 = -5 \cdot 6$, and $-5 + 6 = 1$,
so
$$\begin{aligned}15x^2 + x - 2 &= 15x^2 - 5x + 6x - 2\\ &= 5x(3x-1) + 2(3x-1)\\ &= (3x-1)(5x+2).\end{aligned}$$

31. $8x^2 + 20xy + 8y^2 = 4(2x^2 + 5xy + 2y^2)$
$2 \cdot 2y^2 = 4y^2$, $4y^2 = 4y \cdot y$, and $4y + y = 5y$,
so
$$\begin{aligned}&4(2x^2 + 5xy + 2y^2)\\ &= 4(2x^2 + 4xy + xy + 2y^2)\\ &= 4[2x(x+2y) + y(x+2y)]\\ &= 4(x+2y)(2x+y).\end{aligned}$$

33. $6(-3y^2) = -18y^2$, $-18y^2 = -2y \cdot 9y$, and $-2y + 9y = 7y$, so
$$\begin{aligned}6x^2 + 7xy - 3y^2 &= 6x^2 + 9xy - 2xy - 3y^2\\ &= 3x(2x+3y) - y(2x+3y)\\ &= (2x+3y)(3x-y).\end{aligned}$$

35. $7 \cdot 3y^2 = 21y^2$, $21y^2 = -3y(-7y)$, and $-3y + (-7y) = -10y$, so
$$\begin{aligned}7x^2 - 10xy + 3y^2 &= 7x^2 - 7xy + 3xy + 3y^2\\ &= 7x(x-y) - 3y(x-y)\\ &= (x-y)(7x-3y).\end{aligned}$$

37. $15(-2y^2) = -30y^2$, $-30y^2 = -6y \cdot 5y$, and $-6y + 5y = -y$, so
$$\begin{aligned}15x^2 - xy - 2y^2 &= 15x^2 - 6xy + 5xy - 2y^2\\ &= 3x(5x-2y) + y(5x-2y)\\ &= (5x-2y)(3x+y).\end{aligned}$$

39. $15x^2 - 2xy - 2y^2$

$15(-2y^2) = -30y^2$

It's impossible to find two numbers with product $-30y^2$ and sum $-2y$. The polynomial is not factorable (prime).

41. $12(-5) = -60$, $-60 = -3 \cdot 20$, and $-3 + 20 = 17$, so

$$\begin{aligned} 12r^2 + 17r - 5 &= 12r^2 + 20r - 3r - 5 \\ &= 4r(3r+5) - 1(3r+5) \\ &= (3r+5)(4r-1). \end{aligned}$$

43. $22(-6) = -132$, $-132 = -33 \cdot 4$, and $-33 + 4 = -29$, so

$$\begin{aligned} 22t^2 - 29t - 6 &= 22t^2 - 33t + 4t - 6 \\ &= 11t(2t-3) + 2(2t-3) \\ &= (2t-3)(11t+2). \end{aligned}$$

45. $18x^2 - 21x + 6 = 3(6x^2 - 7x + 2)$

$6 \cdot 2 = 12$, $12 = -4(-3)$, and $-4 + (-3) = -7$, so

$$\begin{aligned} 3(6x^2 - 7x + 2) &= 3(6x^2 - 4x - 3x + 2) \\ &= 3[2x(3x-2) - 1(3x-2)] \\ &= 3(3x-2)(2x-1). \end{aligned}$$

47. $6ab^2 + 5ab + a = a(6b^2 + 5b + 1)$

$6 \cdot 1 = 6$, $6 = 2 \cdot 3$, and $2 + 3 = 5$, so

$$\begin{aligned} a(6b^2 + 5b + 1) &= a(6b^2 + 2b + 3b + 1) \\ &= a[2b(3b+1) + 1(3b+1)] \\ &= a(3b+1)(2b+1). \end{aligned}$$

49. $6x^5y + 25x^4y^2 + 4x^3y^3$

$= x^3y(6x^2 + 25xy + 4y^2)$

$6 \cdot 4y^2 = 24y^2$, $24y^2 = 24y \cdot y$, and $24y + y = 25y$, so

$$\begin{aligned} &x^3y(6x^2 + 25xy + 4y^2) \\ &= x^3y(6x^2 + 24xy + xy + 4y^2) \\ &= x^3y[6x(x+4y) + y(x+4y)] \\ &= x^3y(x+4y)(6x+y). \end{aligned}$$

51. $-6x^2 - 7x - 2 = -(6x^2 + 7x + 2)$
$= -(3x+2)(2x+1)$

53. $-9x^2 - 3x + 2 = -(9x^2 + 3x - 2)$
$= -(3x+2)(3x-1)$

55. $-8m^2 + 10mn + 3n^2 = -(8m^2 - 10mn - 3n^2)$
$= -(4m+n)(2m-3n)$

57. $-8x^2 + 9xy - y^2 = -(8x^2 - 9xy + y^2)$
$= -(8x-y)(x-y)$

59. $-x^3 - 5x^2 - 6x = -x(x^2 + 5x + 6)$
$= -x(x+3)(x+2)$

61. $2g^2 + g - 36 = (2g+9)(g-4)$

63. $2R^2 - 3R + 1 = (2R-1)(R-1)$

65. **a.** $B(t) \cdot P(t) = 0.004t^2 + 0.2544t + 2.72$
$= (0.02t+1)(0.2t+2.72)$

b. $B(0) \cdot P(0) = (0+1)(0+2.72) = \$2.72$

c. $B(5) \cdot P(5) = [0.02(5)+1][0.2(5)+2.72]$
$= (1.1)(3.72) \approx \$4.09$

67. $(x+8)^2 = (x+8)(x+8)$
$= x^2 + 8x + 8x + 64$
$= x^2 + 16x + 64$

69. $(3x-2)^2 = (3x-2)(3x-2)$
$= 9x^2 - 6x - 6x + 4$
$= 9x^2 - 12x + 4$

71. $(2x+3y)^2 = (2x+3y)(2x+3y)$
$= 4x^2 + 6xy + 6xy + 9y^2$
$= 4x^2 + 12xy + 9y^2$

73. $(3x+5y)(3x-5y)$
$= 9x^2 - 15xy + 15xy - 25y^2$
$= 9x^2 - 25y^2$

75. $(x^2+4)(x^2-4) = x^4 - 4x^2 + 4x^2 - 16$
$= x^4 - 16$

77. $2L^2 - 9L + 9 = (2L - 3)(L - 3)$

79. $5t^2 - 12t + 7 = (5t - 7)(t - 1)$

81. Sample answer:
$$\begin{aligned}(2 - x)(1 - 3x) &= [(-1)(x - 2)][(-1)(3x - 1)] \\ &= (-1)^2(x - 2)(3x - 1) \\ &= (x - 2)(3x - 1)\end{aligned}$$
and $(x - 2)(3x - 1) = (3x - 1)(x - 2)$ by the commutative property of multiplication. Both students are correct.

83. $3x^2 - 4 - 4x = 3x^2 - 4x - 4$
$3(-4) = -12$, $-12 = -6 \cdot 2$, and $-6 + 2 = -4$, so
$$\begin{aligned}3x^2 - 4x - 4 &= 3x^2 - 6x + 2x - 4 \\ &= 3x(x - 2) + 2(x - 2) \\ &= (x - 2)(3x + 2).\end{aligned}$$

85. $2(-6y^2) = -12y^2$, $-12y^2 = -4y \cdot 3y$, and $-4y + 3y = -y$, so
$$\begin{aligned}2x^2 - xy - 6y^2 &= 2x^2 - 4xy + 3xy - 6y^2 \\ &= 2x(x - 2y) + 3y(x - 2y) \\ &= (x - 2y)(2x + 3y).\end{aligned}$$

87. $16x^2 + 4x - 2 = 2(8x^2 + 2x - 1)$
$8(-1) = -8$, $-8 = -2 \cdot 4$, and $-2 + 4 = 2$, so
$$\begin{aligned}2(8x^2 + 2x - 1) &= 2(8x^2 + 4x - 2x - 1) \\ &= 2[4x(2x + 1) - 1(2x + 1)] \\ &= 2(2x + 1)(4x - 1).\end{aligned}$$

89. $3x^4 + 5x^3 - 3x^2 = x^2(3x^2 + 5x - 3)$
$3(-3) = -9$
It's impossible to find two numbers with product -9 and sum 5. The polynomial cannot be factored further.

5.4 Factoring Squares of Binomials

Problems 5.4

1. a. $y^2 + 9y + 9$

1. y^2 and $9 = 3^2$ are perfect squares.
2. There are no minus signs before y^2 or 9.
3. $2 \cdot 3 \cdot y = 6y \neq 9y$

The expression is not the square of a binomial.

b. $y^2 + 4y + 4$

1. y^2 and $4 = 2^2$ are perfect squares.
2. There are no minus signs before y^2 or 4.
3. $2 \cdot y \cdot 2 = 4y$

The expression is the square of a binomial.

c. $y^2 + 8y - 16$

1. y^2 and $16 = 4^2$ are perfect squares.
2. There is a minus sign before 16.

The expression is not the square of a binomial.

d. $9x^2 - 12xy + 4y^2$

1. $9x^2 = (3x)^2$ and $4y^2 = (2y)^2$ are perfect squares.
2. There are no minus signs before $9x^2$ or $4y^2$.
3. $-2 \cdot 3x \cdot 2y = -12xy$

The expression is the square of a binomial.

2. a.
$$\begin{aligned}y^2 + 6y + 9 &= y^2 + 2 \cdot (3 \cdot y) + 3^2 \\ &= (y + 3)^2\end{aligned}$$

b. $4x^2 + 28xy + 49y^2$
$$\begin{aligned}&= (2x)^2 + 2 \cdot (7y \cdot 2x) + (7y)^2 \\ &= (2x + 7y)^2\end{aligned}$$

c. $9y^2+12y+4=(3y)^2+2\cdot(2\cdot 3y)+2^2$
$=(3y+2)^2$

3. a. $y^2-4y+4=y^2-2\cdot(2\cdot y)+2^2$
$=(y-2)^2$

b. $9y^2-12y+4=(3y)^2-2\cdot(3y\cdot 2)+2^2$
$=(3y-2)^2$

c. $9x^2-30xy+25y^2$
$=(3x)^2-2\cdot(3x\cdot 5y)+(5y)^2$
$=(3x-5y)^2$

4. a. $y^2-1=(y)^2-(1)^2=(y+1)(y-1)$

b. $9y^2-25=(3y)^2-(5)^2$
$=(3y+5)(3y-5)$

c. $9y^2-25x^2=(3y)^2-(5x)^2$
$=(3y+5x)(3y-5x)$

d. $y^4-81=(y^2)^2-(9)^2$
$=(y^2+9)(y^2-9)$
$=(y^2+9)(y+3)(y-3)$

e. $\frac{1}{9}y^2-\frac{1}{4}=\left(\frac{1}{3}y\right)^2-\left(\frac{1}{2}\right)^2$
$=\left(\frac{1}{3}y+\frac{1}{2}\right)\left(\frac{1}{3}y-\frac{1}{2}\right)$

f. $5y^3-45y=5y(y^2-9)$
$=5y[(y)^2-(3)^2]$
$=5y(y+3)(y-3)$

Exercises 5.4

1. $x^2+14x+49$
1. x^2 and $49=7^2$ are perfect squares.
2. There are no minus signs before x^2 or 49.
3. $2\cdot 7\cdot x=14x$

The expression is a perfect square trinomial.

3. $25x^2+10x-1$
1. $25x^2=(5x)^2$ and $1=1^2$ are perfect squares.
2. There is a minus sign before 1.

The expression is not a perfect square trinomial.

5. $25x^2+10x+1$
1. $25x^2=(5x)^2$ and $1=1^2$ are perfect squares.
2. There are no minus signs before $25x^2$ or 1.
3. $2\cdot 1\cdot 5x=10x$

The expression is a perfect square trinomial.

7. y^2-4y-4
1. y^2 and $4=2^2$ are perfect squares.
2. There is a minus sign before 4.

The expression is not a perfect square trinomial.

9. $16y^2-40yz+25z^2$
1. $16y^2=(4y)^2$ and $25z^2=(5z)^2$ are perfect squares.
2. There are no minus signs before $16y^2$ or $25z^2$.
3. $-2\cdot 5z\cdot 4y=-40yz$

The expression is a perfect square trinomial.

11. $x^2+2x+1=x^2+2\cdot(1\cdot x)+1^2=(x+1)^2$

13. $3x^2+30x+75=3(x^2+10x+25)$
$=3[x^2+2\cdot(5\cdot x)+5^2]$
$=3(x+5)^2$

15. $9x^2+6x+1=(3x)^2+2\cdot(1\cdot 3x)+1^2$
$=(3x+1)^2$

17. $9x^2+12x+4=(3x)^2+2\cdot(2\cdot 3x)+2^2$
$=(3x+2)^2$

19. $16x^2+40xy+25y^2$
$=(4x)^2+2\cdot(5y\cdot 4x)+(5y)^2$
$=(4x+5y)^2$

21. $25x^2+20xy+4y^2$
$=(5x)^2+2\cdot(2y\cdot 5x)+(2y)^2$
$=(5x+2y)^2$

23. $y^2-2y+1=y^2-2\cdot(1\cdot y)+1^2=(y-1)^2$

25. $3y^2-24y+48=3(y^2-8y+16)$
$=3[y^2-2\cdot(4\cdot y)+4^2]$
$=3(y-4)^2$

27. $9x^2-6x+1=(3x)^2-2\cdot(1\cdot 3x)+1^2$
$=(3x-1)^2$

29. $16x^2-56x+49=(4x)^2-2\cdot(7\cdot 4x)+7^2$
$=(4x-7)^2$

31. $9x^2-12xy+4y^2$
$=(3x)^2-2\cdot(2y\cdot 3x)+(2y)^2$
$=(3x-2y)^2$

33. $25x^2-10xy+y^2=(5x)^2-2\cdot(y\cdot 5x)+y^2$
$=(5x-y)^2$

35. $x^2-49=(x)^2-(7)^2=(x+7)(x-7)$

37. $9x^2-49=(3x)^2-(7)^2=(3x+7)(3x-7)$

39. $25x^2-81y^2=(5x)^2-(9y)^2$
$=(5x+9y)(5x-9y)$

41. $x^4-1=(x^2)^2-(1)^2$
$=(x^2+1)(x^2-1)$
$=(x^2+1)(x+1)(x-1)$

43. $16x^4-1=(4x^2)^2-(1)^2$
$=(4x^2+1)(4x^2-1)$
$=(4x^2+1)(2x+1)(2x-1)$

45. $\frac{1}{9}x^2-\frac{1}{16}=\left(\frac{1}{3}x\right)^2-\left(\frac{1}{4}\right)^2$
$=\left(\frac{1}{3}x+\frac{1}{4}\right)\left(\frac{1}{3}x-\frac{1}{4}\right)$

47. $\frac{1}{4}z^2-1=\left(\frac{1}{2}z\right)^2-(1)^2=\left(\frac{1}{2}z+1\right)\left(\frac{1}{2}z-1\right)$

49. $1-\frac{1}{4}s^2=(1)^2-\left(\frac{1}{2}s\right)^2=\left(1+\frac{1}{2}s\right)\left(1-\frac{1}{2}s\right)$

51. $\frac{1}{4}-\frac{1}{9}y^2=\left(\frac{1}{2}\right)^2-\left(\frac{1}{3}y\right)^2$
$=\left(\frac{1}{2}+\frac{1}{3}y\right)\left(\frac{1}{2}-\frac{1}{3}y\right)$

53. $\frac{1}{9}+\frac{1}{4}x^2$ cannot be factored because $\frac{1}{4}x^2$ is preceded by a plus sign.

55. $3x^3-12x=3x(x^2-4)$
$=3x[(x)^2-(2)^2]$
$=3x(x+2)(x-2)$

57. $5t^3-20t=5t(t^2-4)$
$=5t[(t)^2-(2)^2]$
$=5t(t+2)(t-2)$

59. $5t-20t^3=5t(1-4t^2)$
$=5t[(1)^2-(2t)^2]$
$=5t(1+2t)(1-2t)$

61. $49x^2+28x+4=(7x)^2+2\cdot(2\cdot 7x)+2^2$
$=(7x+2)^2$

63. $x^2-100=(x)^2-(10)^2=(x+10)(x-10)$

65. $x^2+20x+100 = x^2+2\cdot(10\cdot x)+10^2$
$= (x+10)^2$

67. $9-16m^2 = (3)^2-(4m)^2 = (3+4m)(3-4m)$

69. $9x^2-30xy+25y^2$
$= (3x)^2-2\cdot(5y\cdot 3x)+(5y)^2$
$= (3x-5y)^2$

71. $x^4-16 = (x^2)^2-(4)^2$
$= (x^2+4)(x^2-4)$
$= (x^2+4)(x+2)(x-2)$

73. $3x^3-75x = 3x(x^2-25)$
$= 3x[(x)^2-(5)^2]$
$= 3x(x+5)(x-5)$

75. $(R+r)(R-r) = R^2-Rr+Rr-r^2$
$= R^2-r^2$

77. $6x^2-18x-24 = 6(x^2-3x-4)$
$1(-4) = -4$ and $-4 + 1 = -3$, so
$6(x^2-3x-4) = 6(x+1)(x-4)$.

79. $2x^2-18 = 2(x^2-9)$
$= 2[(x)^2-(3)^2]$
$= 2(x+3)(x-3)$

81. $D(x) = x^2-14x+49$
$= x^2-2\cdot(7\cdot x)+7^2$
$= (x-7)^2$

83. $C(x) = x^2+12x+36$
$= x^2+2\cdot(6\cdot x)+6^2$
$= (x+6)^2$

85. Sample answer: a^2+b^2 cannot be factored, but the sum of two squares can sometimes be factored. For example,
$(5a)^2+(25b)^2 = 25a^2+625b^2$
$= 25(a^2+25b^2)$.

87. A perfect square trinomial is formed by multiplying a binomial by itself, so $(x + 2)$ multiplied by $(x + 2)$ gives the perfect square trinomial x^2+4x+4.

89. $x^2-1 = (x)^2-(1)^2 = (x+1)(x-1)$

91. $9x^2-25y^2 = (3x)^2-(5y)^2$
$= (3x+5y)(3x-5y)$

93. $9x^2-24xy+16y^2$
$= (3x)^2-2\cdot(4y\cdot 3x)+(4y)^2$
$= (3x-4y)^2$

95. $16x^2+24xy+9y^2$
$= (4x)^2+2\cdot(3y\cdot 4x)+(3y)^2$
$= (4x+3y)^2$

97. $9x^2+30x+25 = (3x)^2+2\cdot(5\cdot 3x)+5^2$
$= (3x+5)^2$

99. $9x^2+4$ cannot be factored because 4 is preceded by a plus sign.

101. $\frac{1}{81}-\frac{1}{4}x^2 = \left(\frac{1}{9}\right)^2-\left(\frac{1}{2}x\right)^2$
$= \left(\frac{1}{9}+\frac{1}{2}x\right)\left(\frac{1}{9}-\frac{1}{2}x\right)$

103. $18x^3-50xy^2 = 2x(9x^2-25y^2)$
$= 2x[(3x)^2-(5y)^2]$
$= 2x(3x+5y)(3x-5y)$

105. x^2+6x+9

1. x^2 and $9=3^2$ are perfect squares.
2. There are no minus signs before x^2 or 9.
3. $2\cdot 3\cdot x = 6x$

The expression is a perfect square trinomial.
$x^2+6x+9 = x^2+2\cdot(3\cdot x)+3^2 = (x+3)^2$

107. x^2+6x-9

1. x^2 and $9=3^2$ are perfect squares.
2. There is a minus sign before 9.

The expression is not a perfect square trinomial.

109. $4x^2-20xy+25y^2$

1. $4x^2=(2x)^2$ and $25y^2=(5y)^2$ are perfect squares.
2. There are no minus signs before $4x^2$ or $25y^2$.
3. $-2\cdot 5y\cdot 2x=-20xy$

The expression is a perfect square trinomial.

$$4x^2-20xy+25y^2$$
$$=(2x)^2-2\cdot(5y\cdot 2x)+(5y)^2$$
$$=(2x-5y)^2$$

5.5 A General Factoring Strategy

Problems 5.5

1. a. $y^3+8=(y)^3+(2)^3$
$$=(y+2)(y^2-2y+2^2)$$
$$=(y+2)(y^2-2y+4)$$

b. $27y^3+8=(3y)^3+(2)^3$
$$=(3y+2)[(3y)^2-2(3y)+2^2]$$
$$=(3y+2)(9y^2-6y+4)$$

c. $y^3-27z^3=(y)^3-(3z)^3$
$$=(y-3z)[y^2+y(3z)+(3z)^2]$$
$$=(y-3z)(y^2+3yz+9z^2)$$

d. $8a^3-27b^3$
$$=(2a)^3-(3b)^3$$
$$=(2a-3b)[(2a)^2+(2a)(3b)+(3b)^2]$$
$$=(2a-3b)(4a^2+6ab+9b^2)$$

2. a. $7a^2-14a-21=7(a^2-2a-3)$

$-3=-3\cdot 1$ and $-3+1=-2$, so

$7(a^2-2a-3)=7(a+1)(a-3)$.

b. $5a^4+10a^3+25a^2=5a^2(a^2+2a+5)$

3. $3a^3+6a^2+a+2=3a^2(a+2)+1(a+2)$
$$=(a+2)(3a^2+1)$$

4. $kt_1^2-2kt_1t_2+kt_2^2=k(t_1^2-2t_1t_2+t_2^2)$
$$=k(t_1-t_2)^2$$

5. $m^4-n^4=(m^2)^2-(n^2)^2$
$$=(m^2+n^2)(m^2-n^2)$$
$$=(m^2+n^2)(m+n)(m-n)$$

6. a. $27a^5-a^2b^3$
$$=a^2(27a^3-b^3)$$
$$=a^2[(3a)^3-(b)^3]$$
$$=a^2(3a-b)[(3a)^2+(3a)(b)+(b)^2]$$
$$=a^2(3a-b)(9a^2+3ab+b^2)$$

b. $64a^5+a^3b^2=a^3(64a^2+b^2)$

7. a. $-a^2-6a-9=-1\cdot(a^2+6a+9)$
$$=-1\cdot(a+3)^2$$
$$=-(a+3)^2$$

b. $-9a^2+12ab-4b^2$
$$=-1\cdot(9a^2-12ab+4b^2)$$
$$=-1\cdot(3a-2b)^2$$
$$=-(3a-2b)^2$$

c. $-4a^2-12ab+9b^2$
$$=-1\cdot(4a^2+12ab-9b^2)$$
$$=-(4a^2+12ab-9b^2)$$

d. $-4a^4+9a^2=-a^2\cdot(4a^2-9)$
$=-a^2\cdot(2a+3)(2a-3)$
$=-a^2(2a+3)(2a-3)$

Exercises 5.5

1. $x^3+8=(x)^3+(2)^3$
$=(x+2)[x^2-x(2)+2^2]$
$=(x+2)(x^2-2x+4)$

3. $8m^3-27=(2m)^3-(3)^3$
$=(2m-3)[(2m)^2+(2m)(3)+(3)^2]$
$=(2m-3)(4m^2+6m+9)$

5. $27m^3-8n^3$
$=(3m)^3-(2n)^3$
$=(3m-2n)[(3m)^2+(3m)(2n)+(2n)^2]$
$=(3m-2n)(9m^2+6mn+4n^2)$

7. $64s^3-s^6=s^3(64-s^3)$
$=s^3[(4)^3-(s)^3]$
$=s^3(4-s)[4^2+4(s)+s^2]$
$=s^3(4-s)(16+4s+s^2)$

9. $27x^4+8x^7=x^4(27+8x^3)$
$=x^4[(3)^3+(2x)^3]$
$=x^4(3+2x)[3^2-3(2x)+(2x)^2]$
$=x^4(3+2x)(9-6x+4x^2)$

11. $3x^2-3x-18=3(x^2-x-6)$
$=3(x-3)(x+2)$

13. $5x^2+11x+2=(5x+1)(x+2)$

15. $3x^3+6x^2+21x=3x(x^2+2x+7)$

17. $2x^4-4x^3-10x^2=2x^2(x^2-2x-5)$

19. $4x^4+12x^3+18x^2=2x^2(2x^2+6x+9)$

21. $3x^3+6x^2+x+2=3x^2(x+2)+1(x+2)$
$=(x+2)(3x^2+1)$

23. $3x^3+3x^2+2x+2=3x^2(x+1)+2(x+1)$
$=(x+1)(3x^2+2)$

25. $2x^3+2x^2-x-1=2x^2(x+1)-1(x+1)$
$=(x+1)(2x^2-1)$

27. $3x^2+24x+48=3(x^2+8x+16)=3(x+4)^2$

29. $kx^2+4kx+4k=k(x^2+4x+4)=k(x+2)^2$

31. $4x^2-24x+36=4(x^2-6x+9)=4(x-3)^2$

33. $kx^2-12kx+36k=k(x^2-12x+36)$
$=k(x-6)^2$

35. $3x^3+12x^2+12x=3x(x^2+4x+4)$
$=3x(x+2)^2$

37. $18x^3+12x^2+2x=2x(9x^2+6x+1)$
$=2x(3x+1)^2$

39. $12x^4-36x^3+27x^2=3x^2(4x^2-12x+9)$
$=3x^2(2x-3)^2$

41. $x^4-1=(x^2)^2-1^2$
$=(x^2+1)(x^2-1)$
$=(x^2+1)(x+1)(x-1)$

43. $x^4-y^4=(x^2)^2-(y^2)^2$
$=(x^2+y^2)(x^2-y^2)$
$=(x^2+y^2)(x+y)(x-y)$

45. $x^4-16y^4=(x^2)^2-(4y^2)^2$
$=(x^2+4y^2)(x^2-4y^2)$
$=(x^2+4y^2)(x+2y)(x-2y)$

47. $-x^2-6x-9=-1\cdot(x^2+6x+9)$
$=-1\cdot(x+3)^2$
$=-(x+3)^2$

49. $-x^2-4x-4=-1\cdot(x^2+4x+4)$
$=-1\cdot(x+2)^2$
$=-(x+2)^2$

51. $-4x^2-4xy-y^2=-1\cdot(4x^2+4xy+y^2)$
$=-1\cdot(2x+y)^2$
$=-(2x+y)^2$

53. $-9x^2-12xy-4y^2=-1\cdot(9x^2+12xy+4y^2)$
$=-1\cdot(3x+2y)^2$
$=-(3x+2y)^2$

55. $-4x^2+12xy-9y^2=-1\cdot(4x^2-12xy+9y^2)$
$=-1\cdot(2x-3y)^2$
$=-(2x-3y)^2$

57. $-18x^3-24x^2y-8xy^2$
$=-2x\cdot(9x^2+12xy+4y^2)$
$=-2x(3x+2y)^2$

59. $-18x^3-60x^2y-50xy^2$
$=-2x\cdot(9x^2+30xy+25y^2)$
$=-2x(3x+5y)^2$

61. $-x^3+x=-x\cdot(x^2-1)=-x(x+1)(x-1)$

63. $-x^4+4x^2=-x^2\cdot(x^2-4)$
$=-x^2(x+2)(x-2)$

65. $-4x^4+9x^2=-x^2\cdot(4x^2-9)$
$=-x^2(2x+3)(2x-3)$

67. $-2x^4+16x=-2x\cdot(x^3-8)$
$=-2x[(x)^3-(2)^3]$
$=-2x(x-2)[x^2+x(2)+2^2]$
$=-2x(x-2)(x^2+2x+4)$

69. $-16x^5-2x^2$
$=-2x^2\cdot(8x^3+1)$
$=-2x^2[(2x)^3+(1)^3]$
$=-2x^2(2x+1)[(2x)^2-(2x)(1)+1^2]$
$=-2x^2(2x+1)(4x^2-2x+1)$

71. $10(-3)=-30$, $-30=-2\cdot 15$, and $-2+15=13$, so
$10x^2+13x-3=10x^2+15x-2x-3$
$=5x(2x+3)-1(2x+3)$
$=(2x+3)(5x-1).$

73. $2(-3)=-6$, $-6=-6\cdot 1$, and $-6+1=-5$, so
$2x^2-5x-3=2x^2-6x+x-3$
$=2x(x-3)+1(x-3)$
$=(x-3)(2x+1).$

75. $\frac{2\pi A}{360}R_1+\frac{2\pi A}{360}Kt=\frac{2\pi A}{360}\cdot R_1+\frac{2\pi A}{360}\cdot Kt$
$=\frac{2\pi A}{360}(R_1+Kt)$

77. $\frac{3Sd^2}{2bd^3}-\frac{12Sz^2}{2bd^3}=\frac{3S}{2bd^3}\cdot d^2-\frac{3S}{2bd^3}\cdot 4z^2$
$=\frac{3S}{2bd^3}(d^2-4z^2)$
$=\frac{3S}{2bd^3}(d+2z)(d-2z)$

79. Sample answer: Factor using
$X^3+A^3=(X+A)(X^2-AX+A^2)$.

81. x^2-9 can be factored as $(x+3)(x-3)$.

83. $5x^4-10x^3+20x^2=5x^2(x^2-2x+4)$

85. $6x^2-x-35=(3x+7)(2x-5)$

87. $27t^3 - 64 = (3t)^3 - (4)^3$
$= (3t-4)[(3t)^2 + (3t)(4) + 4^2]$
$= (3t-4)(9t^2 + 12t + 16)$

89. $x^4 - 81 = (x^2)^2 - (9)^2$
$= (x^2+9)(x^2-9)$
$= (x^2+9)(x+3)(x-3)$

91. $-9x^2 - 30xy - 25y^2$
$= -1 \cdot (9x^2 + 30xy + 25y^2)$
$= -(3x+5y)^2$

93. $64y^3 + 27x^3$
$= (4y)^3 + (3x)^3$
$= (4y+3x)[(4y)^2 - (4y)(3x) + (3x)^2]$
$= (4y+3x)(16y^2 - 12xy + 9x^2)$

95. $-x^5 - x^2y^3 = -x^2 \cdot (x^3 + y^3)$
$= -x^2(x+y)(x^2 - xy + y^2)$

5.6 Solving Quadratic Equations by Factoring

Problems 5.6

1. a.
$$3x^2 + 8x - 3 = 0$$
$$3x^2 + 9x - x - 3 = 0$$
$$3x(x+3) - 1(x+3) = 0$$
$$(3x-1)(x+3) = 0$$
$3x - 1 = 0$ or $x + 3 = 0$
$3x = 1$ $\quad x = -3$
$x = \frac{1}{3}$ or $x = -3$

b.
$$6x^2 - 7x - 3 = 0$$
$$6x^2 - 9x + 2x - 3 = 0$$
$$3x(2x-3) + 1(2x-3) = 0$$
$$(3x+1)(2x-3) = 0$$
$3x + 1 = 0$ or $2x - 3 = 0$
$3x = -1$ $\quad 2x = 3$
$x = -\frac{1}{3}$ or $x = \frac{3}{2}$

2.
$$6x^2 + 13x = 5$$
$$6x^2 + 13x - 5 = 0$$
$$6x^2 + 15x - 2x - 5 = 0$$
$$3x(2x+5) - 1(2x+5) = 0$$
$$(2x+5)(3x-1) = 0$$
$2x + 5 = 0$ or $3x - 1 = 0$
$2x = -5$ $\quad 3x = 1$
$x = -\frac{5}{2}$ or $x = \frac{1}{3}$

3.
$$(4n+1)(n-2) = 4(n+1) - 3$$
$$4n^2 - 7n - 2 = 4n + 1$$
$$4n^2 - 11n - 3 = 0$$
$$4n^2 - 12n + n - 3 = 0$$
$$4n(n-3) + 1(n-3) = 0$$
$$(n-3)(4n+1) = 0$$
$n - 3 = 0$ or $4n + 1 = 0$
$n = 3$ $\quad 4n = -1$
$n = 3$ or $n = -\frac{1}{4}$

4. $(3m+1)(3m+2) = 5(m+1) - 2m - 4$
$$9m^2 + 9m + 2 = 3m + 1$$
$$9m^2 + 6m + 1 = 0$$
$$(3m+1)^2 = 0$$
$$3m + 1 = 0$$
$$3m = -1$$
$$m = -\frac{1}{3}$$

5. a.
$$16x^2-9=0$$
$$(4x+3)(4x-3)=0$$
$$4x+3=0 \quad \text{or} \quad 4x-3=0$$
$$4x=-3 \qquad 4x=3$$
$$x=-\frac{3}{4} \quad \text{or} \quad x=\frac{3}{4}$$

b.
$$y(2y+3)=-1$$
$$2y^2+3y+1=0$$
$$(2y+1)(y+1)=0$$
$$2y+1=0 \quad \text{or} \quad y+1=0$$
$$2y=-1 \qquad y=-1$$
$$y=-\frac{1}{2} \quad \text{or} \quad y=-1$$

c.
$$n^2=4n$$
$$n^2-4n=0$$
$$n(n-4)=0$$
$$n=0 \quad \text{or} \quad n-4=0$$
$$n=0 \quad \text{or} \quad n=4$$

6.
$$(m-3)(m^2-m-2)=0$$
$$(m-3)(m-2)(m+1)=0$$
$$m-3=0 \quad \text{or} \quad m-2=0 \quad \text{or} \quad m+1=0$$
$$m=3 \quad \text{or} \quad m=2 \quad \text{or} \quad m=-1$$

Exercises 5.6

1.
$$2x^2+7x+3=0$$
$$(2x+1)(x+3)=0$$
$$2x+1=0 \quad \text{or} \quad x+3=0$$
$$2x=-1 \qquad x=-3$$
$$x=-\frac{1}{2} \quad \text{or} \quad x=-3$$

3.
$$2x^2+x-3=0$$
$$(2x+3)(x-1)=0$$
$$2x+3=0 \quad \text{or} \quad x-1=0$$
$$2x=-3 \qquad x=1$$
$$x=-\frac{3}{2} \quad \text{or} \quad x=1$$

5.
$$3y^2-11y+6=0$$
$$3y^2-9y-2y+6=0$$
$$3y(y-3)-2(y-3)=0$$
$$(y-3)(3y-2)=0$$
$$y-3=0 \quad \text{or} \quad 3y-2=0$$
$$y=3 \qquad 3y=2$$
$$y=3 \quad \text{or} \quad y=\frac{2}{3}$$

7.
$$3y^2-2y-1=0$$
$$(3y+1)(y-1)=0$$
$$3y+1=0 \quad \text{or} \quad y-1=0$$
$$3y=-1 \qquad y=1$$
$$y=-\frac{1}{3} \quad \text{or} \quad y=1$$

9.
$$6x^2+11x=-4$$
$$6x^2+11x+4=0$$
$$6x^2+8x+3x+4=0$$
$$2x(3x+4)+1(3x+4)=0$$
$$(3x+4)(2x+1)=0$$
$$3x+4=0 \quad \text{or} \quad 2x+1=0$$
$$3x=-4 \qquad 2x=-1$$
$$x=-\frac{4}{3} \quad \text{or} \quad x=-\frac{1}{2}$$

11.
$$3x^2-5x=2$$
$$3x^2-5x-2=0$$
$$3x^2-6x+x-2=0$$
$$3x(x-2)+1(x-2)=0$$
$$(x-2)(3x+1)=0$$
$$x-2=0 \quad \text{or} \quad 3x+1=0$$
$$x=2 \qquad 3x=-1$$
$$x=2 \quad \text{or} \quad x=-\frac{1}{3}$$

13.
$$\begin{aligned} 5x^2+6x&=8\\ 5x^2+6x-8&=0\\ 5x^2+10x-4x-8&=0\\ 5x(x+2)-4(x+2)&=0\\ (x+2)(5x-4)&=0 \end{aligned}$$
$$x+2=0 \quad \text{or} \quad 5x-4=0$$
$$x=-2 \qquad 5x=4$$
$$x=-2 \quad \text{or} \quad x=\frac{4}{5}$$

15.
$$\begin{aligned} 5x^2-13x&=-8\\ 5x^2-13x+8&=0\\ 5x^2-5x-8x+8&=0\\ 5x(x-1)-8(x-1)&=0\\ (x-1)(5x-8)&=0 \end{aligned}$$
$$x-1=0 \quad \text{or} \quad 5x-8=0$$
$$x=1 \qquad 5x=8$$
$$x=1 \quad \text{or} \quad x=\frac{8}{5}$$

17.
$$\begin{aligned} 3y^2&=17y-10\\ 3y^2-17y+10&=0\\ 3y^2-15y-2y+10&=0\\ 3y(y-5)-2(y-5)&=0\\ (y-5)(3y-2)&=0 \end{aligned}$$
$$y-5=0 \quad \text{or} \quad 3y-2=0$$
$$y=5 \qquad 3y=2$$
$$y=5 \quad \text{or} \quad y=\frac{2}{3}$$

19.
$$\begin{aligned} 2y^2&=-5y-2\\ 2y^2+5y+2&=0\\ 2y^2+4y+y+2&=0\\ 2y(y+2)+1(y+2)&=0\\ (y+2)(2y+1)&=0 \end{aligned}$$
$$y+2=0 \quad \text{or} \quad 2y+1=0$$
$$y=-2 \qquad 2y=-1$$
$$y=-2 \quad \text{or} \quad y=-\frac{1}{2}$$

21.
$$\begin{aligned} 9x^2+6x+1&=0\\ (3x+1)^2&=0\\ 3x+1&=0\\ 3x&=-1\\ x&=-\frac{1}{3} \end{aligned}$$

23.
$$\begin{aligned} y^2-8y&=-16\\ y^2-8y+16&=0\\ (y-4)^2&=0\\ y-4&=0\\ y&=4 \end{aligned}$$

25.
$$\begin{aligned} 9x^2+12x&=-4\\ 9x^2+12x+4&=0\\ (3x+2)^2&=0\\ 3x+2&=0\\ 3x&=-2\\ x&=-\frac{2}{3} \end{aligned}$$

27.
$$\begin{aligned} 4y^2-20y&=-25\\ 4y^2-20y+25&=0\\ (2y-5)^2&=0\\ 2y-5&=0\\ 2y&=5\\ y&=\frac{5}{2} \end{aligned}$$

29.
$$\begin{aligned} x^2&=-10x-25\\ x^2+10x+25&=0\\ (x+5)^2&=0\\ x+5&=0\\ x&=-5 \end{aligned}$$

31.
$$\begin{aligned} (2x-1)(x-3)&=3x-5\\ 2x^2-7x+3&=3x-5\\ 2x^2-10x+8&=0\\ x^2-5x+4&=0\\ (x-4)(x-1)&=0 \end{aligned}$$
$$x-4=0 \quad \text{or} \quad x-1=0$$
$$x=4 \quad \text{or} \quad x=1$$

33. $(2x+3)(x+4)=2(x-1)+4$
$2x^2+11x+12=2x+2$
$2x^2+9x+10=0$
$(2x+5)(x+2)=0$
$2x+5=0$ or $x+2=0$
$2x=-5$ $\quad x=-2$
$x=-\frac{5}{2}$ or $x=-2$

35. $(2x-1)(x-1)=x-1$
$2x^2-3x+1=x-1$
$2x^2-4x+2=0$
$x^2-2x+1=0$
$(x-1)^2=0$
$x-1=0$
$x=1$

37. $4x^2-1=0$
$(2x+1)(2x-1)=0$
$2x+1=0$ or $2x-1=0$
$2x=-1$ $\quad 2x=1$
$x=-\frac{1}{2}$ or $x=\frac{1}{2}$

39. $4y^2-25=0$
$(2y+5)(2y-5)=0$
$2y+5=0$ or $2y-5=0$
$2y=-5$ $\quad 2y=5$
$y=-\frac{5}{2}$ or $y=\frac{5}{2}$

41. $z^2=9$
$z^2-9=0$
$(z+3)(z-3)=0$
$z+3=0$ or $z-3=0$
$z=-3$ or $z=3$

43. $25x^2=49$
$25x^2-49=0$
$(5x+7)(5x-7)=0$
$5x+7=0$ or $5x-7=0$
$5x=-7$ $\quad 5x=7$
$x=-\frac{7}{5}$ or $x=\frac{7}{5}$

45. $m^2=5m$
$m^2-5m=0$
$m(m-5)=0$
$m=0$ or $m-5=0$
$m=0$ or $m=5$

47. $2n^2=10n$
$2n^2-10n=0$
$2n(n-5)=0$
$2n=0$ or $n-5=0$
$n=0$ or $n=5$

49. $y(y+11)=-24$
$y^2+11y+24=0$
$(y+8)(y+3)=0$
$y+8=0$ or $y+3=0$
$y=-8$ or $y=-3$

51. $y(y-16)=-63$
$y^2-16y+63=0$
$(y-7)(y-9)=0$
$y-7=0$ or $y-9=0$
$y=7$ or $y=9$

53. $(v-2)(v^2+3v+2)=0$
$(v-2)(v+2)(v+1)=0$
$v-2=0$ or $v+2=0$ or $v+1=0$
$v=2$ or $v=-2$ or $v=-1$

55. $(m^2-3m+2)(m-4)=0$
$(m-2)(m-1)(m-4)=0$
$m-2=0$ or $m-1=0$ or $m-4=0$
$m=2$ or $m=1$ or $m=4$

57. $(n^2-3n-4)(n+2)=0$
$(n-4)(n+1)(n+2)=0$
$n-4=0$ or $n+1=0$ or $n+2=0$
$n=4$ or $n=-1$ or $n=-2$

59. $(x^2+2x-3)(x-1)=0$
$(x+3)(x-1)(x-1)=0$
$x+3=0$ or $x-1=0$
$x=-3$ or $x=1$

61. $H^2+(3+H)^2=H^2+9+6H+H^2$
$=2H^2+6H+9$

63. $(H-9)(H+3)=H^2+3H-9H-27$
$=H^2-6H-27$

65. $D(x)=x^2-14x+49=0$
$(x-7)^2=0$
$x-7=0$
$x=7$

67. Answers may vary.

69. Sample answer:
$x^2+2x+1=0$
$(x+1)^2=0$
$x+1=0$
$x=-1$ one solution
$x=-1$ is called a *double root* because $(x+1)^2=(x+1)(x+1)$.

71. $(3x-2)(x-1)=2(x+3)+2$
$3x^2-5x+2=2x+8$
$3x^2-7x-6=0$
$3x^2-9x+2x-6=0$
$3x(x-3)+2(x-3)=0$
$(x-3)(3x+2)=0$
$x-3=0$ or $3x+2=0$
$x=3$ $\quad 3x=-2$
$x=3$ or $x=-\frac{2}{3}$

73. $5x^2+9x-2=0$
$5x^2+10x-x-2=0$
$5x(x+2)-1(x+2)=0$
$(x+2)(5x-1)=0$
$x+2=0$ or $5x-1=0$
$x=-2$ $\quad 5x=1$
$x=-2$ or $x=\frac{1}{5}$

75. $x(x-1)=0$
$x=0$ or $x-1=0$
$x=0$ or $x=1$

77. $(x-4)(x^2+4x-5)=0$
$(x-4)(x+5)(x-1)=0$
$x-4=0$ or $x+5=0$ or $x-1=0$
$x=4$ or $x=-5$ or $x=1$

79. $m(3m+5)=-2$
$3m^2+5x+2=0$
$(3m+2)(m+1)=0$
$3m+2=0$ or $m+1=0$
$3m=-2$ $\quad m=-1$
$m=-\frac{2}{3}$ or $m=-1$

5.7 Applications of Quadratics

Problems 5.7

1. $n(n+2)=10+5n$
$n^2+2n=10+5n$
$n^2-3n-10=0$
$(n-5)(n+2)=0$
$n-5=0$ or $n+2=0$
$n=5$ or $n=-2$
-2 is not odd. The numbers are 5 and 7.

2. $W(W+4)=(4W+8)+56$
$W^2+4W=4W+64$
$W^2-64=0$
$(W-8)(W+8)=0$
$W-8=0$ or $W+8=0$
$W=8$ or $W=-8$

Since a rectangle can't have a negative width, discard the –8. The width is 8 feet and the length is 12 feet.

3. a. $H(H+3)=108$
$H^2+3H-108=0$
$(H+12)(H-9)=0$
$H+12=0$ or $H-9=0$
$H=-12$ or $H=9$

Height is positive, so the height is 9 inches and the width his 12 inches.

b. $H(H+6)=2(108)$
$H^2+6H-216=0$
$(H+18)(H-12)=0$
$H+18=0$ or $H-12=0$
$H=-18$ or $H=12$

Height is positive, so the height is 12 inches and the width is 18 inches.

4. $W^2+(W+2)^2=(W+4)^2$
$W^2+W^2+4W+4=W^2+8W+16$
$W^2-4W-12=0$
$(W-6)(W+2)=0$
$W-6=0$ or $W+2=0$
$W=6$ or $W=-2$

Width is positive, so the width is 6 inches, the height is 8 inches.

5. $0.06v^2=150$
$6v^2=15{,}000$
$6v^2-15{,}000=0$
$6(v^2-2500)=0$
$6(v+50)(v-50)=0$
$v+50=0$ or $v-50=0$
$v=-50$ or $v=50$

The velocity was positive, so the car was going 50 mi/hr.

Exercises 5.7

1. $n(n+1)=-8+10n$
$n^2+n=-8+10n$
$n^2-9n+8=0$
$(n-8)(n-1)=0$
$n-8=0$ or $n-1=0$
$n=8$ or $n=1$

The integers are 8 and 9 or 1 and 2.

3. $n(n+2)=7+(n+n+2)$
$n^2+2n=7+2n+2$
$n^2-9=0$
$(n+3)(n-3)=0$
$n=-3$ or $n=3$

The integers are 3 and 5 or –3 and –1.

5. $(n+2)^2=10n-4$
$n^2+4n+4=10n-4$
$n^2-6n+8=0$
$(n-4)(n-2)=0$
$n-4=0$ or $n-2=0$
$n=4$ or $n=2$

The integers are 4 and 6 or 2 and 4.

7. $A=\frac{1}{2}bh$
$40=\frac{1}{2}(h+2)h$
$80=h^2+2h$
$0=h^2+2h-80$
$0=(h+10)(h-8)$
$h+10=0$ or $h-8=0$
$h=-10$ or $h=8$

Height is positive, so h = 8 in. and b = 10 in.

9. $A = Lh$

$150 = (h+5)h$

$150 = h^2 + 5h$

$0 = h^2 + 5h - 150$

$0 = (h+15)(h-10)$

$h+15=0$ or $h-10=0$

$h=-15$ or $h=10$

Height is positive, so h = 10 in. and L = 15 in.

11. $A = \pi r^2$

$49\pi = \pi(x+3)^2$

$49 = x^2 + 6x + 9$

$0 = x^2 + 6x - 40$

$0 = (x+10)(x-4)$

$x+10=0$ or $x-4=0$

$x=-10$ or $x=4$

$x > 0$, so $x = 4$ and $r = 7$ units.

13. $A = LW$

$250 = L(L-45)$

$250 = L^2 - 45L$

$0 = L^2 - 45L - 250$

$0 = (L-50)(L+5)$

$L-50=0$ or $L+5=0$

$L=50$ or $L=-5$

Length is positive, so L = 50 ft and W = 5 ft.

15. $L^2 + H^2 = D^2$

$150^2 + H^2 = 289^2$

$22{,}500 + H^2 = 83{,}521$

$H^2 = 61{,}021$

$H \approx \pm 247$

Height is positive, so the height was about 247 ft.

17. $a^2 + b^2 = c^2$

$(c-4)^2 + (c-2)^2 = c^2$

$c^2 - 8c + 16 + c^2 - 4c + 4 = c^2$

$c^2 - 12c + 20 = 0$

$(c-10)(c-2) = 0$

$c-10=0$ or $c-2=0$

$c=10$ or $c=2$

The dimensions must be positive, so they are 6 in., 8 in., and 10 in.

19. $a^2 + b^2 = c^2$

$a^2 + (a+3)^2 = (a+6)^2$

$a^2 + a^2 + 6a + 9 = a^2 + 12a + 36$

$a^2 - 6a - 27 = 0$

$(a-9)(a+3) = 0$

$a-9=0$ or $a+3=0$

$a=9$ or $a=-3$

Length is positive, so the dimensions are 9 in., 12 in., and 15 in.

21. $0.06v^2 = b$

$0.06v^2 = 54$

$6v^2 = 5400$

$6v^2 - 5400 = 0$

$6(v^2 - 900) = 0$

$6(v+30)(v-30) = 0$

$v+30=0$ or $v-30=0$

$v=-30$ or $v=30$

The speed was positive, so the car was traveling at 30 mi/hr.

23. a. $0.06v^2 = 73.5$

$6v^2 = 7350$

$6v^2 - 7350 = 0$

$6(v^2 - 1225) = 0$

$6(v+35)(v-35) = 0$

$v+35=0$ or $v-35=0$

$v=-35$ or $v=35$

The speed was positive, so the car was traveling at 35 mi/hr.

b. 35 > 30, so, yes, the car was speeding.

25.
$$\begin{aligned} d &= 1.5tv + 0.06v^2 \\ 120 &= 1.5(0.4)v + 0.06v^2 \\ 12{,}000 &= 60v + 6v^2 \\ 0 &= 6v^2 + 60v - 12{,}000 \\ 0 &= 6(v^2 + 10v - 2000) \\ 0 &= 6(v+50)(v-40) \end{aligned}$$
$v + 50 = 0$ or $v - 40 = 0$
$v = -50$ or $v = 40$
The speed was positive, so Pedro was going 40 mi/hr.

27. $\frac{5}{6}\cdot\frac{3}{3} = \frac{5\cdot 3}{6\cdot 3} = \frac{15}{18}$

29. $18x - 36y = 18\cdot x - 18\cdot 2y = 18(x - 2y)$

31. $-6 = -1\cdot 6$ and $-1 + 6 = 5$, so $x^2 + 5x - 6 = (x+6)(x-1)$.

33.
$$\begin{aligned} H(t) &= -5t^2 - V_0 t + h \\ 0 &= -5t^2 - 5t + 10 \\ 0 &= -5(t^2 + t - 2) \\ 0 &= -5(t+2)(t-1) \end{aligned}$$
$t + 2 = 0$ or $t - 1 = 0$
$t = -2$ or $t = 1$
Time is positive, so it took 1 second for the object to hit the ground.

35.
$$\begin{aligned} H(t) &= -5t^2 - V_0 t + h \\ 0 &= -5t^2 - 10t + 15 \\ 0 &= -5(t^2 + 2t - 3) \\ 0 &= -5(t+3)(t-1) \end{aligned}$$
$t + 3 = 0$ or $t - 1 = 0$
$t = -3$ or $t = 1$
Time is positive, so it takes 1 second for the object to hit the ground.

37. The second formula cannot be used because we do not know the reaction time, t. We can only use the first formula.

39. Sample answer: the speed of the car, the driver's reaction time, and the roughness of the road

41.
$$\begin{aligned} n(n+2) &= 23 - (n + n + 2) \\ n^2 + 2n &= 23 - 2n - 2 \\ n^2 + 4n - 21 &= 0 \\ (n+7)(n-3) &= 0 \end{aligned}$$
$n + 7 = 0$ or $n - 3 = 0$
$n = -7$ or $n = 3$
The integers are 3 and 5 or –7 and –5.

43.
$$\begin{aligned} a^2 + b^2 &= c^2 \\ a^2 + (a+7)^2 &= (a+8)^2 \\ a^2 + a^2 + 14a + 49 &= a^2 + 16a + 64 \\ a^2 - 2a - 15 &= 0 \\ (a-5)(a+3) &= 0 \end{aligned}$$
$a - 5 = 0$ or $a + 3 = 0$
$a = 5$ or $a = -3$
Length is positive, so the dimensions are 5 in., 12 in., and 13 in.

45.
$$\begin{aligned} 150 &= 0.06v^2 \\ 15{,}000 &= 6v^2 \\ 0 &= 6v^2 - 15{,}000 \\ 0 &= 6(v^2 - 2500) \\ 0 &= 6(v+50)(v-50) \end{aligned}$$
$v + 50 = 0$ or $v - 50 = 0$
$v = -50$ or $v = 50$
The speed was positive, so the car was going 50 mi/hr.

Review Exercises

1. a. $60 = 2^2\cdot 3\cdot 5$
$90 = 2\cdot 3^2\cdot 5$
GCF = 2 · 3 · 5 = 30

b. $12 = 2^2\cdot 3$
$18 = 2\cdot 3^2$
GCF = 2 · 3 = 6

c. $27 = 3^3$
$80 = 2^4\cdot 5$
$17 = 17$
GCF = 1

2. a. $24x^7 = 6x^5 \cdot 4x^2$
$18x^5 = 6x^5 \cdot 3$
$-30x^{10} = 6x^5 \cdot (-5x^5)$
$\text{GCF} = 6x^5$

b. $18x^8 = 2x^8 \cdot 9$
$12x^9 = 2x^8 \cdot 6x$
$-20x^{10} = 2x^8 \cdot (-10x^2)$
$\text{GCF} = 2x^8$

c. $x^6y^4 = x^3 \cdot x^3y^4$
$y^7x^5 = x^3 \cdot x^2y^7$
$x^3y^6 = x^3 \cdot y^6$
$x^9 = x^3 \cdot x^6$
$\text{GCF} = x^3$

3. a. $20x^3 - 55x^5 = 5x^3(4 - 11x^2)$

b. $14x^4 - 35x^6 = 7x^4(2 - 5x^2)$

c. $16x^7 - 40x^9 = 8x^7(2 - 5x^2)$

4. a. $\frac{3}{7}x^6 - \frac{5}{7}x^5 + \frac{2}{7}x^4 - \frac{1}{7}x^2$
$= \frac{1}{7}x^2(3x^4 - 5x^3 + 2x^2 - 1)$

b. $\frac{4}{9}x^7 - \frac{2}{9}x^6 + \frac{2}{9}x^5 - \frac{1}{9}x^3$
$= \frac{1}{9}x^3(4x^4 - 2x^3 + 2x^2 - 1)$

c. $\frac{3}{8}x^9 - \frac{7}{8}x^8 + \frac{3}{8}x^7 - \frac{1}{8}x^5$
$= \frac{1}{8}x^5(3x^4 - 7x^3 + 3x^2 - 1)$

5. a. $3x^3 - 21x^2 - x + 7$
$= 3x^2(x - 7) - 1(x - 7)$
$= (x - 7)(3x^2 - 1)$

b. $3x^3 + 18x^2 + x + 6$
$= 3x^2(x + 6) + 1(x + 6)$
$= (x + 6)(3x^2 + 1)$

c. $4x^3 - 8x^2y + x - 2y$
$= 4x^2(x - 2y) + 1(x - 2y)$
$= (x - 2y)(4x^2 + 1)$

6. a. $1 \cdot 7 = 7$ and $1 + 7 = 8$, so
$x^2 + 8x + 7 = (x + 7)(x + 1)$.

b. $1(-9) = -9$ and $1 + (-9) = -8$, so
$x^2 - 8x - 9 = (x - 9)(x + 1)$.

c. $1 \cdot 5 = 5$ and $1 + 5 = 6$, so
$x^2 + 6x + 5 = (x + 5)(x + 1)$.

7. a. $10 = -5(-2)$ and $-5 + (-2) = -7$, so
$x^2 - 7x + 10 = (x - 5)(x - 2)$.

b. $14 = -7(-2)$ and $-7 + (-2) = -9$, so
$x^2 - 9x + 14 = (x - 7)(x - 2)$.

c. $-8 = -2 \cdot 4$ and $-2 + 4 = 2$, so
$x^2 + 2x - 8 = (x + 4)(x - 2)$.

8. a. $6x^2 - 6 + 5x = 6x^2 + 5x - 6$
$6(-6) = -36$, $-36 = -4 \cdot 9$, and
$-4 + 9 = 5$, so
$6x^2 + 5x - 6 = 6x^2 + 9x - 4x - 6$
$= 3x(2x + 3) - 2(2x + 3)$
$= (2x + 3)(3x - 2)$.

b. $6x^2 - 1 + x = 6x^2 + x - 1$
$6(-1) = -6$, $-6 = -2 \cdot 3$, and $-2 + 3 = 1$,
so $6x^2 + x - 1 = 6x^2 + 3x - 2x - 1$
$= 3x(2x + 1) - 1(2x + 1)$
$= (2x + 1)(3x - 1)$.

c. $6x^2 - 5 + 13x = 6x^2 + 13x - 5$
$6(-5) = -30$, $-30 = -2 \cdot 15$, and
$-2 + 15 = 13$, so
$$6x^2 + 13x - 5 = 6x^2 + 15x - 2x - 5$$
$$= 3x(2x + 5) - 1(2x + 5)$$
$$= (2x + 5)(3x - 1).$$

9. a. $6x^2 - 17xy + 5y^2$
$$= 6x^2 - 15xy - 2xy + 5y^2$$
$$= 3x(2x - 5y) - y(2x - 5y)$$
$$= (2x - 5y)(3x - y)$$

b. $6x^2 - 7xy + 2y^2$
$$= 6x^2 - 4xy - 3xy + 2y^2$$
$$= 2x(3x - 2y) - y(3x - 2y)$$
$$= (3x - 2y)(2x - y)$$

c. $6x^2 - 11xy + 4y^2$
$$= 6x^2 - 8xy - 3xy + 4y^2$$
$$= 2x(3x - 4y) - y(3x - 4y)$$
$$= (3x - 4y)(2x - y)$$

10. a. $x^2 + 4x + 4 = x^2 + 2 \cdot 2 \cdot x + 2^2 = (x + 2)^2$

b. $x^2 + 10x + 25 = x^2 + 2 \cdot 5 \cdot x + 5^2$
$$= (x + 5)^2$$

c. $x^2 + 8x + 16 = x^2 + 2 \cdot 4 \cdot x + 4^2$
$$= (x + 4)^2$$

11. a. $9x^2 + 12xy + 4y^2$
$$= (3x)^2 + 2 \cdot 3x \cdot 2y + (2y)^2$$
$$= (3x + 2y)^2$$

b. $9x^2 + 30xy + 25y^2$
$$= (3x)^2 + 2 \cdot 3x \cdot 5y + (5y)^2$$
$$= (3x + 5y)^2$$

c. $9x^2 + 24xy + 16y^2$
$$= (3x)^2 + 2 \cdot 3x \cdot 4y + (4y)^2$$
$$= (3x + 4y)^2$$

12. a. $x^2 - 4x + 4 = x^2 - 2 \cdot 2 \cdot x + 2^2 = (x - 2)^2$

b. $x^2 - 6x + 9 = x^2 - 2 \cdot 3 \cdot x + 3^2 = (x - 3)^2$

c. $x^2 - 12x + 36 = x^2 - 2 \cdot 6 \cdot x + 6^2$
$$= (x - 6)^2$$

13. a. $4x^2 - 12xy + 9y^2$
$$= (2x)^2 - 2 \cdot 2x \cdot 3y + (3y)^2$$
$$= (2x - 3y)^2$$

b. $4x^2 - 20xy + 25y^2$
$$= (2x)^2 - 2 \cdot 2x \cdot 5y + (5y)^2$$
$$= (2x - 5y)^2$$

c. $4x^2 - 28xy + 49y^2$
$$= (2x)^2 - 2 \cdot 2x \cdot 7y + (7y)^2$$
$$= (2x - 7y)^2$$

14. a. $x^2 - 36 = (x)^2 - (6)^2 = (x + 6)(x - 6)$

b. $x^2 - 49 = (x)^2 - (7)^2 = (x + 7)(x - 7)$

c. $x^2 - 81 = (x)^2 - (9)^2 = (x + 9)(x - 9)$

15. a. $16x^2 - 81y^2 = (4x)^2 - (9y)^2$
$$= (4x + 9y)(4x - 9y)$$

b. $25x^2 - 64y^2 = (5x)^2 - (8y)^2$
$$= (5x + 8y)(5x - 8y)$$

c. $9x^2 - 100y^2 = (3x)^2 - (10y)^2$
$$= (3x + 10y)(3x - 10y)$$

16. a. $m^3 + 125 = (m)^3 + (5)^3$
$$= (m + 5)[m^2 - m(5) + 5^2]$$
$$= (m + 5)(m^2 - 5m + 25)$$

b. $n^3 + 64 = (n)^3 + (4)^3$
$$= (n + 4)[n^2 - n(4) + 4^2]$$
$$= (n + 4)(n^2 - 4n + 16)$$

c. $y^3+8=(y)^3+(2)^3$
$=(y+2)[y^2-y(2)+2^2]$
$=(y+2)(y^2-2y+4)$

17. a. $8y^3-27x^3$
$=(2y)^3-(3x)^3$
$=(2y-3x)[(2y)^2+(2y)(3x)+(3x)^2]$
$=(2y-3x)(4y^2+6xy+9x^2)$

b. $64y^3-125x^3$
$=(4y)^3-(5x)^3$
$=(4y-5x)[(4y)^2+(4y)(5x)+(5x)^2]$
$=(4y-5x)(16y^2+20xy+25x^2)$

c. $8m^3-125n^3$
$=(2m)^3-(5n)^3$
$=(2m-5n)[(2m)^2+(2m)(5n)+(5n)^2]$
$=(2m-5n)(4m^2+10mn+25n^2)$

18. a. $3x^3-6x^2+27x=3x(x^2-2x+9)$

b. $3x^3-6x^2+30x=3x(x^2-2x+10)$

c. $4x^3-8x^2+32x=4x(x^2-2x+8)$

19. a. $2x^3-2x^2-4x=2x(x^2-x-2)$
$=2x(x-2)(x+1)$

b. $3x^3-6x^2-9x=3x(x^2-2x-3)$
$=3x(x-3)(x+1)$

c. $4x^3-12x^2-16x=4x(x^2-3x-4)$
$=4x(x-4)(x+1)$

20. a. $2x^3+8x^2+x+4=2x^2(x+4)+1(x+4)$
$=(x+4)(2x^2+1)$

b. $2x^3+10x^2+x+5$
$=2x^2(x+5)+1(x+5)$
$=(x+5)(2x^2+1)$

c. $2x^3+12x^2+x+6$
$=2x^2(x+6)+1(x+6)$
$=(x+6)(2x^2+1)$

21. a. $9kx^2+12kx+4k$
$=k(9x^2+12x+4)$
$=k[(3x)^2+2\cdot 3x\cdot 2+2^2]$
$=k(3x+2)^2$

b. $9kx^2+30kx+25k$
$=k(9x^2+30x+25)$
$=k[(3x)^2+2\cdot 3x\cdot 5+5^2]$
$=k(3x+5)^2$

c. $4kx^2+20kx+25k$
$=k(4x^2+20x+25)$
$=k[(2x)^2+2\cdot 2x\cdot 5+5^2]$
$=k(2x+5)^2$

22. a. $-3x^4+27x^2=-3x^2(x^2-9)$
$=-3x^2(x+3)(x-3)$

b. $-4x^4+64x^2=-4x^2(x^2-16)$
$=-4x^2(x+4)(x-4)$

c. $-5x^4+20x^2=-5x^2(x^2-4)$
$=-5x^2(x+2)(x-2)$

23. a. $-x^3-y^3=-(x^3+y^3)$
$=-(x+y)(x^2-xy+y^2)$

b. $-8m^3-27n^3$
$=-(8m^3+27n^3)$
$=-[(2m)^3+(3n)^3]$
$=-(2m+3n)[(2m)^2-2m(3n)+(3n)^2]$
$=-(2m+3n)(4m^2-6mn+9n^2)$

c. $-64n^3 - m^3$
$= -(64n^3 + m^3)$
$= -[(4n)^3 + m^3]$
$= -(4n + m)[(4n)^2 - (4n)m + m^2]$
$= -(4n + m)(16n^2 - 4mn + m^2)$

24. a. $-y^3 + x^3 = -(y^3 - x^3)$
$= -(y - x)(y^2 + xy + x^2)$

b. $-8m^3 + 27n^3$
$= -(8m^3 - 27n^3)$
$= -[(2m)^3 - (3n)^3]$
$= -(2m - 3n)[(2m)^2 + (2m)(3n) + (3n)^2]$
$= -(2m - 3n)(4m^2 + 6mn + 9n^2)$

c. $-64t^3 + 125s^3$
$= -(64t^3 - 125s^3)$
$= -[(4t)^3 - (5s)^3]$
$= -(4t - 5s)[(4t)^2 + (4t)(5s) + (5s)^2]$
$= -(4t - 5s)(16t^2 + 20st + 25s^2)$

25. a. $-4x^2 - 12xy + 9y^2$
$= -(4x^2 + 12xy - 9y^2)$

b. $-25x^2 - 30xy + 9y^2$
$= -(25x^2 + 30xy - 9y^2)$

c. $-16x^2 - 24xy + 9y^2$
$= -(16x^2 + 24xy - 9y^2)$

26. a. $x^2 - 4x - 5 = 0$
$(x - 5)(x + 1) = 0$
$x - 5 = 0$ or $x + 1 = 0$
$x = 5$ or $x = -1$

b. $x^2 - 5x - 6 = 0$
$(x - 6)(x + 1) = 0$
$x - 6 = 0$ or $x + 1 = 0$
$x = 6$ or $x = -1$

c. $x^2 - 6x - 7 = 0$
$(x - 7)(x + 1) = 0$
$x - 7 = 0$ or $x + 1 = 0$
$x = 7$ or $x = -1$

27. a. $2x^2 + x = 10$
$2x^2 + x - 10 = 0$
$(2x + 5)(x - 2) = 0$

$2x + 5 = 0$ or $x - 2 = 0$

$2x = -5$ $x = 2$

$x = -\frac{5}{2}$ or $x = 2$

b. $2x^2 + 3x = 5$
$2x^2 + 3x - 5 = 0$
$(2x + 5)(x - 1) = 0$

$2x + 5 = 0$ or $x - 1 = 0$

$2x = -5$ $x = 1$

$x = -\frac{5}{2}$ or $x = 1$

c. $2x^2 + x = 3$
$2x^2 + x - 3 = 0$
$(2x + 3)(x - 1) = 0$

$2x + 3 = 0$ or $x - 1 = 0$

$2x = -3$ $x = 1$

$x = -\frac{3}{2}$ or $x = 1$

28. a.
$$\begin{aligned}(3x+1)(x-2)&=2(x-1)-4\\3x^2-5x-2&=2x-2-4\\3x^2-7x+4&=0\\(3x-4)(x-1)&=0\end{aligned}$$

$3x-4=0$ or $x-1=0$

$3x=4$ $\quad x=1$

$x=\frac{4}{3}$ or $x=1$

b.
$$\begin{aligned}(2x+1)(x-4)&=6(x-4)-1\\2x^2-7x-4&=6x-24-1\\2x^2-13x+21&=0\\(2x-7)(x-3)&=0\end{aligned}$$

$2x-7=0$ or $x-3=0$

$2x=7$ $\quad x=3$

$x=\frac{7}{2}$ or $x=3$

c.
$$\begin{aligned}(2x+1)(x-1)&=3(x+2)-1\\2x^2-x-1&=3x+6-1\\2x^2-4x-6&=0\\2(x^2-2x-3)&=0\\2(x-3)(x+1)&=0\end{aligned}$$

$x-3=0$ or $x+1=0$

$x=3$ or $x=-1$

29. a.
$$\begin{aligned}n(n+2)&=4+5n\\n^2+2n&=4+5n\\n^2-3n-4&=0\\(n-4)(n+1)&=0\end{aligned}$$

$n-4=0$ or $n+1=0$

$n=4$ or $n=-1$

-1 is not even. The integers are 4 and 6.

b.
$$\begin{aligned}n(n+2)&=4+2n\\n^2+2n&=4+2n\\n^2-4&=0\\(n+2)(n-2)&=0\end{aligned}$$

$n+2=0$ or $n-2=0$

$n=-2$ $\quad n=2$

The integers are 2 and 4 or -2 and 0.

c.
$$\begin{aligned}n(n+2)&=10+11n\\n^2+2n&=10+11n\\n^2-9n-10&=0\\(n-10)(n+1)&=0\end{aligned}$$

$n-10=0$ or $n+1=0$

$n=10$ or $n=-1$

-1 is not even. The integers are 10 and 12.

30.
$$\begin{aligned}a^2+b^2&=c^2\\(c-12)^2+(c-6)^2&=c^2\\c^2-24c+144+c^2-12c+36&=c^2\\c^2-36c+180&=0\\(c-30)(c-6)&=0\end{aligned}$$

$c-30=0$ or $c-6=0$

$c=30$ or $c=6$

The lengths must be greater than zero, so the dimensions are 18 in., 24 in., and 30 in.

Cumulative Review Chapters 1–5

1. $-\frac{3}{8}+\left(-\frac{1}{6}\right)=-\frac{9}{24}-\frac{4}{24}=-\frac{13}{24}$

2. $6.6-(-9.8)=6.6+9.8=16.4$

3. $(-5.5)(5.7)=-31.35$

4. $-(5^2)=-(25)=-25$

5. $-\frac{5}{6}\div\left(-\frac{5}{18}\right)=-\frac{5}{6}\left(-\frac{18}{5}\right)=\frac{18}{6}=3$

6. $y \div 5 \cdot x - z = 60 \div 5 \cdot 6 - 3$
$= 12 \cdot 6 - 3$
$= 72 - 3$
$= 69$

7. $7 \cdot (6 \cdot 4) = (7 \cdot 6) \cdot 4$ illustrates the Associative Law of Multiplication.

8. $-8xy^4 - (-9xy^4) = -8xy^4 + 9xy^4 = xy^4$

9. $2x - (x+3) - 2(x+4) = 2x - x - 3 - 2x - 8$
$= -x - 11$

10. $\dfrac{d+5e}{f}$

11. $5 = 3(x-1) + 5 - 2x$
$5 = 3x - 3 + 5 - 2x$
$5 = x + 2$
$x = 3$

12. $\dfrac{x}{7} - \dfrac{x}{9} = 2$
$63\left(\dfrac{x}{7} - \dfrac{x}{9}\right) = 2(63)$
$9x - 7x = 126$
$2x = 126$
$x = 63$

13. $8 - \dfrac{x}{3} = \dfrac{6(x+1)}{7}$
$21\left(8 - \dfrac{x}{3}\right) = 21\left(\dfrac{6x+6}{7}\right)$
$168 - 7x = 3(6x+6)$
$168 - 7x = 18x + 18$
$150 = 25x$
$x = 6$

14. $n + (n+25) = 105$
$2n + 25 = 105$
$2n = 80$
$n = 40$
The numbers are 40 and 65.

15. distance of $A = 50(t+2)$
distance of $B = 60t$
Set the distances equal.
$60t = 50(t+2)$
$60t = 50t + 100$
$10t = 100$
$t = 10$ hr

16. $0.11x + 0.12(6000 - x) = 680$
$0.11x + 720 - 0.12x = 680$
$-0.01x = -40$
$x = 4000$
\$4000 is invested in bonds and \$2000 in certificates.

17. $-\dfrac{x}{6} + \dfrac{x}{2} \geq \dfrac{x-2}{2}$
$-x + 3x \geq 3x - 6$
$-x \geq -6$
$x \leq 6$

0 6

18. $C(3, -3)$: $x = 3$ and $y = -3$

y 5 −5 5 x −5 C(3, −3)

19. $5x - y = -18$
$5(-3) - (-3) = -18$
$-15 + 3 = -18$
$-12 = -18$ False
$(-3, -3)$ is not a solution.

20. $2x - 3y = -5$
$2x - 3(3) = -5$
$2x - 9 = -5$
$2x = 4$
$x = 2$

21. $x + y = 4$

$$y = -x + 4$$

y-intercept: $y = 0 + 4 = 4$

x-intercept: $x + 0 = 4$

$$x = 4$$

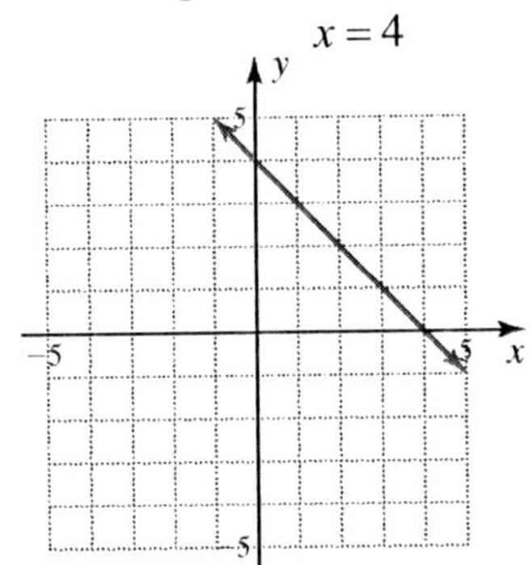

22. $4y - 20 = 0$

$$4y = 20$$

$$y = 5$$

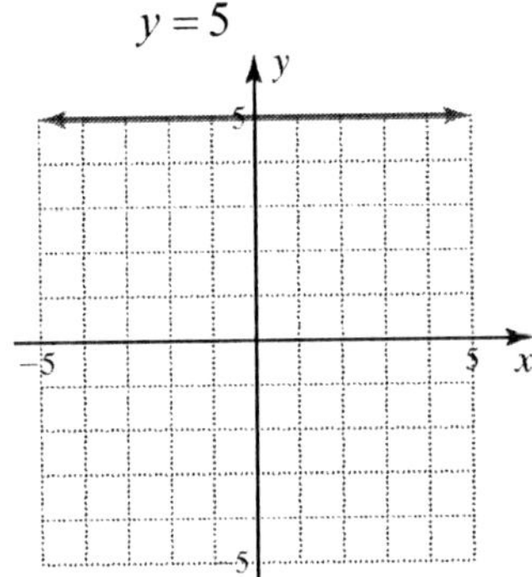

23. $m = \dfrac{y_2 - y_1}{x_2 - x_1} = \dfrac{7-5}{7-(-4)} = \dfrac{2}{11}$

24. $6x - 2y = 15$

$$-2y = -6x + 15$$

$$y = 3x - \frac{15}{2}$$

$$y = mx + b$$

So, the slope is $m = 3$.

25. **(1)** $4y = x - 7$

$$y = \frac{1}{4}x - \frac{7}{4}$$

(2) $7x - 28y = -7$

$$-28y = -7x - 7$$

$$y = \frac{1}{4}x + \frac{1}{4}$$

(3) $28y + 7x = -7$

$$28y = -7x - 7$$

$$y = -\frac{1}{4}x - \frac{1}{4}$$

(1) and (2) have the same slope, therefore, they are parallel.

26. $\dfrac{20x^6y^3}{-5x^4y^7} = -\dfrac{20}{5} \cdot x^{6-4} \cdot y^{3-7}$

$$-4x^2y^{-4}$$

$$= -\frac{4x^2}{y^4}$$

27. $\dfrac{x^{-7}}{x^{-9}} = x^{-7-(-9)} = x^{-7+9} = x^2$

28. $x \cdot x^{-5} = x^{1-5} = x^{-4} = \dfrac{1}{x^4}$

29. $(3x^3y^{-3})^{-3} = 3^{-3} \cdot x^{3(-3)} \cdot y^{-3(-3)}$

$$= \frac{1}{3^3} \cdot x^{-9} \cdot y^9$$

$$= \frac{y^9}{27x^9}$$

30. Move the decimal five places to the right and multiply by 10^{-5}:

$0.000048 = 4.8 \times 10^{-5}$.

31. $(5.98 \times 10^{-3}) \div (1.3 \times 10^2) = \dfrac{5.98}{1.3} \cdot \dfrac{10^{-3}}{10^2}$

$$= 4.6 \times 10^{-5}$$

32. $6x^2 + x + 2$

The greatest exponent is 2, so the degree of the polynomial is 2.

33. $x^3 + 2x^2 + 1 = (-2)^3 + 2(-2)^2 + 1$

$$= -8 + 8 + 1$$

$$= 1$$

34. $(-3x^2 - 2x^3 - 6) + (5x^3 - 7 - 4x^2)$
$= -2x^3 + 5x^3 - 3x^2 - 4x^2 - 6 - 7$
$= 3x^3 - 7x^2 - 13$

35. $-4x^4(8x^2 + 4y) = -32x^6 - 16x^4y$

36. $(3x + 2y)^2 = (3x + 2y)(3x + 2y)$
$= (3x)^2 + 6xy + 6xy + (2y)^2$
$= 9x^2 + 12xy + 4y^2$

37. $(5x - 7y)(5x + 7y)$
$= (5x)^2 + 35xy - 35xy - (7y)^2$
$= 25x^2 - 49y^2$

38. $\left(2x^2 - \frac{1}{5}\right)^2 = \left(2x^2 - \frac{1}{5}\right)\left(2x^2 - \frac{1}{5}\right)$
$= (2x^2)^2 - \frac{2}{5}x^2 - \frac{2}{5}x^2 + \left(\frac{1}{5}\right)^2$
$= 4x^4 - \frac{4}{5}x^2 + \frac{1}{25}$

39. $(5x^2 + 9)(5x^2 - 9)$
$= (5x^2)^2 - 45x^2 + 45x^2 - 9^2$
$= 25x^4 - 81$

40.
$$\begin{array}{r l}
 & 2x^2 \quad + 5x + 5 \quad \text{R1} \\
x - 2 \,\big)\!\! & \overline{2x^3 + \ x^2 \quad - 5x - 9} \\
 & \underline{2x^3 - 4x^2} \\
 & \quad 5x^2 \ - 5x \\
 & \quad \underline{5x^2 - 10x} \\
 & \qquad\quad 5x \ - 9 \\
 & \qquad\quad \underline{5x - 10} \\
 & \qquad\qquad\quad 1
\end{array}$$

41. $12x^6 - 14x^9 = 2x^6 \cdot 6 - 2x^6 \cdot 7x^3$
$= 2x^6(6 - 7x^3)$

42. $\frac{4}{5}x^7 - \frac{3}{5}x^6 + \frac{4}{5}x^5 - \frac{2}{5}x^3$
$= \frac{1}{5}x^3 \cdot 4x^4 - \frac{1}{5}x^3 \cdot 3x^3 + \frac{1}{5}x^3 \cdot 4x^2 - \frac{1}{5}x^3 \cdot 2$
$= \frac{1}{5}x^3(4x^4 - 3x^3 + 4x^2 - 2)$

43. 27 = −3(−9) and −3 + (−9) = −12, so
$x^2 - 12x + 27 = (x - 3)(x - 9)$.

44. $20x^2 - 23xy + 6y^2$
$= 20x^2 - 15xy - 8xy + 6y^2$
$= 5x(4x - 3y) - 2y(4x - 3y)$
$= (4x - 3y)(5x - 2y)$

45. $25x^2 - 49y^2 = (5x)^2 - (7y)^2$
$= (5x + 7y)(5x - 7y)$

46. $-5x^4 + 80x^2 = -(5x^4 - 80x^2)$
$= -5x^2(x^2 - 16)$
$= -5x^2(x + 4)(x - 4)$

47. $3x^3 - 6x^2 - 9x = 3x(x^2 - 2x - 3)$
$= 3x(x + 1)(x - 3)$

48. $2x^2 + 5x + 6x + 15 = x(2x + 5) + 3(2x + 5)$
$= (2x + 5)(x + 3)$

49. $9kx^2 + 6kx + k = k(9x^2 + 6x + 1)$
$= k[(3x)^2 + 2 \cdot 3x \cdot 1 + 1^2]$
$= k(3x + 1)^2$

50. $4x^2 + 17x = 15$
$4x^2 + 17x - 15 = 0$
$(4x - 3)(x + 5) = 0$

$4x - 3 = 0$ or $x + 5 = 0$

$4x = 3$ $\qquad x = -5$

$x = \frac{3}{4}$ or $x = -5$

Chapter 6 Rational Expressions

6.1 Building and Reducing Rational Expressions

Problems 6.1

1. a. $2m+3=0$

$2m=-3$

$m=-\frac{3}{2}$

Thus, $\frac{m}{2m+3}$ is undefined for $m=-\frac{3}{2}$.

b. $y^2+4y-5=(y+5)(y-1)$

$y+5=0$ or $y-1=0$

$y=-5$ $\quad y=1$

Thus, $\frac{y+1}{y^2+4y-5}$ is undefined for $y=-5$ or $y=1$.

c. $n^2+3=0$

Since $n^2\ge 0$, there are no values for which $\frac{n+3}{n^2+3}$ is undefined.

2. a. Since $14=7\cdot 2$, $\frac{4}{7}=\frac{4\cdot 2}{7\cdot 2}=\frac{8}{14}$.

b. Since $14x^3=7x^2\cdot 2x$,

$\frac{2y}{7x^2}=\frac{2y\cdot 2x}{7x^2\cdot 2x}=\frac{4xy}{14x^3}$.

c. Note that $x^2+x-2=(x+2)(x-1)$.

$\frac{2x}{x+2}=\frac{2x(x-1)}{(x+2)(x-1)}=\frac{2x^2-2x}{x^2+x-2}$

3. a. $\frac{-3x^2y}{6xy^4}=\frac{(-1)\cdot 3\cdot x\cdot x\cdot y}{2\cdot 3\cdot x\cdot y\cdot y\cdot y\cdot y}$

$=\frac{(-1)\cdot x}{2\cdot y\cdot y\cdot y}$

$=\frac{-x}{2y^3}$

b. $\frac{-6(m^2-n^2)}{-3(m-n)}$

$=\frac{(-1)\cdot 2\cdot 3\cdot(m-n)(m+n)}{(-1)\cdot 3\cdot(m-n)}$

$=2(m+n)$

4. a. $\frac{3x-12y}{12x-48y}=\frac{3(x-4y)}{12(x-4y)}=\frac{3}{12}=\frac{1}{4}$

b. $-\frac{x}{x+xy}=-\frac{x}{x(1+y)}=\frac{-1}{1+y}$

c. $\frac{y-3}{-(y^2-9)}=\frac{y-3}{-(y-3)(y+3)}$

$=\frac{1}{-(y+3)}$

$=\frac{-1}{y+3}$

5. a. $\frac{y^2-9}{3+y}=\frac{(y+3)(y-3)}{y+3}=y-3$

b. $\frac{y^2-9}{y+3}=\frac{(y+3)(y-3)}{y+3}=y-3$

c. $\frac{y^2+y-12}{3-y}=\frac{(y+4)(y-3)}{-1\cdot(y-3)}$

$=\frac{y+4}{-1}$

$=-(y+4)$

Exercises 6.1

1. $x-7=0$

$x=7$

Thus, $\frac{x}{x-7}$ is undefined for $x=7$.

3. $2y+8=0$

$2y=-8$

$y=-4$

Thus, $\frac{y-5}{2y+8}$ is undefined for $y=-4$.

5. $$x^2 - 9 = 0$$
$$(x+3)(x-3) = 0$$
$$x+3=0 \quad \text{or} \quad x-3=0$$
$$x=-3 \qquad\qquad x=3$$

Thus, $\dfrac{x+9}{x^2-9}$ is undefined for $x = 3$ or $x = -3$.

7. $y^2 + 5 = 0$

Since $y^2 \ge 0$, there are no values for which $\dfrac{y+3}{y^2+5}$ is undefined.

9. $$y^2 - 6y + 8 = 0$$
$$(y-4)(y-2) = 0$$
$$y-4=0 \quad \text{or} \quad y-2=0$$
$$y=4 \qquad\qquad y=2$$

Thus, $\dfrac{2y+9}{y^2-6y+8}$ is undefined for $y = 2$ or $y = 4$.

11. $$x^3 + 8 = 0$$
$$x^3 = -8$$
$$x = -2$$

Thus $\dfrac{x^2-4}{x^3+8}$ is undefined for $x = -2$.

13. $$x^3 - 27 = 0$$
$$x^3 = 27$$
$$x = 3$$

Thus $\dfrac{x^2+4}{x^3-27}$ is undefined for $x = 3$.

15. $$x^2 + 6x + 9 = 0$$
$$(x+3)^2 = 0$$
$$x+3 = 0$$
$$x = -3$$

Thus $\dfrac{x-1}{x^2+6x+9}$ is undefined for $x = -3$.

17. $$y^2 - 6y - 16 = 0$$
$$(y+2)(y-8) = 0$$
$$y+2=0 \quad \text{or} \quad y-8=0$$
$$y=-2 \qquad\qquad y=8$$

Thus $\dfrac{y-3}{y^2-6y-16}$ is undefined for $x = -2$ or $x = 8$.

19. $$x^3 + 2x^2 + x = 0$$
$$x(x^2 + 2x + 1) = 0$$
$$x(x+1)^2 = 0$$
$$x = 0 \quad \text{or} \quad x+1=0$$
$$x = -1$$

Thus, $\dfrac{x+4}{x^3+2x^2+x}$ is undefined for $x = 0$ or $x = -1$.

21. Since $21 = 7 \cdot 3$, $\dfrac{3}{7} = \dfrac{3 \cdot 3}{7 \cdot 3} = \dfrac{9}{21}$.

23. Since $22 = 11 \cdot 2$, $\dfrac{-8}{11} = \dfrac{-8 \cdot 2}{11 \cdot 2} = \dfrac{-16}{22}$.

25. Since $24y^3 = 6y^2 \cdot 4y$,
$$\frac{5x}{6y^2} = \frac{5x \cdot 4y}{6y^2 \cdot 4y} = \frac{20xy}{24y^3}.$$

27. Since $21y^4 = 7y \cdot 3y^3$,
$$\frac{-3x}{7y} = \frac{-3x \cdot 3y^3}{7y \cdot 3y^3} = \frac{-9xy^3}{21y^4}.$$

29. Note that $x^2 - x - 2 = (x+1)(x-2)$.
$$\frac{4x}{x+1} = \frac{4x(x-2)}{(x+1)(x-2)} = \frac{4x(x-2)}{x^2-x-2} \text{ or } \frac{4x^2-8x}{x^2-x-2}$$

31. Note that $x^2+x-6=(x+3)(x-2)$.

$$\frac{-5x}{x+3}=\frac{-5x(x-2)}{(x+3)(x-2)}=\frac{-5x(x-2)}{x^2+x-6} \text{ or } \frac{-5x^2+10x}{x^2+x-6}$$

33. $$\frac{7x^3y}{14xy^4}=\frac{7\cdot x\cdot x\cdot x\cdot y}{2\cdot 7\cdot x\cdot y\cdot y\cdot y\cdot y}=\frac{x\cdot x}{2\cdot y\cdot y\cdot y}=\frac{x^2}{2y^3}$$

35. $$\frac{-9xy^5}{3x^2y}=\frac{(-1)\cdot 3\cdot 3\cdot x\cdot y\cdot y\cdot y\cdot y\cdot y}{3\cdot x\cdot x\cdot y}=\frac{(-1)\cdot 3\cdot y\cdot y\cdot y\cdot y}{x}=\frac{-3y^4}{x}$$

37. $$\frac{-6x^2y}{-12x^3y^4}=\frac{(-1)\cdot 2\cdot 3\cdot x\cdot x\cdot y}{(-1)\cdot 2\cdot 2\cdot 3\cdot x\cdot x\cdot x\cdot y\cdot y\cdot y\cdot y}=\frac{1}{2\cdot x\cdot y\cdot y\cdot y}=\frac{1}{2xy^3}$$

39. $$\frac{-25x^3y^2}{-5x^2y^4}=\frac{(-1)\cdot 5\cdot 5\cdot x\cdot x\cdot x\cdot y\cdot y}{(-1)\cdot 5\cdot x\cdot x\cdot y\cdot y\cdot y\cdot y}=\frac{5\cdot x}{y\cdot y}=\frac{5x}{y^2}$$

41. $$\frac{6(x^2-y^2)}{18(x+y)}=\frac{6\cdot(x+y)\cdot(x-y)}{6\cdot 3\cdot(x+y)}=\frac{x-y}{3}$$

43. $$\frac{-9(x^2-y^2)}{3(x+y)}=\frac{-3\cdot 3\cdot(x+y)\cdot(x-y)}{3\cdot(x+y)}=-3(x-y) \text{ or } 3y-3x$$

45. $$\frac{-6(x+y)}{24(x^2-y^2)}=\frac{-1\cdot 6\cdot(x+y)}{4\cdot 6\cdot(x+y)(x-y)}=\frac{-1}{4(x-y)} \text{ or } \frac{1}{4y-4x}$$

47. $$\frac{-5(x-2)}{-10(x^2-4)}=\frac{-5\cdot(x-2)}{-5\cdot 2\cdot(x-2)(x+2)}=\frac{1}{2(x+2)} \text{ or } \frac{1}{2x+4}$$

49. $$\frac{-3(x-y)}{-3(x^2-y^2)}=\frac{-3\cdot(x-y)}{-3\cdot(x-y)\cdot(x+y)}=\frac{1}{x+y}$$

51. $$\frac{4x-4y}{8x-8y}=\frac{4(x-y)}{8(x-y)}=\frac{4}{8}=\frac{1}{2}$$

53. $$\frac{4x-8y}{12x-24y}=\frac{4(x-2y)}{12(x-2y)}=\frac{4}{12}=\frac{1}{3}$$

55. $$-\frac{6}{6+12y}=\frac{-1\cdot 6}{6(1+2y)}=\frac{-1}{1+2y}$$

57. $$-\frac{x}{x+2xy}=\frac{-1\cdot x}{x(1+2y)}=\frac{-1}{1+2y}$$

59. $$-\frac{6y}{6xy+12y}=\frac{-1\cdot 6y}{6y(x+2)}=\frac{-1}{x+2}$$

61. $$\frac{3x-2y}{2y-3x}=\frac{-1(2y-3x)}{2y-3x}=-1$$

63. $$\frac{x^2+4x-5}{1-x}=\frac{(x-1)(x+5)}{-1(x-1)}=\frac{x+5}{-1}=-(x+5) \text{ or } -x-5$$

65. $$\frac{x^2-6x+8}{4-x}=\frac{(x-4)(x-2)}{-1(x-4)}=\frac{x-2}{-1}=-(x-2) \text{ or } 2-x$$

67. $\dfrac{2-x}{x^2+4x-12}=\dfrac{-1(x-2)}{(x-2)(x+6)}=\dfrac{-1}{x+6}$

69. $-\dfrac{3-x}{x^2-5x+6}=\dfrac{x-3}{(x-3)(x-2)}=\dfrac{1}{x-2}$

71. a. $S(t)=-0.3t^2+10t+50$

$S(0)=-0.3(0)^2+10(0)+50=50$

In 1980, a total of \$50 million was spent.

2000 is 20 years after 1980, so $t=20$.

$S(20)=-0.3(20)^2+10(20)+50=130$

In 2000, a total of \$130 million was spent.

b. $N(t)=-0.13t^2+5t+30$

$N(0)=-0.13(0)^2+5(0)+30=30$

In 1980, the national spending was \$30 million.

$N(20)=-0.13(20)^2+5(20)+30=78$

In 2000, the national spending was \$78 million.

c. $L(t)=-0.17t^2+5t+20$

$L(0)=-0.17(0)^2+5(0)+20=20$

In 1980, the local spending was \$20 million.

$L(20)=-0.17(20)^2+5(20)+20=52$

In 2000, the local spending was \$52 million.

d. $\dfrac{N(t)}{S(t)}$ is the fraction of the total advertising that is spent on national advertising.

73. a. $\dfrac{T(t)}{S(t)}=\dfrac{-0.07t^2+2t+11}{-0.3t^2+10t+50}$

2010 is 30 years after 1980, so $t=30$.

$\dfrac{T(30)}{S(30)}=\dfrac{-0.07(30)^2+2(30)+11}{-0.3(30)^2+10(30)+50}=0.1$

Therefore, 10% of the total amount spent annually on advertising will be spent on television in the year 2010.

b. $\dfrac{L(t)}{S(t)}=\dfrac{-0.17t^2+5t+20}{-0.3t^2+10t+50}$

$\dfrac{L(30)}{S(30)}=\dfrac{-0.17(30)^2+5(30)+20}{-0.3(30)^2+10(30)+50}$

$=0.2125$

Therefore, 21.25% of the total amount spent annually on advertising will be spent on local advertising in the year 2010.

c. $\dfrac{T(t)}{S(t)}$ is the fraction of the total advertising that is spent on TV advertising.

75. $\dfrac{3}{2}\cdot\dfrac{4}{9}=\dfrac{3\cdot 4}{2\cdot 9}=\dfrac{3\cdot 2\cdot 2}{2\cdot 3\cdot 3}=\dfrac{2}{3}$

77. $x^2+2x-3=(x+3)(x-1)$

79. $x^2-7x+10=(x-5)(x-2)$

81. $\dfrac{40}{60}=\dfrac{2}{3}$

83. a. $\dfrac{2000}{500}=\dfrac{4}{1}$

The reduced transmission ratio is 4 to 1.

b. $\dfrac{5}{1}=\dfrac{?}{500}$

$\dfrac{5\times 500}{1\times 500}=\dfrac{2500}{500}$

Thus, the engine speed is 2500 rpm.

85. a. Yes; answers may vary. Sample answer: x^2 is always ≥ 0. So, if a is positive, $\dfrac{1}{x^2+a}$ is always defined.

b. No; answers may vary. Sample answer: Consider the case where $x=1$ and $a=-1$. Then $\dfrac{1}{x^2+a}=\dfrac{1}{1^2-1}=\dfrac{1}{0}$ which is undefined.

87. If $\dfrac{P(x)}{Q(x)} = -1$, then $P(x) = -Q(x)$.

89. $\dfrac{x^2-9}{x+3} = \dfrac{(x+3)(x-3)}{x+3} = x-3$

91. $\dfrac{10x-15y}{4x-6y} = \dfrac{5(2x-3y)}{2(2x-3y)} = \dfrac{5}{2}$

93. $\dfrac{x+4}{-(x^2-16)} = \dfrac{x+4}{-1\cdot(x+4)(x-4)}$
$= \dfrac{1}{-1\cdot(x-4)}$
$= \dfrac{-1}{x-4}$ or $\dfrac{1}{4-x}$

95. $\dfrac{-6(x^2-y^2)}{-3(x-y)} = \dfrac{-3\cdot 2(x-y)(x+y)}{-3\cdot(x-y)}$
$= 2(x+y)$ or $2x+2y$

97. Since $24y^3 = 8y^2\cdot 3y$,
$\dfrac{3x}{8y^2} = \dfrac{3x\cdot 3y}{8y^2\cdot 3y} = \dfrac{9xy}{24y^3}$.

99. $x^2-4=0$
$(x-2)(x+2)=0$
$x-2=0$ or $x+2=0$
$x=2$ $\quad x=-2$

Thus $\dfrac{x^2+1}{x^2-4}$ is undefined for $x = 2$ or $x = -2$.

6.2 Multiplication and Division or Rational Expressions

Problems 6.2

1. a. $\dfrac{m}{4}\cdot\dfrac{5}{n} = \dfrac{5m}{4n}$

b. $\dfrac{5y^3}{2}\cdot\dfrac{4x}{15y} = \dfrac{20xy^3}{30y}$
$= \dfrac{10\cdot 2\cdot x\cdot y\cdot y^2}{10\cdot 3\cdot y}$
$= \dfrac{2xy^2}{3}$

2. a. $\dfrac{-7m}{5n^2}\cdot\dfrac{15n}{21m^2} = \dfrac{-7m}{5n^2}\cdot\dfrac{5n}{7m^2}$
$= \dfrac{-1\cdot 7m}{5n\cdot n}\cdot\dfrac{5n}{7m\cdot m}$
$= \dfrac{-1}{mn}$

b. $5x^2\cdot\dfrac{7y}{10x^2} = \dfrac{5x^2}{1}\cdot\dfrac{7y}{5x^2\cdot 2} = \dfrac{7y}{2}$

3. a. $(m+2)\cdot\dfrac{m+3}{m^2-4} = \dfrac{m+2}{1}\cdot\dfrac{m+3}{(m+2)(m-2)}$
$= \dfrac{m+3}{m-2}$

b. $\dfrac{y^2-y-12}{y-2}\cdot\dfrac{2-y}{y+3}$
$= \dfrac{(y+3)(y-4)}{y-2}\cdot\dfrac{-1\cdot(y-2)}{y+3}$
$= -1\cdot(y-4)$
$= 4-y$

4. a. $\frac{y^2-9}{y+5} \div (y-3) = \frac{y^2-9}{y+5} \div \frac{y-3}{1}$
$= \frac{y^2-9}{y+5} \cdot \frac{1}{y-3}$
$= \frac{(y+3)(y-3)}{y+5} \cdot \frac{1}{y-3}$
$= \frac{y+3}{y+5}$

b. $\frac{y+4}{y-4} \div \frac{y^2-16}{4-y} = \frac{y+4}{y-4} \cdot \frac{4-y}{y^2-16}$
$= \frac{y+4}{y-4} \cdot \frac{-1 \cdot (y-4)}{(y+4)(y-4)}$
$= \frac{-1}{y-4}$ or $\frac{1}{4-y}$

5. a. $\frac{y^2-3y+2}{y^2-4y+3} \div \frac{y^2-49}{y^2-5y-14}$
$= \frac{y^2-3y+2}{y^2-4y+3} \cdot \frac{y^2-5y-14}{y^2-49}$
$= \frac{(y-2)(y-1)}{(y-3)(y-1)} \cdot \frac{(y-7)(y+2)}{(y-7)(y+7)}$
$= \frac{(y-2)(y+2)}{(y-3)(y+7)}$
$= \frac{y^2-4}{y^2+4y-21}$

b. $\frac{y^2-1}{y^2-y-6} \div \frac{y^2-3y+2}{y^2-9}$
$= \frac{y^2-1}{y^2-y-6} \cdot \frac{y^2-9}{y^2-3y+2}$
$= \frac{(y-1)(y+1)}{(y-3)(y+2)} \cdot \frac{(y-3)(y+3)}{(y-2)(y-1)}$
$= \frac{(y+1)(y+3)}{(y+2)(y-2)}$
$= \frac{y^2+4y+3}{y^2-4}$

Exercises 6.2

1. $\frac{x}{3} \cdot \frac{8}{y} = \frac{8x}{3y}$

3. $\frac{-6x^2}{7} \cdot \frac{14y}{9x} = \frac{-2 \cdot 3 \cdot x \cdot x}{7} \cdot \frac{7 \cdot 2y}{3 \cdot 3 \cdot x} = \frac{-4xy}{3}$

5. $7x^2 \cdot \frac{3y}{14x^2} = \frac{7x^2}{1} \cdot \frac{3y}{14x^2} = \frac{7x^2}{1} \cdot \frac{3y}{7x^2 \cdot 2} = \frac{3y}{2}$

7. $\frac{-4y}{7x^2} \cdot 14x^3 = \frac{-4y}{7x^2} \cdot \frac{14x^3}{1}$
$= \frac{-4y}{7x^2} \cdot \frac{7x^2 \cdot 2x}{1}$
$= -8xy$

9. $(x-7) \cdot \frac{x+1}{x^2-49} = \frac{x-7}{1} \cdot \frac{x+1}{x^2-49}$
$= \frac{x-7}{1} \cdot \frac{x+1}{(x-7)(x+7)}$
$= \frac{x+1}{x+7}$

11. $-2(x+2) \cdot \frac{x-1}{x^2-4} = \frac{-2(x+2)}{1} \cdot \frac{x-1}{x^2-4}$
$= \frac{-2(x+2)}{1} \cdot \frac{x-1}{(x+2)(x-2)}$
$= \frac{-2(x-1)}{x-2}$
$= \frac{-2x+2}{x-2}$
$= \frac{2-2x}{x-2}$

13. $\frac{3}{x-5} \cdot \frac{x^2-25}{x+1} = \frac{3}{x-5} \cdot \frac{(x-5)(x+5)}{x+1}$
$= \frac{3(x+5)}{x+1}$
$= \frac{3x+15}{x+1}$

15. $\frac{x^2-x-6}{x-2}\cdot\frac{2-x}{x-3}$
$=\frac{(x-3)(x+2)}{x-2}\cdot\frac{-1\cdot(x-2)}{x-3}$
$=-1\cdot(x+2)$
$=-(x+2)$ or $-x-2$

17. $\frac{x-1}{3-x}\cdot\frac{x+3}{1-x}=\frac{x-1}{-1\cdot(x-3)}\cdot\frac{x+3}{-1\cdot(x-1)}$
$=\frac{x+3}{x-3}$

19. $\frac{3(x-5)}{14(4-x)}\cdot\frac{7(x-4)}{6(5-x)}$
$=\frac{3(x-5)}{-2\cdot7(x-4)}\cdot\frac{7(x-4)}{-2\cdot3(x-5)}$
$=\frac{1}{-2}\cdot\frac{1}{-2}$
$=\frac{1}{4}$

21. $\frac{6x^3}{x^2-16}\cdot\frac{x^2-5x+4}{3x^2}$
$=\frac{3x^2\cdot2x}{(x+4)(x-4)}\cdot\frac{(x-4)(x-1)}{3x^2}$
$=\frac{2x(x-1)}{x+4}$
$=\frac{2x^2-2x}{x+4}$

23. $\frac{y^2+2y-3}{y-5}\cdot\frac{y^2-3y-10}{y^2+5y-6}$
$=\frac{(y-1)(y+3)}{y-5}\cdot\frac{(y-5)(y+2)}{(y+6)(y-1)}$
$=\frac{(y+3)(y+2)}{y+6}$
$=\frac{y^2+5y+6}{y+6}$

25. $\frac{2y^2+y-3}{6-11y-10y^2}\cdot\frac{5y^3-2y^2}{3y^2-5y+2}$
$=\frac{(2y+3)(y-1)}{-1(2y+3)(5y-2)}\cdot\frac{y^2(5y-2)}{(3y-2)(y-1)}$
$=\frac{y^2}{-1(3y-2)}$
$=\frac{y^2}{2-3y}$

27. $\frac{15x^2-x-2}{2x^2+5x-18}\cdot\frac{2x^2+x-36}{3x^2-11x-4}$
$=\frac{(5x-2)(3x+1)}{(2x+9)(x-2)}\cdot\frac{(2x+9)(x-4)}{(3x+1)(x-4)}$
$=\frac{5x-2}{x-2}$

29. $\frac{27y^3+8}{6y^2+19y+10}\cdot\frac{4y^2-25}{9y^2-6y+4}$
$=\frac{(3y+2)(9y^2-6y+4)}{(3y+2)(2y+5)}\cdot\frac{(2y-5)(2y+5)}{9y^2-6y+4}$
$=2y-5$

31. $\frac{x^2-1}{x+2}\div(x+1)=\frac{x^2-1}{x+2}\div\frac{x+1}{1}$
$=\frac{x^2-1}{x+2}\cdot\frac{1}{x+1}$
$=\frac{(x+1)(x-1)}{x+2}\cdot\frac{1}{x+1}$
$=\frac{x-1}{x+2}$

33. $\frac{x^2-25}{x-3}\div5(x+5)=\frac{x^2-25}{x-3}\div\frac{5(x+5)}{1}$
$=\frac{x^2-25}{x-3}\cdot\frac{1}{5(x+5)}$
$=\frac{(x+5)(x-5)}{x-3}\cdot\frac{1}{5(x+5)}$
$=\frac{x-5}{5(x-3)}$
$=\frac{x-5}{5x-15}$

35. $$(x+3)\div\frac{x^2-9}{x+4}=\frac{x+3}{1}\div\frac{x^2-9}{x+4}$$
$$=\frac{x+3}{1}\cdot\frac{x+4}{x^2-9}$$
$$=\frac{x+3}{1}\cdot\frac{x+4}{(x+3)(x-3)}$$
$$=\frac{x+4}{x-3}$$

37. $$\frac{-3}{x-4}\div\frac{6(x+3)}{5(x^2-16)}=\frac{-3}{x-4}\cdot\frac{5(x^2-16)}{6(x+3)}$$
$$=\frac{-1\cdot 3}{x-4}\cdot\frac{5(x+4)(x-4)}{2\cdot 3(x+3)}$$
$$=\frac{-1\cdot 5\cdot(x+4)}{2(x+3)}$$
$$=\frac{-5x-20}{2x+6}$$

39. $$\frac{-4(x+1)}{3(x+2)}\div\frac{-8(x^2-1)}{6(x^2-4)}$$
$$=\frac{-4(x+1)}{3(x+2)}\cdot\frac{6(x^2-4)}{-8(x^2-1)}$$
$$=\frac{-4(x+1)}{3(x+2)}\cdot\frac{3\cdot 2(x-2)(x+2)}{-4\cdot 2\cdot(x-1)(x+1)}$$
$$=\frac{x-2}{x-1}$$

41. $$\frac{x+3}{x-3}\div\frac{x^2-1}{3-x}=\frac{x+3}{x-3}\cdot\frac{3-x}{x^2-1}$$
$$=\frac{x+3}{x-3}\cdot\frac{-1(x-3)}{(x-1)(x+1)}$$
$$=\frac{-(x+3)}{(x-1)(x+1)}$$
$$=\frac{x+3}{-(x^2-1)}$$
$$=\frac{x+3}{1-x^2}$$

43. $$\frac{x^2-4}{7(x^2-9)}\div\frac{x+2}{14(x+3)}$$
$$=\frac{x^2-4}{7(x^2-9)}\cdot\frac{14(x+3)}{x+2}$$
$$=\frac{(x+2)(x-2)}{7(x+3)(x-3)}\cdot\frac{2\cdot 7(x+3)}{x+2}$$
$$=\frac{2(x-2)}{x-3}$$
$$=\frac{2x-4}{x-3}$$

45. $$\frac{3(x^2-36)}{14(5-x)}\div\frac{6(6-x)}{7(x^2-25)}$$
$$=\frac{3(x^2-36)}{14(5-x)}\cdot\frac{7(x^2-25)}{6(6-x)}$$
$$=\frac{3(x-6)(x+6)}{7(-2)(x-5)}\cdot\frac{7(x-5)(x+5)}{(-2)(3)(x-6)}$$
$$=\frac{(x+6)(x+5)}{(-2)(-2)}$$
$$=\frac{x^2+11x+30}{4}$$

47. $$\frac{x+2}{x-1}\div\frac{x^2+5x+6}{x^2-4x+4}=\frac{x+2}{x-1}\cdot\frac{x^2-4x+4}{x^2+5x+6}$$
$$=\frac{x+2}{x-1}\cdot\frac{(x-2)(x-2)}{(x+3)(x+2)}$$
$$=\frac{(x-2)(x-2)}{(x-1)(x+3)}$$
$$=\frac{x^2-4x+4}{x^2+2x-3}$$

49. $$\frac{x-5}{x+3}\div\frac{5(x-5)}{x^2+9x+18}=\frac{x-5}{x+3}\cdot\frac{x^2+9x+18}{5(x-5)}$$
$$=\frac{x-5}{x+3}\cdot\frac{(x+6)(x+3)}{5(x-5)}$$
$$=\frac{x+6}{5}$$

51. $\frac{x^2+2x-3}{x-5} \div \frac{x^2+6x+9}{x^2-2x-15}$
$= \frac{x^2+2x-3}{x-5} \cdot \frac{x^2-2x-15}{x^2+6x+9}$
$= \frac{(x+3)(x-1)}{x-5} \cdot \frac{(x-5)(x+3)}{(x+3)(x+3)}$
$= x-1$

53. $\frac{x^2-1}{x^2+3x-10} \div \frac{x^2-3x-4}{x^2-25}$
$= \frac{x^2-1}{x^2+3x-10} \cdot \frac{x^2-25}{x^2-3x-4}$
$= \frac{(x+1)(x-1)}{(x+5)(x-2)} \cdot \frac{(x-5)(x+5)}{(x-4)(x+1)}$
$= \frac{(x-1)(x-5)}{(x-2)(x-4)}$
$= \frac{x^2-6x+5}{x^2-6x+8}$

55. $\frac{x^2+3x-4}{x^2+7x+12} \div \frac{x^2+x-2}{x^2+5x+6}$
$= \frac{x^2+3x-4}{x^2+7x+12} \cdot \frac{x^2+5x+6}{x^2+x-2}$
$= \frac{(x+4)(x-1)}{(x+4)(x+3)} \cdot \frac{(x+3)(x+2)}{(x-1)(x+2)}$
$= 1$

57. $\frac{x^2-y^2}{x^2-2xy} \div \frac{x^2+xy-2y^2}{x^2-4y^2}$
$= \frac{x^2-y^2}{x^2-2xy} \cdot \frac{x^2-4y^2}{x^2+xy-2y^2}$
$= \frac{(x+y)(x-y)}{x(x-2y)} \cdot \frac{(x-2y)(x+2y)}{(x-y)(x+2y)}$
$= \frac{x+y}{x}$

59. $\frac{x^2+2xy-3y^2}{y^2-7y+10} \div \frac{x^2+5xy-6y^2}{y^2-3y-10}$
$= \frac{x^2+2xy-3y^2}{y^2-7y+10} \cdot \frac{y^2-3y-10}{x^2+5xy-6y^2}$
$= \frac{(x+3y)(x-y)}{(y-5)(y-2)} \cdot \frac{(y-5)(y+2)}{(x+6y)(x-y)}$
$= \frac{(x+3y)(y+2)}{(y-2)(x+6y)}$
$= \frac{xy+2x+6y+3y^2}{xy-2x-12y+6y^2}$

61. $\frac{2x^2-x-28}{3x^2-x-2} \div \frac{4x^2+16x+7}{3x^2+11x+6}$
$= \frac{2x^2-x-28}{3x^2-x-2} \cdot \frac{3x^2+11x+6}{4x^2+16x+7}$
$= \frac{(2x+7)(x-4)}{(3x+2)(x-1)} \cdot \frac{(3x+2)(x+3)}{(2x+7)(2x+1)}$
$= \frac{(x-4)(x+3)}{(x-1)(2x+1)}$
$= \frac{x^2-x-12}{2x^2-x-1}$

63. $\frac{(a^3-27)(a^2-9)}{(a-3)^2(a+3)^3} \div \frac{a^2+3a+9}{a^2+3a} = \frac{(a^3-27)(a^2-9)}{(a-3)^2(a+3)^3} \cdot \frac{a^2+3a}{a^2+3a+9}$

$= \frac{(a-3)(a^2+3a+9)(a+3)(a-3)}{(a-3)(a-3)(a+3)(a+3)(a+3)} \cdot \frac{a(a+3)}{a^2+3a+9}$

$= \frac{a}{a+3}$

65. $\frac{7}{8}+\frac{2}{5}=\frac{7\cdot 5}{8\cdot 5}+\frac{2\cdot 8}{5\cdot 8}=\frac{35}{40}+\frac{16}{40}=\frac{35+16}{40}=\frac{51}{40}$

67. $\frac{7}{8}-\frac{2}{5}=\frac{7\cdot 5}{8\cdot 5}-\frac{2\cdot 8}{5\cdot 8}=\frac{35}{40}-\frac{16}{40}=\frac{35-16}{40}=\frac{19}{40}$

69. $\frac{5}{2}-\frac{1}{6}=\frac{5\cdot 3}{2\cdot 3}-\frac{1}{6}=\frac{15}{6}-\frac{1}{6}=\frac{15-1}{6}=\frac{14}{6}=\frac{7}{3}$

71. $R\cdot\frac{R_T}{R-R_T}=\frac{R}{1}\cdot\frac{R_T}{R-R_T}=\frac{RR_T}{R-R_T}$

73. $C_R = N\cdot C = \frac{3000}{x}\cdot(20+3x)=\frac{3000}{x}\cdot\frac{20+3x}{1}=\frac{3000(20+3x)}{x}=\frac{60,000+9000x}{x}$

75. Answers may vary.

77. Answers may vary.

79. $\frac{x^2-1}{x^2-x-6} \div \frac{x^2-3x+2}{x^2-9} = \frac{x^2-1}{x^2-x-6}\cdot\frac{x^2-9}{x^2-3x+2}$

$= \frac{(x-1)(x+1)}{(x-3)(x+2)}\cdot\frac{(x-3)(x+3)}{(x-2)(x-1)}$

$= \frac{(x+1)(x+3)}{(x+2)(x-2)}$

$= \frac{x^2+4x+3}{x^2-4}$

81. $\frac{x+6}{x-6} \div \frac{x^2-36}{6-x} = \frac{x+6}{x-6}\cdot\frac{6-x}{x^2-36} = \frac{x+6}{x-6}\cdot\frac{-1(x-6)}{(x-6)(x+6)} = \frac{-1}{x-6} = \frac{1}{6-x}$

83. $\frac{x^3-27}{x+3} \div \frac{x^2+3x+9}{x^2-9} = \frac{x^3-27}{x+3}\cdot\frac{x^2-9}{x^2+3x+9}$

$= \frac{(x-3)(x^2+3x+9)}{x+3}\cdot\frac{(x+3)(x-3)}{x^2+3x+9}$

$= (x-3)(x-3)$

$= x^2-6x+9$

85. $\frac{x^2+x-6}{x-3}\cdot\frac{3-x}{x+3}=\frac{(x+3)(x-2)}{x-3}\cdot\frac{-1(x-3)}{x+3}$
$=-(x-2)$
$=2-x$

87. $\frac{6x}{11y^2}\cdot 22y^2=\frac{6x}{11y^2}\cdot\frac{22y^2}{1}$
$=\frac{6x}{11y^2}\cdot\frac{11y^2\cdot 2}{1}$
$=6x\cdot 2$
$=12x$

89. $\frac{2x^2-x-28}{3x^2-x-2}\cdot\frac{3x^2+11x+6}{4x^2+16x+7}$
$=\frac{(2x+7)(x-4)}{(3x+2)(x-1)}\cdot\frac{(3x+2)(x+3)}{(2x+1)(2x+7)}$
$=\frac{(x-4)(x+3)}{(x-1)(2x+1)}$
$=\frac{x^2-x-12}{2x^2-x-1}$

6.3 Addition and Subtraction of Rational Expressions

Problems 6.3

1. a. $\frac{4}{5(y-1)}+\frac{1}{5(y-1)}=\frac{4+1}{5(y-1)}$
$=\frac{5}{5(y-1)}$
$=\frac{1}{y-1}$

b. $\frac{9}{7(y+2)}-\frac{2}{7(y+2)}=\frac{9-7}{7(y+2)}$
$=\frac{7}{7(y+2)}$
$=\frac{1}{y+2}$

2. $12=2^2\cdot 3^1$
$18=2^1\cdot 3^2$
$\text{LCD}=2^2\cdot 3^2=36$
$\frac{5}{12}=\frac{5\cdot 3}{12\cdot 3}=\frac{15}{36}$
$\frac{7}{18}=\frac{7\cdot 2}{18\cdot 2}=\frac{14}{36}$
$\frac{5}{12}+\frac{7}{18}=\frac{15}{36}+\frac{14}{36}=\frac{15+14}{36}=\frac{29}{36}$

3. $15=3\cdot 5$
$12=2^2\cdot 3$
$\text{LCD}=2^2\cdot 3\cdot 5=60$
$\frac{13}{15}=\frac{13\cdot 4}{15\cdot 4}=\frac{52}{60}$
$\frac{7}{12}=\frac{7\cdot 5}{12\cdot 5}=\frac{35}{60}$
$\frac{13}{15}-\frac{7}{12}=\frac{52}{60}-\frac{35}{60}=\frac{52-35}{60}=\frac{17}{60}$

4. a. Since 5 and y don't have any common factors, the LCD is $5y$.
$\frac{3}{5}=\frac{3\cdot y}{5\cdot y}=\frac{3y}{5y}$
$\frac{2}{y}=\frac{2\cdot 5}{y\cdot 5}=\frac{10}{5y}$
$\frac{3}{5}+\frac{2}{y}=\frac{3y}{5y}+\frac{10}{5y}=\frac{3y+10}{5y}$

b. Since $(y-2)$ and $(y+1)$ don't have any common factors, the LCD is $(y-2)(y+1)$.
$\frac{2}{y-2}=\frac{2(y+1)}{(y-2)(y+1)}$
$\frac{1}{y+1}=\frac{1\cdot(y-2)}{(y+1)(y-2)}=\frac{y-2}{(y-2)(y+1)}$

$\frac{2}{y-2}-\frac{1}{y+1}$
$=\frac{2(y+1)}{(y-2)(y+1)}-\frac{y-2}{(y-2)(y+1)}$
$=\frac{2(y+1)-(y-2)}{(y-2)(y+1)}$
$=\frac{2y+2-y+2}{(y-2)(y+1)}$
$=\frac{y+4}{(y-2)(y+1)}$

5. $(x+1)(x-2)$
$x^2-4=(x-2)(x+2)$
$\text{LCD}=(x+1)(x-2)(x+2)$
$\frac{x-3}{(x+1)(x-2)}=\frac{(x-3)(x+2)}{(x+1)(x-2)(x+2)}$
$\frac{x+3}{x^2-4}=\frac{(x+3)(x+1)}{(x+2)(x-2)(x+1)}$
$\frac{x-3}{(x+1)(x-2)}-\frac{x+3}{x^2-4}$
$=\frac{(x-3)(x+2)}{(x+1)(x-2)(x+2)}-\frac{(x+3)(x+1)}{(x+1)(x-2)(x+2)}$
$=\frac{(x-3)(x+2)-(x+3)(x+1)}{(x+1)(x-2)(x+2)}$
$=\frac{x^2-x-6-x^2-4x-3}{(x+1)(x-2)(x+2)}$
$=\frac{-5x-9}{(x+1)(x-2)(x+2)}$

6. a. $\frac{T(t)}{F(t)}=\frac{-0.4t^2+10t+360}{-0.1t^2+6t+125}$

b. 2020:
$\frac{F(35)}{T(35)}=\frac{-0.1(35)^2+6(35)+125}{-0.4(35)^2+10(35)+360}$
≈ 0.97 or 97%
2021:
$\frac{F(36)}{T(36)}=\frac{-0.1(36)^2+6(36)+125}{-0.4(36)^2+10(36)+360}$
≈ 1.05 or 105%
No; in 2020 the percent of female students would be 97% and in 2021 it would be 105%.

Exercises 6.3

1. a. $\frac{2}{7}+\frac{3}{7}=\frac{2+3}{7}=\frac{5}{7}$

b. $\frac{3}{x}+\frac{8}{x}=\frac{3+8}{x}=\frac{11}{x}$

3. a. $\frac{8}{9}-\frac{2}{9}=\frac{8-2}{9}=\frac{6}{9}=\frac{2}{3}$

b. $\frac{6}{x}-\frac{2}{x}=\frac{6-2}{x}=\frac{4}{x}$

5. a. $\frac{6}{7}+\frac{8}{7}=\frac{6+8}{7}=\frac{14}{7}=2$

b. $\frac{3}{2x}+\frac{7}{2x}=\frac{3+7}{2x}=\frac{10}{2x}=\frac{5}{x}$

7. a. $\frac{8}{3}-\frac{2}{3}=\frac{8-2}{3}=\frac{6}{3}=2$

b. $\frac{11}{3(x+1)}-\frac{9}{3(x+1)}=\frac{11-9}{3(x+1)}=\frac{2}{3(x+1)}$

9. a. $\frac{8}{9}+\frac{4}{9}=\frac{8+4}{9}=\frac{12}{9}=\frac{4}{3}$

b. $\frac{7x}{4(x+1)}+\frac{3x}{4(x+1)}=\frac{7x+3x}{4(x+1)}$
$=\frac{10x}{4(x+1)}$
$=\frac{5x}{2(x+1)}$

11. a. Since 4 and 3 don't have any common factors, the LCD is $4\cdot 3=12$.
$\frac{3}{4}=\frac{3\cdot 3}{4\cdot 3}=\frac{9}{12}$
$\frac{1}{3}=\frac{1\cdot 4}{3\cdot 4}=\frac{4}{12}$
$\frac{3}{4}-\frac{1}{3}=\frac{9}{12}-\frac{4}{12}=\frac{9-4}{12}=\frac{5}{12}$

b. Since x and 8 don't have any common factors, the LCD is $x \cdot 8 = 8x$.

$\frac{7}{x} = \frac{7 \cdot 8}{x \cdot 8} = \frac{56}{8x}$

$\frac{3}{8} = \frac{3 \cdot x}{8 \cdot x} = \frac{3x}{8x}$

$\frac{7}{x} - \frac{3}{8} = \frac{56}{8x} - \frac{3x}{8x} = \frac{56 - 3x}{8x}$

13. a. Since 5 and 7 don't have any common factors, the LCD is $5 \cdot 7 = 35$.

$\frac{1}{5} = \frac{1 \cdot 7}{5 \cdot 7} = \frac{7}{35}$

$\frac{1}{7} = \frac{1 \cdot 5}{7 \cdot 5} = \frac{5}{35}$

$\frac{1}{5} + \frac{1}{7} = \frac{7}{35} + \frac{5}{35} = \frac{7+5}{35} = \frac{12}{35}$

b. Since x and 9 don't have any common factors, the LCD is $9x$.

$\frac{4}{x} = \frac{4 \cdot 9}{x \cdot 9} = \frac{36}{9x}$

$\frac{x}{9} = \frac{x \cdot x}{9 \cdot x} = \frac{x^2}{9x}$

$\frac{4}{x} + \frac{x}{9} = \frac{36}{9x} + \frac{x^2}{9x} = \frac{36 + x^2}{9x}$ or $\frac{x^2 + 36}{9x}$

15. a. 5

$15 = 3 \cdot 5$

$\text{LCD} = 3 \cdot 5 = 15$

$\frac{2}{5} = \frac{2 \cdot 3}{5 \cdot 3} = \frac{6}{15}$

$\frac{2}{5} - \frac{4}{15} = \frac{6}{15} - \frac{4}{15} = \frac{6-4}{15} = \frac{2}{15}$

b. $7(x - 1)$

$14(x - 1) = 2 \cdot 7(x - 1)$

$\text{LCD} = 2 \cdot 7(x - 1) = 14(x - 1)$

$\frac{4}{7(x-1)} = \frac{4 \cdot 2}{2 \cdot 7(x-1)} = \frac{8}{14(x-1)}$

$\frac{4}{7(x-1)} - \frac{3}{14(x-1)}$

$= \frac{8}{14(x-1)} - \frac{3}{14(x-1)}$

$= \frac{8-3}{14(x-1)}$

$= \frac{5}{14(x-1)}$

17. a. Since 7 and 8 don't have any common factors, the LCD is $7 \cdot 8 = 56$.

$\frac{4}{7} = \frac{4 \cdot 8}{7 \cdot 8} = \frac{32}{56}$

$\frac{3}{8} = \frac{3 \cdot 7}{8 \cdot 7} = \frac{21}{56}$

$\frac{4}{7} + \frac{3}{8} = \frac{32}{56} + \frac{21}{56} = \frac{32+21}{56} = \frac{53}{56}$

b. Since $(x + 1)$ and $(x - 2)$ don't have any common factors, the LCD is $(x + 1)(x - 2)$.

$\frac{3}{x+1} = \frac{3(x-2)}{(x+1)(x-2)}$

$\frac{5}{x-2} = \frac{5(x+1)}{(x-2)(x+1)}$

$\frac{3}{x+1} + \frac{5}{x-2}$

$= \frac{3(x-2)}{(x+1)(x-2)} + \frac{5(x+1)}{(x-2)(x+1)}$

$= \frac{3(x-2) + 5(x+1)}{(x+1)(x-2)}$

$= \frac{3x - 6 + 5x + 5}{(x+1)(x-2)}$

$= \frac{8x - 1}{(x+1)(x-2)}$

19. a. Since 8 and 3 don't have any common factors, the LCD is $8 \cdot 3 = 24$.

$\frac{7}{8} = \frac{7 \cdot 3}{5 \cdot 3} = \frac{21}{24}$

$\frac{1}{3} = \frac{1 \cdot 8}{3 \cdot 8} = \frac{8}{24}$

$\frac{7}{8} - \frac{1}{3} = \frac{21}{24} - \frac{8}{24} = \frac{21-8}{24} = \frac{13}{24}$

b. Since $(x-2)$ and $(x+1)$ don't have any common factors, the LCD is $(x-2)(x+1)$.

$$\frac{6}{x-2}=\frac{6(x+1)}{(x-2)(x+1)}$$
$$\frac{3}{x+1}=\frac{3(x-2)}{(x+1)(x-2)}$$
$$\frac{6}{x-2}-\frac{3}{x+1}$$
$$=\frac{6(x+1)}{(x-2)(x+1)}-\frac{3(x-2)}{(x-2)(x+1)}$$
$$=\frac{6(x+1)-3(x-2)}{(x-2)(x+1)}$$
$$=\frac{6x+6-3x+6}{(x-2)(x+1)}$$
$$=\frac{3x+12}{(x-2)(x+1)}$$

21. $x^2+3x-4=(x+4)(x-1)$

$x^2-16=(x+4)(x-4)$

LCD $=(x+4)(x-4)(x-1)$

$$\frac{x+1}{x^2+3x-4}=\frac{(x+1)(x-4)}{(x+4)(x-1)(x-4)}$$
$$\frac{x+2}{x^2-16}=\frac{(x+2)(x-1)}{(x+4)(x-4)(x-1)}$$
$$\frac{x+1}{x^2+3x-4}+\frac{x+2}{x^2-16}$$
$$=\frac{(x+1)(x-4)}{(x+4)(x-4)(x-1)}+\frac{(x+2)(x-1)}{(x+4)(x-4)(x-1)}$$
$$=\frac{(x+1)(x-4)+(x+2)(x-1)}{(x+4)(x-4)(x-1)}$$
$$=\frac{x^2-3x-4+x^2+x-2}{(x+4)(x-4)(x-1)}$$
$$=\frac{2x^2-2x-6}{(x+4)(x-4)(x-1)}$$

23. $x^2+3x-10=(x+5)(x-2)$

$x^2+x-6=(x+3)(x-2)$

LCD $=(x+5)(x-2)(x+3)$

$$\frac{3x}{x^2+3x-10}=\frac{3x(x+3)}{(x+5)(x-2)(x+3)}$$
$$\frac{2x}{x^2+x-6}=\frac{2x(x+5)}{(x+3)(x-2)(x+5)}$$
$$\frac{3x}{x^2+3x-10}+\frac{2x}{x^2+x-6}$$
$$=\frac{3x(x+3)}{(x+5)(x-2)(x+3)}+\frac{2x(x+5)}{(x+5)(x-2)(x+3)}$$
$$=\frac{3x(x+3)+2x(x+5)}{(x+5)(x-2)(x+3)}$$
$$=\frac{3x^2+9x+2x^2+10x}{(x+5)(x-2)(x+3)}$$
$$=\frac{5x^2+19x}{(x+5)(x-2)(x+3)}$$

25. $x^2-y^2=(x+y)(x-y)$

$(x+y)^2$

LCD $=(x+y)^2(x-y)$

$$\frac{1}{x^2-y^2}=\frac{1\cdot(x+y)}{(x+y)(x-y)(x+y)}$$
$$\frac{5}{(x+y)^2}=\frac{5(x-y)}{(x+y)^2(x-y)}$$
$$\frac{1}{x^2-y^2}+\frac{5}{(x+y)^2}$$
$$=\frac{(x+y)}{(x+y)^2(x-y)}+\frac{5(x-y)}{(x+y)^2(x-y)}$$
$$=\frac{(x+y)+5(x-y)}{(x+y)^2(x-y)}$$
$$=\frac{x+y+5x-5y}{(x+y)^2(x-y)}$$
$$=\frac{6x-4y}{(x+y)^2(x-y)}$$

27. $x-5$

$x^2-25=(x-5)(x+5)$

$\text{LCD}=(x-5)(x+5)$

$$\frac{2}{x-5}=\frac{2(x+5)}{(x-5)(x+5)}$$

$$\frac{2}{x-5}-\frac{3x}{x^2-25}$$
$$=\frac{2(x+5)}{(x-5)(x+5)}-\frac{3x}{(x+5)(x-5)}$$
$$=\frac{2(x+5)-3x}{(x+5)(x-5)}$$
$$=\frac{2x+10-3x}{(x+5)(x-5)}$$
$$=\frac{10-x}{(x+5)(x-5)}$$

29. $x^2+3x+2=(x+2)(x+1)$

$x^2+5x+6=(x+2)(x+3)$

$\text{LCD}=(x+2)(x+1)(x+3)$

$$\frac{x-1}{x^2+3x+2}=\frac{(x-1)(x+3)}{(x+2)(x+1)(x+3)}$$
$$\frac{x+7}{x^2+5x+6}=\frac{(x+7)(x+1)}{(x+2)(x+3)(x+1)}$$
$$\frac{x-1}{x^2+3x+2}-\frac{x+7}{x^2+5x+6}$$
$$=\frac{(x-1)(x+3)}{(x+2)(x+1)(x+3)}-\frac{(x+7)(x+1)}{(x+2)(x+1)(x+3)}$$
$$=\frac{(x-1)(x+3)-(x+7)(x+1)}{(x+2)(x+1)(x+3)}$$
$$=\frac{x^2+2x-3-x^2-8x-7}{(x+2)(x+1)(x+3)}$$
$$=\frac{-6x-10}{(x+2)(x+1)(x+3)}$$

31. $y^2-1=(y+1)(y-1)$

$y+1$

$\text{LCD}=(y+1)(y-1)$

$$\frac{y}{y+1}=\frac{y(y-1)}{(y+1)(y-1)}$$
$$\frac{y}{y^2-1}+\frac{y}{y+1}=\frac{y}{(y+1)(y-1)}+\frac{y(y-1)}{(y+1)(y-1)}$$
$$=\frac{y+y(y-1)}{(y+1)(y-1)}$$
$$=\frac{y+y^2-y}{(y+1)(y-1)}$$
$$=\frac{y^2}{(y+1)(y-1)}$$

33. $y^2-16=(y-4)(y+4)$

$y-4$

$\text{LCD}=(y-4)(y+4)$

$$\frac{2y-1}{y-4}=\frac{(2y-1)(y+4)}{(y-4)(y+4)}$$
$$\frac{3y+1}{y^2-16}-\frac{2y-1}{y-4}$$
$$=\frac{3y+1}{(y-4)(y+4)}-\frac{(2y-1)(y+4)}{(y-4)(y+4)}$$
$$=\frac{3y+1-(2y-1)(y+4)}{(y-4)(y+4)}$$
$$=\frac{3y+1-2y^2-7y+4}{(y-4)(y+4)}$$
$$=\frac{5-4y-2y^2}{(y-4)(y+4)}$$

35. $x^2-x-2=(x-2)(x+1)$

$x^2+2x+1=(x+1)^2$

$\text{LCD}=(x-2)(x+1)^2$

$$\frac{x+1}{x^2-x-2}=\frac{(x+1)(x+1)}{(x-2)(x+1)(x+1)}$$
$$\frac{x-1}{x^2+2x+1}=\frac{(x-1)(x-2)}{(x+1)(x+1)(x-2)}$$

$$\frac{x+1}{x^2-x-2}+\frac{x-1}{x^2+2x+1}=\frac{(x+1)(x+1)}{(x-2)(x+1)^2}+\frac{(x-1)(x-2)}{(x-2)(x+1)^2}$$
$$=\frac{(x+1)(x+1)+(x-1)(x-2)}{(x-2)(x+1)^2}$$
$$=\frac{x^2+2x+1+x^2-3x+2}{(x-2)(x+1)^2}$$
$$=\frac{2x^2-x+3}{(x-2)(x+1)^2}$$

37. $a-w$

$a+w$

$a^2-w^2=(a-w)(a+w)$

$\text{LCD}=(a-w)(a+w)$

$$\frac{a}{a-w}=\frac{a(a+w)}{(a-w)(a+w)}$$
$$\frac{w}{a+w}=\frac{w(a-w)}{(a+w)(a-w)}$$
$$\frac{a}{a-w}-\frac{w}{a+w}-\frac{a^2+w^2}{a^2-w^2}=\frac{a(a+w)}{(a-w)(a+w)}-\frac{w(a-w)}{(a-w)(a+w)}-\frac{a^2+w^2}{a^2-w^2}$$
$$=\frac{a(a+w)-w(a-w)-(a^2+w^2)}{(a-w)(a+w)}$$
$$=\frac{a^2+aw-aw+w^2-a^2-w^2}{(a-w)(a+w)}$$
$$=\frac{0}{(a-w)(a+w)}$$
$$=0$$

39. $a^3+8=(a+2)(a^2-2a+4)$

a^2-2a+4

$\text{LCD}=(a+2)(a^2-2a+4)$

$$\frac{a+1}{a^2-2a+4}=\frac{(a+1)(a+2)}{(a^2-2a+4)(a+2)}$$

$$\begin{aligned}\frac{1}{a^3+8}+\frac{a+1}{a^2-2a+4} &= \frac{1}{(a+2)(a^2-2a+4)}+\frac{(a+1)(a+2)}{(a+2)(a^2-2a+4)}\\ &= \frac{1+(a+1)(a+2)}{(a+2)(a^2-2a+4)}\\ &= \frac{1+a^2+3a+2}{(a+2)(a^2-2a+4)}\\ &= \frac{a^2+3a+3}{(a+2)(a^2-2a+4)}\end{aligned}$$

41. a. $\dfrac{P(t)}{R(t)}=\dfrac{(0.04t^3-0.28t^2+3.2t+23)\text{ billion}}{(0.01t^3-0.12t^2+0.28t+22)\text{ million}}$

b. $\dfrac{P(0)}{R(0)}=\dfrac{23\text{ billion}}{22\text{ million}}\approx 1045.45$

In 1980, each recipient got \$1045.45.

c. $$\begin{aligned}\frac{P(20)}{R(20)} &= \frac{[0.04(20)^3-0.28(20)^2+3.2(20)+23]\text{ billion}}{[0.01(20)^3-0.12(20)^2+0.28(20)+22]\text{ million}}\\ &= \frac{295\text{ billion}}{59.6\text{ million}}\\ &\approx 4949.66\end{aligned}$$

In 2000, each recipient will get \$4949.66.

$$\begin{aligned}\frac{P(30)}{R(30)} &= \frac{[0.04(30)^3-0.28(30)^2+3.2(30)+23]\text{ billion}}{[0.01(30)^3-0.12(30)^2+0.28(30)+22]\text{ million}}\\ &= \frac{947\text{ billion}}{192.4\text{ million}}\\ &\approx 4922.04\end{aligned}$$

In 2010, each recipient will get \$4922.04.

43. \$1045.45 – \$428.57 = \$616.88

This average amount in 1980 was \$616.88.

45. $1\div\dfrac{20}{9}=1\cdot\dfrac{9}{20}=\dfrac{9}{20}$

47. $$\begin{aligned}12x\left(\frac{2}{x}+\frac{3}{2x}\right) &= 12x\cdot\frac{2}{x}+12x\cdot\frac{3}{2x}\\ &= 12\cdot 2+6\cdot 3\\ &= 24+18\\ &= 42\end{aligned}$$

49. $x^2\left(1-\dfrac{1}{x^2}\right)=x^2\cdot 1-x^2\cdot\dfrac{1}{x^2}=x^2-1$

51. $1+\frac{1}{2}=1+0.5=1.5$

53. $1+\frac{1}{2+\frac{1}{2+\frac{1}{2}}}=1+\frac{1}{1+1.4}=1.41\overline{6}$

55. 1.4167 – 1.4142 = 0.0025

57. Answers may vary.

59.
$$\frac{R(t)-P(t)}{G(t)}=\frac{(0.04t^2-0.59t+7.42)-(0.02t^2-0.25t+6)}{0.04t^2+2.34t+90}$$
$$=\frac{0.04t^2-0.59t+7.42-0.02t^2+0.25t-6}{0.04t^2+2.34t+90}$$
$$=\frac{0.02t^2-0.34t+1.42}{0.04t^2+2.34t+90}$$

61. Since $(x-1)$ and $(x+3)$ have no common factors, the LCD is $(x-1)(x+3)$.
$$\frac{5}{x-1}=\frac{5(x+3)}{(x-1)(x+3)}$$
$$\frac{3}{x+3}=\frac{3(x-1)}{(x+3)(x-1)}$$
$$\frac{5}{x-1}-\frac{3}{x+3}=\frac{5(x+3)}{(x-1)(x+3)}-\frac{3(x-1)}{(x+3)(x-1)}=\frac{5(x+3)-3(x-1)}{(x-1)(x+3)}=\frac{5x+15-3x+3}{(x-1)(x+3)}=\frac{2x+18}{(x-1)(x+3)}$$

63.
$$\frac{4}{5(x-2)}+\frac{6}{5(x-2)}=\frac{4+6}{5(x-2)}=\frac{10}{5(x-2)}=\frac{2}{x-2}$$

65. $x^2-x-2=(x-2)(x+1)$

$x^2-4=(x-2)(x+2)$

$\text{LCD}=(x+2)(x-2)(x+1)$
$$\frac{x+3}{x^2-x-2}=\frac{(x+3)(x+2)}{(x-2)(x+1)(x+2)}$$
$$\frac{x-3}{x^2-4}=\frac{(x-2)(x+1)}{(x-2)(x+2)(x+1)}$$

$$\frac{x+3}{x^2-x-2}-\frac{x-3}{x^2-4}$$
$$=\frac{(x+3)(x+2)}{(x+2)(x-2)(x+1)}-\frac{(x-3)(x+1)}{(x+2)(x-2)(x+1)}$$
$$=\frac{(x+3)(x+2)-(x-3)(x+1)}{(x+2)(x-2)(x+1)}$$
$$=\frac{x^2+5x+6-x^2+2x+3}{(x+2)(x-2)(x+1)}$$
$$=\frac{7x+9}{(x+2)(x-2)(x+1)}$$

6.4 Complex Fractions

Problems 6.4

1. Method 1: The LCD of a and b is ab.

$$\frac{\frac{2}{a}-\frac{3}{b}}{\frac{1}{a}+\frac{2}{b}}=\frac{ab\cdot\left(\frac{2}{a}-\frac{3}{b}\right)}{ab\cdot\left(\frac{1}{a}+\frac{2}{b}\right)}$$
$$=\frac{ab\cdot\frac{2}{a}-ab\cdot\frac{3}{b}}{ab\cdot\frac{1}{a}+ab\cdot\frac{2}{b}}$$
$$=\frac{2b-3a}{b+2a}$$

Method 2:

$$\frac{\frac{2}{a}-\frac{3}{b}}{\frac{1}{a}+\frac{2}{b}}=\frac{\frac{2\cdot b}{a\cdot b}-\frac{3a}{a\cdot b}}{\frac{1\cdot b}{a\cdot b}+\frac{2\cdot a}{a\cdot b}}$$
$$=\frac{\frac{2b-3a}{ab}}{\frac{b+2a}{ab}}$$
$$=\frac{2b-3a}{ab}\div\frac{b+2a}{ab}$$
$$=\frac{2b-3a}{ab}\cdot\frac{ab}{b+2a}$$
$$=\frac{2b-3a}{b+2a}$$

2. The LCD of $4x$, $3x$, $2x$, and x is $12x$.

$$\frac{\frac{1}{4x}+\frac{2}{3x}}{\frac{3}{2x}-\frac{1}{x}}=\frac{12x\left(\frac{1}{4x}+\frac{2}{3x}\right)}{12x\left(\frac{3}{2x}-\frac{1}{x}\right)}$$
$$=\frac{12x\cdot\frac{1}{4x}+12x\cdot\frac{2}{3x}}{12x\cdot\frac{3}{2x}-12x\cdot\frac{1}{x}}$$
$$=\frac{3\cdot1+4\cdot2}{6\cdot3-12\cdot1}$$
$$=\frac{3+8}{18-12}$$
$$=\frac{11}{6}$$

3. The LCD is x^2.

$$\frac{1-\frac{1}{x^2}}{1-\frac{1}{x}}=\frac{x^2\cdot\left(1-\frac{1}{x^2}\right)}{x^2\cdot\left(1-\frac{1}{x}\right)}$$
$$=\frac{x^2\cdot1-x^2\cdot\frac{1}{x^2}}{x^2\cdot1-x^2\cdot\frac{1}{x}}$$
$$=\frac{x^2-1}{x^2-x}$$
$$=\frac{(x-1)(x+1)}{x(x-1)}$$
$$=\frac{x+1}{x}$$

Exercises 6.4

1. The LCD is 2.

$$\frac{\frac{1}{2}}{2+\frac{1}{2}}=\frac{2\cdot\frac{1}{2}}{2\cdot\left(2+\frac{1}{2}\right)}=\frac{1}{2\cdot2+2\cdot\frac{1}{2}}=\frac{1}{4+1}=\frac{1}{5}$$

3. The LCD is 2.

$$\frac{\frac{1}{2}}{2-\frac{1}{2}}=\frac{2\cdot\frac{1}{2}}{2\cdot\left(2-\frac{1}{2}\right)}=\frac{1}{2\cdot2-2\cdot\frac{1}{2}}=\frac{1}{4-1}=\frac{1}{3}$$

5. The LCD is b.

$$\frac{a-\frac{a}{b}}{1+\frac{a}{b}}=\frac{b\cdot\left(a-\frac{a}{b}\right)}{b\cdot\left(1+\frac{a}{b}\right)}=\frac{b\cdot a-b\cdot\frac{a}{b}}{b\cdot1+b\cdot\frac{a}{b}}=\frac{ab-a}{b+a}$$

7. The LCD is ab.

$$\frac{\frac{1}{a}+\frac{1}{b}}{\frac{1}{a}-\frac{1}{b}}=\frac{ab\cdot\left(\frac{1}{a}+\frac{1}{b}\right)}{ab\cdot\left(\frac{1}{a}-\frac{1}{b}\right)}=\frac{ab\cdot\frac{1}{a}+ab\cdot\frac{1}{b}}{ab\cdot\frac{1}{a}-ab\cdot\frac{1}{b}}=\frac{b+a}{b-a}$$

9. The LCD is $12ab$.

$$\frac{\frac{1}{2a}+\frac{1}{3b}}{\frac{4}{a}-\frac{3}{4b}}=\frac{12ab\cdot\left(\frac{1}{2a}+\frac{1}{3b}\right)}{12ab\cdot\left(\frac{4}{a}-\frac{3}{4b}\right)}$$
$$=\frac{12ab\cdot\frac{1}{2a}+12ab\cdot\frac{1}{3b}}{12ab\cdot\frac{4}{a}-12ab\cdot\frac{3}{4b}}$$
$$=\frac{6b\cdot 1+4a\cdot 1}{12b\cdot 4-3a\cdot 3}$$
$$=\frac{6b+4a}{48b-9a}$$

11. The LCD is 24.

$$\frac{\frac{1}{3}+\frac{3}{4}}{\frac{3}{8}-\frac{1}{6}}=\frac{24\cdot\left(\frac{1}{3}+\frac{3}{4}\right)}{24\cdot\left(\frac{3}{8}-\frac{1}{6}\right)}$$
$$=\frac{24\cdot\frac{1}{3}+24\cdot\frac{3}{4}}{24\cdot\frac{3}{8}-24\cdot\frac{1}{6}}$$
$$=\frac{8\cdot 1+6\cdot 3}{3\cdot 3-4\cdot 1}$$
$$=\frac{8+18}{9-4}$$
$$=\frac{26}{5}$$

13. The LCD is x^2.

$$\frac{2+\frac{1}{x}}{4-\frac{1}{x^2}}=\frac{x^2\cdot\left(2+\frac{1}{x}\right)}{x^2\cdot\left(4-\frac{1}{x^2}\right)}$$
$$=\frac{x^2\cdot 2+x^2\cdot\frac{1}{x}}{x^2\cdot 4-x^2\cdot\frac{1}{x^2}}$$
$$=\frac{2x^2+x}{4x^2-1}$$
$$=\frac{x(2x+1)}{(2x-1)(2x+1)}$$
$$=\frac{x}{2x-1}$$

15. The LCD is x.

$$\frac{2+\frac{2}{x}}{1+\frac{1}{x}}=\frac{x\cdot\left(2+\frac{2}{x}\right)}{x\cdot\left(1+\frac{1}{x}\right)}$$
$$=\frac{x\cdot 2+x\cdot\frac{2}{x}}{x\cdot 1+x\cdot\frac{1}{x}}$$
$$=\frac{2x+2}{x+1}$$
$$=\frac{2(x+1)}{x+1}$$
$$=2$$

17. The LCD is xy.

$$\frac{\frac{1}{y}+\frac{1}{x}}{\frac{x}{y}-\frac{y}{x}}=\frac{xy\left(\frac{1}{y}+\frac{1}{x}\right)}{xy\left(\frac{x}{y}-\frac{y}{x}\right)}$$
$$=\frac{xy\cdot\frac{1}{y}+xy\cdot\frac{1}{x}}{xy\cdot\frac{x}{y}-xy\cdot\frac{y}{x}}$$
$$=\frac{x+y}{x^2-y^2}$$
$$=\frac{x+y}{(x+y)(x-y)}$$
$$=\frac{1}{x-y}$$

19. The LCD is x.

$$\frac{x-2-\frac{8}{x}}{x-3-\frac{4}{x}}=\frac{x\cdot\left(x-2-\frac{8}{x}\right)}{x\cdot\left(x-3-\frac{4}{x}\right)}$$
$$=\frac{x\cdot x-x\cdot 2-x\cdot\frac{8}{x}}{x\cdot x-x\cdot 3-x\cdot\frac{4}{x}}$$
$$=\frac{x^2-2x-8}{x^2-3x-4}$$
$$=\frac{(x+2)(x-4)}{(x+1)(x-4)}$$
$$=\frac{x+2}{x+1}$$

21. The LCD is $(x^2-25)=(x+5)(x-5)$.

$$\frac{\frac{1}{x+5}}{\frac{4}{x^2-25}}=\frac{(x+5)(x-5)\cdot\frac{1}{x+5}}{(x^2-25)\cdot\frac{4}{x^2-25}}=\frac{x-5}{4}$$

23. The LCD is $(x^2-16)=(x+4)(x-4)$.

$$\frac{\frac{1}{x^2-16}}{\frac{2}{x+4}}=\frac{(x^2-16)\cdot\frac{1}{x^2-16}}{(x+4)(x-4)\cdot\frac{2}{x+4}}=\frac{1}{2(x-4)}$$

25.
$$19w=2356$$
$$\frac{19w}{19}=\frac{2356}{19}$$
$$w=124$$

27.
$$9x+24=x$$
$$9x-9x+24=x-9x$$
$$24=-8x$$
$$\frac{24}{-8}=\frac{-8x}{-8}$$
$$-3=x \text{ or } x=-3$$

29.
$$5x=4x+3$$
$$5x-4x=4x-4x+3$$
$$x=3$$

31. The LCD is 6.

$$\frac{1}{4+\frac{1}{6}}=\frac{6\cdot1}{6\cdot\left(4+\frac{1}{6}\right)}=\frac{6}{6\cdot4+6\cdot\frac{1}{6}}=\frac{6}{24+1}=\frac{6}{25}$$

It takes Mercury $\frac{6}{25}$ year to go around the sun.

33. The LCD is 43.

$$11+\frac{1}{1+\frac{7}{43}}=11+\frac{43\cdot1}{43\cdot\left(1+\frac{7}{43}\right)}=11+\frac{43}{43\cdot1+43\cdot\frac{7}{43}}=11+\frac{43}{43+7}=11+\frac{43}{50}=11\frac{43}{50}$$

It takes Jupiter $11\frac{43}{50}$ years to go around the sun.

35. A complex fraction is a fraction that has one or more fractions in its numerator, denominator, or both.

37. Answers may vary.

39. Answers may vary.

41. The LCD is $12x$.

$$\frac{\frac{1}{4x}+\frac{2}{3x}}{\frac{3}{2x}-\frac{1}{x}}=\frac{12x\cdot\left(\frac{1}{4x}+\frac{2}{3x}\right)}{12x\cdot\left(\frac{3}{2x}-\frac{1}{x}\right)}=\frac{12x\cdot\frac{1}{4x}+12x\cdot\frac{2}{3x}}{12x\cdot\frac{3}{2x}-12x\cdot\frac{1}{x}}=\frac{3\cdot1+4\cdot2}{6\cdot3-12\cdot1}=\frac{3+8}{18-12}=\frac{11}{6}$$

43. The LCD is $w-5$.

$$\frac{w+2-\frac{18}{w-5}}{w-1-\frac{12}{w-5}}=\frac{(w-5)\cdot\left(w+2-\frac{18}{w-5}\right)}{(w-5)\cdot\left(w-1-\frac{12}{w-5}\right)}$$
$$=\frac{(w-5)\cdot w+(w-5)\cdot2-(w-5)\cdot\frac{18}{w-5}}{(w-5)\cdot w-(w-5)\cdot1-(w-5)\cdot\frac{12}{w-5}}$$
$$=\frac{w(w-5)+2(w-5)-18}{w(w-5)-(w-5)-12}$$
$$=\frac{w^2-5w+2w-10-18}{w^2-5w-w+5-12}$$
$$=\frac{w^2-3w-28}{w^2-6w-7}$$
$$=\frac{(w-7)(w+4)}{(w-7)(w+1)}$$
$$=\frac{w+4}{w+1}$$

45. The LCD is $(m-3)(m-4)(m+5)$.

$$\frac{\frac{3}{m-4}-\frac{16}{m-3}}{\frac{2}{m-3}-\frac{15}{m+5}}=\frac{(m-3)(m-4)(m+5)\left(\frac{3}{m-4}-\frac{16}{m-3}\right)}{(m-3)(m-4)(m+5)\left(\frac{2}{m-3}-\frac{15}{m+5}\right)}$$

$$=\frac{(m-3)(m-4)(m+5)\left(\frac{3}{m-4}\right)-(m-3)(m-4)(m+5)\left(\frac{16}{m-3}\right)}{(m-3)(m-4)(m+5)\left(\frac{2}{m-3}\right)-(m-3)(m-4)(m+5)\left(\frac{15}{m+5}\right)}$$

$$=\frac{3(m-3)(m+5)-16(m-4)(m+5)}{2(m-4)(m+5)-15(m-3)(m-4)}$$

$$=\frac{3m^2+6m-45-16m^2-16m+320}{2m^2+2m-40-15m^2+105m-180}$$

$$=\frac{-13m^2-10m+275}{-13m^2+107m-220}$$

$$=\frac{-1\cdot(13m-55)(m+5)}{-1\cdot(13m-55)(m-4)}$$

$$=\frac{m+5}{m-4}$$

6.5 Solving Equations Containing Rational Expressions

Problems 6.5

1. The LCD is $10x$.

$$\frac{4}{5}+\frac{1}{x}=\frac{3}{10}$$

$$10x\cdot\frac{4}{5}+10x\cdot\frac{1}{x}=10x\cdot\frac{3}{10}$$

$$8x+10=3x$$

$$8x=3x-10$$

$$5x=-10$$

$$x=-2$$

Check:

$$\frac{4}{5}+\frac{1}{x} \stackrel{?}{=} \frac{3}{10}$$

$\frac{4}{5}+\frac{1}{-2}$	$\frac{3}{10}$
$\frac{4\cdot 2}{5\cdot 2}+\frac{1\cdot 5}{-2\cdot 5}$	
$\frac{8}{10}-\frac{5}{10}$	
$\frac{3}{10}$	

The solution is $x = -2$.

2. The LCD is $x - 1$.

$$\frac{8}{x-1}-4=\frac{2x}{x-1}$$
$$(x-1)\cdot\frac{8}{x-1}-4(x-1)=(x-1)\cdot\frac{2x}{x-1}$$
$$8-4x+4=2x$$
$$-4x+12=2x$$
$$12=6x$$

$2 = x$ or $x = 2$

By substitution, the solution $x = 2$ checks.

3. The LCD is $x^2-9=(x+3)(x-3)$.

$$\frac{x}{x^2-9}+\frac{3}{x-3}=\frac{1}{x+3}$$
$$(x+3)(x-3)\cdot\frac{x}{(x+3)(x-3)}+(x+3)(x-3)\cdot\frac{3}{x-3}=(x+3)(x-3)\cdot\frac{1}{x+3}$$
$$x+3(x+3)=x-3$$
$$x+3x+9=x-3$$
$$4x+9=x-3$$
$$4x=x-12$$
$$3x=-12$$
$$x=-4$$

By substitution, the solution $x = -4$ checks.

4. The LCD is $4(x-2)$.

$$\frac{x}{x-2}+\frac{3}{4}=\frac{2}{x-2}$$
$$4(x-2)\cdot\frac{x}{x-2}+4(x-2)\cdot\frac{3}{4}=4(x-2)\cdot\frac{2}{x-2}$$
$$4x+3(x-2)=4\cdot 2$$
$$4x+3x-6=8$$
$$7x-6=8$$
$$7x=14$$
$$x=2$$

If we substitute $x = 2$ in the original equation, we have

$$\frac{2}{2-2}+\frac{3}{4}=\frac{2}{2-2}$$
$$\frac{2}{0}+\frac{3}{4}=\frac{2}{0}$$

Since division by 0 is undefined, there is no solution.

5. The LCD is $(x^2-1)=(x+1)(x-1)$.

$$1-\frac{4}{x^2-1}=\frac{-2}{x-1}$$
$$1\cdot(x+1)(x-1)-(x+1)(x-1)\cdot\frac{4}{(x+1)(x-1)}=(x+1)(x-1)\cdot\frac{-2}{x-1}$$
$$(x+1)(x-1)-4=-2(x+1)$$
$$x^2-1-4=-2x-2$$
$$x^2-5=-2x-2$$
$$x^2+2x-3=0$$
$$(x+3)(x-1)=0$$

$x+3=0$ or $x-1=0$

$x=-3$ $\quad x=1$

Since $x = 1$ makes the denominator $x - 1$ equal to zero, the only possible solution is $x = -3$. This solution can be verified in the original equation.

6. The LCD is $r - 1$.

$$S=\frac{a(r^n-1)}{r-1}$$
$$(r-1)S=(r-1)\cdot\frac{a(r^n-1)}{r-1}$$
$$S(r-1)=a(r^n-1)$$
$$\frac{S(r-1)}{r^n-1}=a \text{ or } a=\frac{S(r-1)}{r^n-1}$$

Exercises 6.5

1. The LCD is 4.

$$\frac{x}{4}=\frac{3}{2}$$
$$4\cdot\frac{x}{4}=4\cdot\frac{3}{2}$$
$$x=6$$

The solution $x = 6$ checks.

3. The LCD is $4x$.

$$\frac{3}{x}=\frac{3}{4}$$
$$4x\cdot\frac{3}{x}=4x\cdot\frac{3}{4}$$
$$12=3x$$
$$4=x$$

The solution $x = 4$ checks.

5. The LCD is $3x$.

$$\frac{-8}{3}=\frac{16}{x}$$
$$3x\cdot\frac{-8}{3}=3x\cdot\frac{16}{x}$$
$$-8x=48$$
$$x=-6$$

The solution $x = -6$ checks.

7. The LCD is 9.

$$\frac{4}{3}=\frac{x}{9}$$
$$9\cdot\frac{4}{3}=9\cdot\frac{x}{9}$$
$$12=x$$

The solution $x = 12$ checks.

9. The LCD is $20x$.

$$\frac{2}{5}+\frac{3}{x}=\frac{23}{20}$$
$$20x\cdot\frac{2}{5}+20x\cdot\frac{3}{x}=20x\cdot\frac{23}{20}$$
$$8x+60=23x$$
$$60=15x$$
$$4=x$$

The solution $x = 4$ checks.

11. The LCD is $35x$.

$$\frac{3}{x}-\frac{2}{7}=\frac{11}{35}$$
$$35x\cdot\frac{3}{x}-35x\cdot\frac{2}{7}=35x\cdot\frac{11}{35}$$
$$105-10x=11x$$
$$105=21x$$
$$\frac{105}{21}=x$$
$$5=x$$

The solution $x = 5$ checks.

13. The LCD is 10.

$$\frac{3}{5}+\frac{7x}{10}=2$$
$$10\cdot\frac{3}{5}+10\cdot\frac{7x}{10}=10\cdot 2$$
$$6+7x=20$$
$$7x=14$$
$$x=2$$

The solution $x = 2$ checks.

15. The LCD is 20.

$$\frac{3x}{4}-\frac{1}{5}=\frac{13}{10}$$
$$20\cdot\frac{3x}{4}-20\cdot\frac{1}{5}=20\cdot\frac{13}{10}$$
$$15x-4=26$$
$$15x=30$$
$$x=2$$

The solution $x = 2$ checks.

17. The LCD is $(x + 2)(x - 1)$.

$$\frac{3}{x+2}=\frac{4}{x-1}$$
$$(x+2)(x-1)\cdot\frac{3}{x+2}=(x+2)(x-1)\cdot\frac{4}{x-1}$$
$$3(x-1)=4(x+2)$$
$$3x-3=4x+8$$
$$3x-11=4x$$
$$-11=x$$

The solution $x = -11$ checks.

19. The LCD is $(x + 1)(x + 5)$.

$$\frac{-1}{x+1}=\frac{3}{x+5}$$
$$(x+1)(x+5)\cdot\frac{-1}{x+1}=(x+1)(x+5)\cdot\frac{3}{x+5}$$
$$-1\cdot(x+5)=3(x+1)$$
$$-x-5=3x+3$$
$$-x=3x+8$$
$$-4x=8$$
$$x=-2$$

The solution $x = -2$ checks.

21. The LCD is $x - 3$.

$$\frac{3x}{x-3}+2=\frac{5x}{x-3}$$
$$(x-3)\cdot\frac{3x}{x-3}+2\cdot(x-3)=(x-3)\cdot\frac{5x}{x-3}$$
$$3x+2x-6=5x$$
$$5x-6=5x$$
$$-6=0$$

Since this is false, there is no solution.

23. The LCD is $x + 1$.

$$\frac{5x}{x+1}-6=\frac{3x}{x+1}$$
$$(x+1)\cdot\frac{5x}{x+1}-6(x+1)=(x+1)\cdot\frac{3x}{x+1}$$
$$5x-6x-6=3x$$
$$-x-6=3x$$
$$-6=4x$$
$$\frac{-3}{2}=x$$

The solution $x=\frac{-3}{2}$ checks.

25. The LCD is $x^2-25=(x-5)(x+5)$.

$$\frac{x}{x^2-25}+\frac{5}{x-5}=\frac{1}{x+5}$$
$$(x-5)(x+5)\cdot\frac{x}{(x-5)(x+5)}+(x-5)(x+5)\cdot\frac{5}{x-5}=(x-5)(x+5)\cdot\frac{1}{x+5}$$
$$x+5(x+5)=1\cdot(x-5)$$
$$x+5x+25=x-5$$
$$6x+25=x-5$$
$$6x=x-30$$
$$5x=-30$$
$$x=-6$$

The solution $x = -6$ checks.

27. The LCD is $x^2-49=(x-7)(x+7)$.

$$\frac{x}{x^2-49}+\frac{7}{x-7}=\frac{1}{x+7}$$

$$(x-7)(x+7)\cdot\frac{x}{(x-7)(x+7)}+(x-7)(x+7)\cdot\frac{7}{x-7}=(x-7)(x+7)\cdot\frac{1}{x+7}$$

$$x+7(x+7)=1\cdot(x-7)$$
$$x+7x+49=x-7$$
$$8x+49=x-7$$
$$8x=x-56$$
$$7x=-56$$
$$x=-8$$

The solution $x=-8$ checks.

29. The LCD is $4(x+3)$.

$$\frac{x}{x+3}+\frac{3}{4}=\frac{-3}{x+3}$$

$$4(x+3)\cdot\frac{x}{x+3}+4(x+3)\cdot\frac{3}{4}=4(x+3)\cdot\frac{-3}{x+3}$$

$$4x+3(x+3)=4(-3)$$
$$4x+3x+9=-12$$
$$7x+9=-12$$
$$7x=-21$$
$$x=-3$$

Since $x=-3$ makes the denominator $x+3$ equal to zero, there is no solution.

31. The LCD is $7(x-4)$.

$$\frac{x}{x-4}-\frac{2}{7}=\frac{4}{x-4}$$

$$7(x-4)\cdot\frac{x}{x-4}-7(x-4)\cdot\frac{2}{7}=7(x-4)\cdot\frac{4}{x-4}$$

$$7x-2(x-4)=7\cdot 4$$
$$7x-2x+8=28$$
$$5x+8=28$$
$$5x=20$$
$$x=4$$

Since $x=4$ makes the denominator $x-4$ equal to zero, there is no solution.

33. The LCD is $(x^2-1)=(x+1)(x-1)$.

$$1+\frac{2}{x-1}=\frac{4}{x^2-1}$$

$$(x+1)(x-1)\cdot 1+(x+1)(x-1)\cdot\frac{2}{x-1}=(x+1)(x-1)\cdot\frac{4}{(x+1)(x-1)}$$

$$(x+1)(x-1)+2(x+1)=4$$

$$x^2-1+2x+2=4$$

$$x^2+2x-3=0$$

$$(x+3)(x-1)=0$$

$x+3=0$ or $x-1=0$

$x=-3$ $\quad x=1$

Since $x = 1$ makes the denominator $x - 1$ equal to zero, the only solution is $x = -3$, which checks.

35. The LCD is $(x^2-1)=(x+1)(x-1)$.

$$2-\frac{6}{x^2-1}=\frac{-3}{x-1}$$

$$(x+1)(x-1)\cdot 2-(x+1)(x-1)\cdot\frac{6}{(x+1)(x-1)}=(x+1)(x-1)\cdot\frac{-3}{x-1}$$

$$2(x^2-1)-6=-3(x+1)$$

$$2x^2-2-6=-3x-3$$

$$2x^2+3x-5=0$$

$$(2x+5)(x-1)=0$$

$2x+5=0$ or $x-1=0$

$x=\frac{-5}{2}$ $\quad x=1$

Since $x = 1$ makes the denominator $x - 1$ equal to zero, the only solution is $x=\frac{-5}{2}$, which checks.

37. The LCD is $(x-1)(x+2)(x-3)$.

$$\frac{4}{x-3}-\frac{2}{x-1}=\frac{2}{x+2}$$

$$(x-1)(x+2)(x-3)\cdot\frac{4}{x-3}-(x-1)(x+2)(x-3)\cdot\frac{2}{x-1}=(x-1)(x+2)(x-3)\cdot\frac{2}{x+2}$$

$$4(x-1)(x+2)-2(x+2)(x-3)=2(x-1)(x-3)$$

$$4(x^2+x-2)-2(x^2-x-6)=2(x^2-4x+3)$$

$$4x^2+4x-8-2x^2+2x+12=2x^2-8x+6$$

$$2x^2+6x+4=2x^2-8x+6$$

$$6x+4=-8x+6$$

$$14x=2$$

$$x=\frac{1}{7}$$

The solution $x=\frac{1}{7}$ checks.

39. The LCD is $(x^2-1)=(x+1)(x-1)$.

$$\frac{2x}{x^2-1}+\frac{4}{x-1}=\frac{1}{x-1}$$
$$(x+1)(x-1)\cdot\frac{2x}{(x+1)(x-1)}+(x+1)(x-1)\cdot\frac{4}{x-1}=(x+1)(x-1)\cdot\frac{1}{x-1}$$
$$2x+4(x+1)=1\cdot(x+1)$$
$$2x+4x+4=x+1$$
$$6x+4=x+1$$
$$5x=-3$$
$$x=\frac{-3}{5}$$

The solution $x=\frac{-3}{5}$ checks.

41. The LCD is $(z-3)(z-6)$.

$$\frac{2z+7}{z-3}+1=\frac{z+5}{z-6}+2$$
$$(z-3)(z-6)\cdot\frac{2z+7}{z-3}+(z-3)(z-6)\cdot 1=(z-3)(z-6)\cdot\frac{z+5}{z-6}+(z-3)(z-6)\cdot 2$$
$$(z-6)(2z+7)+(z-3)(z-6)=(z-3)(z+5)+2(z-3)(z-6)$$
$$2z^2-5z-42+z^2-9z+18=z^2+2z-15+2z^2-18z+36$$
$$3z^2-14z-24=3z^2-16z+21$$
$$-14z-24=-16z+21$$
$$2z=45$$
$$z=\frac{45}{2}$$

The solution $z=\frac{45}{2}$ checks.

43. The LCD is $2(y-1)$.

$$\frac{2y-5}{2}-\frac{1}{y-1}=y+1$$
$$2(y-1)\cdot\frac{2y-5}{2}-2(y-1)\cdot\frac{1}{y-1}=2(y-1)\cdot y+2(y-1)\cdot 1$$
$$(y-1)(2y-5)-2=2y(y-1)+2(y-1)$$
$$2y^2-7y+5-2=2y^2-2y+2y-2$$
$$2y^2-7y+3=2y^2-2$$
$$-7y+3=-2$$
$$-7y=-5$$
$$y=\frac{5}{7}$$

The solution $y=\frac{5}{7}$ checks.

45. The LCD is $v(4v^2-1)=v(2v+1)(2v-1)$.

$$\frac{5}{2v-1}+\frac{2}{v}=\frac{18v}{4v^2-1}$$

$$v(2v+1)(2v-1)\cdot\frac{5}{2v-1}+v(2v+1)(2v-1)\cdot\frac{2}{v}=v(2v+1)(2v-1)\cdot\frac{18v}{(2v+1)(2v-1)}$$

$$5v(2v+1)+2(2v+1)(2v-1)=18v\cdot v$$

$$10v^2+5v+2(4v^2-1)=18v^2$$

$$10v^2+5v+8v^2-2=18v^2$$

$$18v^2+5v-2=18v^2$$

$$5v-2=0$$

$$5v=2$$

$$v=\frac{2}{5}$$

The solution $v=\frac{2}{5}$ checks.

47. The LCD is $(z-1)(z-2)(z-6)$.

$$\frac{z+7}{z-1}-\frac{z+3}{z-2}=\frac{3}{z-6}$$

$$(z-1)(z-2)(z-6)\cdot\frac{z+7}{z-1}-(z-1)(z-2)(z-6)\cdot\frac{z+3}{z-2}=(z-1)(z-2)(z-6)\cdot\frac{3}{z-6}$$

$$(z-2)(z-6)(z+7)-(z-1)(z-6)(z+3)=3(z-1)(z-2)$$

$$(z-2)(z^2+z-42)-(z-1)(z^2-3z-18)=3(z^2-3z+2)$$

$$z^3+z^2-42z-2z^2-2z+84-z^3+3z^2+18z+z^2-3z-18=3z^2-9z+6$$

$$3z^2-29z+66=3z^2-9z+6$$

$$-29z+66=-9z+6$$

$$60=20z$$

$$3=z$$

The solution $z=3$ checks.

49. $(x^2 - 4x + 3) = (x - 3)(x - 1)$
$(x^2 - x - 6) = (x - 3)(x + 2)$
The LCD is $(x - 1)(x + 2)(x - 3)$.

$$\frac{2}{x^2 - 4x + 3} - \frac{5}{x^2 - x - 6} = \frac{x - 7}{(x - 1)(x - 3)(x + 2)}$$

$$(x - 1)(x + 2)(x - 3) \cdot \frac{2}{(x - 3)(x - 1)} - (x - 1)(x + 2)(x - 3) \cdot \frac{5}{(x - 3)(x + 2)} = (x - 1)(x + 2)(x - 3) \cdot \frac{x - 7}{(x - 1)(x - 3)(x + 2)}$$

$$2(x + 2) - 5(x - 1) = x - 7$$
$$2x + 4 - 5x + 5 = x - 7$$
$$-3x + 9 = x - 7$$
$$-4x + 9 = -7$$
$$-4x = -16$$
$$x = 4$$

The solution $x = 4$ checks.

51. $4(x + 3) = 45$
$4x + 12 = 45$
$4x = 33$
$x = \frac{33}{4}$

The solution $x = \frac{33}{4}$ checks.

53. The LCD is 12.
$$\frac{h}{4} + \frac{h}{6} = 1$$
$$12 \cdot \frac{h}{4} + 12 \cdot \frac{h}{6} = 12 \cdot 1$$
$$3h + 2h = 12$$
$$5h = 12$$
$$h = \frac{12}{5}$$

The solution $h = \frac{12}{5}$ checks.

55. $30(R - 5) = 10(R + 15)$
$30R - 150 = 10R + 150$
$20R - 150 = 150$
$20R = 300$
$R = 15$
The solution $R = 15$ checks.

57. The LCD is 2.

$$A = \frac{h(b_1 + b_2)}{2}$$
$$2 \cdot A = 2 \cdot \frac{h(b_1 + b_2)}{2}$$
$$2A = h(b_1 + b_2)$$
$$\frac{2A}{b_1 + b_2} = h \text{ or } h = \frac{2A}{b_1 + b_2}$$

59. The LCD is $Q_2 - Q_1$.

$$\frac{Q_1}{Q_2 - Q_1} = P$$
$$(Q_2 - Q_1) \cdot \frac{Q_1}{Q_2 - Q_1} = (Q_2 - Q_1) \cdot P$$
$$Q_1 = PQ_2 - PQ_1$$
$$Q_1 + PQ_1 = PQ_2$$
$$Q_1(1 + P) = PQ_2$$
$$Q_1 = \frac{PQ_2}{1 + P}$$

61. The LCD is *fab*.

$$\frac{1}{f} = \frac{1}{a} + \frac{1}{b}$$
$$fab \cdot \frac{1}{f} = fab \cdot \frac{1}{a} + fab \cdot \frac{1}{b}$$
$$ab = fb + fa$$
$$ab = f(b + a)$$
$$\frac{ab}{b + a} = f \text{ or } f = \frac{ab}{a + b}$$

63. Answers may vary.

65. The LCD is $(x^2 - 1) = (x + 1)(x - 1)$.

$$1 - \frac{4}{x^2 - 1} = \frac{-2}{x - 1}$$
$$(x+1)(x-1) \cdot 1 - (x+1)(x-1) \cdot \frac{4}{(x+1)(x-1)} = (x+1)(x-1) \cdot \frac{-2}{x - 1}$$
$$(x+1)(x-1) - 4 = -2(x+1)$$
$$x^2 - 1 - 4 = -2x - 2$$
$$x^2 - 5 = -2x - 2$$
$$x^2 + 2x - 3 = 0$$
$$(x+3)(x-1) = 0$$

$$\begin{aligned} x+3=0 \quad &\text{or} \quad x-1=0 \\ x=-3 \quad &\qquad\quad x=1 \end{aligned}$$

Since $x = 1$ makes the denominator $x - 1$ equal to zero, the only solution is $x = -3$, which checks.

67. The LCD is $(x^2-9)=(x+3)(x-3)$.

$$\begin{aligned} \frac{x}{x^2-9}+\frac{3}{x-3}&=\frac{1}{x+3} \\ (x+3)(x-3)\cdot\frac{x}{(x+3)(x-3)}+(x+3)(x-3)\cdot\frac{3}{x-3}&=(x+3)(x-3)\cdot\frac{1}{x+3} \\ x+3(x+3)&=x-3 \\ x+3x+9&=x-3 \\ 4x+9&=x-3 \\ 3x&=-12 \\ x&=-4 \end{aligned}$$

The solution $x = -4$ checks.

69. The LCD is $10x$.

$$\begin{aligned} \frac{4}{5}+\frac{1}{x}&=\frac{3}{10} \\ 10x\cdot\frac{4}{5}+10x\cdot\frac{1}{x}&=10x\cdot\frac{3}{10} \\ 8x+10&=3x \\ 10&=-5x \\ -2&=x \end{aligned}$$

The solution $x = -2$ checks.

71. The LCD is 9.

$$\begin{aligned} C&=\frac{5}{9}(F-32) \\ 9\cdot C&=9\cdot\frac{5}{9}(F-32) \\ 9C&=5(F-32) \\ 9C&=5F-160 \\ 9C+160&=5F \\ \frac{9C+160}{5}&=F \\ F&=\frac{9C}{5}+\frac{160}{5} \\ F&=\frac{9}{5}C+32 \end{aligned}$$

6.6 Ratio, Proportion, and Applications

Problems 6.6

1. The ratio of miles to gallons is $\frac{150}{9}=\frac{50}{3}$.

   Let g = gallons needed.

   $$\frac{900}{g}=\frac{50}{3}$$
   $$3g\cdot\frac{900}{g}=3g\cdot\frac{50}{3}$$
   $$2700=50g$$
   $$g=\frac{2700}{50}=54$$

   Hence, 54 gallons are needed.

2. 1. **Read the problem.**
   We are asked to find the time if both work together.

   2. **Select the unknown.**
   Let h = hours to complete with both workers.

   3. **Think of a plan.**
   The first worker completes $\frac{1}{5}$ of the report in 1 hour. The second worker completes $\frac{1}{8}$ of the report in 1 hour.
   $$\frac{1}{5}+\frac{1}{8}=\frac{1}{h}$$

   4. **Use algebra to solve the problem.**
   $$40h\cdot\frac{1}{5}+40h\cdot\frac{1}{8}=40h\cdot\frac{1}{h}$$
   $$8h+5h=40$$
   $$13h=40$$
   $$h=\frac{40}{13}=3\frac{1}{13}$$
   Thus, both workers working together take $3\frac{1}{13}$ hours to complete the report.

   5. **Verify the solution.**
   The solution checks.

3. 1. **Read the problem.**
   Find the speed of the freight train.

   2. **Select the unknown.**
   Let R = speed of freight train.

   3. **Think of a plan.**
   $D=RT$ or $T=\frac{D}{R}$

	Rate	Time	Distance
Freight	R	T	120
Passenger	$R+5$	T	140

   $$\frac{120}{R}=\frac{140}{R+5}$$

   4. **Use the cross-product rule to solve the problem.**
   $$120(R+5)=140R$$
   $$120R+600=140R$$
   $$600=20R$$
   $$30=R$$
   Thus the speed of the freight train is 30 miles per hour.

   5. **Verify the solution.**
   $$\frac{120}{R}=\frac{140}{R+5}$$
   $$\frac{120}{30}=\frac{140}{30+5}$$
   $$4=4$$
   The solution checks.

4. Solve the same problem as in Example 4 using 162 instead of 152.
   $$\frac{w}{162}=\frac{121}{113}$$
   $$113w=162\cdot121$$
   $$113w=19{,}602$$
   $w=173$ (to the nearest whole number)

   Yes. In 162 games, Thomas would have had 173 walks. However, the record would be for 162 games, not 152 games!

5. $\frac{d}{4}=\frac{4}{6}$
$6d=16$
$d=\frac{16}{6}=\frac{8}{3}=2\frac{2}{3}$
Thus d is $\frac{8}{3}$ or $2\frac{2}{3}$ centimeters.

Exercises 6.6

Students should use the RSTUV method to solve where appropriate.

1. The ratio of gallons to hours is $\frac{7400}{24}=\frac{925}{3}$.
Let g = gallons.
$\frac{g}{30}=\frac{925}{3}$
$3g=30\cdot 925$
$3g=27,750$
$g=9250$
Thus 9250 gallons would be sold in 30 hours.

3. The ratio of tortillas to minutes is $\frac{25}{15}=\frac{5}{3}$.
Let t = number of tortillas.
$\frac{t}{60}=\frac{5}{3}$
$3t=300$
$t=100$
Michael could eat 100 tortillas in 1 hour.

5. The ratio of pounds to square feet is $\frac{4}{100}=\frac{1}{25}$.
Let p = number of pounds.
$\frac{p}{15\times 10}=\frac{1}{25}$
$\frac{p}{150}=\frac{1}{25}$
$25p=150$
$p=6$
You will need 6 pounds of Ironite.

7. The ratio of pounds to square feet is $\frac{2}{100}=\frac{1}{50}$.
Let p = number of pounds.
$\frac{p}{12\times 15}=\frac{1}{50}$
$\frac{p}{180}=\frac{1}{50}$
$50p=180$
$p=3.6$
You will need 3.6 pounds of fertilizer.

9. a. The ratio of persons on street to persons in shelters is $\frac{400}{1200}=\frac{1}{3}$.
Let s = number in shelters.
$\frac{750}{s}=\frac{1}{3}$
$s=2250$
You would expect 2250 persons in shelters.

b. $\frac{600}{s}=\frac{1}{3}$
$s=1800$
To accommodate 1800 persons in shelters, $\frac{1800}{50}=36$ shelters are needed.

11. a. The ratio is 1200 to 400 or 3 to 1.

b. The ratio is 1050 to 30 or 35 to 1.

c. The ratio is higher in Minneapolis, which could have to do with the colder climate.

13. The ratio of foreign to total is $\frac{60,000}{600,000}=\frac{1}{10}$.
Let f = number of foreign-born.
$\frac{f}{700,000}=\frac{1}{10}$
$10f=700,000$
$f=70,000$
You would expect 70,000 persons born in a foreign country.

15. Let h = number of hours working together.

$$\frac{1}{9}+\frac{1}{12}=\frac{1}{h}$$
$$36h\cdot\frac{1}{9}+36h\cdot\frac{1}{12}=36h\cdot\frac{1}{h}$$
$$4h+3h=36$$
$$7h=36$$
$$h=\frac{36}{7}=5\frac{1}{7}$$

It will take $5\frac{1}{7}$ hours.

17. Let m = number of minutes operating together.

$$\frac{1}{5}+\frac{1}{3}=\frac{1}{m}$$
$$15m\cdot\frac{1}{5}+15m\cdot\frac{1}{3}=15m\cdot\frac{1}{m}$$
$$3m+5m=15$$
$$8m=15$$
$$m=\frac{15}{8}=1\frac{7}{8}$$

It will take $1\frac{7}{8}$ minutes.

19. Let h = number of hours for the remaining fax machine.

$$\frac{1}{h}+\frac{1}{3}=\frac{1}{2}$$
$$6h\cdot\frac{1}{h}+6h\cdot\frac{1}{3}=6h\cdot\frac{1}{2}$$
$$6+2h=3h$$
$$6=h$$

It would take 6 hours.

21. $D=R\cdot T$ or $T=\frac{D}{R}$

Let R = speed of boat in still water.

	R	T	D
up	$R+6$	T	16
down	$R-6$	T	4

$$\frac{16}{R+6}=\frac{4}{R-6}$$
$$16(R-6)=4(R+6)$$
$$16R-96=4R+24$$
$$12R=120$$
$$R=10$$

The boat travels 10 miles per hour in still water.

23. $D=R\cdot T$ or $T=\frac{D}{R}$

Let R = speed of *Tokaido* train.

	R	T	D
regular	$R-60$	T	180
Tokaido	R	T	450

$$\frac{180}{R-60}=\frac{450}{R}$$
$$180R=450(R-60)$$
$$180R=450R-27,000$$
$$-270R=-27,000$$
$$R=100$$

The *New Tokaido* travels at 100 miles per hour.

25. a. The ratio of home runs to games is $\frac{40}{111}$.

Let h = number of home runs.

$$\frac{h}{162}=\frac{40}{111}$$
$$111h=40\cdot162$$
$$111h=6480$$
$$h\approx58$$

He would have hit 58 home runs.

b. No, since $58<61$.

27. The ratio of runs to games in $\frac{106}{113}$.

Let g = number of games.

$$\frac{177}{g}=\frac{106}{113}$$
$$106g=177\cdot113$$
$$106g=20,001$$
$$g\approx189$$

He would need 189 games.
No, he could not have broken the record since 189 > 162.

29. $\frac{x}{4}=\frac{4}{3}$
$3x=16$
$x=\frac{16}{3}=5\frac{1}{3}$

$\frac{y}{5}=\frac{4}{3}$
$3y=20$
$y=\frac{20}{3}=6\frac{2}{3}$

31. $\frac{DE}{6}=\frac{16}{7}$
$7DE=96$
$DE=\frac{96}{7}=13\frac{5}{7}$ in.

33. $\frac{x}{36}=\frac{20}{45}$
$45x=720$
$x=\frac{720}{45}$
$x=16$

$\frac{y}{18}=\frac{20}{45}$
$45y=360$
$y=\frac{360}{45}$
$y=8$

35. $\frac{x}{8}=\frac{9}{6}$
$6x=72$
$x=12$

$\frac{y}{8}=\frac{9}{6}$
$6y=72$
$y=12$

37. $\frac{a}{32}=\frac{18}{24}$
$24a=576$
$a=24$

$\frac{b}{40}=\frac{18}{24}$
$24b=720$
$b=30$

39. $\frac{x}{40}=\frac{10}{50}$
$50x=400$
$x=8$

$\frac{y}{44}=\frac{10}{50}$
$50y=440$
$y=\frac{440}{50}=8\frac{4}{5}$

41. $\frac{x}{4}=\frac{20}{8}$
$8x=80$
$x=10$

$\frac{y}{10}=\frac{20}{8}$
$8y=200$
$y=25$

43. $5x-2y=5(-2)-2(-1)=-10+2=-8$

45. $11=4x+3$
$8=4x$
$2=x$ or $x=2$

47. The ratio of inches to pounds is $\frac{3}{7}$.

Let p = pounds of force.
$\frac{8}{p}=\frac{3}{7}$
$3p=56$
$p=\frac{56}{3}=18\frac{2}{3}$

It would require $18\frac{2}{3}$ pounds of force.

49. The ratio of rise to run is $\frac{2}{5}$.

Let s = rise.
$\frac{s}{15}=\frac{2}{5}$
$5s=30$
$s=6$
The rise is 6 feet.

51. The ratio of tagged to total is $\frac{5}{53}$.

Let f = number of fish in lake.
$\frac{250}{f}=\frac{5}{53}$
$5f=13{,}250$
$f=2650$
There are approximately 2650 fish in the lake.

53. Answers may vary.

55. Answers may vary.

57. Let h = hours working together.
$$\frac{1}{5}+\frac{1}{8}=\frac{1}{h}$$
$$40h\cdot\frac{1}{5}+40h\cdot\frac{1}{8}=40h\cdot\frac{1}{h}$$
$$8h+5h=40$$
$$13h=40$$
$$h=\frac{40}{13}=3\frac{1}{13}$$

It will take $3\frac{1}{13}$ hours if both work on the report.

59. The ratio of singles to games is $\frac{160}{120}=\frac{4}{3}$.

Let s = number of singles.

$$\frac{s}{150}=\frac{4}{3}$$
$$3s=600$$
$$s=200$$

He will hit 200 singles.

Review Exercises

1. a. Since $16y^2=8y\cdot 2y$,

$$\frac{5x}{8y}=\frac{5x\cdot 2y}{8y\cdot 2y}=\frac{10xy}{16y^2}.$$

b. Since $16y^3=4y^2\cdot 4y$,

$$\frac{3x}{4y^2}=\frac{3x\cdot 4y}{4y^2\cdot 4y}=\frac{12xy}{16y^3}.$$

c. Since $15x^5=3x^3\cdot 5x^2$,

$$\frac{2y}{3x^3}=\frac{2y\cdot 5x^2}{3x^3\cdot 5x^2}=\frac{10x^2y}{15x^5}.$$

2. a. $$\frac{-9(x^2-y^2)}{3(x+y)}=\frac{-3\cdot 3\cdot(x+y)\cdot(x-y)}{3\cdot(x+y)}$$
$$=-3(x-y)\text{ or }3y-3x$$

b. $$\frac{-10(x^2-y^2)}{5(x+y)}=\frac{-2\cdot 5\cdot(x+y)\cdot(x-y)}{5\cdot(x+y)}$$
$$=-2(x-y)\text{ or }2y-2x$$

c. $$\frac{-16(x^2-y^2)}{-4(x+y)}=\frac{-4\cdot 4\cdot(x+y)\cdot(x-y)}{-4\cdot(x+y)}$$
$$=4(x-y)\text{ or }4x-4y$$

3. a. $$\frac{-x}{x^2+x}=\frac{-1\cdot x}{x(x+1)}=\frac{-1}{x+1}$$
$$x^2+x=0$$
$$x(x+1)=0$$
$$x=0\quad\text{or}\quad x+1=0$$
$$x=-1$$

Thus, $\frac{-x}{x^2+x}$ is undefined for $x = 0$ or $x = -1$.

b. $$\frac{-x}{x^2-x}=\frac{-1\cdot x}{x(x-1)}=\frac{-1}{x-1}\text{ or }\frac{1}{1-x}$$
$$x^2-x=0$$
$$x(x-1)=0$$
$$x=0\quad\text{or}\quad x-1=0$$
$$x=1$$

Thus $\frac{-x}{x^2-x}$ is undefined for $x = 0$ or $x = 1$.

c. $$\frac{-x}{x-x^2}=\frac{-1\cdot x}{x(1-x)}=\frac{-1}{1-x}\text{ or }\frac{1}{x-1}$$
$$x-x^2=0$$
$$x(1-x)=0$$
$$x=0\quad\text{or}\quad 1-x=0$$
$$x=1$$

Thus, $\frac{-x}{x-x^2}$ is undefined for $x = 0$ or $x = 1$.

4. a. $$\frac{x^2-3x-18}{6-x}=\frac{(x-6)(x+3)}{-1\cdot(x-6)}$$
$$=\frac{x+3}{-1}$$
$$=-(x+3)\text{ or }-x-3$$

b. $$\frac{x^2-2x-8}{4-x}=\frac{(x-4)(x+2)}{-1\cdot(x-4)}$$
$$=\frac{x+2}{-1}$$
$$=-(x+2)\text{ or }-x-2$$

c. $$\frac{x^2-2x-15}{5-x}=\frac{(x-5)(x+3)}{-1\cdot(x-5)}$$
$$=\frac{x+3}{-1}$$
$$=-(x+3)\text{ or }-x-3$$

5. a. $\frac{3y^2}{7}\cdot\frac{14x}{9y}=\frac{3y\cdot y}{7}\cdot\frac{2\cdot 7\cdot x}{3\cdot 3y}=\frac{2xy}{3}$

b. $\frac{7y^2}{5}\cdot\frac{15x}{14y}=\frac{7y\cdot y}{5}\cdot\frac{3\cdot 5\cdot x}{2\cdot 7y}=\frac{3xy}{2}$

c. $\frac{6y^3}{7}\cdot\frac{28x}{3y}=\frac{2\cdot 3y\cdot y^2}{7}\cdot\frac{7\cdot 4\cdot x}{3y}$
$=2\cdot y^2\cdot 4\cdot x$
$=8xy^2$

6. a. $(x-3)\cdot\frac{x+2}{x^2-9}=\frac{x-3}{1}\cdot\frac{x+2}{x^2-9}$
$=\frac{x-3}{1}\cdot\frac{x+2}{(x-3)(x+3)}$
$=\frac{x+2}{x+3}$

b. $(x-5)\cdot\frac{x+1}{x^2-25}=\frac{x-5}{1}\cdot\frac{x+1}{x^2-25}$
$=\frac{x-5}{1}\cdot\frac{x+1}{(x-5)(x+5)}$
$=\frac{x+1}{x+5}$

c. $(x-4)\cdot\frac{x+5}{x^2-16}=\frac{x-4}{1}\cdot\frac{x+5}{x^2-16}$
$=\frac{x-4}{1}\cdot\frac{x+5}{(x-4)(x+4)}$
$=\frac{x+5}{x+4}$

7. a. $\frac{x^2-9}{x+2}\div(x+3)=\frac{x^2-9}{x+2}\div\frac{x+3}{1}$
$=\frac{x^2-9}{x+2}\cdot\frac{1}{x+3}$
$=\frac{(x+3)(x-3)}{x+2}\cdot\frac{1}{x+3}$
$=\frac{x-3}{x+2}$

b. $\frac{x^2-16}{x+1}\div(x+4)=\frac{x^2-16}{x+1}\div\frac{x+4}{1}$
$=\frac{x^2-16}{x+1}\cdot\frac{1}{x+4}$
$=\frac{(x+4)(x-4)}{x+1}\cdot\frac{1}{x+4}$
$=\frac{x-4}{x+1}$

c. $\frac{x^2-25}{x+4}\div(x+5)=\frac{x^2-25}{x+4}\div\frac{x+5}{1}$
$=\frac{x^2-25}{x+4}\cdot\frac{1}{x+5}$
$=\frac{(x-5)(x+5)}{x+4}\cdot\frac{1}{x+5}$
$=\frac{x-5}{x+4}$

8. a. $\frac{x+5}{x-5}\div\frac{x^2-25}{5-x}=\frac{x+5}{x-5}\cdot\frac{5-x}{x^2-25}$
$=\frac{x+5}{x-5}\cdot\frac{-1\cdot(x-5)}{(x-5)(x+5)}$
$=\frac{-1}{x-5}$ or $\frac{1}{5-x}$

b. $\frac{x+1}{x-1}\div\frac{x^2-1}{1-x}=\frac{x+1}{x-1}\cdot\frac{1-x}{x^2-1}$
$=\frac{x+1}{x-1}\cdot\frac{-1\cdot(x-1)}{(x-1)(x+1)}$
$=\frac{-1}{x-1}$ or $\frac{1}{1-x}$

c. $\frac{x+2}{x-2}\div\frac{x^2-4}{2-x}=\frac{x+2}{x-2}\cdot\frac{2-x}{x^2-4}$
$=\frac{x+2}{x-2}\cdot\frac{-1\cdot(x-2)}{(x-2)(x+2)}$
$=\frac{-1}{x-2}$ or $\frac{1}{2-x}$

9. a. $\frac{3}{2(x-1)}+\frac{1}{2(x-1)}=\frac{3+1}{2(x-1)}$
$=\frac{4}{2(x-1)}$
$=\frac{2}{x-1}$

b. $\frac{5}{6(x-2)}+\frac{7}{6(x-2)}=\frac{5+7}{6(x-2)}$
$=\frac{12}{6(x-2)}$
$=\frac{2}{x-2}$

c. $\frac{3}{4(x+1)}+\frac{1}{4(x+1)}=\frac{3+1}{4(x+1)}$
$=\frac{4}{4(x+1)}$
$=\frac{1}{x+1}$

10. a. $\frac{7}{2(x+1)}-\frac{3}{2(x+1)}=\frac{7-3}{2(x+1)}$
$=\frac{4}{2(x+1)}$
$=\frac{2}{x+1}$

b. $\frac{11}{5(x+2)}-\frac{1}{5(x+2)}=\frac{11-1}{5(x+2)}$
$=\frac{10}{5(x+2)}$
$=\frac{2}{x+2}$

c. $\frac{17}{7(x+3)}-\frac{3}{7(x+3)}=\frac{17-3}{7(x+3)}$
$=\frac{14}{7(x+3)}$
$=\frac{2}{x+3}$

11. a. Since $(x + 2)$ and $(x - 2)$ do not have any common factors, the LCD is $(x + 2)(x - 2)$.
$\frac{2}{x+2}=\frac{2(x-2)}{(x+2)(x-2)}$
$\frac{1}{x-2}=\frac{1\cdot(x+2)}{(x-2)(x+2)}$
$\frac{2}{x+2}+\frac{1}{x-2}$
$=\frac{2(x-2)}{(x+2)(x-2)}+\frac{(x+2)}{(x+2)(x-2)}$
$=\frac{2(x-2)+(x+2)}{(x+2)(x-2)}$
$=\frac{2x-4+x+2}{(x+2)(x-2)}$
$=\frac{3x-2}{(x+2)(x-2)}$

b. Since $(x + 1)$ and $(x - 1)$ do not have any common factors, the LCD is $(x + 1)(x - 1)$.
$\frac{3}{x+1}=\frac{3(x-1)}{(x+1)(x-1)}$
$\frac{1}{x-1}=\frac{1\cdot(x+1)}{(x-1)(x+1)}$
$\frac{3}{x+1}+\frac{1}{x-1}$
$=\frac{3(x-1)}{(x+1)(x-1)}+\frac{(x+1)}{(x+1)(x-1)}$
$=\frac{3(x-1)+(x+1)}{(x+1)(x-1)}$
$=\frac{3x-3+x+1}{(x+1)(x-1)}$
$=\frac{4x-2}{(x+1)(x-1)}$

c. Since $(x + 3)$ and $(x - 3)$ do not have any common factors, the LCD is $(x + 3)(x - 3)$.
$\frac{4}{x+3}=\frac{4(x-3)}{(x+3)(x-3)}$
$\frac{1}{x-3}=\frac{1\cdot(x+3)}{(x-3)(x+3)}$

$$\frac{4}{x+3}+\frac{1}{x-3}$$
$$=\frac{4(x-3)}{(x+3)(x-3)}+\frac{(x+3)}{(x-3)(x+3)}$$
$$=\frac{4(x-3)+(x+3)}{(x+3)(x-3)}$$
$$=\frac{4x-12+x+3}{(x+3)(x-3)}$$
$$=\frac{5x-9}{(x+3)(x-3)}$$

12. a. $x^2+3x+2=(x+2)(x+1)$

$x^2+5x+6=(x+2)(x+3)$

$\text{LCD}=(x+1)(x+2)(x+3)$

$$\frac{x-1}{x^2+3x+2}=\frac{(x-1)(x+3)}{(x+2)(x+1)(x+3)}$$
$$\frac{x+7}{x^2+5x+6}=\frac{(x+7)(x+1)}{(x+2)(x+3)(x+1)}$$
$$\frac{x-1}{x^2+3x+2}-\frac{x+7}{x^2+5x+6}=\frac{(x-1)(x+3)}{(x+1)(x+2)(x+3)}-\frac{(x+7)(x+1)}{(x+1)(x+2)(x+3)}$$
$$=\frac{(x-1)(x+3)-(x+7)(x+1)}{(x+1)(x+2)(x+3)}$$
$$=\frac{x^2+2x-3-x^2-8x-7}{(x+1)(x+2)(x+3)}$$
$$=\frac{-6x-10}{(x+1)(x+2)(x+3)}$$

b. $x^2-x-2=(x-2)(x+1)$

$x^2+2x+1=(x+1)^2$

$\text{LCD}=(x-2)(x+1)^2$

$$\frac{x+3}{x^2-x-2}=\frac{(x+3)(x+1)}{(x-2)(x+1)(x+1)}$$
$$\frac{x-1}{x^2+2x+1}=\frac{(x-1)(x-2)}{(x+1)(x+1)(x-2)}$$
$$\frac{x+3}{x^2-x-2}-\frac{x-1}{x^2+2x+1}=\frac{(x+3)(x+1)}{(x-2)(x+1)^2}-\frac{(x-1)(x-2)}{(x-2)(x+1)^2}$$
$$=\frac{(x+3)(x+1)-(x-1)(x-2)}{(x-2)(x+1)^2}$$
$$=\frac{x^2+4x+3-x^2+3x-2}{(x-2)(x+1)^2}$$
$$=\frac{7x+1}{(x-2)(x+1)^2}$$

c. $x^2+3x+2=(x+2)(x+1)$

$x^2+x-2=(x+2)(x-1)$

LCD = $(x+2)(x+1)(x-1)$

$$\frac{x-1}{x^2+3x+2}=\frac{(x-1)(x-1)}{(x+2)(x+1)(x-1)}$$

$$\frac{x+1}{x^2+x-2}=\frac{(x+1)(x+1)}{(x+2)(x-1)(x+1)}$$

$$\begin{aligned}\frac{x-1}{x^2+3x+2}-\frac{x+1}{x^2+x-2}&=\frac{(x-1)(x-1)}{(x+2)(x+1)(x-1)}-\frac{(x+1)(x+1)}{(x+2)(x+1)(x-1)}\\&=\frac{(x-1)^2-(x+1)^2}{(x+2)(x+1)(x-1)}\\&=\frac{x^2-2x+1-x^2-2x-1}{(x+2)(x+1)(x-1)}\\&=\frac{-4x}{(x+2)(x+1)(x-1)}\end{aligned}$$

13. a. The LCD is $12x$.

$$\frac{\frac{3}{2x}-\frac{1}{x}}{\frac{2}{3x}+\frac{3}{4x}}=\frac{12x\cdot\left(\frac{3}{2x}-\frac{1}{x}\right)}{12x\cdot\left(\frac{2}{3x}+\frac{3}{4x}\right)}=\frac{12x\cdot\frac{3}{2x}-12x\cdot\frac{1}{x}}{12x\cdot\frac{2}{3x}+12x\cdot\frac{3}{4x}}=\frac{6\cdot3-12\cdot1}{4\cdot2+3\cdot3}=\frac{18-12}{8+9}=\frac{6}{17}$$

b. The LCD is $12x$.

$$\frac{\frac{3}{2x}-\frac{1}{3x}}{\frac{2}{x}+\frac{1}{4x}}=\frac{12x\cdot\left(\frac{3}{2x}-\frac{1}{3x}\right)}{12x\cdot\left(\frac{2}{x}+\frac{1}{4x}\right)}=\frac{12x\cdot\frac{3}{2x}-12x\cdot\frac{1}{3x}}{12x\cdot\frac{2}{x}+12x\cdot\frac{1}{4x}}=\frac{6\cdot3-4\cdot1}{12\cdot2+3\cdot1}=\frac{18-4}{24+3}=\frac{14}{27}$$

c. The LCD is $12x$.

$$\frac{\frac{3}{2x}-\frac{1}{x}}{\frac{3}{4x}+\frac{4}{3x}}=\frac{12x\cdot\left(\frac{3}{2x}-\frac{1}{x}\right)}{12x\cdot\left(\frac{3}{4x}+\frac{4}{3x}\right)}=\frac{12x\cdot\frac{3}{2x}-12x\cdot\frac{1}{x}}{12x\cdot\frac{3}{4x}+12x\cdot\frac{4}{3x}}=\frac{6\cdot3-12\cdot1}{3\cdot3+4\cdot4}=\frac{18-12}{9+16}=\frac{6}{25}$$

14. a. The LCD is $x-1$.

$$\begin{aligned}\frac{2x}{x-1}+3&=\frac{4x}{x-1}\\(x-1)\cdot\frac{2x}{x-1}+3(x-1)&=(x-1)\cdot\frac{4x}{x-1}\\2x+3x-3&=4x\\5x-3&=4x\\-3&=-x\\3&=x\end{aligned}$$

The solution $x=3$ checks.

b. The LCD is $x - 5$.

$$\begin{aligned}\frac{6x}{x-5}+7&=\frac{8x}{x-5}\\(x-5)\cdot\frac{6x}{x-5}+7(x-5)&=(x-5)\cdot\frac{8x}{x-5}\\6x+7x-35&=8x\\13x-35&=8x\\-35&=-5x\\7&=x\end{aligned}$$

The solution $x = 7$ checks.

c. The LCD is $x - 4$.

$$\begin{aligned}\frac{5x}{x-4}+6&=\frac{7x}{x-4}\\(x-4)\cdot\frac{5x}{x-4}+6(x-4)&=(x-4)\cdot\frac{7x}{x-4}\\5x+6x-24&=7x\\11x-24&=7x\\-24&=-4x\\6&=x\end{aligned}$$

The solution $x = 6$ checks.

15. a. $\dfrac{x-3}{x^2-x-6}=\dfrac{x-3}{(x-3)(x+2)}=\dfrac{1}{x+2}$

The LCD is $(x + 2)(x - 2)$.

$$\begin{aligned}\frac{x}{x^2-4}+\frac{2}{x-2}&=\frac{x-3}{x^2-x-6}\\(x+2)(x-2)\cdot\frac{x}{(x+2)(x-2)}+(x+2)(x-2)\cdot\frac{2}{x-2}&=(x+2)(x-2)\cdot\frac{1}{x+2}\\x+2(x+2)&=1\cdot(x-2)\\x+2x+4&=x-2\\3x+4&=x-2\\2x+4&=-2\\2x&=-6\\x&=-3\end{aligned}$$

The solution $x = -3$ checks.

b. $\dfrac{x-5}{x^2-x-20}=\dfrac{x-5}{(x-5)(x+4)}=\dfrac{1}{x+4}$

The LCD is $(x-4)(x+4)$.

$$\begin{aligned}\frac{x}{x^2-16}+\frac{4}{x-4}&=\frac{x-5}{x^2-x-20}\\(x-4)(x+4)\cdot\frac{x}{(x-4)(x+4)}+(x-4)(x+4)\cdot\frac{4}{x-4}&=(x-4)(x+4)\cdot\frac{1}{x+4}\\x+4(x+4)&=1\cdot(x-4)\\x+4x+16&=x-4\\5x+16&=x-4\\4x+16&=-4\\4x&=-20\\x&=-5\end{aligned}$$

The solution $x=-5$ checks.

c. $\dfrac{x-6}{x^2-x-30}=\dfrac{x-6}{(x-6)(x+5)}=\dfrac{1}{x+5}$

The LCD is $(x+5)(x-5)$.

$$\begin{aligned}\frac{x}{x^2-25}+\frac{5}{x-5}&=\frac{x-6}{x^2-x-30}\\(x+5)(x-5)\cdot\frac{x}{(x+5)(x-5)}+(x+5)(x-5)\cdot\frac{5}{x-5}&=(x+5)(x-5)\cdot\frac{1}{x+5}\\x+5(x+5)&=1\cdot(x-5)\\x+5x+25&=x-5\\6x+25&=x-5\\5x+25&=-5\\5x&=30\\x&=-6\end{aligned}$$

The solution $x=-6$ checks.

16. a. The LCD is $7(x+6)$.

$$\begin{aligned}\frac{x}{x+6}-\frac{1}{7}&=\frac{-6}{x+6}\\7(x+6)\cdot\frac{x}{x+6}-7(x+6)\cdot\frac{1}{7}&=7(x+6)\cdot\frac{-6}{x+6}\\7x-(x+6)&=7(-6)\\7x-x-6&=-42\\6x-6&=-42\\6x&=-36\\x&=-6\end{aligned}$$

Since $x=-6$ makes the denominator $x+6$ equal to zero, there is no solution.

b. The LCD is 8(*x* + 7).

$$\frac{x}{x+7}-\frac{1}{8}=\frac{-7}{x+7}$$
$$8(x+7)\cdot\frac{x}{x+7}-8(x+7)\cdot\frac{1}{8}=8(x+7)\cdot\frac{-7}{x+7}$$
$$8x-(x+7)=8(-7)$$
$$8x-x-7=-56$$
$$7x-7=-56$$
$$7x=-49$$
$$x=-7$$

Since $x = -7$ makes the denominator $x + 7$ equal to zero, there is no solution.

c. The LCD is 9(*x* + 8).

$$\frac{x}{x+8}-\frac{1}{9}=\frac{-8}{x+8}$$
$$9(x+8)\cdot\frac{x}{x+8}-9(x+8)\cdot\frac{1}{9}=9(x+8)\cdot\frac{-8}{x+8}$$
$$9x-(x+8)=9(-8)$$
$$9x-x-8=-72$$
$$8x-8=-72$$
$$8x=-64$$
$$x=-8$$

Since $x = -8$ makes the denominator $x + 8$ equal to zero, there is no solution.

17. a. The LCD is $(x^2-16)=(x-4)(x+4)$.

$$3+\frac{5}{x-4}=\frac{50}{x^2-16}$$
$$(x-4)(x+4)\cdot 3+(x-4)(x+4)\cdot\frac{5}{x-4}=(x-4)(x+4)\cdot\frac{50}{(x-4)(x+4)}$$
$$3(x^2-16)+5(x+4)=50$$
$$3x^2-48+5x+20=50$$
$$3x^2+5x-78=0$$
$$(3x-13)(x+6)=0$$

$3x-13=0$ or $x+6=0$

$x=\frac{13}{3}$ $\qquad$ $x=-6$

The solution is $x=\frac{13}{3}$ or $x = -6$, and both check.

b. The LCD is $(x^2-25)=(x-5)(x+5)$.

$$4+\frac{6}{x-5}=\frac{84}{x^2-25}$$

$$(x-5)(x+5)\cdot 4+(x-5)(x+5)\cdot\frac{6}{x-5}=(x-5)(x+5)\cdot\frac{84}{(x-5)(x+5)}$$

$$4(x^2-25)+6(x+5)=84$$

$$4x^2-100+6x+30=84$$

$$4x^2+6x-154=0$$

$$2(2x-11)(x+7)=0$$

$2x-11=0$ or $x+7=0$

$x=\frac{11}{2}$ $\quad x=-7$

The solution is $x=\frac{11}{2}$ or $x=-7$, and both check.

c. The LCD is $(x^2-36)=(x-6)(x+6)$.

$$5+\frac{7}{x-6}=\frac{126}{x^2-36}$$

$$(x-6)(x+6)\cdot 5+(x-6)(x+6)\cdot\frac{7}{x-6}=(x-6)(x+6)\cdot\frac{126}{(x-6)(x+6)}$$

$$5(x^2-36)+7(x+6)=126$$

$$5x^2-180+7x+42=126$$

$$5x^2+7x-264=0$$

$$(5x-33)(x+8)=0$$

$5x-33=0$ or $x+8=0$

$x=\frac{33}{5}$ $\quad x=-8$

The solution is $x=\frac{33}{5}$ or $x=-8$, and both check.

18. a. The LCD is $1-b^n$.

$$A=\frac{a_1(1-b)}{1-b^n}$$

$$(1-b^n)\cdot A=(1-b^n)\cdot\frac{a_1(1-b)}{1-b^n}$$

$$A(1-b^n)=a_1(1-b)$$

$$\frac{A(1-b^n)}{1-b}=a_1 \text{ or } a_1=\frac{A(1-b^n)}{1-b}$$

b. The LCD is $1 - c$.

$$B = \frac{b_1(1-c^n)}{1-c}$$
$$(1-c)\cdot B = (1-c)\cdot\frac{b_1(1-c^n)}{1-c}$$
$$B(1-c) = b_1(1-c^n)$$
$$\frac{B(1-c)}{1-c^n} = b_1 \text{ or } b_1 = \frac{B(1-c)}{1-c^n}$$

c. The LCD is $d + 1$.

$$C = \frac{c_1(1-d)^n}{d+1}$$
$$(d+1)\cdot C = (d+1)\cdot\frac{c_1(1-d)^n}{d+1}$$
$$C(d+1) = c_1(1-d)^n$$
$$\frac{C(d+1)}{(1-d)^n} = c_1 \text{ or } c_1 = \frac{C(d+1)}{(1-d)^n}$$

19. a. The ratio of miles to gallons is $\frac{160}{7}$.

Let g = number of gallons.

$$\frac{240}{g} = \frac{160}{7}$$
$$160g = 1680$$
$$g = 10\frac{1}{2}$$

It will need $10\frac{1}{2}$ gallons.

b. The ratio of miles to gallons is $\frac{180}{9} = \frac{20}{1}$.

Let g = number of gallons.

$$\frac{270}{g} = \frac{20}{1}$$
$$20g = 270$$
$$g = 13\frac{1}{2}$$

It will need $13\frac{1}{2}$ gallons.

c. The ratio of miles to gallons is $\frac{200}{12} = \frac{50}{3}$.

Let g = number of gallons.

$$\frac{300}{g} = \frac{50}{3}$$
$$50g = 900$$
$$g = 18$$

It will need 18 gallons.

20. a.
$$\frac{x+3}{6} = \frac{7}{2}$$
$$2(x+3) = 6\cdot 7$$
$$2x + 6 = 42$$
$$2x = 36$$
$$x = 18$$

b.
$$\frac{x+4}{8} = \frac{9}{5}$$
$$5(x+4) = 8\cdot 9$$
$$5x + 20 = 72$$
$$5x = 52$$
$$x = \frac{52}{5} = 10\frac{2}{5}$$

c.
$$\frac{x+5}{2} = \frac{6}{5}$$
$$5(x+5) = 2\cdot 6$$
$$5x + 25 = 12$$
$$5x = -13$$
$$x = \frac{-13}{5}$$

21. Let h = hours if working together.

a.
$$\frac{1}{6} + \frac{1}{8} = \frac{1}{h}$$
$$24h\cdot\frac{1}{6} + 24h\cdot\frac{1}{8} = 24h\cdot\frac{1}{h}$$
$$4h + 3h = 24$$
$$7h = 24$$
$$h = \frac{24}{7} = 3\frac{3}{7}$$

It will take $3\frac{3}{7}$ hours.

b. $$\frac{1}{10}+\frac{1}{8}=\frac{1}{h}$$
$$40h\cdot\frac{1}{10}+40h\cdot\frac{1}{8}=40h\cdot\frac{1}{h}$$
$$4h+5h=40$$
$$9h=40$$
$$h=\frac{40}{9}=4\frac{4}{9}$$

It will take $4\frac{4}{9}$ hours.

c. $$\frac{1}{9}+\frac{1}{6}=\frac{1}{h}$$
$$18h\cdot\frac{1}{9}+18h\cdot\frac{1}{6}=18h\cdot\frac{1}{h}$$
$$2h=3h=18$$
$$5h=18$$
$$h+\frac{18}{5}=3\frac{3}{5}$$

It will take $3\frac{3}{5}$ hours.

22. $D=R\cdot T$ or $T=\frac{D}{R}$

Let R = speed of boat in still water and f = current flow.

	R	T	D
against	$R-f$	T	10
with	$R+f$	T	30

$$\frac{10}{R-f}=\frac{30}{R+f}$$

a. $f=2$
$$\frac{10}{R-2}=\frac{30}{R+2}$$
$$10(R+2)=30(R-2)$$
$$10R+20=30R-60$$
$$80=20R$$
$$4=R$$

The speed of the boat is 4 miles per hour.

b. $f=4$
$$\frac{10}{R-4}=\frac{30}{R+4}$$
$$10(R+4)=30(R-4)$$
$$10R+40=30R-120$$
$$160=20R$$
$$8=R$$

The speed of the boat is 8 miles per hour.

c. $f=6$
$$\frac{10}{R-6}=\frac{30}{R+6}$$
$$10(R+6)=30(R-6)$$
$$10R+60=30R-180$$
$$240=20R$$
$$12=R$$

The speed of the boat is 12 miles per hour.

23. The ratio of home runs to games is $\frac{30}{120}=\frac{1}{4}$.

Let r = number of home runs.

a. $$\frac{r}{128}=\frac{1}{4}$$
$$4r=128$$
$$r=32$$

He will have 32 home runs.

b. $$\frac{r}{140}=\frac{1}{4}$$
$$4r=140$$
$$r=35$$

He will have 35 home runs.

c. $$\frac{r}{160}=\frac{1}{4}$$
$$4r=160$$
$$r=40$$

He will have 40 home runs.

24. The ratio of items to months is $\frac{1500}{12}=\frac{125}{1}$.

Let $x \doteq$ number of items.

a. $\frac{x}{9}=\frac{125}{1}$
$x=9\cdot 125$
$x=1125$
Thus 1125 items should be produced.

b. $\frac{x}{10}=\frac{125}{1}$
$x=10\cdot 125$
$x=1250$
Thus 1250 items should be produced.

c. $\frac{x}{11}=\frac{125}{1}$
$x=11\cdot 125$
$x=1375$
Thus 1375 items should be produced.

25. a. $\frac{x}{3}=\frac{7}{2}$
$2x=21$
$x=\frac{21}{2}=10\frac{1}{2}$

b. $\frac{b}{8}=\frac{12}{9}$
$9b=96$
$b=\frac{96}{9}=10\frac{2}{3}$

c. $\frac{s}{4}=\frac{6}{9}$
$9s=24$
$s=\frac{24}{9}=2\frac{2}{3}$

Cumulative Review Chapters 1–6

1. The LCD is $6\cdot 5=30$.

$$\begin{aligned}-\frac{1}{6}+\left(-\frac{2}{5}\right)&=-\frac{1\cdot 5}{6\cdot 5}+\left(-\frac{2\cdot 6}{5\cdot 6}\right)\\&=-\frac{5}{30}+\left(-\frac{12}{30}\right)\\&=\frac{-5+(-12)}{30}\\&=-\frac{17}{30}\end{aligned}$$

2. $-5.6-(-8.3)=-5.6+8.3=2.7$

3. $(-5)^2=(-5)(-5)=25$

4. $-\frac{7}{8}\div\left(-\frac{7}{16}\right)=-\frac{7}{8}\cdot\left(-\frac{16}{7}\right)=2$

5.
$$\begin{aligned}y\div 5\cdot x-z&=60\div 5\cdot 6-3\\&=12\cdot 6-3\\&=72-3\\&=69\end{aligned}$$

6.
$$\begin{aligned}&x+4(x-3)+(x-2)\\&=x+4x-12+x-2\\&=(x+4x+x)+(-12-2)\\&=6x-14\end{aligned}$$

7. "The quotient" means divide.
$\frac{a-b}{c}$

8.
$$\begin{aligned}5&=3(x-1)+4-2x\\5&=3x-3+4-2x\\5&=(3x-2x)+(-3+4)\\5&=x+1\\5-1&=x+1-1\\4&=x \text{ or } x=4\end{aligned}$$

9. The LCD is 63.

$$\frac{x}{7}-\frac{x}{9}=2$$
$$63\cdot\frac{x}{7}-63\cdot\frac{x}{9}=63\cdot 2$$
$$9x-7x=126$$
$$2x=126$$
$$\frac{2x}{2}=\frac{126}{2}$$
$$x=63$$

10. Let n = first number.
Then $n + 35$ = other number.

$$n+(n+35)=95$$
$$2n+35=95$$
$$2n=60$$
$$\frac{2n}{2}=\frac{60}{2}$$
$$n=30$$

$n + 35 = 30 + 35 = 65$
The numbers are 30 and 65.

11.

	P	$\times$ r	$=$ I
bonds	x	0.08	$0.08x$
CDs	$12{,}000 - x$	0.10	$0.10(12{,}000 - x)$

$$0.08x+0.10(12{,}000-x)=1020$$
$$0.08x+1200-0.10x=1020$$
$$1200-0.02x=1020$$
$$-0.02x=-180$$
$$x=9000$$

$12{,}000 - x = 12{,}000 - 9000 = 3000$
Susan invested \$9000 in bonds and \$3000 in certificates of deposit.

12.

$$-\frac{x}{2}+\frac{x}{4}\geq\frac{x-4}{4}$$
$$4\cdot\left(-\frac{x}{2}\right)+4\cdot\frac{x}{4}\geq 4\cdot\frac{x-4}{4}$$
$$-2x+x\geq x-4$$
$$-x\geq x-4$$
$$-2x\geq -4$$
$$\frac{-2x}{-2}\leq\frac{-4}{-2}$$
$$x\leq 2$$

Any number less than or equal to 2 is a solution.

0 2

13. Start at the origin. To reach the point $C(-1, -3)$, go 1 unit to the left and 3 units down.

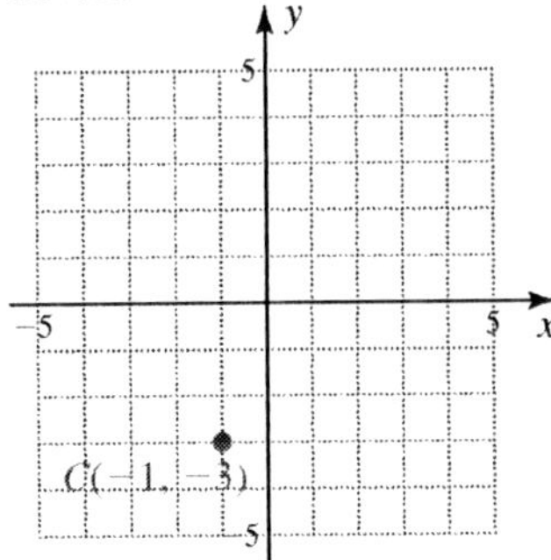

14. In the ordered pair $(-3, -3)$, $x = -3$ and $y = -3$. Substituting in $5x - y = -12$, we get $5(-3) - (-3) = -15 + 3 = -12$, which is true. Yes, $(-3, -3)$ is a solution.

15. In the ordered pair $(x, -2)$, $y = -2$. Substituting -2 for y in the given equation, we have

$$\begin{aligned} 3x - y &= 11 \\ 3x - (-2) &= 11 \\ 3x + 2 &= 11 \\ 3x &= 9 \\ x &= 3 \end{aligned}$$

Thus, $x = 3$ and the ordered pair is $(3, -2)$.

16. Graph two points and join them with a line.

For $x = 0$, $\quad 2x + y = 2$

$$\begin{aligned} 2(0) + y &= 2 \\ y &= 2 \end{aligned}$$

Thus $(0, 2)$ is on the graph.

For $y = 0$, $\quad 2x + y = 2$

$$\begin{aligned} 2x + 0 &= 2 \\ 2x &= 2 \\ x &= 1 \end{aligned}$$

Thus $(1, 0)$ is on the graph.

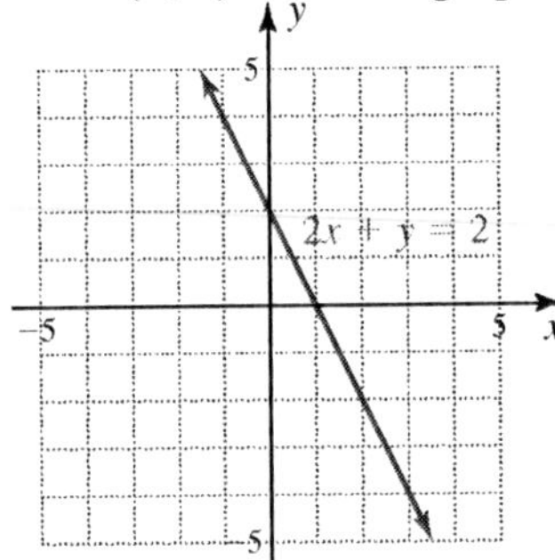

17. Solve for y.

$$\begin{aligned} 3y + 6 &= 0 \\ 3y &= -6 \\ y &= -2 \end{aligned}$$

The graph of $y = -2$ is a horizontal line crossing the y-axis at -2.

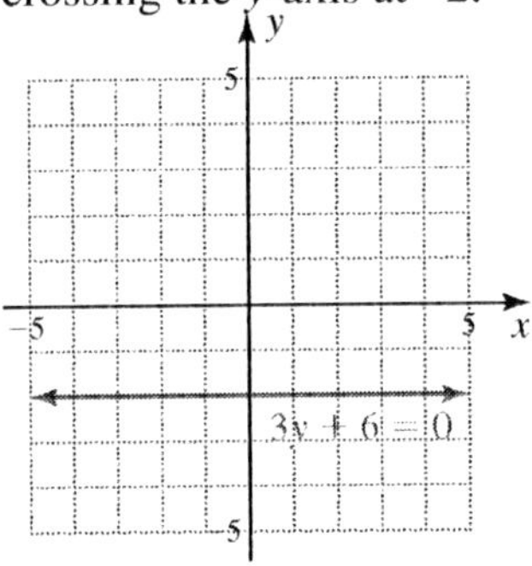

18. Let $(x_1, y_1) = (-7, -9)$ and $(x_2, y_2) = (9, -2)$.

$$m = \frac{y_2 - y_1}{x_2 - x_1} = \frac{-2 - (-9)}{9 - (-7)} = \frac{-2 + 9}{9 + 7} = \frac{7}{16}$$

19. Solve for y.

$$\begin{aligned} 6x - 3y &= -10 \\ -3y &= -6x - 10 \\ y &= 2x + \frac{10}{3} \end{aligned}$$

Since the coefficient of x is 2, the slope is 2.

20. Find the slope of each line by solving for y.

(1) $15y + 20x = 8$

$$\begin{aligned} 15y &= -20x + 8 \\ y &= -\frac{4}{3}x + \frac{8}{15} \end{aligned}$$

The slope is $-\frac{4}{3}$.

(2) $20x - 15y = 8$

$$\begin{aligned} -15y &= -20x + 8 \\ y &= \frac{4}{3}x - \frac{8}{15} \end{aligned}$$

The slope is $\frac{4}{3}$.

(3) $3y = 4x + 8$

$$y = \frac{4}{3}x + \frac{8}{3}$$

The slope is $\frac{4}{3}$.

Since the slopes of (2) and (3) are equal and the y-intercepts are different, the lines are parallel.

21. $\frac{x^{-8}}{x^{-9}} = x^{-8-(-9)} = x^{-8+9} = x^1 = x$

22. $x^5 \cdot x^{-7} = x^{5+(-7)} = x^{-2} = \frac{1}{x^2}$

23. $(3x^3y^{-2})^{-4} = 3^{-4}x^{3(-4)}y^{-2(-4)}$

$$= \frac{1}{81} \cdot x^{-12}y^8$$

$$= \frac{y^8}{81x^{12}}$$

24. $0.00036 = 3.6 \times 10^{-4}$

25. $(14.04 \times 10^{-3}) \div (7.8 \times 10^2)$

$= (14.04 \div 7.8) \times (10^{-3} \div 10^2)$

$= 1.80 \times 10^{-3-2}$

$= 1.80 \times 10^{-5}$

26. Let $x = -2$.

$x^2 + 3x - 2 = (-2)^2 + 3(-2) - 2$

$= 4 - 6 - 2$

$= -4$

27. $(2x^4 - 6x^6 - 7) + (-6x^6 + 3 + 7x^4)$

$= [-6x^6 + (-6x^6)] + (2x^4 + 7x^4) + (-7 + 3)$

$= -12x^6 + 9x^4 - 4$

28. $\left(5x^2 - \frac{1}{5}\right)^2 = (5x^2)^2 + 2(5x^2)\left(-\frac{1}{5}\right) + \left(\frac{1}{5}\right)^2$

$$= 25x^4 - 2x^2 + \frac{1}{25}$$

29. $(9x^2 + 2)(9x^2 - 2) = (9x^2)^2 - 2^2 = 81x^4 - 4$

30.

$$\begin{array}{r} 3x^2 - x + 4 \\ x+4 \overline{\big)\ 3x^3 + 11x^2 + 0x + 18} \\ (-)\underline{3x^3 + 12x^2} \\ -x^2 + 0x \\ (-)\underline{-x^2 - 4x} \\ 4x + 18 \\ (-1)\underline{4x + 16} \\ 2 \end{array}$$

Thus

$(3x^3 + 11x^2 + 18) \div (x + 4)$

$= (3x^2 - x + 4)$ R2 or $3x^2 - x + 4 + \frac{2}{x+4}$

31. $6x^2 - 9x^4 = 3x^2 \cdot 2 - 3x^2 \cdot 3x^2$

$= 3x^2(2 - 3x^2)$

32. $\frac{2}{5}x^7 - \frac{4}{5}x^6 + \frac{4}{5}x^5 - \frac{1}{5}x^3$

$= \frac{1}{5}x^3 \cdot 2x^4 - \frac{1}{5}x^3 \cdot 4x^3 + \frac{1}{5}x^3 \cdot 4x^2 - \frac{1}{5}x^3 \cdot 1$

$= \frac{1}{5}x^3(2x^4 - 4x^3 + 4x^2 - 1)$

33.

Factors	Sum
−1, −56	−57
−2, −28	−30
−4, −14	−18
−7, −8	−15

$x^2 - 15x + 56 = (x - 7)(x - 8)$

34. $15x^2 - 37xy + 20y^2$

$= 15x^2 - 12xy - 25xy + 20y^2$

$= (15x^2 - 12xy) + (-25xy + 20y^2)$

$= 3x(5x - 4y) - 5y(5x - 4y)$

$= (5x - 4y)(3x - 5y)$

35. $16x^2 - 9y^2 = (4x)^2 - (3y)^2$

$= (4x + 3y)(4x - 3y)$

36. $-4x^4+4x^2=-4x^2(x^2-1)$
$=-4x^2(x+1)(x-1)$

37. $3x^3-6x^2-9x=3x(x^2-2x-3)$
$=3x(x+1)(x-3)$

38. $4x^2+4x+3x+3=(4x^2+4x)+(3x+3)$
$=4x(x+1)+3(x+1)$
$=(x+1)(4x+3)$

39. $16kx^2-8kx+k=k(16x^2-8x+1)$
$=k[(4x)^2-2(4x)(1)+1^2]$
$=k(4x-1)^2$

40. $2x^2+x=15$
$2x^2+x-15=0$
$(2x-5)(x+3)=0$
$2x-5=0$ or $x+3=0$
$x=\frac{5}{2}$ $\quad x=-3$

The solution is $x=\frac{5}{2}$ or $x=-3$.

41. Since $12y^3=6y\cdot 2y^2$, then
$\frac{7x}{6y}=\frac{7x\cdot 2y^2}{6y\cdot 2y^2}=\frac{14xy^2}{12y^3}$.

42. $\frac{-8(x^2-y^2)}{4(x-y)}=\frac{-2\cdot 4(x-y)(x+y)}{4(x-y)}$
$=-2(x+y)$ or $-2x-2y$

43. $\frac{x^2+4x-5}{1-x}=\frac{(x-1)(x+5)}{-1(x-1)}$
$=\frac{(x+5)}{-1}$
$=-(x+5)$ or $-x-5$

44. $(x-7)\cdot\frac{x+3}{x^2-49}=\frac{x-7}{1}\cdot\frac{x+3}{x^2-49}$
$=\frac{x-7}{1}\cdot\frac{x+3}{(x-7)(x+7)}$
$=\frac{x+3}{x+7}$

45. $\frac{x+3}{x-3}\div\frac{x^2-9}{3-x}=\frac{x+3}{x-3}\cdot\frac{3-x}{x^2-9}$
$=\frac{x+3}{x-3}\cdot\frac{-1(x-3)}{(x-3)(x+3)}$
$=\frac{-1}{x-3}$ or $\frac{1}{3-x}$

46. $\frac{3}{2(x+8)}+\frac{9}{2(x+8)}=\frac{3+9}{2(x+8)}$
$=\frac{12}{2(x+8)}$
$=\frac{6}{x+8}$

47. $x^2+x-30=(x+6)(x-5)$
$x^2-25=(x+5)(x-5)$
The LCD is $(x+6)(x+5)(x-5)$
$\frac{x+5}{x^2+x-30}=\frac{(x+5)(x+5)}{(x+6)(x+5)(x-5)}$
$\frac{x+6}{x^2-25}=\frac{(x+6)(x+6)}{(x+6)(x+5)(x-5)}$
$\frac{x+5}{x^2+x-30}-\frac{x+6}{x^2-25}$
$=\frac{(x+5)^2}{(x+6)(x+5)(x-5)}-\frac{(x+6)^2}{(x+6)(x+5)(x-5)}$
$=\frac{(x+5)^2-(x+6)^2}{(x+6)(x+5)(x-5)}$
$=\frac{(x^2+10x+25)-(x^2+12x+36)}{(x+6)(x+5)(x-5)}$
$=\frac{x^2+10x+25-x^2-12x-36}{(x+6)(x+5)(x-5)}$
$=\frac{-2x-11}{(x+6)(x+5)(x-5)}$

48. The LCD is $12x$.

$$\frac{\frac{2}{x}+\frac{3}{2x}}{\frac{1}{3x}-\frac{1}{4x}}=\frac{12x\cdot\frac{2}{x}+12x\cdot\frac{3}{2x}}{12x\cdot\frac{1}{3x}-12x\cdot\frac{1}{4x}}$$
$$=\frac{12\cdot 2+6\cdot 3}{4-3}$$
$$=\frac{24+18}{1}$$
$$=42$$

49.
$$\frac{3x}{x-2}+3=\frac{4x}{x-2}$$
$$(x-2)\cdot\frac{3x}{x-2}+3(x-2)=(x-2)\cdot\frac{4x}{x-2}$$
$$3x+3x-6=4x$$
$$6x-6=4x$$
$$-6=-2x$$
$$3=x$$
The solution $x = 3$ checks.

50.
$$\frac{x}{x^2-49}+\frac{7}{x-7}=\frac{1}{x+7}$$
$$(x+7)(x-7)\cdot\frac{x}{(x+7)(x-7)}+(x+7)(x-7)\cdot\frac{7}{x-7}=(x+7)(x-7)\cdot\frac{1}{x+7}$$
$$x+7(x+7)=1\cdot(x-7)$$
$$x+7x+49=x-7$$
$$8x+49=x-7$$
$$7x=-56$$
$$x=-8$$
The solution $x = -8$ checks.

51.
$$\frac{x}{x+4}-\frac{1}{5}=\frac{-4}{x+4}$$
$$5(x+4)\cdot\frac{x}{x+4}-5(x+4)\cdot\frac{1}{5}=5(x+4)\cdot\frac{-4}{x+4}$$
$$5x-(x+4)=5(-4)$$
$$5x-x-4=-20$$
$$4x-4=-20$$
$$4x=-16$$
$$x=-4$$
Since $x = -4$ makes the denominator $x + 4$ equal to zero, there is no solution.

52. $$1+\frac{2}{x-3}=\frac{12}{x^2-9}$$
$$(x+3)(x-3)\cdot 1+(x+3)(x-3)\cdot\frac{2}{x-3}=(x+3)(x-3)\cdot\frac{12}{(x+3)(x-3)}$$
$$(x+3)(x-3)+2(x+3)=12$$
$$x^2-9+2x+6=12$$
$$x^2+2x-15=0$$
$$(x-3)(x+5)=0$$

$x-3=0$ or $x+5=0$

$x=3$ $\quad x=-5$

Since $x = 3$ makes the denominator $x - 3$ equal to zero, there is only one solution, $x = -5$, which checks.

53. The ratio of miles to gallons is $\frac{120}{6}=\frac{20}{1}$.

Let g = number of gallons.

$$\frac{460}{g}=\frac{20}{1}$$
$$20g=460$$
$$g=23$$

It needs 23 gallons.

54. $$\frac{x+8}{9}=\frac{2}{11}$$
$$11(x+8)=9\cdot 2$$
$$11x+88=18$$
$$11x=-70$$
$$x=-\frac{70}{11}$$

55. Let h = hours working together.

$$\frac{1}{6}+\frac{1}{7}=\frac{1}{h}$$
$$42h\cdot\frac{1}{6}+42h\cdot\frac{1}{7}=42h\cdot\frac{1}{h}$$
$$7h+6h=42$$
$$13h=42$$
$$h=\frac{42}{13}=3\frac{3}{13}$$

It will take $3\frac{3}{13}$ hours.

Chapter 7 Graphs, Slopes, Inequalities, and Applications

7.1 Graphing Lines Using Points and Slopes

Problems 7.1

1. a. Use the point-slope form.

$$y - y_1 = m(x - x_1)$$
$$y - (-4) = -3(x - 2)$$
$$y + 4 = -3x + 6$$
$$y = -3x + 2$$

b.

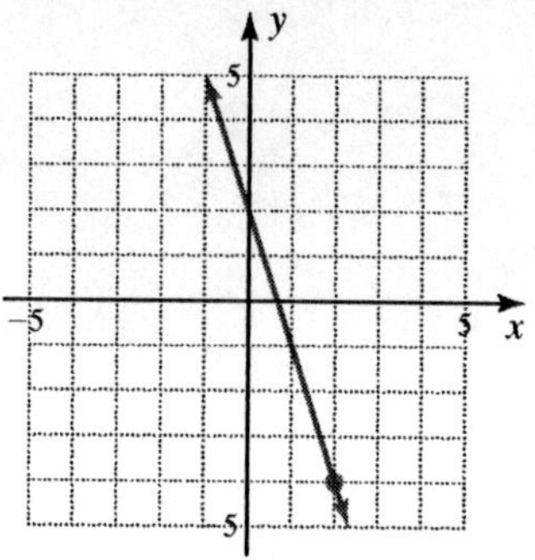

2. a. Use the slope-intercept form.

$y = mx + b$
$y = 3x + (-6)$
$y = 3x - 6$

b.

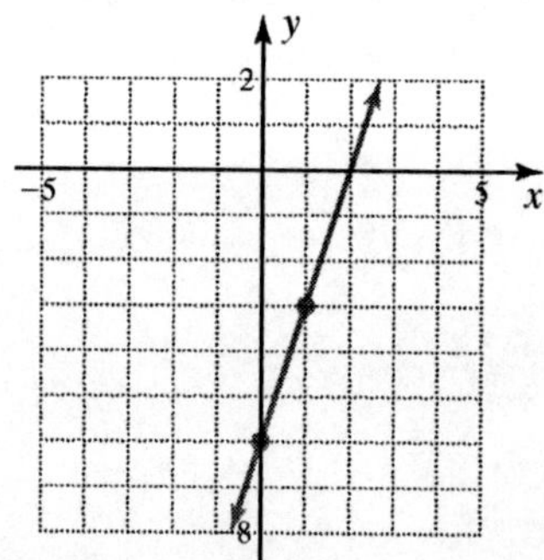

3. a. Find the slope.

$$m = \frac{4-(-6)}{-3-5} = \frac{10}{-8} = -\frac{5}{4}$$

Use the point (5, –6) in the point-slope form.

$$y - (-6) = -\frac{5}{4}(x - 5)$$
$$y + 6 = -\frac{5}{4}x + \frac{25}{4}$$
$$y = -\frac{5}{4}x + \frac{1}{4}$$

b.

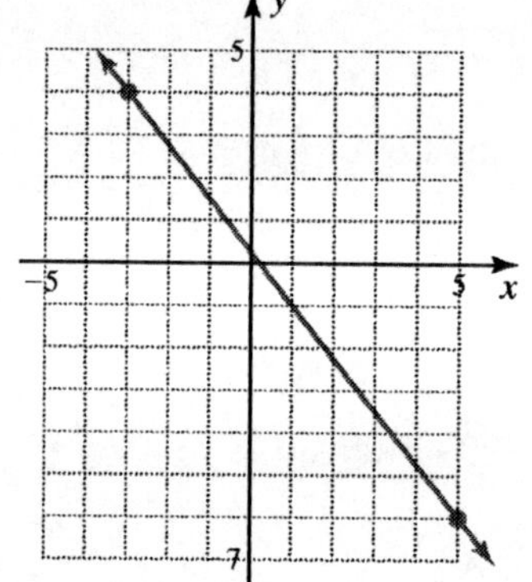

Exercises 7.1

1. $y - 2 = \frac{1}{2}(x - 1)$

$$y - 2 = \frac{1}{2}x - \frac{1}{2}$$
$$y = \frac{1}{2}x + \frac{3}{2}$$

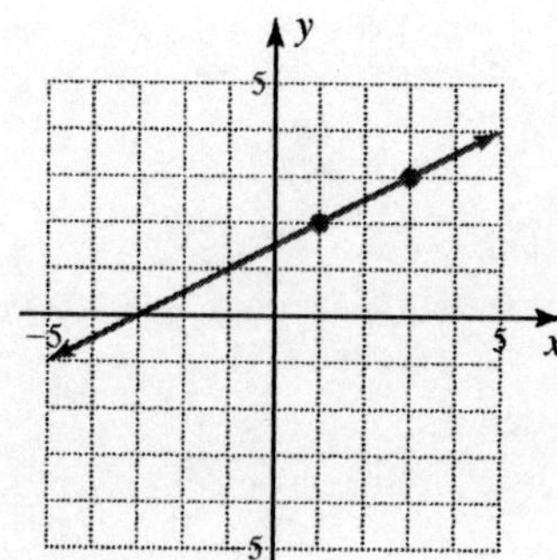

3. $y-2=-1(x-4)$
$y-2=-x+4$
$y=-x+6$

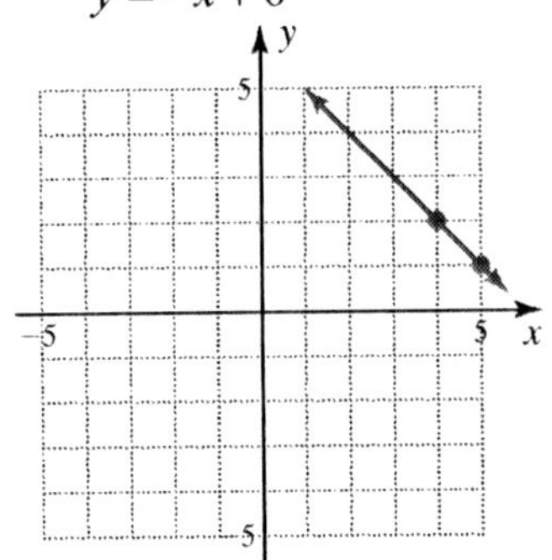

5. $y-5=0(x-4)$
$y-5=0$
$y=5$

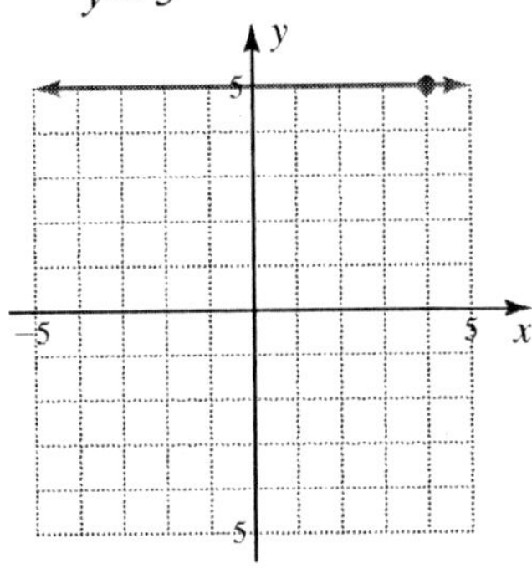

7. $m=2;\ b=-3$
$y=2x-3$

9. $m=-4;\ b=6$
$y=-4x+6$

11. $m=\frac{3}{4};\ b=\frac{7}{8}$
$y=\frac{3}{4}x+\frac{7}{8}$

13. $m=2.5;\ b=-4.7$
$y=2.5x-4.7$

15. $m=-3.5;\ b=5.9$
$y=-3.5x+5.9$

17. $y=\frac{1}{4}x+3$

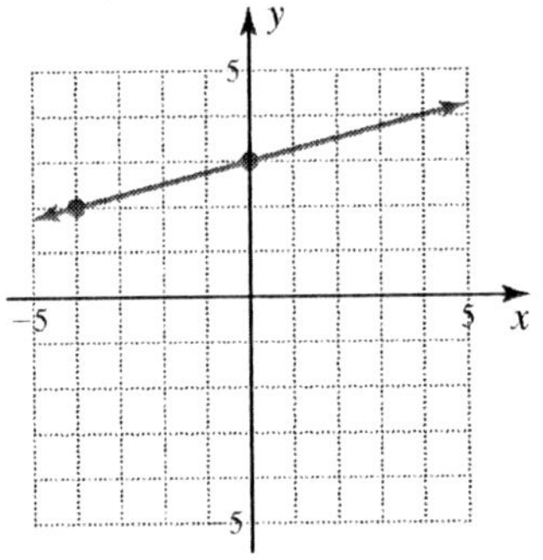

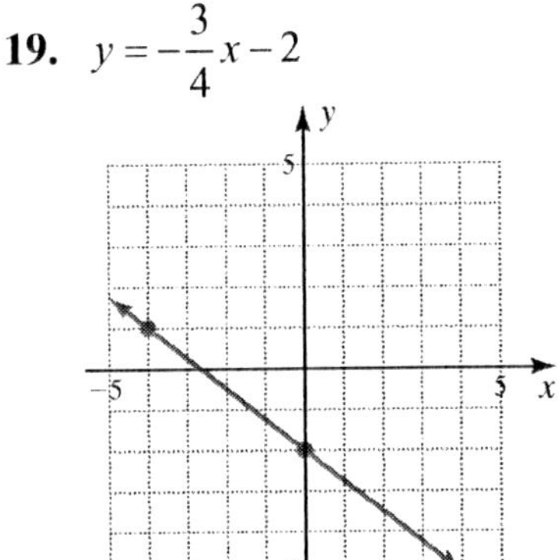

19. $y=-\frac{3}{4}x-2$

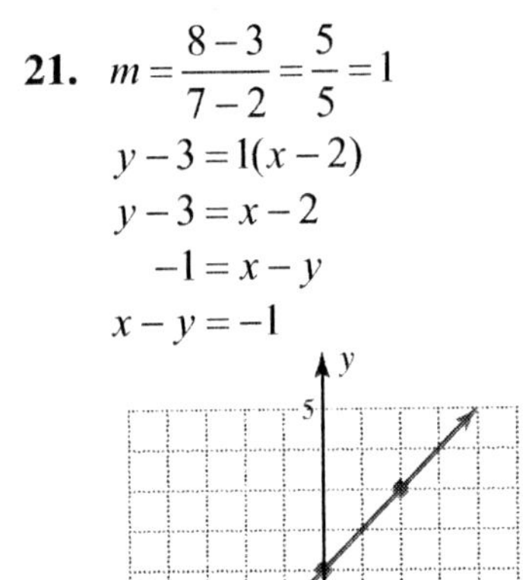

21. $m=\frac{8-3}{7-2}=\frac{5}{5}=1$
$y-3=1(x-2)$
$y-3=x-2$
$-1=x-y$
$x-y=-1$

23. $m = \dfrac{-1-2}{1-2} = \dfrac{-3}{-1} = 3$

$y - 2 = 3(x - 2)$
$y - 2 = 3x - 6$
$4 = 3x - y$
$3x - y = 4$

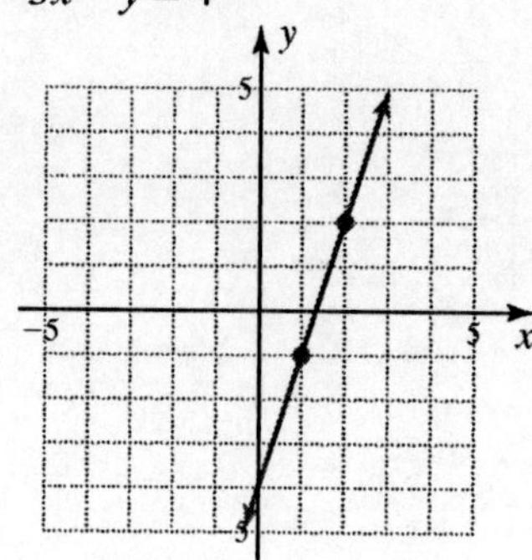

25. $m = \dfrac{4-0}{0-3} = \dfrac{4}{-3} = -\dfrac{4}{3}$

$y - 0 = -\dfrac{4}{3}(x - 3)$
$y = -\dfrac{4}{3}x + 4$
$3y = -4x + 12$
$4x + 3y = 12$

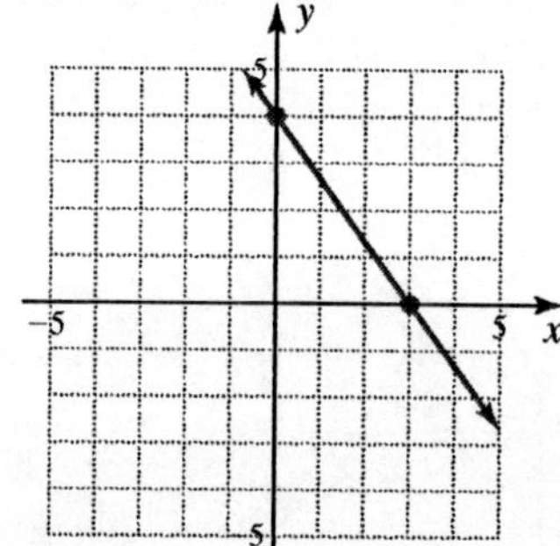

27. $m = \dfrac{2-0}{3-3} = \dfrac{2}{0} = \text{undefined}$

A vertical line through (3, 0).
$x = 3$

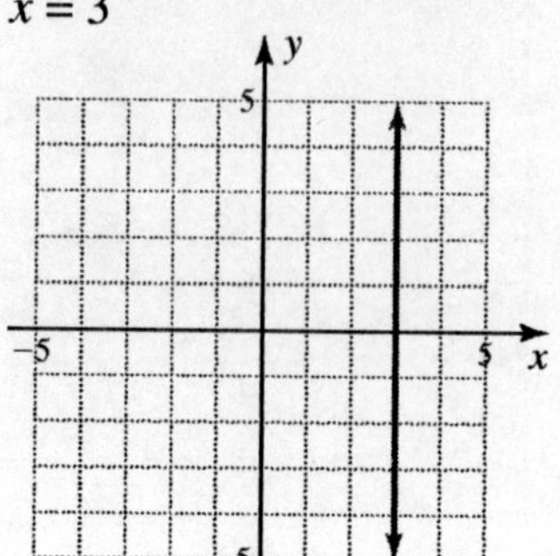

29. $m = \dfrac{-3-(-3)}{1-(-2)} = \dfrac{0}{3} = 0$

$y - (-3) = 0(x - 1)$
$y + 3 = 0$
$y = -3$

31. $m = \dfrac{7-3}{4-0} = \dfrac{4}{4} = 1$

$y - 3 = 1(x - 0)$
$y - 3 = x$
$y = x + 3$

33. $m = 2$

$y - 2 = 2(x - 1)$
$y - 2 = 2x - 2$
$y = 2x$

35. $m = -2$; $b = -3$

$y = -2x - 3$

37. The production cost, m, is \$2 per unit and the fixed cost, b, is \$50.

$y = 2x + 50$

39. **a.** $m = 2$, so the production cost is \$2 per unit.

b. b is 75, so the fixed cost is \$75.

41. The vertical line through the origin is the y-axis.

43. There is no y component in the equation of a vertical line, so the standard form is $Ax + 0y = C$.

45. Answers may vary.

47. $m = -\frac{3}{4}$; $b = 2$

$$y = -\frac{3}{4}x + 2$$

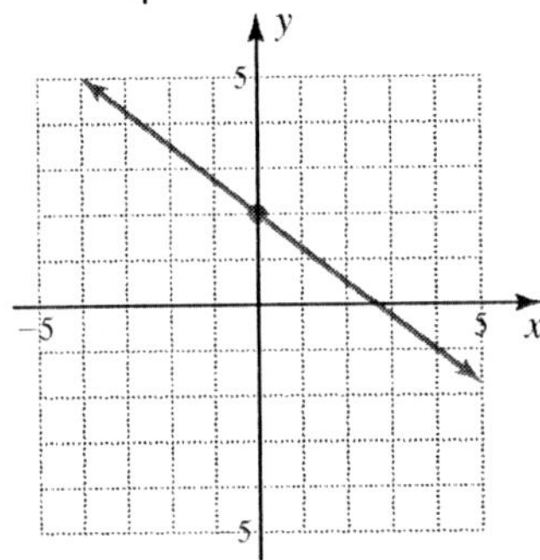

49. $m = \frac{2}{3}$

$$\begin{aligned} y-(-1) &= \frac{2}{3}(x-3) \\ y+1 &= \frac{2}{3}x-2 \\ y &= \frac{2}{3}x-3 \\ 3y &= 2x-9 \\ 9 &= 2x-3y \\ 2x-3y &= 9 \end{aligned}$$

51. $m = \frac{4-(-6)}{-3-(-1)} = \frac{10}{-2} = -5$

$$\begin{aligned} y-(-6) &= -5[x-(-1)] \\ y+6 &= -5(x+1) \\ y+6 &= -5x-5 \\ y &= -5x-11 \\ 5x+y &= -11 \end{aligned}$$

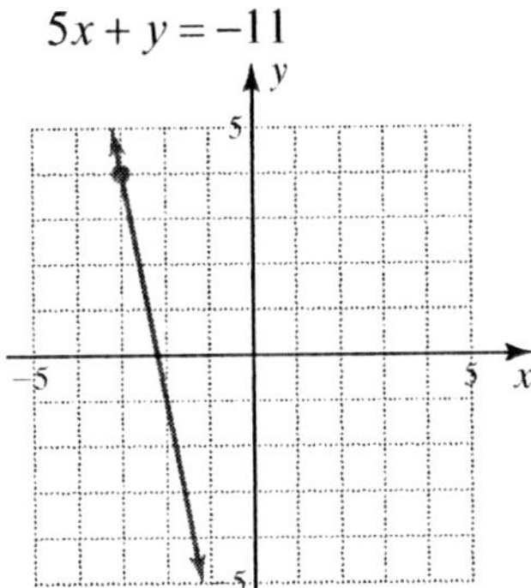

7.2 Applications of Equations of Lines

Problems 7.2

1. We have the point (10, 22) and the slope 2.

$$\begin{aligned} C-22 &= 2(m-10) \\ C &= 2m-20+22 \\ C &= 2m+2 \end{aligned}$$

For a 25-mile ride, $m = 25$ and $C = 2(25) + 2 = 52$. The cost is \$52.

2. Since the cost for no extra minutes is \$40, the y-intercept is 40. Since the cost is \$0.50 for each additional minute, the slope is 0.5.
$C = 0.5m + 40$, for $m > 900$
If she uses 1000 minutes, she will have to pay for $1000 - 900 = 100$ of them, and the cost will be
$C = 0.5(100) + 40 = 50 + 40 = 90$
The cost for 1000 minutes is \$90.

3. Use $(h_1, C_1) = (6, 17.45)$ and $(h_2, C_2) = (9, 24.95)$ to find m.

$$m = \frac{24.95-17.45}{9-6} = \frac{7.5}{3} = 2.5$$

Use (6, 17.45) in the point-slope form to get the equation.

$C - 17.45 = 2.5(h - 6)$
$C = 2.5h - 15 + 17.45$
$C = 2.5h + 2.45$
When 10 hours are used, $h = 10$.
$C = 2.5(10) + 2.45 = 25 + 2.45 = 27.45$
The cost for 10 hours is \$27.45.

Exercises 7.2

1. We have the point (10, 17.30) and the slope 1.7.
$C - 17.3 = 1.7(m - 10)$
$C = 1.7m - 17 + 17.3$
$C = 1.7m + 0.3$
For a 30-mile ride, $m = 30$ and
$C = 1.7(30) + 0.3 = 51 + 0.3 = 51.3$.
A 30-mile ride costs \$51.30.

3. The equation from Problem 1 is $C = 1.7m + 0.3$, while that from Problem 2 is $C = 1.5m + 1.5$. Set the two expressions for C equal to find the number of miles, m.
$1.7m + 0.3 = 1.5m + 1.5$
$0.2m + 0.3 = 1.5$
$0.2m = 1.2$
$m = \frac{1.2}{0.2} = 6$
They both rode 6 miles.

5. a. The cost for the first mile is given as \$2.

b. The number of miles after the first mile is $m - 1$.

c. The cost per mile after the first is given as \$1.70.

d. The cost of all the miles after the first is the cost per mile times the number of miles, or \$1.70(m – 1).

e. The total cost, C, is the cost of the first mile, \$2, plus the cost of all the additional miles, \$1.70(m – 1).
$C = 2 + 1.7(m - 1)$
$C = 2 + 1.7m - 1.7$
$C = 1.7m + 0.3$
This is the same as in Problem 1.

7. a. The cost is \$175, which includes 60 minutes of calls to New York, so $C = 175$, $m \le 60$.

b. The weekly charge is \$175, so $C = 175$, $m \le 60$.

c. The cost for m minutes, $m \le 60$, is \$175, so the per-minute charge is $C = \frac{175}{m}$, $m \le 60$.

9. a. Using (20, 13) and (25, 19),
$m = \frac{19-13}{25-20} = \frac{6}{5}$.

b. Using (25, 19) and (20, 25),
$m = \frac{25-19}{30-25} = \frac{6}{5}$.

c. Yes, the slopes are the same $\left(\text{both } \frac{6}{5}\right)$.

d. Use $m = \frac{6}{5}$ and (20, 13).
$y - 13 = \frac{6}{5}(x - 20)$
$y = \frac{6}{5}x - 24 + 13$
$y = \frac{6}{5}x - 11$

e. Use $m = \frac{6}{5}$ and (30, 25).
$y - 25 = \frac{6}{5}(x - 30)$
$y = \frac{6}{5}x - 36 + 25$
$y = \frac{6}{5}x - 11$
This is equivalent to the equation in (d).

f. For a temperature of 5°F, $x = 5$.
$y = \frac{6}{5}(5) - 11 = 6 - 11 = -5$
The wind chill is –5.

11. a. The total cost, C, is the cost of the service call, \$100, plus the product of the hourly rate, \$37.50, and the number of hours, h.
$C = 37.5h + 100$

b. $212.50 = 37.5h + 100$
$112.50 = 37.5h$
$3 = h$
A bill of \$212.50 is for 3 hours.

13. a. The cost, C, is the per-week rate, \$40, plus the product of the rate per minute, \$2.30, and the number of minutes, m.
$C = 2.3m + 40$

b. $201 = 2.3m + 40$
$161 = 2.3m$
$70 = m$
A total cost of \$201 is for 70 minutes of calls.

c. The cost, C, is the per-minute rate, \$7.20, times the number of minutes, m.
$C = 7.2m$

d. $7.2m = 2.3m + 40$
$4.9m = 40$
$m = \frac{40}{4.9} \approx 8$
The charges are the same for about 8 minutes of calls.

15. a. The cost, C, is the base charge of \$50, plus the product of the charge for each additional minute, \$0.35, and the number of additional minutes, m.
$C = 0.35m + 50$

b. $88.5 = 0.35m + 50$
$38.5 = 0.35m$
$110 = m$
110 additional minutes were used.

17. From Problem 15: $C = 0.35m + 50$
From Problem 16: $C = 0.4m + 45$
$0.4m + 45 = 0.35m + 50$
$0.05m + 45 = 50$
$0.05m = 5$
$m = 100$
The charges are the same for 100 additional minutes.
The plan in Problem 15 is cheaper after 100 additional minutes, since it has a lower per-minute charge.

19. a. $y = 1.95x + 29$

b. The slope is $m = 1.95$.

c. The y-intercept is $b = 29$.

21. The slope is $m = -0.1$ (a decrease of 0.1 day per year).
$x = 0$ in 1970, when $y = 8$, so the y-intercept is $b = 8$.
$y = -0.1x + 8$
In 2000, $x = 2000 - 1970 = 30$.
$y = -0.1(30) + 8 = -3 + 8 = 5$
The average stay in 2000 would be 5 days.
In 2010, $x = 2010 - 1970 = 40$.
$y = -0.1(40) + 8 = -4 + 8 = 4$
The average stay in 2010 would be 4 days.

23. a. $(h, d) = (3, 30)$ means that 3 human years correspond to 30 dog years.
$(h, d) = (9, 60)$ means that 9 human years correspond to 60 dog years.

b. $m = \frac{60-30}{9-3} = \frac{30}{6} = 5$

c. Use $m = 5$ and $(h, d) = (3, 30)$.
$d - 30 = 5(h - 3)$
$d = 5h - 15 + 30$
$d = 5h + 15$

d. When a dog is 4 human years old, $h = 4$ and $d = 5(4) + 15 = 20 + 15 = 35$.
The dog is 35 years old in dog years.

e. Find h when $d = 65$.
$65 = 5h + 15$
$50 = 5h$
$10 = h$
The retirement age for dogs is 10 human years.

f. Find h when $d = 21$.
$21 = 5h + 15$
$6 = 5h$
$1.2 = h$
The drinking age for dogs is 1.2 human years.

25. a. $m = \frac{0.103 - 0.052}{5 - 3} = \frac{0.051}{2} = 0.0255$

b. Use $m = 0.0255$ and (3, 0.052).
$c - 0.052 = 0.0255(b - 3)$
$c = 0.0255b - 0.0765 + 0.052$
$c = 0.0255b - 0.0245$

c. $b = 4$:
$c = 0.0255(4) - 0.0245$
$= 0.102 - 0.0245$
$= 0.0775$
4 beers in 1 hour gives a BAC of 0.0755.
$b = 6$:
$c = 0.0255(6) - 0.0245$
$= 0.153 - 0.0245$
$= 0.1285$
6 beers in 1 hour gives a BAC of 0.1285.

d. Find b when $c = 0.08$.
$0.08 = 0.0255b - 0.0245$
$0.1045 = 0.0255b$
$4 \approx b$
You will be legally drunk after 4 beers in 1 hour.

27. Since 0 is to the right of –8 on a number line, $0 > -8$.

29. Since 10 is to the right of 0 on a number line, $10 > 0$.

31. Since –1 is to the left of 0 on a number line, $-1 < 0$.

33. Answers may vary.

35. Answers may vary.

37–39.

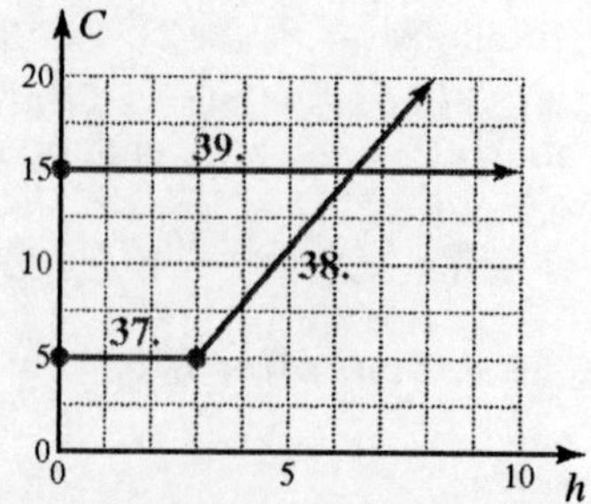

41. The total cost, C, is the daily charge, \$40, plus the product of the per-mile rate, \$0.15, and the number of miles, m.
$C = 0.15m + 40$
$m = 450$:
$C = 0.15(450) + 40 = 67.5 + 40 = 107.5$
The cost for 450 miles in one day is \$107.50.

43. a. (30, 24) means that 30 hours of use cost \$24.

b. (33, 26.70) means that 33 hours of use cost \$26.70.

c. $m = \frac{26.70 - 24}{33 - 30} = \frac{2.70}{3} = 0.9$
$C - 24 = 0.9(x - 30)$
$C = 0.9x - 27 + 24$
$C = 0.9x - 3$

7.3 Graphing Inequalities in Two Variables

Problems 7.3

1. Graph $3x - 2y = -6$ with a dashed line.
Test (0, 0), which is below the line.
$3x - 2y < -6$
$3(0) - 2(0) < -6$
$0 < -6$ False
Shade the points above the line.

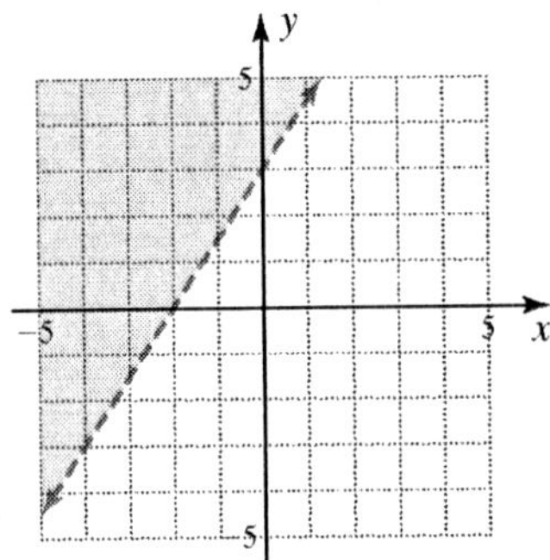

2. Graph $y = -3x + 6$ with a solid line. Test (0, 0), which is below the line:
$y \le -3x + 6$
$0 \le -3(0) + 6$
$0 \le 6$ True
Shade the points below the line.

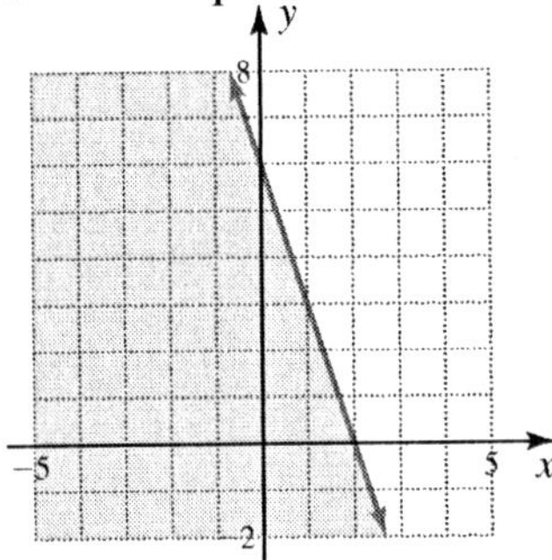

3. Graph $y + 3x = 0$ with a dashed line. (0, 0) is on the line, so we test (1, 1), which is above the line:
$y + 3x > 0$
$1 + 3(1) > 0$
$4 > 0$ True
Shade the points above the line.

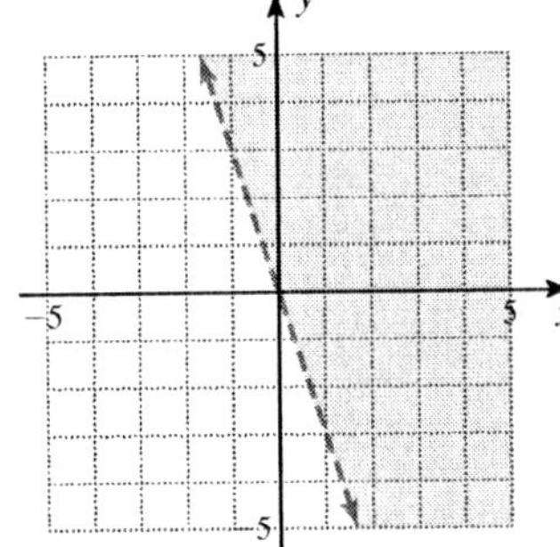

4. a. Graph the vertical line $x = 2$ as solid. Test (0, 0), which is to the left of the line:
$x \ge 2$
$0 \ge 2$ False
Shade the points to the right of the line.

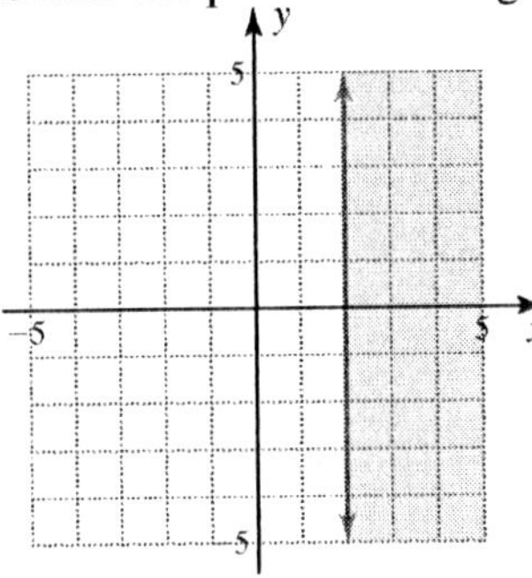

b. Graph the horizontal line $y + 2 = 0$ ($y = -2$) as dashed. Test (0, 0), which is above the line.
$y + 2 > 0$
$0 + 2 > 0$
$2 > 0$ True
Shade the points above the line.

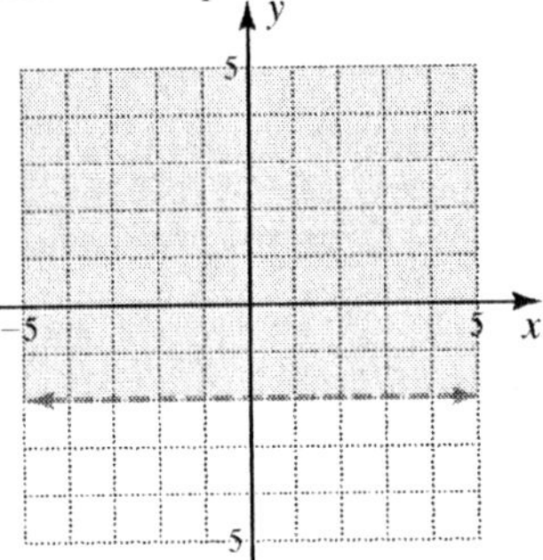

Exercises 7.3

1. Graph $2x + y = 4$ with a dashed line. Test (0, 0), which is below the line.
$2x + y > 4$
$2(0) + 0 > 4$
$0 > 4$ False
Shade the points above the line.

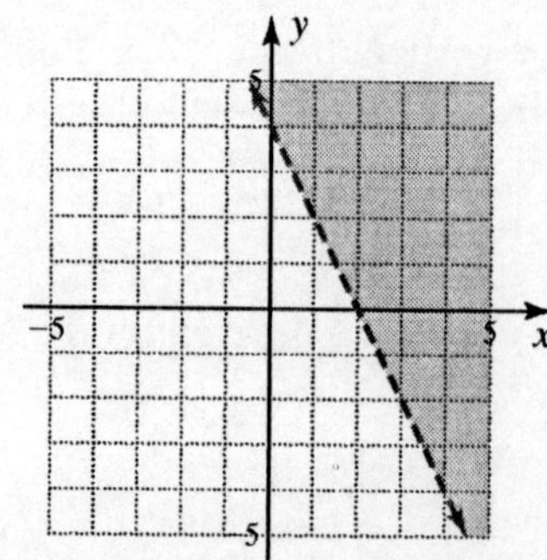

3. Graph $-2x - 5y = 10$ with a solid line. Test (0, 0), which is above the line:
$$-2x - 5y \le 10$$
$$-2(0) - 5(0) \le 10$$
$$0 \le 10 \quad \text{True}$$
Shade the points above the line.

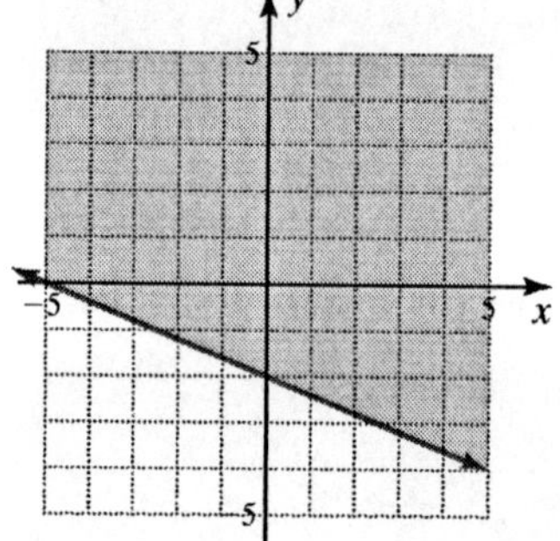

5. Graph $y = 3x - 3$ with a solid line. Test (0, 0), which is above the line:
$$y \ge 3x - 3$$
$$0 \ge 3(0) - 3$$
$$0 \ge -3 \quad \text{True}$$
Shade the points above the line.

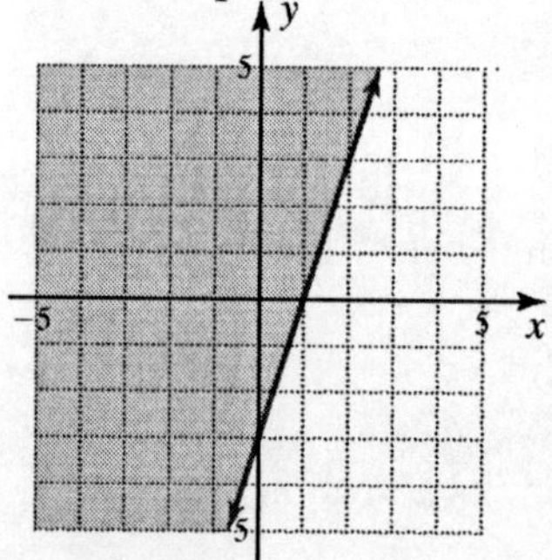

7. Graph $6 = 3x - 6y$ with a dashed line. Test (0, 0), which is above the line:
$$6 < 3x - 6y$$
$$6 < 3(0) - 6(0)$$
$$6 < 0 \quad \text{False}$$
Shade the points below the line.

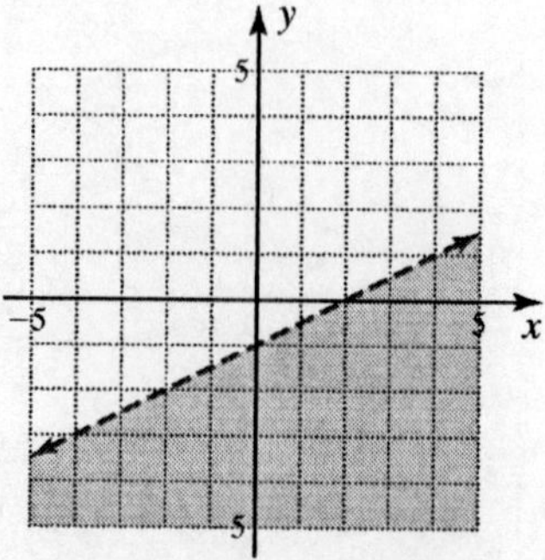

9. Graph $3x + 4y = 12$ with a solid line. Test (0, 0), which is below the line:
$$3x + 4y \ge 12$$
$$3(0) + 4(0) \ge 12$$
$$0 \ge 12 \quad \text{False}$$
Shade the points above the line.

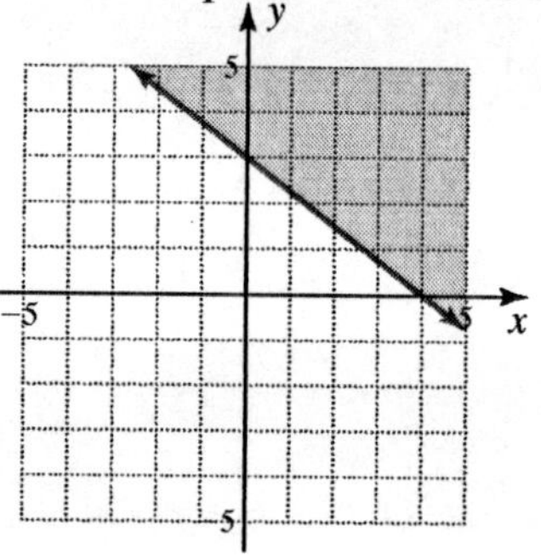

11. Graph $10 = -2x + 5y$ with a dashed line. Test (0, 0), which is below the line:
$$10 < -2x + 5y$$
$$10 < -2(0) + 5(0)$$
$$10 < 0 \quad \text{False}$$
Shade the points above the line.

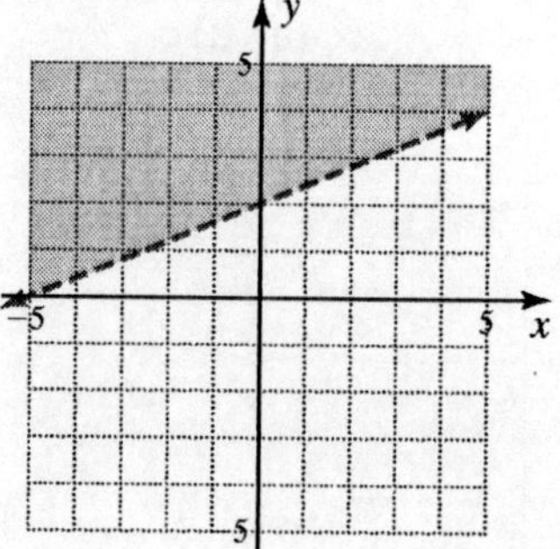

13. Graph $x = 2y - 4$ with a solid line. Test (0, 0), which is below the line:

$x \geq 2y - 4$

$0 \geq 2(0) - 4$

$0 \geq -4$ True

Shade the points below the line.

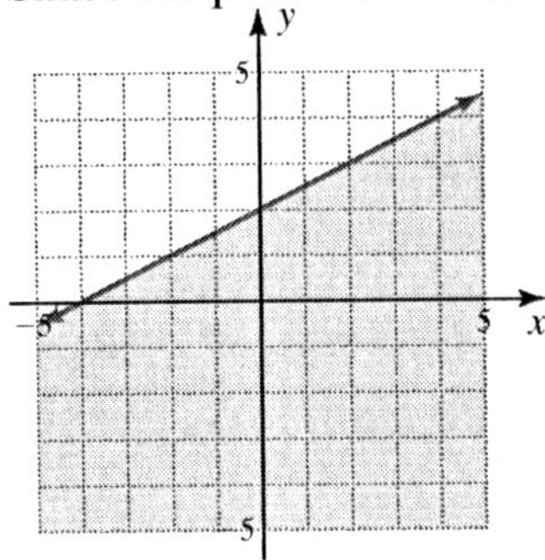

15. Graph $y = -x + 5$ with a dashed line. Test (0, 0), which is below the line:

$y < -x + 5$

$0 < -0 + 5$

$0 < 5$ True

Shade the points below the line.

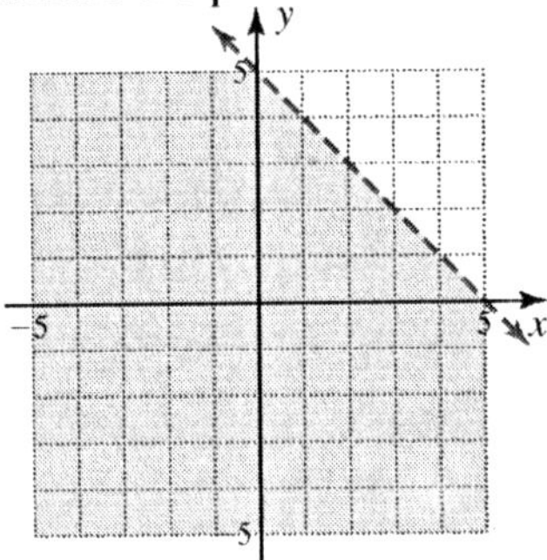

17. Graph $2y = 4x + 5$ with a dashed line. Test (0, 0), which is below the line.

$2y < 4x + 5$

$2(0) < 4(0) + 5$

$0 < 5$ True

Shade the points below the line.

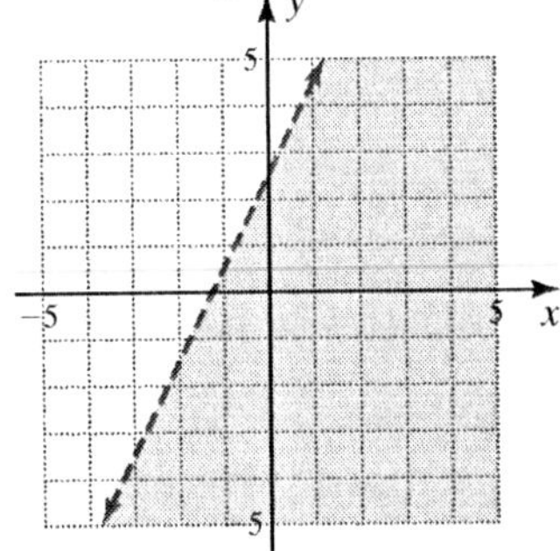

19. Graph the vertical line $x = 1$ as dashed. Test (0, 0), which is to the left of the line:

$x > 1$

$0 > 1$ False

Shade the points to the right of the line.

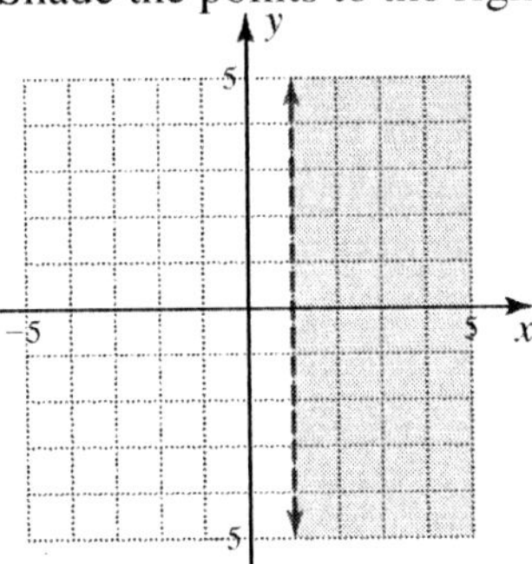

21. Graph the vertical line $x = \frac{5}{2}$ as solid. Test (0, 0), which is to the left of the line.

$x \leq \frac{5}{2}$

$0 \leq \frac{5}{2}$ True

Shade the points to the left of the line.

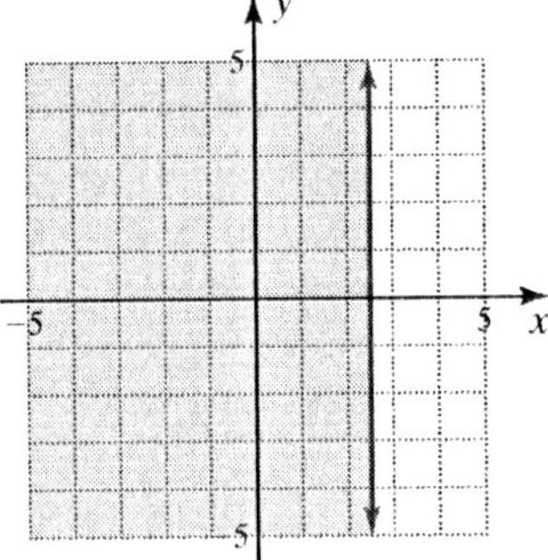

23. Graph the horizontal line $y = -\frac{3}{2}$ as solid.

Test (0, 0), which is above the line:

$y \leq -\frac{3}{2}$

$0 \leq -\frac{3}{2}$ False

Shade the points below the line.

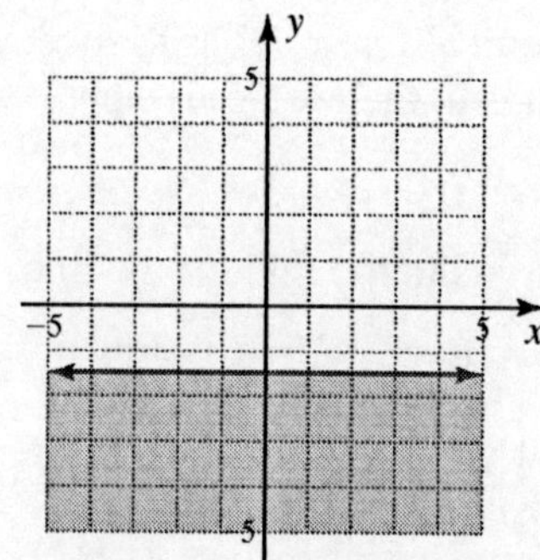

25. Graph the vertical line $x-\frac{2}{3}=0$, or $x=\frac{2}{3}$, as dashed. Test (0, 0), which is to the left of the line:

$$x-\frac{2}{3}>0$$
$$0-\frac{2}{3}>0$$
$$-\frac{2}{3}>0 \quad \text{False}$$

Shade the points to the right of the line.

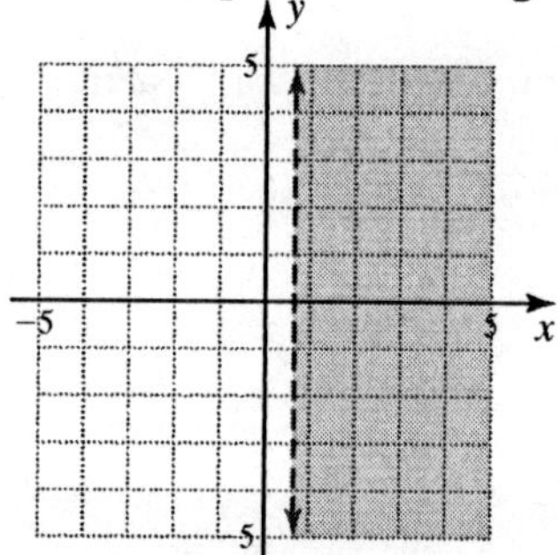

27. Graph the horizontal line $y+\frac{1}{3}=\frac{2}{3}$, or $y=\frac{1}{3}$, as solid.

Test (0, 0), which is below the line:

$$y+\frac{1}{3}\geq\frac{2}{3}$$
$$0+\frac{1}{3}\geq\frac{2}{3}$$
$$\frac{1}{3}\geq\frac{2}{3} \quad \text{False}$$

Shade the points above the line.

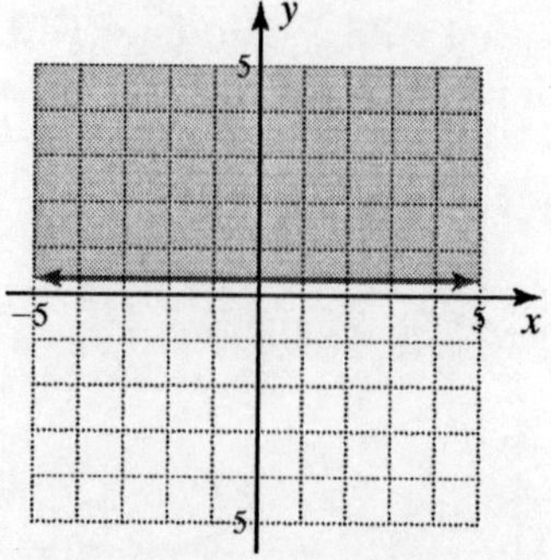

29. Graph $2x + y = 0$ with a dashed line. Since (0, 0) is on the line, test (1, 1), which is above the line:

$$2x+y<0$$
$$2(1)+1<0$$
$$3<0 \quad \text{False}$$

Shade the points below the line.

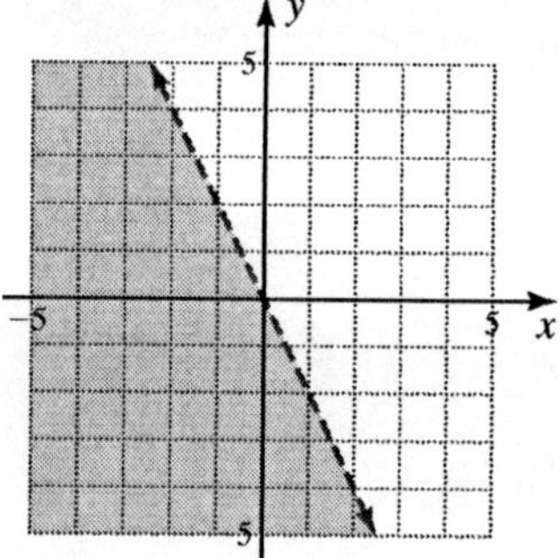

31. Graph $y - 3x = 0$ with a dashed line. Since (0, 0) is on the line, test (1, 1), which is below the line:

$$y-3x>0$$
$$1-3(1)>0$$
$$-2>0 \quad \text{False}$$

Shade the points above the line.

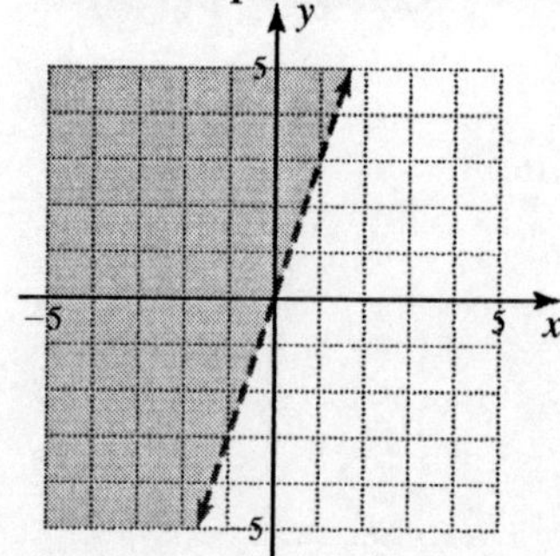

33. Graph the following lines as solid:

$-x = -3$ or $x = 3$

$-y = -4$ or $y = 4$

$x = 4$

$y = 2$

$-x \le -3$ is $x \ge 3$, and $-y \ge -4$ is $y \le 4$.

$x \ge 3$ and $x \le 4$ is $3 \le x \le 4$.

$y \le 4$ and $y \ge 2$ is $2 \le y \le 4$.

Shade the region between the pairs of horizontal and vertical lines to get a rectangle.

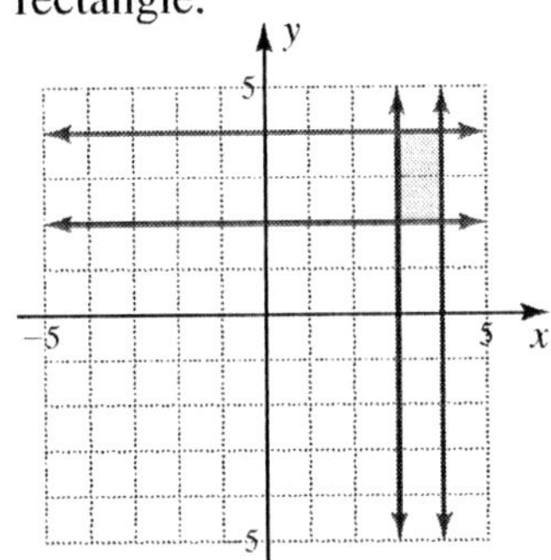

35.

$$0.06 = \frac{k}{130}$$
$$130(0.06) = k$$
$$7.8 = k$$

37.

$$0.103 = 4k$$
$$\frac{0.103}{4} = k$$
$$0.02575 = k$$

39.

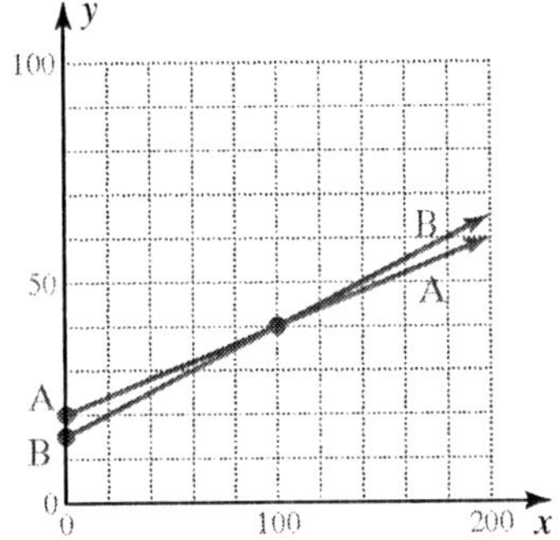

41. Company A: $y = 0.20x + 20$

Company B: $y = 0.25x + 15$

$$0.25x + 15 = 0.20x + 20$$
$$0.05x + 15 = 20$$
$$0.05x = 5$$
$$x = 100$$

The cost is the same for 100 miles.

43. The cost for company B is less when traveling less than 100 miles.

45. Answers may vary.

47. A point on the boundary does not determine whether the points above or below the line should be shaded, so it does not help as a test point.

49. Graph the horizontal line $y = -4$ as solid. Test (0, 0), which is above the line:

$y \ge -4$

$0 \ge -4$ True

Shade the points above the line.

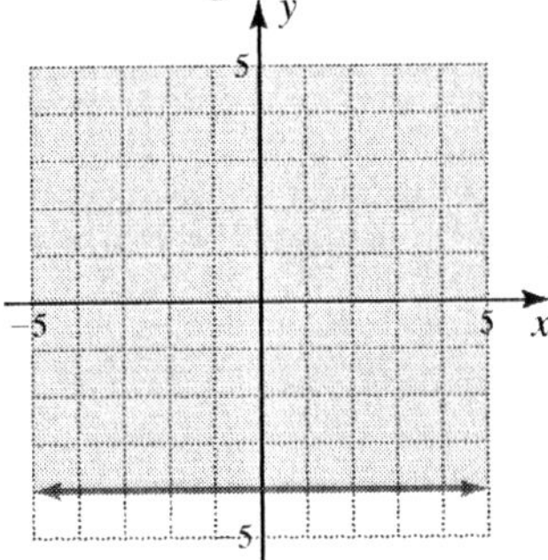

51. Graph the horizontal line $y - 3 = 0$, or $y = 3$, as dashed. Test (0, 0), which is below the line:

$y - 3 > 0$

$0 - 3 > 0$

$-3 > 0$ False

Shade the points above the line.

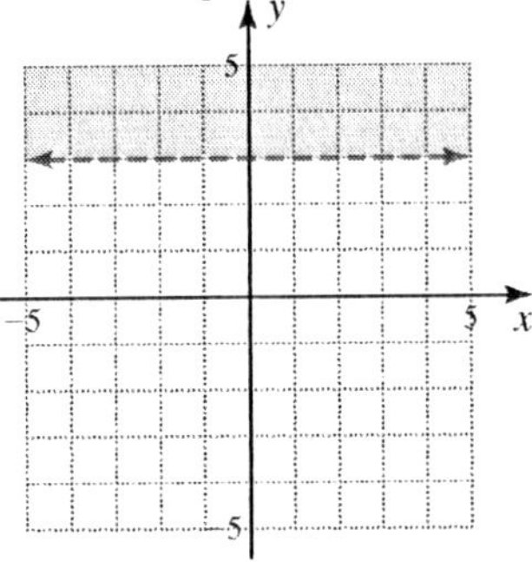

53. Graph $y = -4x + 8$ with a solid line. Test (0, 0), which is below the line:

$y \le -4x + 8$

$0 \le -4(0) + 8$

$0 \le 8$ True

Shade the points below the line.

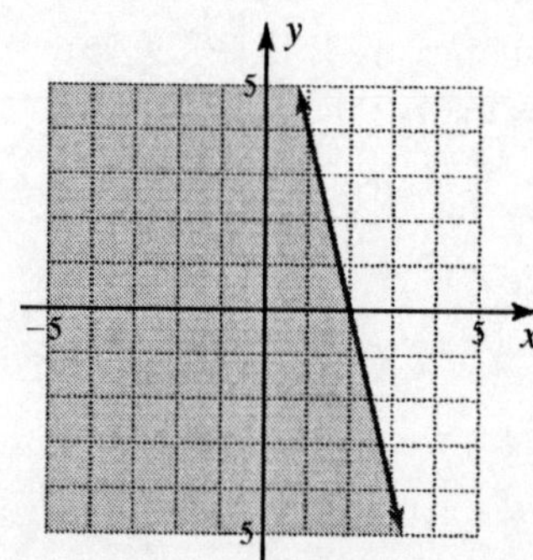

55. Graph $y = 3x$ with a dashed line. Since (0, 0) is on the line, test (1, 1), which is below the line:

$y < 3x$
$1 < 3(1)$
$1 < 3$ True

Shade the points below the line.

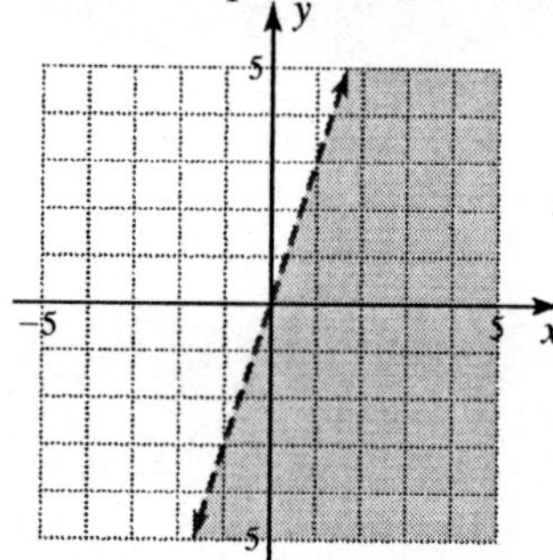

57. Graph $3x + y = 0$ with a solid line. Since (0, 0) is on the line, test (1, 1), which is above the line:

$3x + y \le 0$
$3(1) + 1 \le 0$
$4 \le 0$ False

Shade the points below the line.

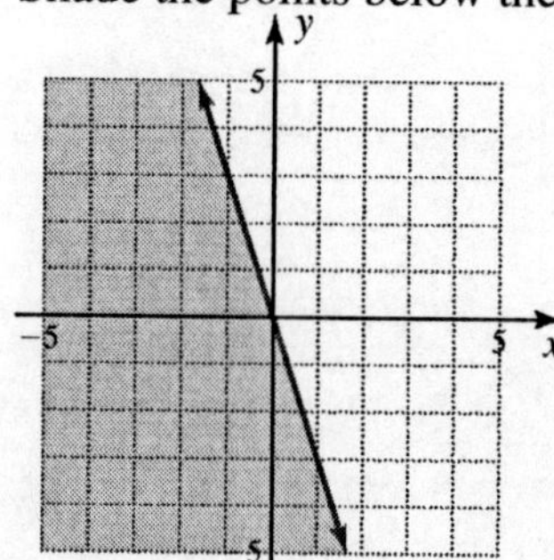

7.4 Direct and Inverse Variation

Problems 7.4

1. a. If C varies directly as n, then $C = kn$.

b. $C = 590$, when $n = 101$:

$590 = k(101)$

$\frac{590}{101} = k$

$k = \frac{590}{101} \approx 5.8$

c. Find n when $C = 100$:

$C = 5.8n$

$100 = 5.8n$

$\frac{100}{5.8} = n$

$n = \frac{100}{5.8} \approx 17$

For 100 calories, you would eat about 17 fries.

2. a. Since C is proportional to t, then $C = kt$.

b. $C = 105$ when $t = 15$:

$105 = k(15)$

$\frac{105}{15} = k$

$k = 7$

c. Find C when $t = 20$:

$C = 7t$

$C = 7(20) = 140$

You would use 140 calories by jogging for 20 minutes.

d. Find t when $C = 3500$:

$C = 7t$

$3500 = 7t$

$\frac{3500}{7} = t$

$t = 500$

You would have to jog for 500 minutes, or 8 hours and 20 minutes, to lose 1 pound.

3. a. Since v is inversely proportional to t,
$v = \frac{k}{t}$.
$v = 55$ when $t = 2$:
$55 = \frac{k}{2}$
$110 = k$
$k = 110$ and represents the distance traveled.

b. $v = \frac{110}{t}$

4. a. Since L is inversely proportional to d^2,
$L = \frac{k}{d^2}$.
$L = 80$ when $d = 5$:
$80 = \frac{k}{5^2}$
$80 = \frac{k}{25}$
$2000 = k$

b. $L = \frac{2000}{d^2}$

c. Find L when $d = 10$:
$L = \frac{2000}{10^2} = \frac{2000}{100} = 20$
The loudness is 20 dB when you are 10 feet away.

5. Since g is directly proportional to A, then $g = kA$.
$g = 60$ when $A = 40$:
$60 = k(40)$
$\frac{60}{40} = k$
$k = \frac{3}{2}$
For Boston, the equation of variation is
$g = \frac{3}{2}A$.

Exercises 7.4

1. a. $S = \frac{W}{16}$

b. Since S is directly proportional to W and $S = \frac{W}{16}$, then $k = \frac{1}{16}$.

c. $W = 160$:
$S = \frac{160}{16} = 10$
If you weigh 160 pounds, your skin weighs 10 pounds.

3. a. $R = \frac{W}{10}$

b. Since R is directly proportional to W and $R = \frac{W}{10}$, then $k = \frac{1}{10}$.

c. $W = 160$:
$R = \frac{160}{10} = 16$
If you weigh 160 pounds, your basal metabolic rate is 16.

5. a. Since R varies directly with t, then $R = kt$.

b. $R = 112.5$ when $t = 2\frac{1}{2} = 2.5$:
$112.5 = k(2.5)$
$\frac{112.5}{2.5} = k$
$45 = k$

c. $R = 108$:
$R = 45t$
$108 = 45t$
$\frac{108}{45} = t$
$t = \frac{12}{5} = 2.4$
A record that makes 108 revolutions takes 2.4 minutes to play.

7. a. Since weight, thus the threshold weight T, varies directly as h^3, then $T = kh^3$.

b. $T = 196$ when $h = 70$:

$$196 = k(70)^3$$
$$196 = 343{,}000k$$
$$\frac{196}{343{,}000} = k$$
$$k = \frac{196}{343{,}000} \approx 0.00057$$

c. $h = 75$:

$$T = 0.00057h^3$$
$$T = 0.00057(75)^3 \approx 240$$

The threshold weight for a person 75 inches tall is 240 pounds.

9. a. Since f varies inversely as d, then $f = \frac{k}{d}$.

b. $f = 8$ when $d = \frac{1}{2}$:

$$8 = \frac{k}{\frac{1}{2}}$$
$$8\left(\frac{1}{2}\right) = k$$
$$4 = k$$

c. $d = \frac{1}{4}$:

$$f = \frac{4}{d}$$
$$f = \frac{4}{\frac{1}{4}} = 4\left(\frac{4}{1}\right) = 16$$

the f-stop is 16 when the aperture is $\frac{1}{4}$.

11. Since P varies inversely as V, then $P = \frac{k}{V}$.

$P = 24$ when $V = 18$:

$$24 = \frac{k}{18}$$
$$432 = k$$

$$P = \frac{432}{V}$$

$P = 40$:

$$40 = \frac{432}{V}$$
$$V = \frac{432}{40} = 10.8$$

The volume is 10.8 in^3.

13. a. Since w is inversely proportional to s, then $w = \frac{k}{s}$.

b. $w = 600$ when $s = 12$:

$$600 = \frac{k}{12}$$
$$7200 = k$$

c. $s = 10$:

$$w = \frac{7200}{s}$$
$$w = \frac{7200}{10} = 720$$

720 words could be typed on the page if a 10-point font were used.

15. a. Since b is inversely proportional to a, then $b = \frac{k}{a}$.

b. $b = 110$ when $a = 27$:

$$110 = \frac{k}{27}$$
$$2970 = k$$

c. $a = 33$:

$$b = \frac{2970}{a}$$
$$b = \frac{2970}{33} = 90$$

90 births per 1000 women are expected for 33-year-old women.

17. a. Since d is directly proportional to s, then $d = ks$.

b. $d = 501$ when $s = 28.41$:

$$501 = k(28.41)$$
$$\frac{501}{28.41} = k$$
$$k = \frac{501}{28.41} \approx 17.63$$

c. Since d represents distance in miles and s represents speed in mi/hr, k represents the time it took to drive d miles at s miles per hour.

19. a. Since C is directly proportional to $F - 37$, then $C = k(F - 37)$.
$C = 80$ when $F = 57$:

$$80 = k(57 - 37)$$
$$80 = k(20)$$
$$4 = k$$

$C = 4(F - 37)$ or $C = 4F - 148$

b. $F = 90$
$C = 4(90) - 148 = 360 - 148 = 212$
212 chirps per minute would be expected at a temperature of 90°F.

21. a. Let n be the number of years after 1930, then I is proportional to n^2, so $I = kn^2$.

b. In 1940, $n = 1940 - 1930 = 10$, and $I = 5$:

$$5 = k(10)^2$$
$$5 = 100k$$
$$0.05 = k$$

c. In the year 2000, $n = 2000 - 1930 = 70$.

$$I = 0.05n^2$$
$$I = 0.05(70)^2 = 0.05(4900) = 245$$

The cumulative increase in 2000 would be 245%.

23. a. Since BAC is directly proportional to $N - 1$, then $\text{BAC} = k(N - 1)$.

b. $\text{BAC} = 0.052$ when $N = 3$:

$$0.052 = k(3 - 1)$$
$$0.052 = 2k$$
$$0.026 = k$$

c. $N = 5$:
$\text{BAC} = 0.026(N - 1)$
$\text{BAC} = 0.026(5 - 1) = 0.026(4) = 0.104$
The man's BAC after 5 beers is 0.104.

d. $\text{BAC} = 0.08$:

$$0.08 = 0.026(N - 1)$$
$$\frac{0.08}{0.026} = N - 1$$
$$\frac{0.08}{0.026} + 1 = N$$
$$N \approx 4$$

The man can drink 4 beers in an hour before going over the legal limit.

25. a. Since BAC is inversely proportional to W, then $\text{BAC} = \frac{k}{W}$.
$\text{BAC} = 0.06$ when $W = 130$:

$$0.06 = \frac{k}{130}$$
$$7.8 = k$$
$$\text{BAC} = \frac{7.8}{W}$$

b. $W = 260$:

$$\text{BAC} = \frac{7.8}{260} = 0.03$$

The BAC of a 260-pound male after 3 beers is 0.03.

c. $\text{BAC} = 0.08$:

$$0.08 = \frac{7.8}{W}$$
$$W = \frac{7.8}{0.08} = 97.5$$

A male whose BAC is exactly 0.08 after drinking 3 beers in the last hour weighs 97.5 pounds.

d. A male weighing more than 97.5 pounds would have a BAC of less than 0.08 after drinking 3 beers in the last hour.

27. $x+2y=4$

$2y=-x+4$

$y=-\frac{1}{2}x+2$

$y=0$: $x+2(0)=4$

$x=4$

The line has slope $m=-\frac{1}{2}$, y-intercept $b = 2$, and x-intercept $x = 4$.

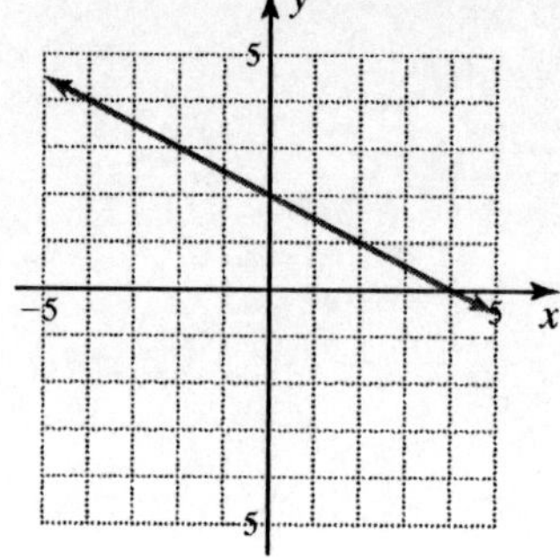

29. $y-2x=4$

$y=2x+4$

$y=0$: $0-2x=4$

$-2x=4$

$x=-2$

The line has slope $m = 2$, y-intercept $b = 4$, and x-intercept $x = -2$.

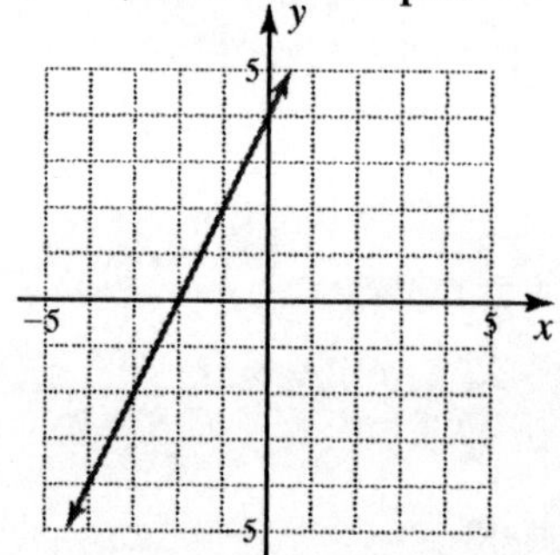

31. $x+y=4$

$y=-x+4$

slope: $m_1=-1$

$2x-y=-1$

$-y=-2x-1$

$y=2x+1$

slope: $m_2=2$

Since $m_1 \neq m_2$, the lines are not parallel.

33. **a.** $T=15S$

b. 1 hour = 60 minutes, so $T = 60$.

$60=15S$

$4=S$

You need an SPF of 4 to stay out for 1 hour.

35. Answers may vary.

37. Answers may vary.

Review Exercises

1. Use the point-slope form: $y-y_1=m(x-x_1)$ with $(x_1, y_1)=(3, -5)$.

a. $m=-2$:

$y-(-5)=-2(x-3)$

$y+5=-2x+6$

$y=-2x+1$

$2x+y=1$

b. $m=-3$:

$y-(-5)=-3(x-3)$

$y+5=-3x+9$

$y=-3x+4$

$3x+y=4$

c. $m=-4$:

$y-(-5)=-4(x-3)$

$y+5=-4x+12$

$y=-4x+7$

$4x+y=7$

2. Use the slope-intercept form: $y = mx + b$.

a. $m = 5$; $b = -2$

$y = 5x - 2$

b. $m = 4$; $b = 7$

$y = 4x + 7$

c. $m = 6$; $b = -4$

$y = 6x - 4$

3. a. $(x_1, y_1) = (-1, 2); (x_2, y_2) = (4, 7)$

$$m = \frac{7-2}{4-(-1)} = \frac{5}{5} = 1$$

$$y - 2 = 1[x - (-1)]$$
$$y = x + 1 + 2$$
$$y = x + 3$$
$$-3 = x - y$$

or $x - y = -3$

b. $(x_1, y_1) = (-3, 1); (x_2, y_2) = (7, 6)$

$$m = \frac{6-1}{7-(-3)} = \frac{5}{10} = \frac{1}{2}$$

$$y - 1 = \frac{1}{2}[x - (-3)]$$
$$y = \frac{1}{2}x + \frac{3}{2} + 1$$
$$y = \frac{1}{2}x + \frac{5}{2}$$
$$-\frac{5}{2} = \frac{1}{2}x - y$$

or $x - 2y = -5$

c. $(x_1, y_1) = (1, 2); (x_2, y_2) = (7, -2)$

$$m = \frac{-2-2}{7-1} = \frac{-4}{6} = -\frac{2}{3}$$

$$y - 2 = -\frac{2}{3}(x - 1)$$
$$y = -\frac{2}{3}x + \frac{2}{3} + 2$$
$$y = -\frac{2}{3}x + \frac{8}{3}$$
$$\frac{2}{3}x + y = \frac{8}{3}$$

or $2x + 3y = 8$

4. a. We have the point (10, 5) and the slope 0.20.

$$C - 5 = 0.2(m - 10)$$
$$C = 0.2m - 2 + 5$$
$$C = 0.2m + 3$$

$m = 15$:

$C = 0.2(15) + 3 = 3 + 3 = 6$

A 15-minute call costs $6.

b. We have the point (10, 7) and the slope 0.20.

$$C - 7 = 0.2(m - 10)$$
$$C = 0.2m - 2 + 7$$
$$C = 0.2m + 5$$

$m = 15$:

$C = 0.2(15) + 5 = 3 + 5 = 8$

A 15-minute call costs $8.

c. We have the point (10, 8) and the slope 0.30.

$$C - 8 = 0.3(m - 10)$$
$$C = 0.3m - 3 + 8$$
$$C = 0.3m + 5$$

$m = 15$:

$C = 0.3(15) + 5 = 4.5 + 5 = 9.5$

A 15-minute call costs $9.50.

5. a. We have the slope 0.40 and the y-intercept 30.

$C = 0.4m + 30$

When 800 minutes are used, the number of additional minutes is

$m = 800 - 500 = 300$:

$C = 0.4(300) + 30 = 120 + 30 = 150$

The cost is $150 when 800 minutes are used.

b. We have the slope 0.30 and the y-intercept 40.

$C = 0.3m + 40$

When 600 minutes are used, the number of additional minutes is

$m = 600 - 500 = 100$:

$C = 0.3(100) + 40 = 30 + 40 = 70$

The cost is $70 when 600 minutes are used.

c. We have the slope 0.20 and the y-intercept 50.

$C = 0.2m + 50$

When 900 minutes are used, the number of additional minutes is

$m = 900 - 500 = 400$:

$C = 0.2(400) + 50 = 80 + 50 = 130$

The cost is $130 when 900 minutes are used.

6. a. We have the points (5, 3) and (10, 5).

$$\text{slope} = \frac{5-3}{10-5} = \frac{2}{5} = 0.4$$

Using (5, 3):

$$C - 3 = 0.4(m-5)$$
$$C = 0.4m - 2 + 3$$
$$C = 0.40m + 1$$

b. We have the points (5, 4) and (10, 6).

$$\text{slope} = \frac{6-4}{10-5} = \frac{2}{5} = 0.4$$

Using (5, 4):

$$C - 4 = 0.4(m-5)$$
$$C = 0.4m - 2 + 4$$
$$C = 0.40m + 2$$

c. We have the points (5, 5) and (10, 7).

$$\text{slope} = \frac{7-5}{10-5} = \frac{2}{5} = 0.4$$

Using (5, 5):

$$C - 5 = 0.4(m-5)$$
$$C = 0.4m - 2 + 5$$
$$C = 0.40m + 3$$

7. a. Graph $2x - 4y = -8$ with a dashed line.
Test (0, 0), which is below the line:

$$2x - 4y < -8$$
$$2(0) - 4(0) < -8$$
$$0 < -8 \quad \text{False}$$

Shade the points above the line.

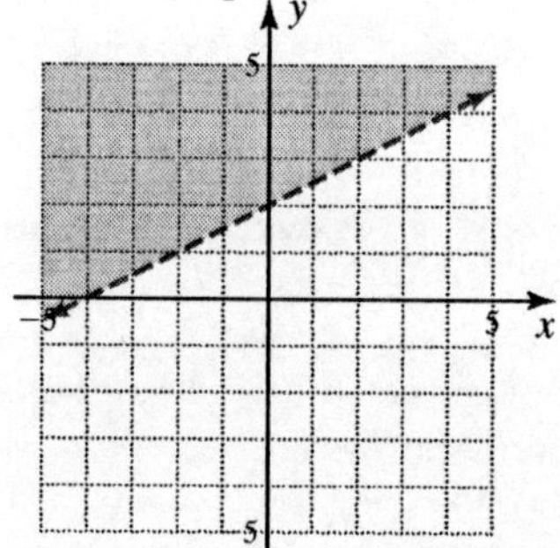

b. Graph $3x - 6y = -12$ with a dashed line.
Test (0, 0), which is below the line:

$$3x - 6y < -12$$
$$3(0) - 6(0) < -12$$
$$0 < -12 \quad \text{False}$$

Shade the points above the line.

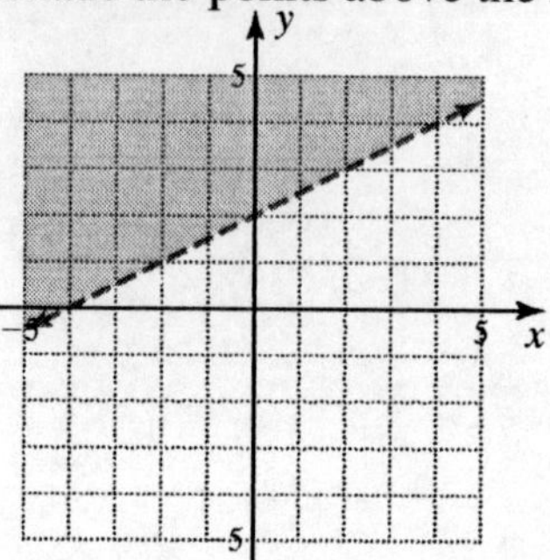

c. Graph $4x - 2y = -8$ with a dashed line.
Test (0, 0), which is below the line:

$$4x - 2y < -8$$
$$4(0) - 2(0) < -8$$
$$0 < -8 \quad \text{False}$$

Shade the points above the line.

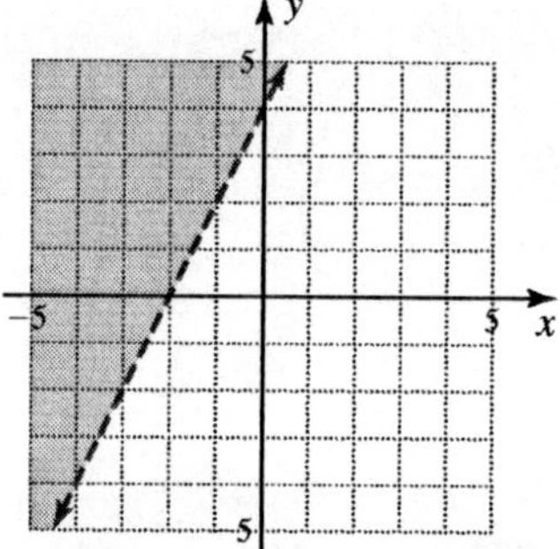

8. a. Graph $-y = -2x + 2$ with a solid line.
Test (0, 0), which is above the line:

$$-y \le -2x + 2$$
$$-0 \le -2(0) + 2$$
$$0 \le 4 \quad \text{True}$$

Shade the points above the line.

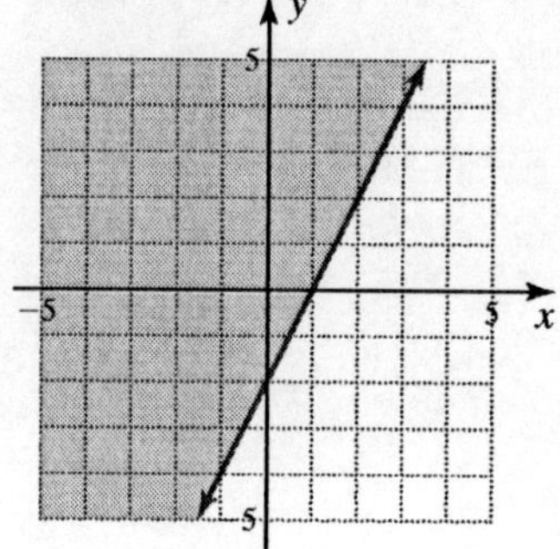

b. Graph $-y = -2x + 4$ with a solid line. Test (0, 0), which is above the line:

$-y \le -2x+4$

$-0 \le -2(0)+4$

$0 \le 4$ True

Shade the points above the line.

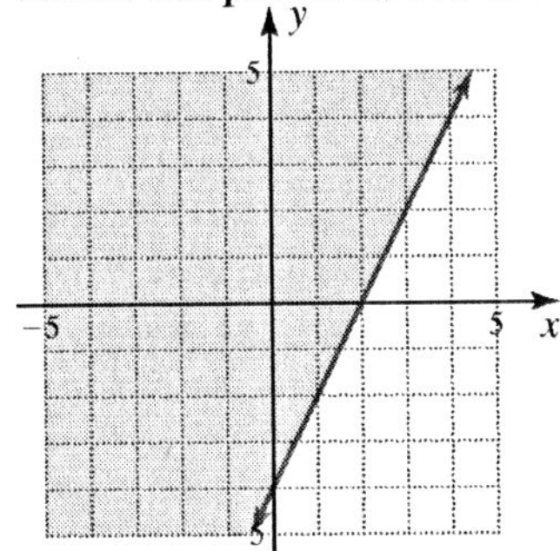

c. Graph $-y = -x + 3$ with a solid line. Test (0, 0), which is above the line:

$y \le -x+3$

$-0 \le -0+3$

$0 \le 3$ True

Shade the points above the line.

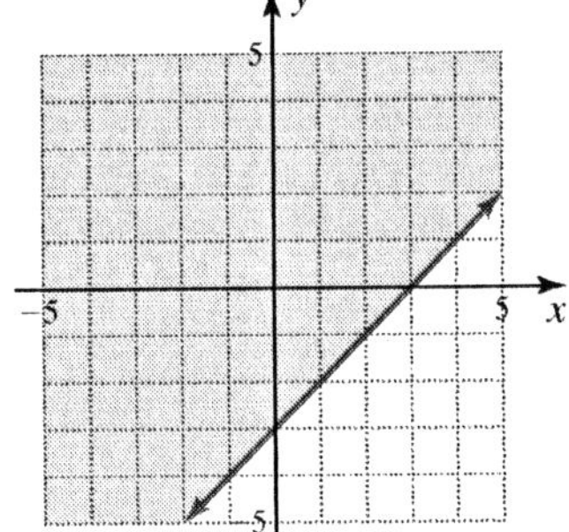

9. a. Graph $2x + y = 0$ with a dashed line. Since (0, 0) is on the line, test (1, 1), which is above the line:

$2x + y > 0$

$2(1)+1>0$

$3>0$ True

Shade the points above the line.

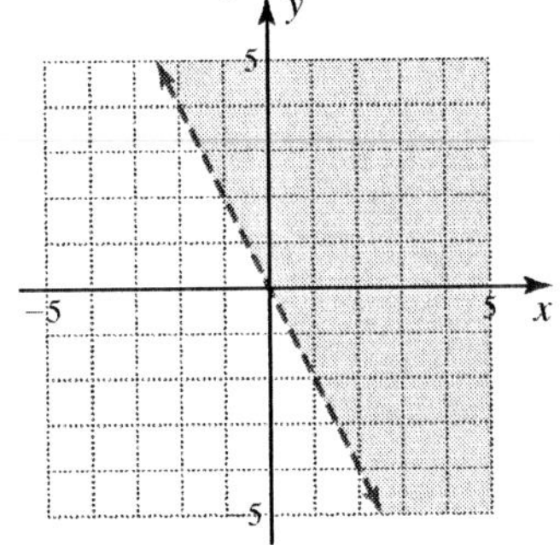

b. Graph $3x + y = 0$ with a dashed line. Since (0, 0) is on the line, test (1, 1), which is above the line:

$3x + y > 0$

$3(1)+1>0$

$4>0$ True

Shade the points above the line.

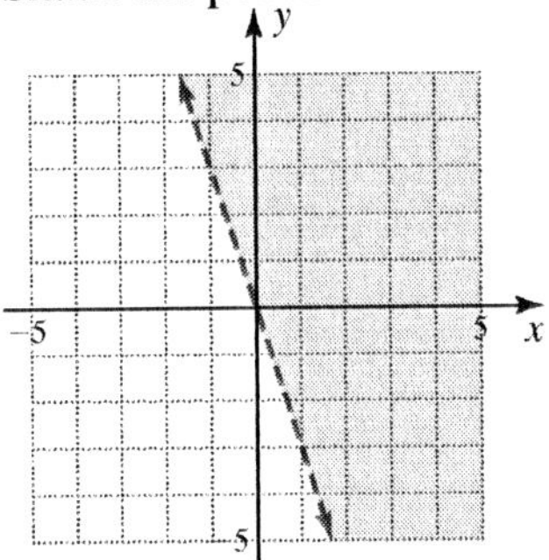

c. Graph $3x - y = 0$ with a dashed line. Since (0, 0) is on the line, test (1, 1), which is below the line:

$3x - y < 0$

$3(1)-1<0$

$2<0$ False

Shade the points above the line.

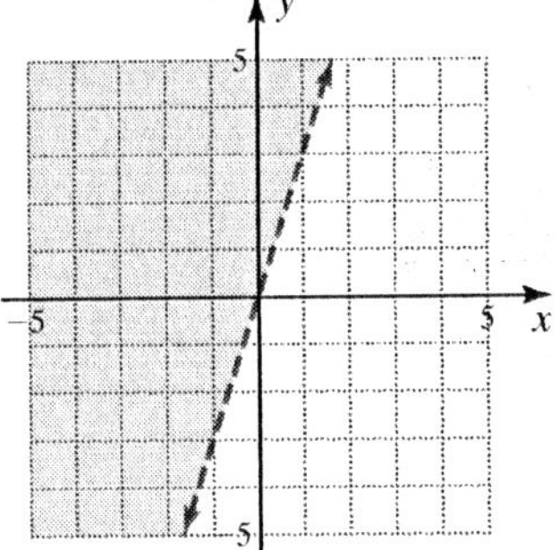

10. a. Graph the vertical line $x = -4$ as solid. Test (0, 0), which is to the right of the line:

$x \ge -4$

$0 \ge -4$ True

Shade the points to the right of the line.

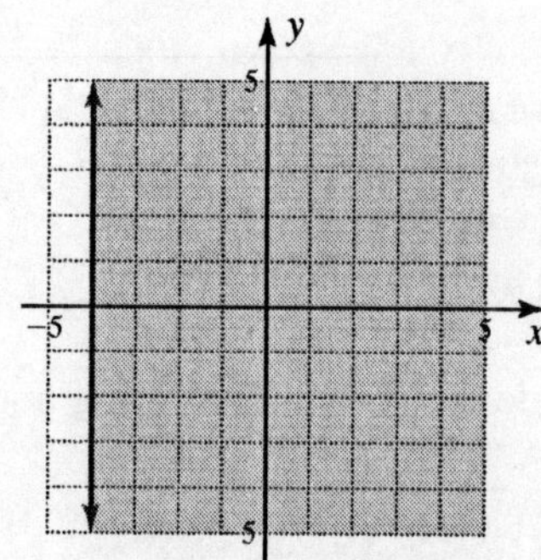

b. Graph the horizontal line $y - 4 = 0$, or $y = 4$, as dashed. Test (0, 0), which is below the line:

$y - 4 < 0$
$0 - 4 < 0$
$-4 < 0$ True

Shade the points below the line.

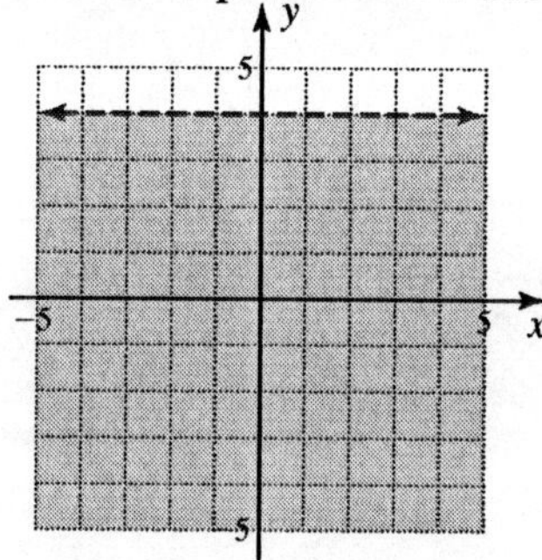

c. Graph the horizontal line $2y - 4 = 0$, or $y = 2$, as solid. Test (0, 0), which is below the line:

$2y - 4 \geq 0$
$2(0) - 4 \geq 0$
$-4 \geq 0$ False

Shade the points above the line.

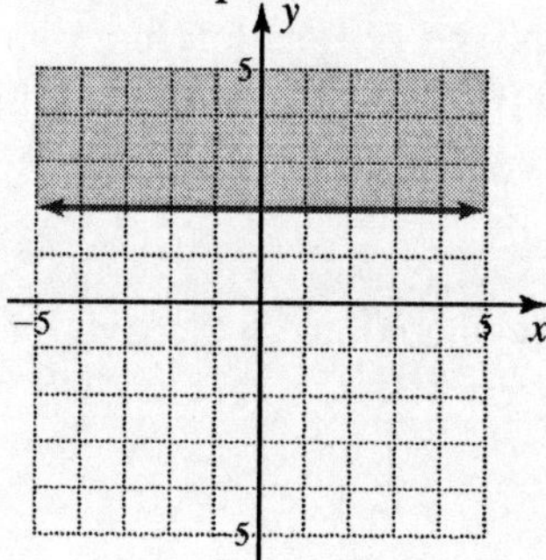

11. a. Since C is proportional to m, then $C = km$.
$C = 30$ when $m = 12$:
$30 = k(12)$
$\frac{30}{12} = k$
$\frac{5}{2} = k$
$C = \frac{5}{2}m$
One hour is $m = 60$ minutes:
$C = \frac{5}{2}(60) = 150$
You use 150 calories when you walk for an hour.

b. Since C is proportional to m, then $C = km$.
$C = 140$ when $m = 20$:
$140 = k(20)$
$7 = k$
$C = 7m$
One hour is $m = 60$ minutes:
$C = 7(60) = 420$
You use 420 calories when you row for an hour.

c. Since C is proportional to m, then $C = km$.
$C = 180$ when $m = 180$:
$180 = k(180)$
$1 = k$
$C = m$
$m = 45$:
$C = 45$
You use 45 calories when you play tennis for 45 minutes.

12. a. Since F varies inversely with L and $k = 30$ the equation of variation is $F = \frac{30}{L}$.
$L = 6$:
$F = \frac{30}{6} = 5$
A force of 5 is needed when the handle is 6 inches long.

b. $F = \dfrac{30}{L}$ and $L = 10$:

$F = \dfrac{30}{10} = 3$

A force of 3 is needed when the handle is 10 inches long.

c. $F = \dfrac{30}{L}$ and $L = 15$:

$F = \dfrac{30}{15} = 2$

A force of 2 is needed when the handle is 15 inches long.

Cumulative Review Chapters 1–7

1. The LCD is $9 \cdot 8 = 72$.

$$\begin{aligned} -\frac{2}{9}+\left(-\frac{1}{8}\right) &= -\frac{2\cdot 8}{9\cdot 8}+\left(-\frac{1\cdot 9}{8\cdot 9}\right) \\ &= -\frac{16}{72}+\left(-\frac{9}{72}\right) \\ &= \frac{-16+(-9)}{72} \\ &= \frac{-25}{72} \\ &= -\frac{25}{72} \end{aligned}$$

2. $4.3 - (-3.9) = 4.3 + 3.9 = 8.2$

3. $(-2)^4 = (-2)(-2)(-2)(-2) = 16$

4. $-\dfrac{1}{5} \div \left(-\dfrac{1}{10}\right) = -\dfrac{1}{5} \cdot \left(-\dfrac{10}{1}\right) = \dfrac{10}{5} = 2$

5.
$$\begin{aligned} y \div 2 \cdot x - z &= 8 \div 2 \cdot 2 - 3 \\ &= 4 \cdot 2 - 3 \\ &= 8 - 3 \\ &= 5 \end{aligned}$$

6.
$$\begin{aligned} x + 4(x-2) + (x-3) &= x + 4x - 8 + x - 3 \\ &= x + 4x + x - 8 - 3 \\ &= 6x - 11 \end{aligned}$$

7. $\dfrac{d+4e}{f}$

8.
$$\begin{aligned} 5 &= 3(x-1) + 5 - 2x \\ 5 &= 3x - 3 + 5 - 2x \\ 5 &= x + 2 \\ 5 - 2 &= x + 2 - 2 \\ 3 &= x \end{aligned}$$

9.
$$\begin{aligned} \frac{x}{7} - \frac{x}{9} &= 2 \\ 63\left(\frac{x}{7} - \frac{x}{9}\right) &= 63(2) \\ 63\left(\frac{x}{7}\right) - 63\left(\frac{x}{9}\right) &= 126 \\ 9x - 7x &= 126 \\ 2x &= 126 \\ x &= 63 \end{aligned}$$

10. Let x be one of the numbers. Then $x + 40$ is the other number.

$$\begin{aligned} x + (x + 40) &= 110 \\ 2x + 40 &= 110 \\ 2x &= 70 \\ x &= 35 \end{aligned}$$

$x + 40 = 75$

The numbers are 35 and 75.

11. Let b be the amount invested in bonds. Then $9000 - b$ is invested in certificates of deposit.

$$\begin{aligned} 0.11b + 0.14(9000 - b) &= 1140 \\ 0.11b + 1260 - 0.14b &= 1140 \\ 1260 - 0.03b &= 1140 \\ -0.03b &= -120 \\ b &= 4000 \end{aligned}$$

$9000 - b = 5000$

She has \$4000 invested in bonds and \$5000 in certificates of deposit.

12.
$$-\frac{x}{7}+\frac{x}{4}\ge\frac{x-4}{4}$$
$$28\left(-\frac{x}{7}+\frac{x}{4}\right)\ge 28\left(\frac{x-4}{4}\right)$$
$$28\left(-\frac{x}{7}\right)+28\left(\frac{x}{4}\right)\ge 7(x-4)$$
$$-4x+7x\ge 7x-28$$
$$3x\ge 7x-28$$
$$-4x\ge -28$$
$$x\le 7$$

0 7

13.

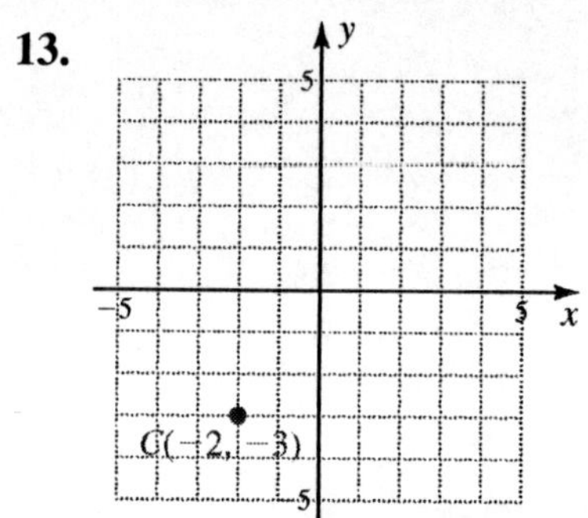

14. $5x-y=-19;\ (-3,-4)$
$$5(-3)-(-4)\overset{?}{=}-19$$
$$-15+4\overset{?}{=}-19$$
$$-11\overset{?}{=}-19\quad\text{False}$$
(–3, –4) is not a solution.

15. $4x+2y=-10;\ (x,-1)$
$$4x+2(-1)=-10$$
$$4x-2=-10$$
$$4x=-8$$
$$x=-2$$

16. $3x+y=3$
$$y=-3x+3$$
The line has slope $m=-3$ and y-intercept $b=3$.

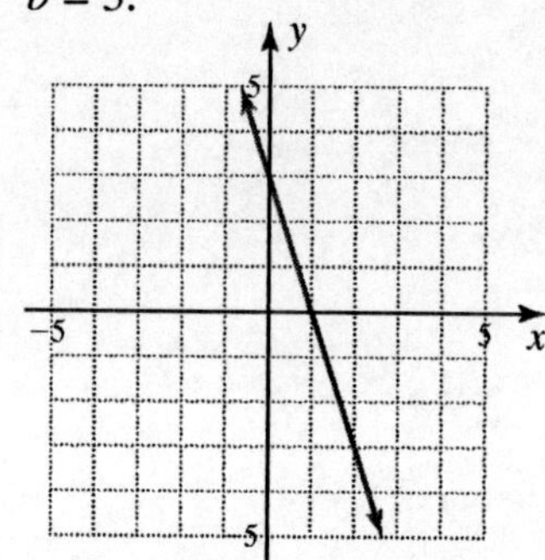

17. $3x+9=0$
$$3x=-9$$
$$x=-3$$
This is a vertical line with x-intercept (–3, 0).

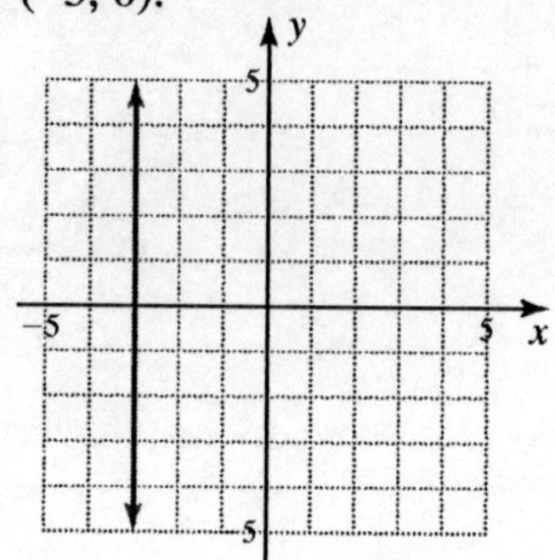

18. $m=\frac{1-(-6)}{-7-(-4)}=\frac{1+6}{-7+4}=\frac{7}{-3}=-\frac{7}{3}$

19. $12x-4y=-14$
$$-4y=-12x-14$$
$$y=\frac{-12x-14}{-4}$$
$$y=3x+\frac{7}{2}$$
The slope is $m=3$.

20. **(1)** $20x-5y=2$
$$-5y=-20x+2$$
$$y=4x-\frac{2}{5}$$
$$m_1=4$$

(2) $5x+20y=2$
$$20y=-5x+2$$
$$y=-\frac{1}{4}x+\frac{1}{10}$$
$$m_2=-\frac{1}{4}$$

(3) $-y=-4x+2$
$$y=4x-2$$
$$m_3=4$$

Since $m_1=m_3$, lines (1) and (3) are parallel.

21. $\frac{x^{-4}}{x^{-8}} = x^{-4-(-8)} = x^{-4+8} = x^4$

22. $x^{-7} \cdot x^{-2} = x^{-7+(-2)} = x^{-9} = \frac{1}{x^9}$

23. $(2x^3y^{-5})^{-4} = 2^{-4}(x^3)^{-4}(y^{-5})^{-4}$
$= \frac{1}{2^4}x^{3\cdot(-4)}y^{-5\cdot(-4)}$
$= \frac{1}{16}x^{-12}y^{20}$
$= \frac{y^{20}}{16x^{12}}$

24. $3{,}400{,}000 = 3.4 \times 10^6$

25. $(8.33 \times 10^3) \div (1.7 \times 10^3) = \frac{8.33}{1.7} \times \frac{10^3}{10^3}$
$= 4.9 \times 10^{3-3}$
$= 4.9 \times 10^0$
$= 4.9$

26. $3x^3 + 2x^2 - 5;\ x = -3$
$3(-3)^3 + 2(-3)^2 - 5 = 3(-27) + 2(9) - 5$
$= -81 + 18 - 5$
$= -68$

27. $(-7x^2 - 7x^5 - 6) + (3x^5 + 6 + 8x^2)$
$= -7x^2 - 7x^5 - 6 + 3x^5 + 6 + 8x^2$
$= -7x^5 + 3x^5 - 7x^2 + 8x^2 - 6 + 6$
$= -4x^5 + x^2$

28. $\left(3x^2 - \frac{1}{2}\right)^2 = (3x^2)^2 - 2(3x^2)\left(\frac{1}{2}\right) + \left(\frac{1}{2}\right)^2$
$= 9x^4 - 3x^2 + \frac{1}{4}$

29. $(5x^2 + 2)(5x^2 - 2) = (5x^2)^2 - 2^2 = 25x^4 - 4$

30.
$$\begin{array}{r} 3x^2 - 4x - 5 \\ x-2\overline{\big)\ 3x^3 - 10x^2 + 3x + 12} \\ \underline{-(3x^3 - 6x^2)} \\ -4x^2 + 3x \\ \underline{-(-4x^2 + 8x)} \\ -5x + 12 \\ \underline{-(-5x + 10)} \\ 2 \end{array}$$

$\frac{3x^3 - 10x^2 + 3x + 12}{x - 2} = (3x^2 - 4x - 5)$ R 2

31. $15x^4 - 35x^7 = 5x^4(3) - 5x^4(7x^3)$
$= 5x^4(3 - 7x^3)$

32. $\frac{5}{7}x^8 - \frac{2}{7}x^7 + \frac{2}{7}x^6 - \frac{3}{7}x^4$
$= \frac{1}{7}x^4(5x^4 - 2x^3 + 2x^2 - 3)$

33. $x^2 - 3x + 2 = (x - 1)(x - 2)$

34. $9x^2 - 24xy + 16y^2$
$= (3x)^2 - 2(3x)(4y) + (4y)^2$
$= (3x - 4y)^2$
$= (3x - 4y)(3x - 4y)$

35. $4x^2 - 9y^2 = (2x)^2 - (3y)^2$
$= (2x + 3y)(2x - 3y)$

36. $-6x^4 + 24x^2 = -6x^2(x^2 - 4)$
$= -6x^2[(x)^2 - (2)^2]$
$= -6x^2(x + 2)(x - 2)$

37. $2x^3 - 2x^2 - 4x = 2x(x^2 - x - 2)$
$= 2x(x + 1)(x - 2)$

38. $4x^2 + 3x + 4x + 3 = x(4x + 3) + 1(4x + 3)$
$= (4x + 3)(x + 1)$

39. $25kx^2+10kx+k=k(25x^2+10x+1)$
$=k[(5x)^2+2(5x)(1)+1^2]$
$=k(5x+1)^2$

40. $4x^2+5x=6$
$4x^2+5x-6=0$
$(x+2)(4x-3)=0$
$x+2=0$ or $4x-3=0$
$x=-2$ or $x=\frac{3}{4}$

41. $8y^3=2y(4y^2)$
$$\frac{3x}{2y}=\frac{3x(4y^2)}{2y(4y^2)}=\frac{12xy^2}{8y^3}$$

42. $$\frac{-6(x^2-y^2)}{3(x-y)}=\frac{-2\cdot 3(x+y)(x-y)}{3(x-y)}=\frac{-2(x+y)}{1}=-2(x+y)$$

43. $$\frac{x^2+5x-14}{2-x}=\frac{(x-2)(x+7)}{-(x-2)}=\frac{x+7}{-1}=-(x+7)$$

44. $$(x-9)\cdot\frac{x-4}{x^2-81}=\frac{x-9}{1}\cdot\frac{x-4}{(x+9)(x-9)}=\frac{x-4}{x+9}$$

45. $$\frac{x+9}{x-9}\div\frac{x^2-81}{9-x}=\frac{x+9}{x-9}\cdot\frac{9-x}{x^2-81}=\frac{x+9}{x-9}\cdot\frac{-(x-9)}{(x+9)(x-9)}=\frac{-1}{x-9}=\frac{1}{9-x}$$

46. $$\frac{4}{3(x-2)}+\frac{8}{3(x-2)}=\frac{4+8}{3(x-2)}=\frac{12}{3(x-2)}=\frac{4}{x-2}$$

47. $x^2+x-6=(x+3)(x-2)$
$x^2-4=(x+2)(x-2)$
$\text{LCD}=(x+3)(x-2)(x+2)$
$$\frac{x+2}{x^2+x-6}-\frac{x+3}{x^2-4}$$
$$=\frac{(x+2)(x+2)}{(x+3)(x-2)(x+2)}-\frac{(x+3)(x+3)}{(x+3)(x-2)(x+2)}$$
$$=\frac{x^2+4x+4-(x^2+6x+9)}{(x+3)(x+2)(x-2)}$$
$$=\frac{-2x-5}{(x+3)(x+2)(x-2)}$$

48. $$\frac{\frac{4}{x}-\frac{1}{2x}}{\frac{1}{3x}-\frac{3}{4x}}=\frac{12x\left(\frac{4}{x}-\frac{1}{2x}\right)}{12x\left(\frac{1}{3x}-\frac{3}{4x}\right)}=\frac{12x\left(\frac{4}{x}\right)-12x\left(\frac{1}{2x}\right)}{12x\left(\frac{1}{3x}\right)-12x\left(\frac{3}{4x}\right)}=\frac{48-6}{4-9}=\frac{42}{-5}=-\frac{42}{5}$$

49. $$\frac{5x}{x-5}-2=\frac{x}{x-5}$$
$$(x-5)\left(\frac{5x}{x-5}-2\right)=(x-5)\left(\frac{x}{x-5}\right)$$
$5x-2(x-5)=x$
$5x-2x+10=x$
$3x+10=x$
$10=-2x$
$-5=x$

Since $x=-5$ does not make a denominator 0 in the original equation, the solution is $x=-5$.

50. The LCD is $x^2-64=(x+8)(x-8)$.

$$\begin{aligned}\frac{x}{x^2-64}+\frac{8}{x-8}&=\frac{1}{x+8}\\(x^2-64)\left(\frac{x}{x^2-64}+\frac{8}{x-8}\right)&=(x^2-64)\left(\frac{1}{x+8}\right)\\x+8(x+8)&=x-8\\x+8x+64&=x-8\\9x+64&=x-8\\8x&=-72\\x&=-9\end{aligned}$$

Since $x=-9$ does not make a denominator 0 in the original equation, the solution is $x=-9$.

51.

$$\begin{aligned}\frac{x}{x+1}-\frac{1}{2}&=\frac{-1}{x+1}\\2(x+1)\left(\frac{x}{x+1}-\frac{1}{2}\right)&=2(x+1)\left(\frac{-1}{x+1}\right)\\2x-(x+1)&=-2\\2x-x-1&=-2\\x-1&=-2\\x&=-1\end{aligned}$$

Since $x=-1$ makes a denominator 0 in the original equation, it is an extraneous solution. The equation has no solution.

52. The LCD is $x^2-16=(x+4)(x-4)$.

$$\begin{aligned}1+\frac{4}{x-4}&=\frac{32}{x^2-16}\\(x^2-16)\left(1+\frac{4}{x-4}\right)&=(x^2-16)\left(\frac{32}{x^2-16}\right)\\x^2-16+4(x+4)&=32\\x^2-16+4x+16&=32\\x^2+4x&=32\\x^2+4x-32&=0\\(x+8)(x-4)&=0\end{aligned}$$

$x+8=0$ or $x-4=0$

$x=-8$ or $x=4$

Since $x=4$ makes a denominator 0 in the original equation, it is an extraneous solution. The only solution is $x=-8$.

53. $\frac{120 \text{ mi}}{8 \text{ gal}} = \frac{210 \text{ mi}}{x \text{ gal}}$

$$\frac{120}{8} = \frac{210}{x}$$
$$120x = 210(8)$$
$$120x = 1680$$
$$x = 14$$

The van will need 14 gallons of gas.

54. $\frac{x-6}{3} = \frac{3}{10}$

$$10(x-6) = 3(3)$$
$$10x - 60 = 9$$
$$10x = 69$$
$$x = \frac{69}{10}$$

55. Let h be the number of hours it takes Janet and James to paint the kitchen together.

Janet paints $\frac{1}{3}$ of the kitchen in 1 hour.

James paints $\frac{1}{5}$ of the kitchen in 1 hour.

Together they paint $\frac{1}{h}$ of the kitchen in 1 hour.

$$\frac{1}{3} + \frac{1}{5} = \frac{1}{h}$$

The LCD is $3 \cdot 5 \cdot h = 15h$.

$$15h \cdot \frac{1}{3} + 15h \cdot \frac{1}{5} = 15h \cdot \frac{1}{h}$$
$$5h + 3h = 15$$
$$8h = 15$$
$$h = \frac{15}{8} = 1\frac{7}{8}$$

It would take them $1\frac{7}{8}$ hours working together.

56. $y - y_1 = m(x - x_1)$

$$y - 0 = -3(x-2)$$
$$y = -3x + 6$$

57. $y = mx + b;\ m = 3,\ b = -1$

$$y = 3x - 1$$

58. Graph $6x - y = -6$ with a dashed line. Test (0, 0), which is below the line:

$$6x - y < -6$$
$$6(0) - 0 < -6$$
$$0 < -6 \quad \text{False}$$

Shade the points above the line.

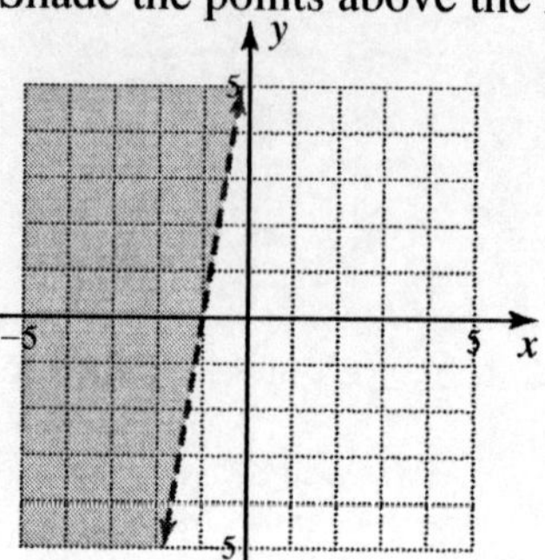

59. Graph $-y = -6x - 6$ with a solid line. Test (0, 0), which is below the line:

$$-y \geq -6x - 6$$
$$-0 \geq -6(0) - 6$$
$$0 \geq -6 \quad \text{True}$$

Shade the points below the line.

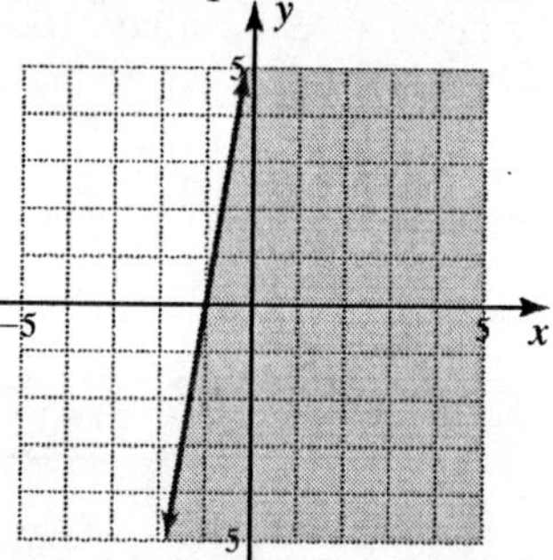

60. Since P is directly proportional to T, then $P = kT$.

$P = 5$ when $T = 250$:

$$5 = k(250)$$
$$\frac{5}{250} = k$$
$$\frac{1}{50} = k$$

61. Since P varies inversely as V, then $P = \frac{k}{V}$.

$P = 1960$ when $V = 7$:

$$1960 = \frac{k}{7}$$
$$7(1960) = k$$
$$k = 13{,}720$$

Chapter 8 Solving Systems of Linear Equations and Inequalities

8.1 Solving Systems of Equations by Graphing

Problems 8.1

1. Graph the equation $x + 2y = 4$ using the points (0, 2) and (4, 0). Then graph the equation $2x - 4y = 0$ using the points (0, 0) and (2, 1).

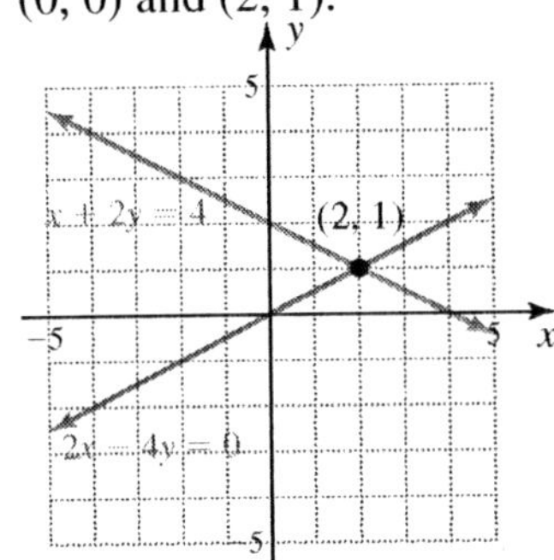

The graphs intersect at (2, 1), which is the solution of the system of equations.

2. Graph the equation $y - 3x = 3$ using the points (0, 3) and (−1, 0). Then graph the equation $2y - 6x = 12$ using the points (−1, 3) and (−2, 0).

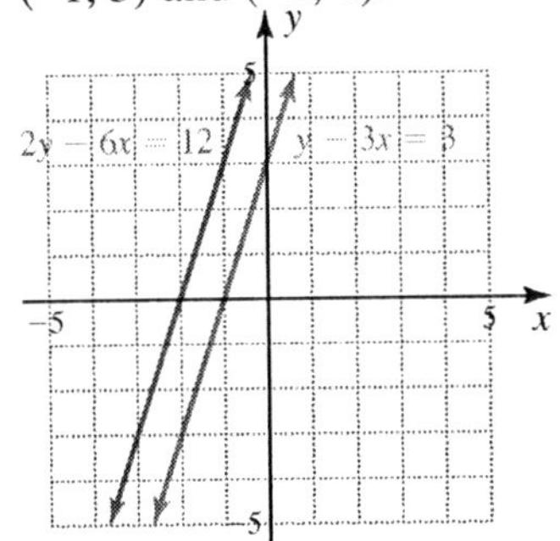

The lines are parallel; there is no solution.

3. Graph the equation $x + 2y = 4$ using the points (0, 2) and (4, 0). Then graph the equation $4y + 2x = 8$ using the points (0, 2) and (4, 0).

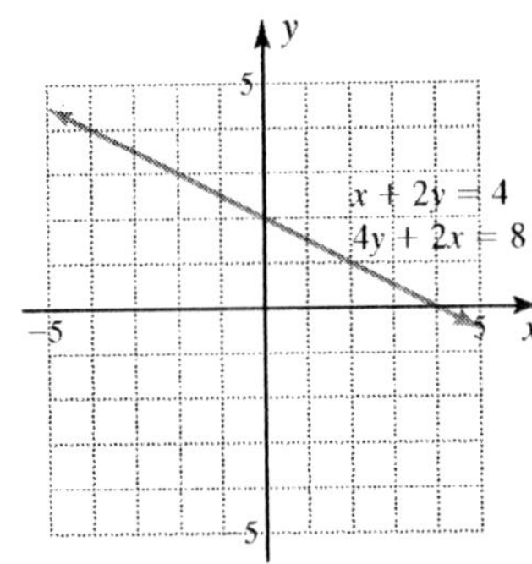

The lines coincide; there are infinitely many solutions.

4. **a.** Graph the equation $x + y = 4$ using the points (0, 4) and (4, 0). Then graph the equation $2y - x = 2$ using the points (0, 1) and (−2, 0).

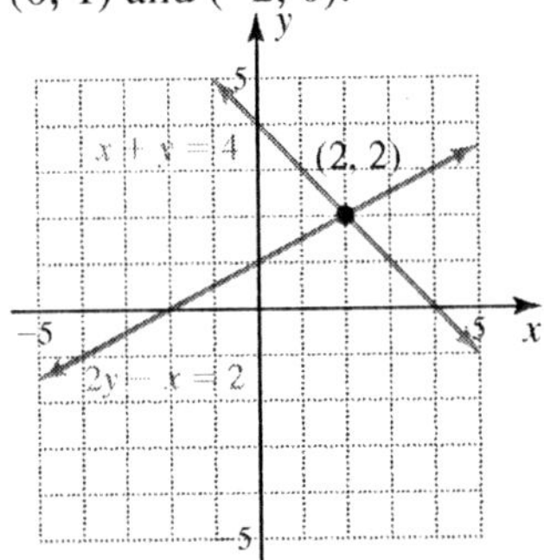

The system is consistent; the solution is (2, 2).

b. Graph the equation $2x + y = 4$ using the points (0, 4) and (2, 0). Then graph the equation $2y + 4x = 6$ using the points (0, 3) and $\left(\frac{3}{2}, 0\right)$.

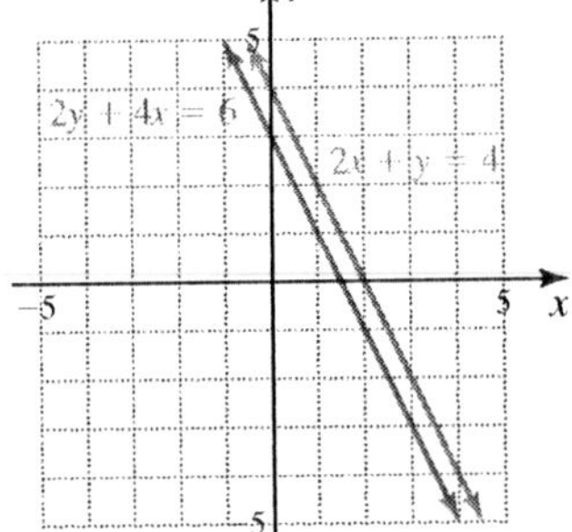

The system is inconsistent; there is no solution.

c. Graph the equation $2x + y = 4$ using the points (0, 4) and (2, 0). Then graph the equation $2y + 4x = 8$ using the points (0, 4) and (2, 0).

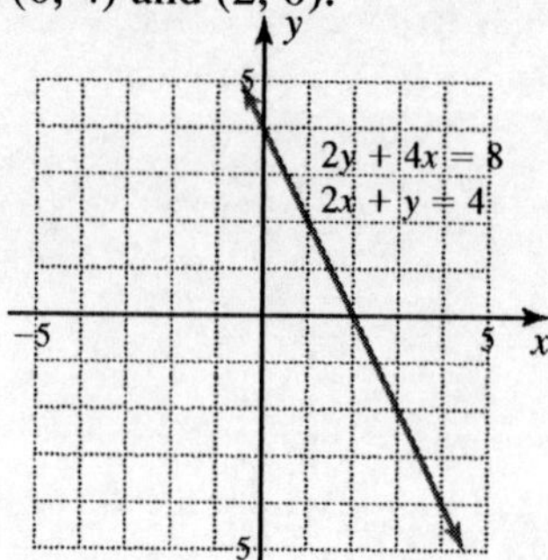

The system is dependent; there are infinitely many solutions.

5. For the first Internet provider, plot (0, 20), (5, 20), (10, 20), (15, 20), (20, 20), (25, 27.50), (30, 35), and (35, 42.50). For the second Internet provider, plot (0, 15), (5, 15), (10, 15), (15, 15), (20, 15), (25, 25), (30, 35), and (35, 45).

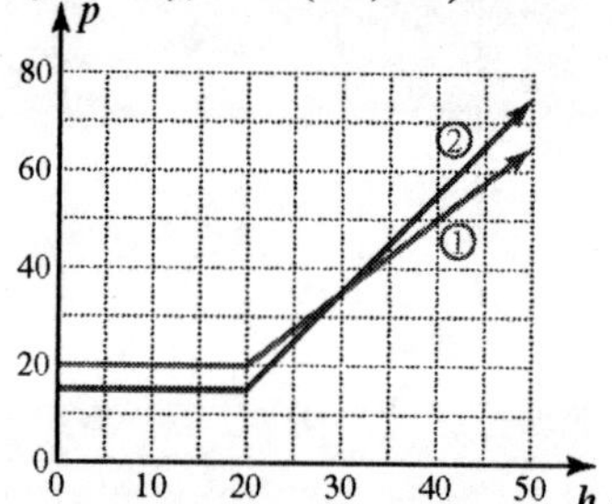

Exercises 8.1

1. Graph the equation $x + y = 4$ using the points (0, 4) and (4, 0). Then graph the equation $x - y = -2$ using the points (0, 2) and (–2, 0).

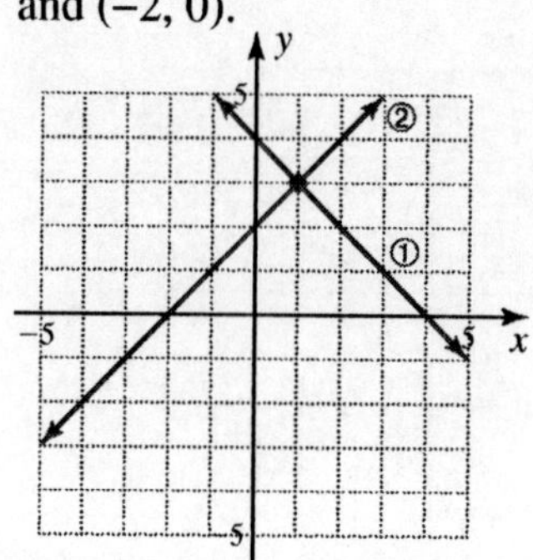

The system is consistent; the solution is (1, 3).

3. Graph the equation $x + 2y = 0$ using the points (0, 0) and (2, –1). Then graph the equation $x - y = -3$ using the points (0, 3) and (–3, 0).

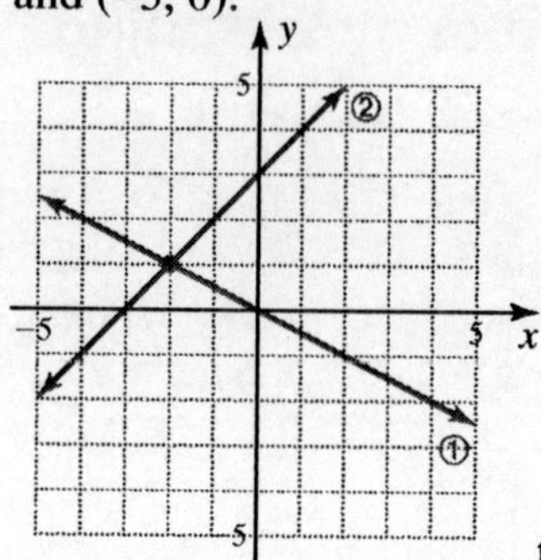

The system is consistent; the solution is (–2, 1).

5. Graph the equation $3x - 2y = 6$ using the points (0, –3) and (2, 0). Then graph the equation $6x - 4y = 12$ using the points (0, –3) and (2, 0).

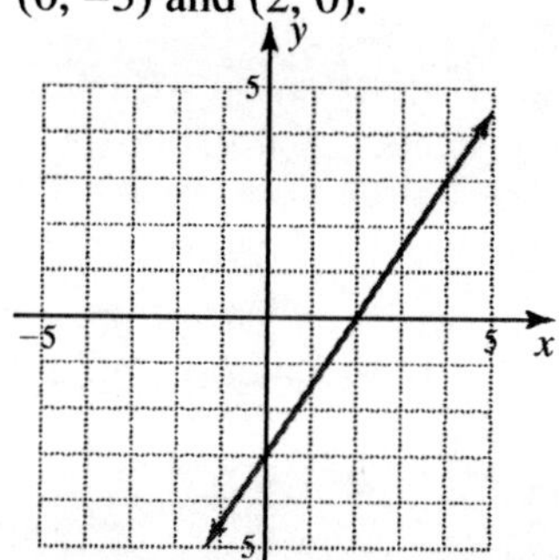

The system is dependent; there are infinitely many solutions.

7. Graph the equation $3x - y = -3$ using the points (0, 3) and (–1, 0). Then graph the equation $y - 3x = 3$ using the points (0, 3) and (–1, 0).

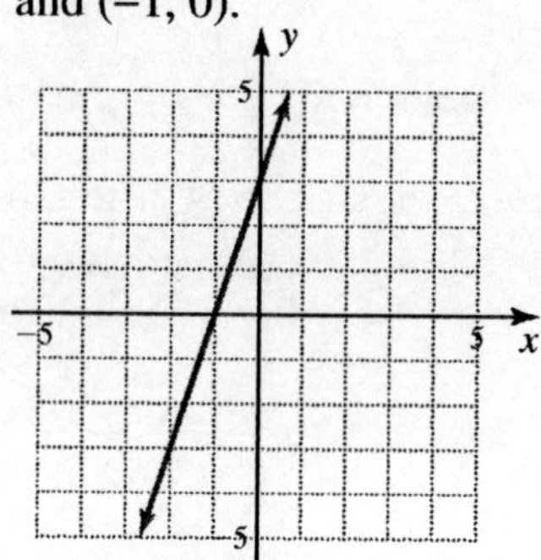

The system is dependent; there are infinitely many solutions.

9. Graph the equation $2x - y = -2$ using the points (0, 2) and (−1, 0). Then graph the equation $y = 2x + 4$ using the points (0, 4) and (−2, 0).

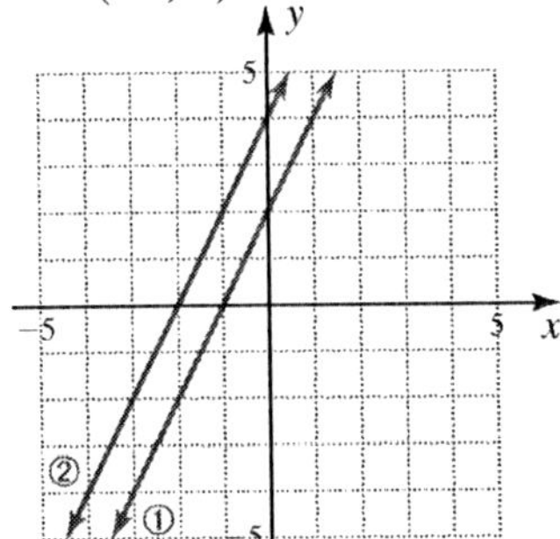

The system is inconsistent; there is no solution.

11. Graph the equation $y = -2$ using the points (0, −2) and (2, −2). Then graph the equation $2y = x - 2$ using the points (0, −1) and (2, 0).

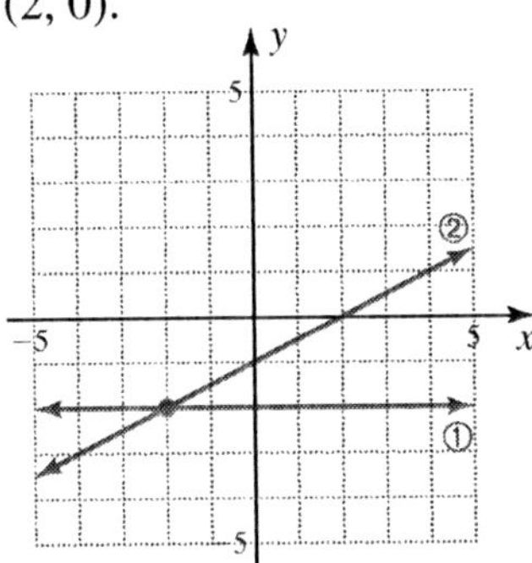

The system is consistent; the solution is (−2, −2).

13. Graph the equation $x = 3$ using the points (3, 0) and (3, 2). Then graph the equation $y = 2x - 4$ using the points (0, −4) and (2, 0).

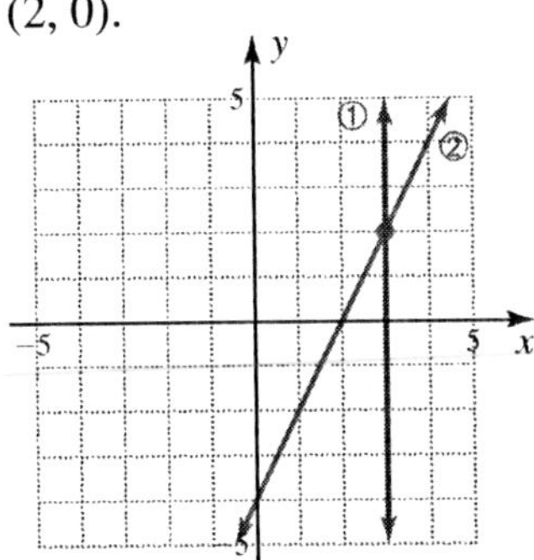

The system is consistent; the solution is (3, 2).

15. Graph the equation $y + x = 3$ using the points (0, 3) and (3, 0). Then graph the equation $2x - y = 0$ using the points (0, 0) and (2, 4).

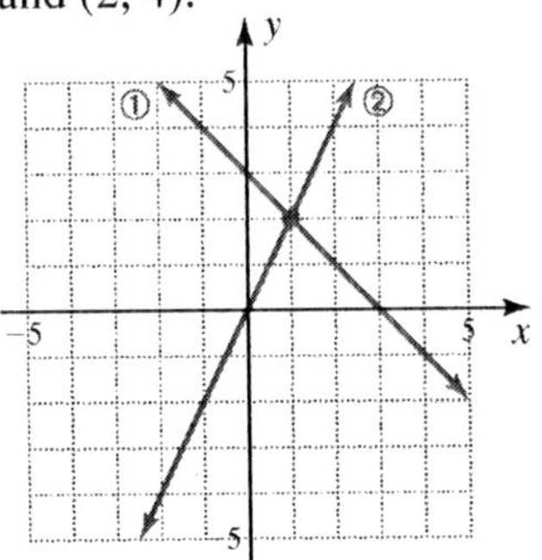

The system is consistent; the solution is (1, 2).

17. Graph the equation $5x + y = 5$ using the points (0, 5) and (1, 0). Then graph the equation $5x = 15 - 3y$ using the points (0, 5) and (3, 0).

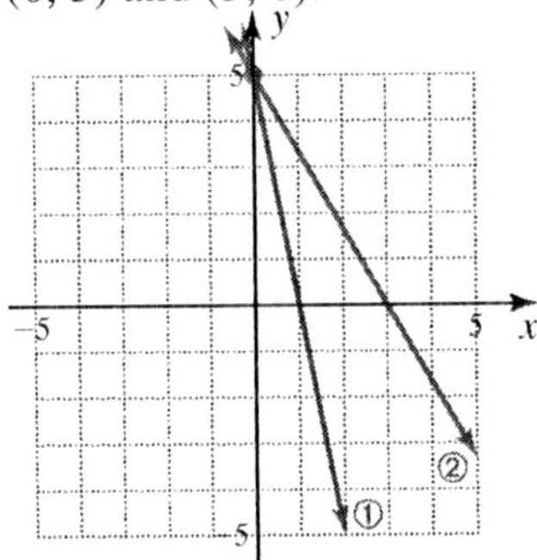

The system is consistent; the solution is (0, 5).

19. Graph the equation $3x + 4y = 12$ using the points (0, 3) and (4, 0). Then graph the equation $8y = 24 - 6x$ using the points (0, 3) and (4, 0).

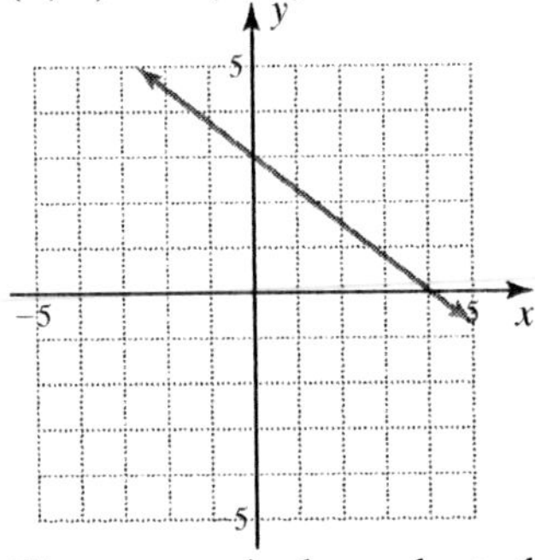

The system is dependent; there are infinitely many solutions.

21. Graph the equation $y = x + 3$ using the points (0, 3) and (–3, 0). Then graph the equation $y = -x + 3$ using the points (0, 3) and (3, 0).

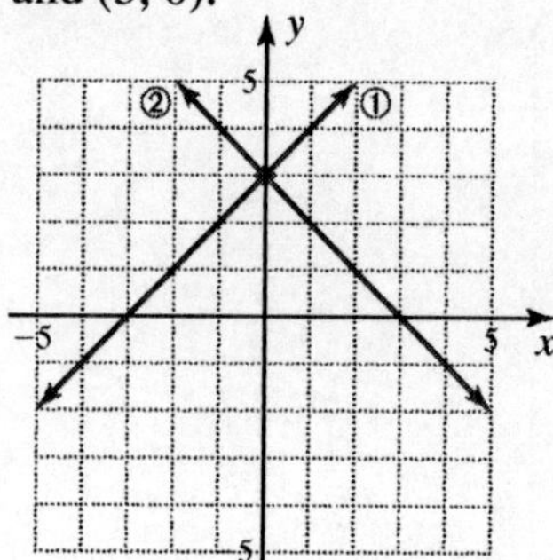

The system is consistent; the solution is (0, 3).

23. Graph the equation $y = 2x - 2$ using the points (0, –2) and (1, 0). Then graph the equation $y = -3x + 3$ using the points (0, 3) and (1, 0).

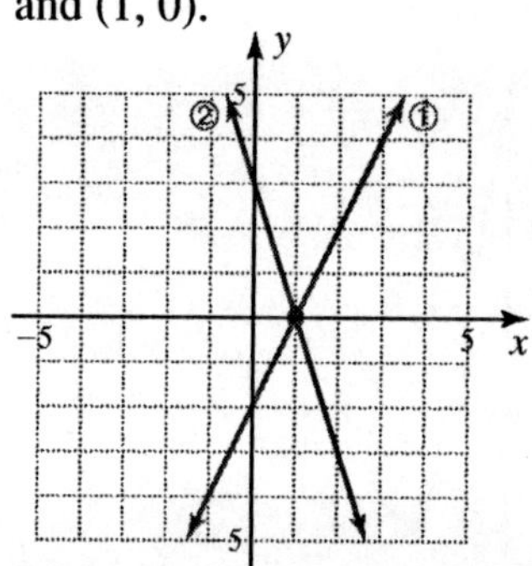

The system is consistent; the solution is (1, 0).

25. Graph the equation $-2x = 4$ using the points (–2, 0) and (–2, 2). Then graph the equation $y = -3$ using the points (0, –3) and (2, –3).

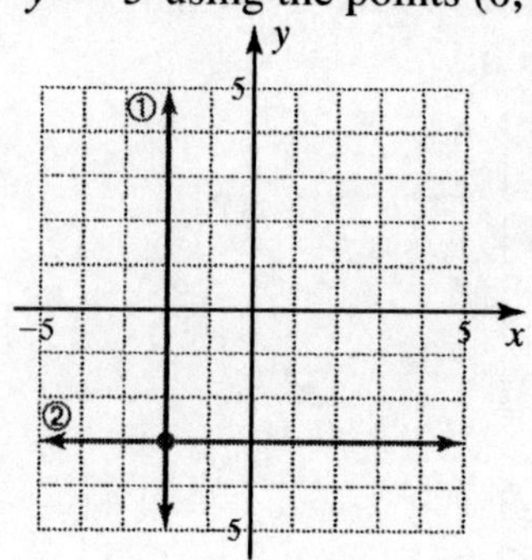

The system is consistent; the solution is (–2, –3).

27. Graph the equation $y = -3$ using the points (0, –3) and (2, –3). Then graph the equation $y = -3x + 6$ using the points (1, 3) and (2, 0).

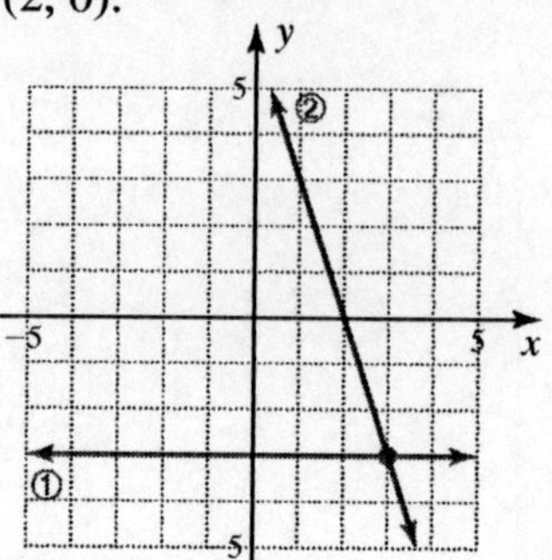

The system is consistent; the solution is (3, –3).

29. Graph the equation $x + 4y = 4$ using the points (0, 1) and (4, 0). Then graph the equation $y = -\frac{1}{4}x + 2$ using the points (0, 2) and (4, 1).

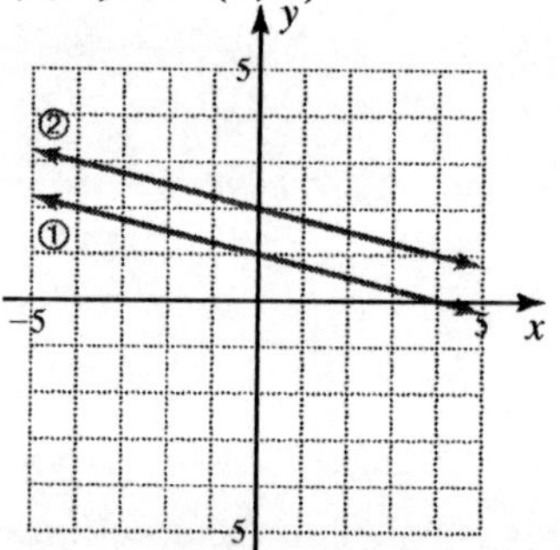

The system is inconsistent; there is no solution.

31. **a.** The cost *C* is the cost of installation, \$20, plus the cost per month, \$35, times the number of months, *m*. The equation is $C = 20 + 35m$.

b. For $m = 6$, $C = 20 + 35 \cdot 6 = 230$.
For $m = 12$, $C = 20 + 35 \cdot 12 = 440$.
For $m = 18$, $C = 20 + 35 \cdot 18 = 650$.

c. Graph the equation from part **a** using the points (6, 230), (12, 440), and (18, 650).

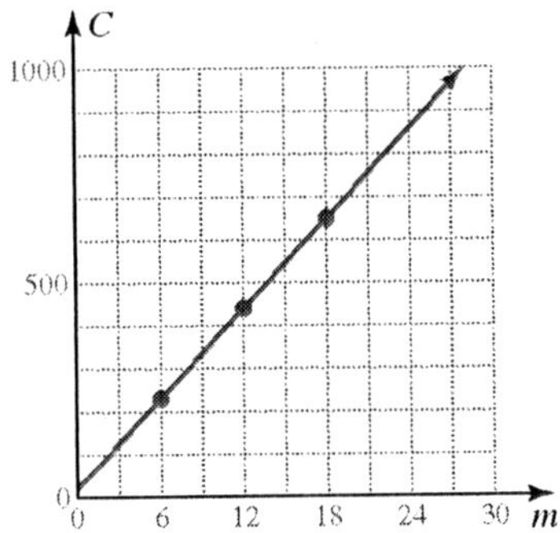

33. Draw the graphs from Problems 31 and 32 on the same coordinate axes.

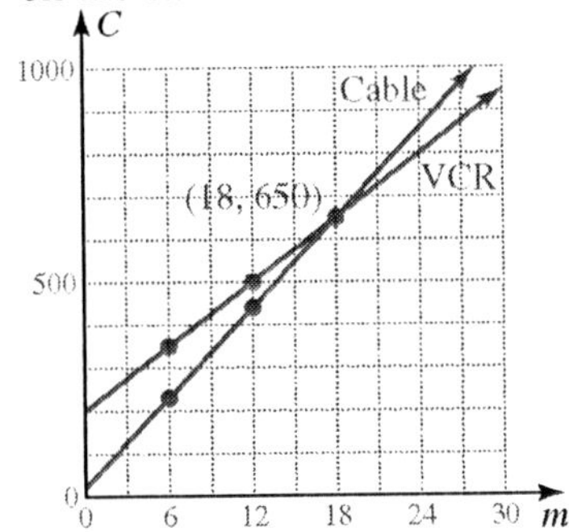

35. The graph for the VCR is lower than the graph for cable to the right of 18 on the horizontal axis. The VCR and rental option is cheaper if used more than 18 months. (The options are equal if used for 18 months.)

37. **a.** The weekly wages W are \$100 plus \$3 per table times the number of tables, t. The equation is $W = 100 + 3t$.

b. For $t = 5$, $W = 100 + 3 \cdot 5 = 115$.
For $t = 10$, $W = 100 + 3 \cdot 10 = 130$.
For $t = 15$, $W = 100 + 3 \cdot 15 = 145$.
For $t = 20$, $W = 100 + 3 \cdot 20 = 160$.

c. Graph the equation from part **a** using the points (5, 115), (10, 130), (15, 145), and (20, 160).

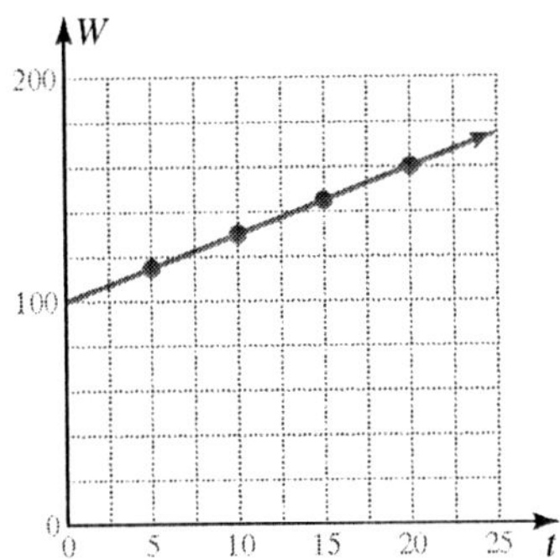

39. **a.** The cost C of plan A is the cost of the cell phone, \$0, plus \$0.60 per minute times the number of minutes, m. The equation is $C = 0.60m$.

b. The cost C of plan B is the cost of the cell phone, \$45, plus \$0.45 per minute times the number of minutes, m. The equation is $C = 45 + 0.45m$ or $C = 0.45m + 45$.

c. Graph the equation for plan A using the points (0, 0) and (500, 300). Then graph the equation for plan B using (0, 45) and (500, 270).

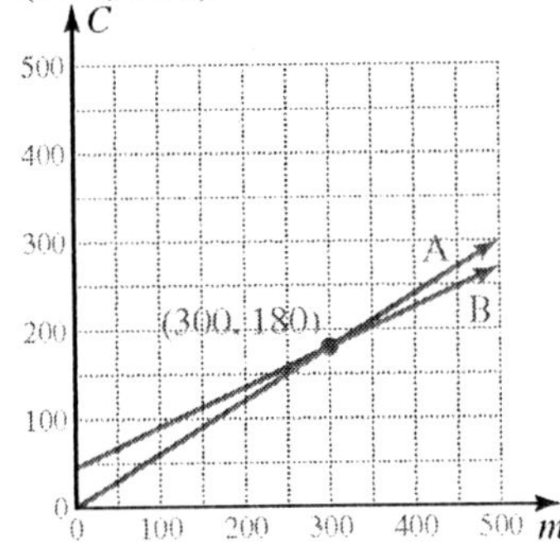

41.
$$3x + 72 = 2.5x + 74$$
$$0.5x + 72 = 74$$
$$0.5x = 2$$
$$x = 4$$

43.
$$2x + y = 11$$
$$2(3) + 5 = 11$$
$$6 + 5 = 11$$
Yes, (3, 5) is a solution of the equation.

45.
$$2x - y = 0$$
$$2(-1) - 2 = 0$$
$$-2 - 2 = 0$$
No, (−1, 2) is not a solution of the equation.

47. Graph the points (28, 17), (35, 15), and (42, 13).
See graph in problem 49.

49. Graph the points (26, 12), (35, 15), and (44, 18) on the same coordinate system used in Problems 48 and 49.

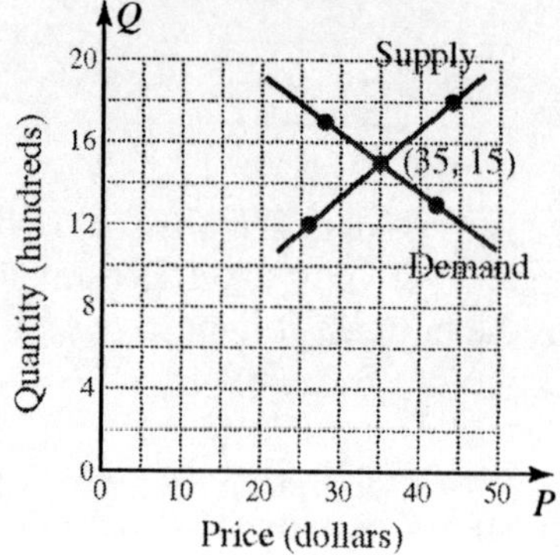

51. The lines in the graph for Problems 47–50 intersect at the point (35, 15).

53. In the ordered pair (35, 15), the second coordinate, 15, represents the quantity of books, in hundreds. The quantity of books is 1500.

55. Answers may vary. Sample answer: The lines intersect if the system is consistent. The lines are parallel for an inconsistent system. The lines are identical for a dependent system.

57. Answers may vary. Sample answer: Since the slopes are the same but the y-intercepts are different, the lines are parallel and the system has no solution.

59. Answers may vary. Sample answer: Since the lines have different slopes, they will intersect in exactly one point and the system has one solution.

61. Graph the equation $y-3x=3$ using the points (0, 3) and (–1, 0). Then graph the equation $2y=6x+12$ using the points (–1, 3) and (–2, 0).

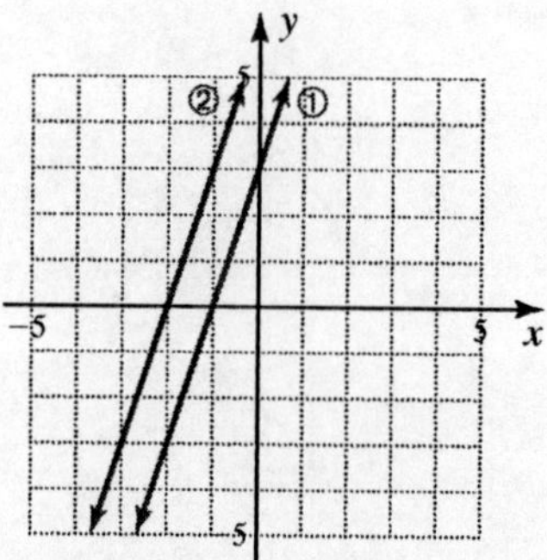

The lines are parallel; there is no solution.

63. Graph the equation $x+y=4$ using the points (0, 4) and (4, 0). Then graph the equation $2x-y=2$ using the points (0, –2) and (1, 0).

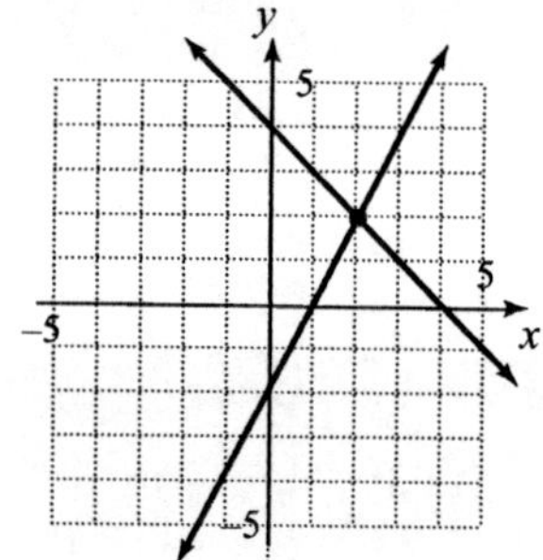

The system is consistent; the solution is (2, 2).

65. Graph the equation $2x+y=4$ using the points (0, 4) and (2, 0). Then graph the equation $2y+4x=8$ using the points (0, 4) and (2, 0).

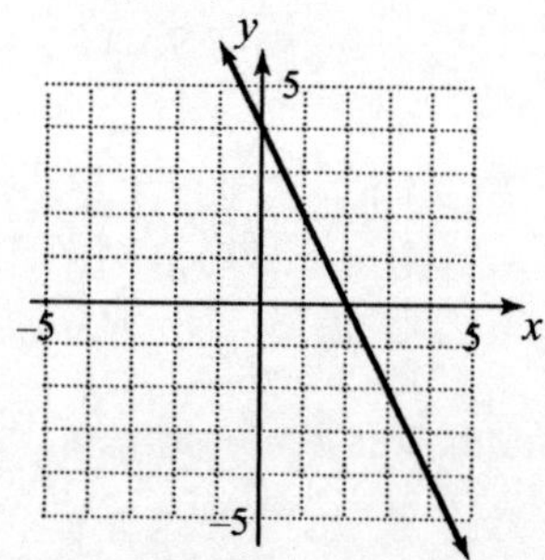

The system is dependent; there are infinitely many solutions.

8.2 Solving Systems of Equations by Substitution

Problems 8.2

1. 1. Solve the first equation for y: $y = 5 - x$.

2. Substitute $5 - x$ for y in $2x - 4y = -8$: $2x - 4(5 - x) = -8$.

3. Solve the new equation for x.

$$\begin{aligned} 2x - 4(5 - x) &= -8 \\ 2x - 20 + 4x &= -8 \\ 6x - 20 &= -8 \\ 6x &= 12 \\ x &= 2 \end{aligned}$$

4. Substitute 2 for x in $x + y = 5$. Then solve for y.

$$\begin{aligned} 2 + y &= 5 \\ y &= 3 \end{aligned}$$

The solution is the ordered pair (2, 3).

5. CHECK When $x = 2$ and $y = 3$,

$$\begin{aligned} x + y &= 5 \\ 2 + 3 &= 5 \\ 5 &= 5 \end{aligned} \quad \text{and} \quad \begin{aligned} 2x - 4y &= -8 \\ 2(2) - 4(3) &= -8 \\ 4 - 12 &= -8 \\ -8 &= -8 \end{aligned}$$

which are both true. Thus, the solution (2, 3) is correct.

2. 1. Solve the first equation for x: $x = 6 + 3y$.

2. Substitute $6 + 3y$ for x in $2x - 6y = 8$: $2(6 + 3y) - 6y = 8$.

$$\begin{aligned} 2(6 + 3y) - 6y &= 8 \\ 12 + 6y - 6y &= 8 \\ 12 &= 8 \end{aligned}$$

3. The result, $12 = 8$, is never true. The system has no solution; it is inconsistent.

4. We do not need step 4.

5. CHECK If you divide both sides of the second equation by –2, you get $x - 3y = 4$ which contradicts the first equation, $x - 3y = 6$.

3. 1. Solve the first equation for x: $x = 6 + 3y$.

2. Substitute $6 + 3y$ for x in $6y - 2x = -12$: $6y - 2(6 + 3y) = -12$.

$$\begin{aligned} 6y - 2(6 + 3y) &= -12 \\ 6y - 12 - 6y &= -12 \\ -12 &= -12 \end{aligned}$$

3. The result, $-12 = -12$, is always true. The system has infinitely many solutions; it is dependent.

4. We do not need step 4.

5. CHECK Note that if you divide the second equation by –2 and rearrange, you get $x - 3y = 6$, which is identical to the first equation.

4. Simplify the second equation.

$$\begin{aligned} 6 - 3x + y &= -5x + 2 \\ 2x + y &= -4 \end{aligned}$$

Now we have an equivalent system.

$$\begin{aligned} -3x &= -y + 6 \\ 2x + y &= -4 \end{aligned}$$

Solving the second equation for y, we get $y = -2x - 4$. Substituting $-2x - 4$ for y in the first equation, we have

$$\begin{aligned} -3x &= -y + 6 \\ -3x &= -(-2x - 4) + 6 \\ -3x &= 2x + 4 + 6 \\ -5x &= 10 \\ x &= -2 \end{aligned}$$

Since $-3x = -y + 6$ and $x = -2$, we have

$$\begin{aligned} -3(-2) &= -y + 6 \\ 6 &= -y + 6 \\ y &= 0 \end{aligned}$$

The system is consistent, and its solution is (–2, 0). You can verify this by substituting –2 for x and 0 for y in the two original equations.

5. Multiply both sides of the first equation by 3 and both sides of the second equation by 12 (the LCM of 4 and 6).

$3\left(2x+\frac{y}{3}\right)=3(-1)$ or $6x+y=-3$

$12\left(\frac{x}{4}+\frac{y}{6}\right)=12\left(\frac{1}{4}\right)$ or $3x+2y=3$

Solving the first equation for y, we have $y=-6x-3$. Now we substitute $-6x-3$ for y in $3x+2y=3$.

$$3x+2(-6x-3)=3$$
$$3x-12x-6=3$$
$$-9x=9$$
$$x=-1$$

Substituting -1 for x in $6x+y=-3$, we get $6(-1)+y=-3$ or $y=3$. The system is consistent, and its solution is (–1, 3). You should verify the solution.

6. The equation for Quickie Internet is $p=25+2.5(h-10)$. Substitute $25+2.5(h-10)$ for p in the equation for AOL, $p=20+3(h-20)$.

$$25+2.5(h-10)=20+3(h-20)$$
$$25+2.5h-25=20+3h-60$$
$$2.5h=3h-40$$
$$-0.5h=-40$$
$$h=80$$

If you use 80 hours, the price is the same for both services.

CHECK The price for 80 hours of Quickie Internet is

$p=25+2.5(80-10)=25+175=200$

The price for 80 hours of AOL is

$p=20+3(80-20)=20+180=200$

When 80 hours are used, the price is the same for both services, $200.

Exercises 8.2

1. The first equation, $y=2x-4$, is already solved for y. Substitute $2x-4$ for y in the second equation, $-2x=y-4$.

$$-2x=(2x-4)-4$$
$$-2x=2x-8$$
$$-4x=-8$$
$$x=2$$

Substitute 2 for x in $y=2x-4$.

$y=2(2)-4=4-4=0$

The system is consistent, and its solution is (2, 0).

3. Solve the first equation, $x+y=5$, for y: $y=5-x$. Substitute $5-x$ for y in the second equation, $3x+y=9$.

$$3x+(5-x)=9$$
$$2x+5=9$$
$$2x=4$$
$$x=2$$

Substitute 2 for x in $y=5-x$.

$y=5-2=3$

The system is consistent, and its solution is (2, 3).

5. The second equation, $y=2x+2$, is already solved for y. Substitute $2x+2$ for y in the first equation, $y-4=2x$.

$$(2x+2)-4=2x$$
$$2x-2=2x$$
$$-2=0$$

The result, $-2=0$, is never true. The system is inconsistent; it has no solution.

7. The first equation, $x=8-2y$, is already solved for x. Substitute $8-2y$ for x in the second equation, $x+2y=4$.

$$(8-2y)+2y=4$$
$$8=4$$

The result, $8=4$, is never true. The system is inconsistent; it has no solution.

9. The second equation, $x=-2y+4$, is already solved for *x*. Substitute $-2y+4$ for *x* in the first equation, $x+2y=4$.

$(-2y+4)+2y=4$
$4=4$

The result, $4=4$, is always true. The system is dependent; it has infinitely many solutions.

11. The second equation, $y=2x+1$, is already solved for *y*. Substitute $2x+1$ for *y* in the first equation, $x=2y+1$.

$x=2(2x+1)+1$
$x=4x+2+1$
$-3x=3$
$x=-1$

Substitute -1 for *x* in $y=2x+1$.

$y=2(-1)+1=-2+1=-1$

The system is consistent, and its solution is $(-1, -1)$.

13. Solve the first equation, $2x-y=-4$, for *y*: $-y=-2x-4$ or $y=2x+4$. Substitute $2x+4$ for *y* in the second equation, $4x=4+2y$.

$4x=4+2(2x+4)$
$4x=4+4x+4$
$0=8$

The result, $0=8$, is never true. The system is inconsistent; it has no solution.

15. The first equation, $x=5-y$, is already solved for *x*. Substitute $5-y$ for *x* in the second equation, $0=x-4y$.

$0=(5-y)-4y$
$0=5-5y$
$5y=5$
$y=1$

Substitute 1 for *y* in $x=5-y$.

$x=5-1=4$

The system is consistent, and its solution is (4, 1).

17. Simplify both equations.

$x+1=y+3 \qquad x-3=3y-7$
$x=y+2 \qquad x=3y-4$

Substitute $y+2$ for *x* in the equation $x=3y-4$.

$y+2=3y-4$
$-2y=-6$
$y=3$

Substitute 3 for *y* in $x=y+2$.

$x=3+2=5$

The system is consistent, and its solution is (5, 3).

19. Simplify the second equation.

$8+x-4y=-2y+4$
$x=2y-4$

Substitute $2y-4$ for *x* in the equation $2y=-x+4$.

$2y=-(2y-4)+4$
$2y=-2y+4+4$
$4y=8$
$y=2$

Substitute 2 for *y* in $x=2y-4$.

$x=2(2)-4=4-4=0$

The system is consistent, and its solution is (0, 2).

21. Simplify both equations.

$3x+y-5=7x+2 \qquad y+3=4x-2$
$y=4x+7 \qquad y=4x-5$

Substitute $4x+7$ for *y* in the equation $y=4x-5$.

$4x+7=4x-5$
$7=-5$

The result, $7=-5$, is never true. The system is inconsistent; it has no solution.

23. Simplify both equations.

$4x-2y-1=3x-1 \qquad x+2=6-2y$
$x=2y \qquad x=4-2y$

Substitute $2y$ for *x* in the equation $x=4-2y$.

$2y=4-2y$
$4y=4$
$y=1$

Substitute 1 for *y* in $x=2y$.

$x=2(1)=2$

The system is consistent, and its solution is (2, 1).

25. Multiply each side of the first equation by 6 (the LCM of 6 and 2).

$6\left(\frac{x}{6}+\frac{y}{2}\right)=6(1)$

$x+3y=6$

Then solve this equation for x.

$x=6-3y$

Substitute $6-3y$ for x in the equation $5x-2y=13$.

$5(6-3y)-2y=13$

$30-15y-2y=13$

$-17y=-17$

$y=1$

Substitute 1 for y in $x=6-3y$.

$x=6-3(1)=6-3=3$

The system is consistent, and its solution is (3, 1).

27. Multiply each side of the second equation by 6 (the LCM of 2 and 6).

$6\left(-\frac{x}{2}+\frac{y}{6}\right)=6(-2)$

$-3x+y=-12$

Then solve this equation for y.

$y=3x-12$

Substitute $3x-12$ for y in the equation $3x-y=12$.

$3x-(3x-12)=12$

$12=12$

The result, $12=12$, is always true. The system is dependent; it has infinitely many solutions.

29. Multiply each side of the first equation by 8 (the LCM of 4 and 8).

$8\left(\frac{y}{4}+x\right)=8\left(\frac{3}{8}\right)$

$2y+8x=3$

The second equation, $y=8-4x$, is already solved for y.

Substitute $8-4x$ for y in the equation $2y+8x=3$.

$2(8-4x)+8x=3$

$16-8x+8x=3$

$16=3$

The result, $16=3$, is never true. The system is inconsistent; it has no solution.

31. Multiply each side of the first equation by 3 and each side of the second equation by 6 (the LCM of 2 and 3).

$3\left(3x+\frac{y}{3}\right)=3(5) \quad 6\left(\frac{x}{2}-\frac{2y}{3}\right)=6(3)$

$9x+y=15 \quad 3x-4y=18$

Then solve the equation $9x+y=15$ for y.

$y=15-9x$

Substitute $15-9x$ for y in the equation $3x-4y=18$.

$3x-4(15-9x)=18$

$3x-60+36x=18$

$39x=78$

$x=2$

Substitute 2 for x in $y=15-9x$.

$y=15-9(2)=15-18=-3$

The system is consistent, and its solution is (2, –3).

33. a. The equation for The Information Network is $p=20+3(h-15)$ when $h>15$.

b. The equation for InterServe Communications is $p=20+2(h-15)$ when $h>15$.

c. Substitute $20+3(h-15)$ for p in the equation $p=20+2(h-15)$.

$20+3(h-15)=20+2(h-15)$

$20+3h-45=20+2h-30$

$3h-25=2h-10$

$h=15$

The price of both services is the same, \$20, when $h\le 15$ hours.

35. The equation for company A is $C=20+0.6m$, and the equation for company B is $C=50+0.4m$. Substitute $20+0.6m$ for C in the equation for company B.

$20 + 0.6m = 50 + 0.4m$
$0.2m = 30$
$m = 150$
The cost for both companies is the same when 150 minutes are used.

37. The equation for Le Bon Ton is $W = 50 + 10t$, and the equation for Le Magnifique is $W = 100 + 5t$. Substitute $50 + 10t$ for W in the equation for Le Magnifique.
$50 + 10t = 100 + 5t$
$5t = 50$
$t = 10$
The weekly income is the same at both restaurants when 10 tables are served.

39. The equation for the first company is $C = 35 + 20m$, and the equation for the second company is $C = 20 + 35m$, where m is the number of months. Substitute $35 + 20m$ for C in the equation for the second company.
$35 + 20m = 20 + 35m$
$-15m = -15$
$m = 1$
The cost for both companies is the same at the end of the 1st month.

41. The temperatures are the same when $F = C$. Substitute C for F in the equation
$F = \frac{9}{5}C + 32$.
$C = \frac{9}{5}C + 32$
$-\frac{4}{5}C = 32$
$C = -40$
The temperatures are the same at $-40°$.

43. Substitute $4x$ for y in the supply equation, $y = 2x + 8$.
$4x = 2x + 8$
$2x = 8$
$x = 4$
The supply will equal demand in 4 days.

45. Substitute $7x$ for y in the supply equation, $y = 5x + 10$.
$7x = 5x + 10$
$2x = 10$
$x = 5$
The supply will equal demand in 5 days.

47. $-0.5x = -4$
$\frac{-0.5x}{-0.5} = \frac{-4}{-0.5}$
$x = 8$

49. $9x = -9$
$\frac{9x}{9} = \frac{-9}{9}$
$x = -1$

51. Substitute $4L_1$ for L_2 in the equation $L_1 = -L_2 + 400$.
$L_1 = -4L_1 + 400$
$5L_1 = 400$
$L_1 = 80$
Substitute 80 for L_1 in the equation $L_2 = 4L_1$.
$L_2 = 4(80) = 320$
The regeneration for the oscillator circuit is correct when $L_1 = 80$ and $L_2 = 320$.

53. a. Substitute $5x$ for R and $4x + 500$ for C in the equation $R = C$. The equation is $5x = 4x + 500$.

b. Solve the equation from part **a**.
$5x = 4x + 500$
$x = 500$
The manufacturer must produce and sell 500 units to break even.

55. Answers may vary.

57. Solve the first equation, $x - 3y = 6$, for x: $x = 3y + 6$. Substitute $3y + 6$ for x in the second equation, $2x - 6y = 8$.
$2(3y + 6) - 6y = 8$
$6y + 12 - 6y = 8$
$12 = 8$
The result, $12 = 8$, is never true. The system is inconsistent.

59. Solve the first equation, $x+y=5$, for *y*: $y=5-x$. Substitute $5-x$ for *y* in the second equation, $2x-3y=-5$.

$$2x-3(5-x)=-5$$
$$2x-15+3x=-5$$
$$5x=10$$
$$x=2$$

Substitute 2 for *x* in $y=5-x$.

$$y=5-2=3$$

The system is consistent, and its solution is (2, 3).

61. Simplify the first equation.

$$3x-3y+1=5+2x$$
$$x-3y=4$$

Solve this equation for *x*: $x=3y+4$. Substitute $3y+4$ for *x* in the second equation, $x-3y=4$.

$$(3y+4)-3y=4$$
$$4=4$$

The result, $4=4$, is always true. The system is dependent.

63. Multiply each side of the first equation by 4 (the LCM of 2 and 4).

$$4\left(\frac{x}{2}-\frac{y}{4}\right)=4(-1)$$
$$2x-y=-4$$

Then solve the second equation, $2x=2+y$ for *y*: $2x-2=y$. Substitute $2x-2$ for *y* in the equation $2x-y=-4$.

$$2x-(2x-2)=-4$$
$$2=-4$$

The result, $2=-4$, is never true. The system is inconsistent.

65. The equation for the Sony system is $p=300+50m$, and the equation for the RCA system is $p=500+30m$, where *p* is the price and *m* is the number of months. Substitute $300+50m$ for *p* in the equation for RCA.

$$300+50m=500+30m$$
$$20m=200$$
$$m=10$$

The prices for both systems are the same for 10 months.

8.3 Solving Systems of Equations by Elimination

Problems 8.3

1. Multiply the first equation by 4. Leave the second equation as is.

$$\begin{array}{rcrcr} 12x & + & 4y & = & 4 \\ 3x & - & 4y & = & 11 \\ \hline 15x & + & 0 & = & 15 \\ & & 15x & = & 15 \\ & & x & = & 1 \end{array}$$

Substitute 1 for *x* in $3x+y=1$.

$$3(1)+y=1$$
$$y=-2$$

The solution of the system is (1, –2).

2. Multiply the first equation by –2. Leave the second equation as is.

$$\begin{array}{rcrcr} -6x & - & 4y & = & -2 \\ 6x & + & 4y & = & 12 \\ \hline 0 & + & 0 & = & 10 \\ & & 0 & = & 10 \end{array}$$

There is no solution; the system is inconsistent.

3. Multiply the second equation by 3. Leave the first equation as is.

$$\begin{array}{rcrcr} 3x & - & 6y & = & 9 \\ -3x & + & 6y & = & -9 \\ \hline 0 & + & 0 & = & 0 \\ & & 0 & = & 0 \end{array}$$

There are infinitely many solutions; the system is dependent.

4. Write the system in standard form.

$$\begin{array}{rcrcr} 5x & + & 4y & = & 6 \\ -4x & + & 3y & = & -11 \end{array}$$

Multiply the first equation by 4 and the second equation by 5.

$$\begin{array}{rcrcr} 20x & + & 16y & = & 24 \\ -20x & + & 15y & = & -55 \\ \hline 0 & + & 31y & = & -31 \\ & & 31y & = & -31 \\ & & y & = & -1 \end{array}$$

Substitute –1 for *y* in $5x+4y=6$.

$5x+4(-1)=6$
$5x=10$
$x=2$
The solution of the system is (2, –1).

5. Write the system of equations.

$$\begin{array}{rcrcr} 0.50p & + & 0.40n & = & 44 \\ p & + & n & = & 100 \end{array}$$

Multiply the second equation by –0.40. Leave the first equation as is.

$$\begin{array}{rcrcr} 0.50p & + & 0.40n & = & 44 \\ -0.40p & - & 0.40n & = & -40 \\ \hline 0.10p & & & = & 4 \end{array}$$

$$p = \frac{4}{0.10}$$
$$p = 40$$

Substitute 40 for p in $p+n=100$.

$40+n=100$
$n=60$

40 minutes of peak time and 60 minutes of off-peak time were used.

Exercises 8.3

1. Leave both equations as is.

$$\begin{array}{rcrcr} x & + & y & = & 3 \\ x & - & y & = & -1 \\ \hline 2x & & & = & 2 \end{array}$$
$$x = 1$$

Substitute 1 for x in $x+y=3$.

$1+y=3$
$y=2$

The solution of the system is (1, 2).

3. Leave both equations as is.

$$\begin{array}{rcrcr} x & + & 3y & = & 6 \\ x & - & 3y & = & -6 \\ \hline 2x & & & = & 0 \end{array}$$
$$x = 0$$

Substitute 0 for x in $x+3y=6$.

$0+3y=6$
$y=2$

The solution of the system is (0, 2).

5. Multiply the first equation by –2. Leave the second equation as is.

$$\begin{array}{rcrcr} -4x & - & 2y & = & -8 \\ 4x & + & 2y & = & 0 \\ \hline & & 0 & = & -8 \end{array}$$

There is no solution; the system is inconsistent.

7. Multiply the first equation by –2. Leave the second equation as is.

$$\begin{array}{rcrcr} -4x & - & 6y & = & -12 \\ 4x & + & 6y & = & 2 \\ \hline & & 0 & = & -10 \end{array}$$

There is no solution; the system is inconsistent.

9. Leave both equations as is.

$$\begin{array}{rcrcr} x & - & 5y & = & 15 \\ x & + & 5y & = & 5 \\ \hline 2x & & & = & 20 \end{array}$$
$$x = 10$$

Substitute 10 for x in $x+5y=5$.

$10+5y=5$
$5y=-5$
$y=-1$

The solution of the system is (10, –1).

11. Multiply the first equation by –2. Leave the second equation as is.

$$\begin{array}{rcrcr} -2x & - & 4y & = & -4 \\ 2x & + & 3y & = & -10 \\ \hline & & -y & = & -14 \\ & & y & = & 14 \end{array}$$

Substitute 14 for y in $x+2y=2$.

$x+2(14)=2$
$x+28=2$
$x=-26$

The solution of the system is (–26, 14).

13. Multiply the second equation by 2. Leave the first equation as is.

$$\begin{array}{rcrcr} 3x & - & 4y & = & 10 \\ 10x & + & 4y & = & 68 \\ \hline 13x & & & = & 78 \end{array}$$
$$x = 6$$

Substitute 6 for x in $5x+2y=34$.

$5(6)+2y=34$
$30+2y=34$
$2y=4$
$y=2$

The solution of the system is (6, 2).

15. Multiply the first equation by 8 and the second equation by 3.

$$\begin{array}{rcrcr} 88x & - & 24y & = & 200 \\ 15x & + & 24y & = & 6 \\ \hline 103x & & & = & 206 \\ & & x & = & 2 \end{array}$$

Substitute 2 for x in $5x+8y=2$.

$5(2)+8y=2$
$10+8y=2$
$8y=-8$
$y=-1$

The solution of the system is (2, −1).

17. Write the system in standard form.

$$\begin{array}{rcrcr} 2x & + & 3y & = & 21 \\ 3x & - & y & = & 4 \end{array}$$

Multiply the second equation by 3. Leave the first equation as is.

$$\begin{array}{rcrcr} 2x & + & 3y & = & 21 \\ 9x & - & 3y & = & 12 \\ \hline 11x & & & = & 33 \\ & & x & = & 3 \end{array}$$

Substitute 3 for x in $2x+3y=21$.

$2(3)+3y=21$
$6+3y=21$
$3y=15$
$y=5$

The solution of the system is (3, 5).

19. Write the system in standard form.

$$\begin{array}{rcrcr} x & - & 2y & = & 1 \\ -x & - & y & = & 5 \end{array}$$

Leave both equations as is.

$$\begin{array}{rcrcr} x & - & 2y & = & 1 \\ -x & - & y & = & 5 \\ \hline & & -3y & = & 6 \\ & & y & = & -2 \end{array}$$

Substitute −2 for y in $x=1+2y$.

$x=1+2(-2)=1-4=-3$

The solution of the system is (−3, −2).

21. Multiply the first equation by 12 and the second equation by 6.

$$\begin{array}{rcrcr} 3x & + & 4y & = & 48 \\ 3x & - & y & = & 18 \end{array}$$

Multiply the second equation by −1. Leave the first equation as is.

$$\begin{array}{rcrcr} 3x & + & 4y & = & 48 \\ -3x & + & y & = & -18 \\ \hline & & 5y & = & 30 \\ & & y & = & 6 \end{array}$$

Substitute 6 for y in $3x-y=18$.

$3x-6=18$
$3x=24$
$x=8$

The solution of the system is (8, 6).

23. Multiply the first equation by 12 and the second equation by 5.

$$\begin{array}{rcrcr} 3x & - & 4y & = & -5 \\ x & + & 2y & = & 5 \end{array}$$

Multiply the second equation by 2. Leave the first equation as is.

$$\begin{array}{rcrcr} 3x & - & 4y & = & -5 \\ 2x & + & 4y & = & 10 \\ \hline 5x & & & = & 5 \\ & & x & = & 1 \end{array}$$

Substitute 1 for x in $x+2y=5$.

$1+2y=5$
$2y=4$
$y=2$

The solution of the system is (1, 2).

25. Multiply the first equation by 8 and the second equation by 2.

$$\begin{array}{rcrcr} x & + & y & = & 8 \\ x & - & y & = & 2 \end{array}$$

Leave both equations as is.

$$\begin{array}{rcrcr} x & + & y & = & 8 \\ x & - & y & = & 2 \\ \hline 2x & & & = & 10 \\ & & x & = & 5 \end{array}$$

Substitute 5 for x in $x+y=8$.

$5+y=8$
$y=3$

The solution of the system is (5, 3).

27. Multiply the first equation by 6 and the second equation by 2.

$$3x - 2y = 6$$
$$x + y = 7$$

Multiply the second equation by 2. Leave the first equation as is.

$$\begin{array}{rcrcr} 3x & - & 2y & = & 6 \\ 2x & + & 2y & = & 14 \\ \hline 5x & & & = & 20 \\ & & x & = & 4 \end{array}$$

Substitute 4 for x in $x+y=7$.

$$4+y=7$$
$$y=3$$

The solution of the system is (4, 3).

29. Multiply the first equation by 18 and the second equation by 4.

$$4x - 9y = -18$$
$$4x - 9y = -18$$

Multiply the second equation by −1. Leave the first equation as is.

$$\begin{array}{rcrcr} 4x & - & 9y & = & -18 \\ -4x & + & 9y & = & 18 \\ \hline & & 0 & = & 0 \end{array}$$

There are infinitely many solutions; the system is dependent.

31. Write the system of equations.

$$8x + 9y = 8.2$$
$$x + y = 1$$

(x represents the amount of Costa Rican coffee, and y represents the amount of Indian Mysore coffee, both in pounds.)

Multiply the second equation by −8. Leave the first equation as is.

$$\begin{array}{rcrcr} 8x & + & 9y & = & 8.2 \\ -8x & - & 8y & = & -8 \\ \hline & & y & = & 0.2 \end{array}$$

Substitute 0.2 for y in $x+y=1$.

$$x+0.2=1$$
$$x=0.8$$

0.8 pound of Costa Rican coffee and 0.2 pound of Indian Mysore coffee should go into each pound.

33. Write the system of equations.

$$19x + 4y = 350$$
$$x + y = 50$$

(x represents the amount of oolong tea, and y represents the amount of regular tea, both in pounds. $350 is the total cost of 50 pounds of tea costing $7 per pound.)

Multiply the second equation by −4. Leave the first equation as is.

$$\begin{array}{rcrcr} 19x & + & 4y & = & 350 \\ -4x & - & 4y & = & -200 \\ \hline 15x & & & = & 150 \\ & & x & = & 10 \end{array}$$

Substitute 10 for x in $x+y=50$.

$$10+y=50$$
$$x=40$$

10 pounds of oolong tea and 40 pounds of regular tea should be used.

35. The expression is $n+d=300$.

37. The expression is $4(x-y)=48$.

39. The expression is $m=n-3$.

41. Write the system of equations.

$$2x + y = 361$$
$$x + 2y = 360$$

(x represents Tweedledee's weight, and y represents Tweedledum's weight, both in pounds.)

Multiply the second equation by −2. Leave the first equation as is.

$$\begin{array}{rcrcr} 2x & + & y & = & 361 \\ -2x & - & 4y & = & -720 \\ \hline & & -3y & = & -359 \\ & & y & = & \frac{359}{3} \end{array}$$

Substitute $\frac{359}{3}$ for y in $x + 2y = 360$.

$$x+2\left(\frac{359}{3}\right)=360$$
$$x=360-\frac{718}{3}$$
$$x=\frac{362}{3}$$

Tweedledee weighs $\frac{362}{3}$ or $120\frac{2}{3}$ pounds, and Tweedledum weighs $\frac{359}{3}$ or $119\frac{1}{3}$ pounds.

43. Answers may vary.

45. Multiply the first equation by –2. Leave the second equation as is.

$$\begin{aligned} -4x - 10y &= -18 \\ 4x - 3y &= 11 \\ \hline -13y &= -7 \\ y &= \frac{7}{13} \end{aligned}$$

Substitute $\frac{7}{13}$ for y in $2x+5y=9$.

$$2x+5\left(\frac{7}{13}\right)=9$$
$$2x=9-\frac{35}{13}$$
$$2x=\frac{82}{13}$$
$$x=\frac{41}{13}$$

The solution of the system is $\left(\frac{41}{13},\frac{7}{13}\right)$.

47. Write the system in standard form.

$$\begin{aligned} 2x - 5y &= 5 \\ x - y &= 4 \end{aligned}$$

Multiply the second equation by –2. Leave the first equation as is.

$$\begin{aligned} 2x - 5y &= 5 \\ -2x + 2y &= -8 \\ \hline -3y &= -3 \\ y &= 1 \end{aligned}$$

Substitute 1 for y in $x-y=4$.

$$x-1=4$$
$$x=5$$

The solution of the system is (5, 1).

49. Multiply the first equation by 6 and the second equation by 4.

$$\begin{aligned} x - 3y &= 6 \\ -x + 3y &= -3 \\ \hline 0 &= 3 \end{aligned}$$

There is no solution; the system is inconsistent.

8.4 Coin, General, Motion, and Investment Problems

Problems 8.4

2. 1. We are asked to find the numbers of nickels and dimes.

2. Let n be the number of nickels and d the number of dimes.

3. Translate the problem into a system of equations.
$$0.05n+0.10d=1.50$$
$$d=2n$$

4. Substitute $2n$ for d in the first equation.
$$0.05n+0.10(2n)=1.50$$
$$0.05n+0.20n=1.50$$
$$0.25n=1.50$$
$$n=6$$
Substitute 6 for n in the equation $d=2n$.
$$d=2(6)=12$$
Jill has 6 nickels and 12 dimes.

5. Since 6 nickels are \$0.30 and 12 dimes are \$1.20, Jill does have \$1.50. She also has twice as many dimes, 12, as nickels, 6.

3. 1. We are asked to find the weight of each of the astronauts.

2. Let m be the weight of the male astronaut and f be the weight of the female astronaut.

3. Translate the problem into a system of equations.
   $m+f=320$
   $m-f=60$

4. Add the equations to eliminate f.

$$\begin{array}{rcrcr} m & + & f & = & 320 \\ m & - & f & = & 60 \\ \hline 2m & & & = & 380 \\ & & m & = & 190 \end{array}$$

   Substitute 190 for m in the equation $m+f=320$.
   $190+f=320$
   $f=130$
   The man weighed 190 pounds, and the woman weighed 130 pounds.

5. If the man weighed 190 pounds and the woman weighed 130 pounds, their combined weight was 320 pounds and the woman was 60 pounds lighter than the man.

4. 1. We are asked to find the speed of the wind and the speed of the plane in still air.

2. Let x be the speed of the plane in still air and y be the speed of the wind.

3. Make a chart.

	R	$\times$ T	$=$ D
Against the wind:	$x-y$	4	800
With the wind:	$x+y$	2	800

   Write the system of equations.
   $4(x-y)=800$
   $2(x+y)=800$

4. Solve the system of equations. Divide the first equation by 4, and divide the second equation by 2.

$$\begin{array}{rcrcr} x & - & y & = & 200 \\ x & + & y & = & 400 \\ \hline 2x & & & = & 600 \\ & & x & = & 300 \end{array}$$

   Substitute 300 for x in the equation $2(x+y)=800$.
   $2(300+y)=800$
   $300+y=400$
   $y=100$
   The speed of the plane in still air is 300 miles per hour, and the speed of the wind is 100 miles per hour.

5. You should verify this solution.

5. 1. We are asked to find the amount owed on each card.

2. Let x be the amount owed on the card at 9% and y the amount owed on the card at 12%.

3. Make a chart.

	P	$\times$ R	$=$ I
Card at 9%:	x	0.09	$0.09x$
Card at 12%:	y	0.12	$0.12y$

   Write the system of equations
   $x+y=11{,}000$
   $0.09x+0.12y=1140$

4. Multiply the first equation by −0.09.

$$\begin{array}{rcrcr} -0.09x & - & 0.09y & = & -990 \\ 0.09x & + & 0.12y & = & 1140 \\ \hline & & 0.03y & = & 150 \\ & & y & = & 5000 \end{array}$$

   Substitute 5000 for y in the equation $x+y=11{,}000$.
   $x+5000=11{,}000$
   $x=6000$
   Dorothy owes \$6000 on the card at 9% and \$5000 on the card at 12%.

5. Since
   $0.09\cdot\$6000+0.12\cdot\$5000$
   $=\$540+\$600$
   $=\$1140,$
   the amounts of \$6000 and \$5000 are correct.

Exercises 8.4

1. We are asked to find the numbers of nickels and dimes. Let n be the number of nickels and d the number of dimes. Translate the problem into a system of equations.
   $0.05n + 0.10d = 2.25$
   $d = 4n$
   Substitute $4n$ for d in the first equation.
   $0.05n + 0.10(4n) = 2.25$
   $0.05n + 0.40n = 2.25$
   $0.45n = 2.25$
   $n = 5$
   Substitute 5 for n in the equation $d = 4n$.
   $d = 4(5) = 20$
   Mida has 5 nickels and 20 dimes.

3. We are asked to find the numbers of nickels and dimes. Let n be the number of nickels and d the number of dimes. Translate the problem into a system of equations.
   $n + d = 20$
   $0.05n + 0.10d = 0.05d + 0.10n - 0.50$
   Write the system in standard form.
   $n + d = 20$
   $-0.05n + 0.05d = -0.50$
   Multiply the first equation by 0.05.
   $$\begin{array}{rcrcr} 0.05n & + & 0.05d & = & 1 \\ -0.05n & + & 0.05d & = & -0.50 \\ \hline & & 0.10d & = & 0.50 \\ & & d & = & 5 \end{array}$$
   Substitute 5 for d in the equation $n + d = 20$.
   $n + 5 = 20$
   $n = 15$
   Mongo has 15 nickels and 5 dimes.

5. We are asked to find the numbers of \$1 bills and \$5 bills. Let x be the number of \$1 bills and y the number of \$5 bills. Translate the problem into a system of equations.
   $x + y = 10$
   $x + 5y = 26$
   Multiply the first equation by -1.
   $$\begin{array}{rcrcr} -x & - & y & = & -10 \\ x & + & 5y & = & 26 \\ \hline & & 4y & = & 16 \\ & & y & = & 4 \end{array}$$
   Substitute 4 for y in the equation $x + y = 10$.
   $x + 4 = 10$
   $x = 6$
   Don had 6 \$1 bills and 4 \$5 bills.

7. Let x and y be the numbers.
   $$\begin{array}{rcrcr} x & + & y & = & 102 \\ x & - & y & = & 16 \\ \hline 2x & & & = & 118 \\ & & x & = & 59 \end{array}$$
   Substitute 59 for x in $x + y = 102$.
   $59 + y = 102$
   $y = 43$
   The numbers are 59 and 43.

9. Let m and n be the integers.
   $m + n = 126$
   $m = 5n$
   Substitute $5n$ for m in $m + n = 126$.
   $5n + n = 126$
   $6n = 126$
   $n = 21$
   Substitute 21 for n in $m = 5n$.
   $m = 5(21) = 105$
   The integers are 105 and 21.

11. Let x and y be the numbers.
   $x - y = 16$
   $x = y + 4$
   Write the system in standard form.
   $x - y = 16$
   $x - y = 4$
   Multiply the second equation by -1.
   $$\begin{array}{rcrcr} x & - & y & = & 16 \\ -x & + & y & = & -4 \\ \hline & & 0 & = & 12 \end{array}$$
   It is impossible to have two numbers satisfying the given conditions.

13. Let x be the elevation of Longs Peak and y the elevation of Pikes Peak, both in feet.
   $x - y = 145$
   $x + y + 637 = 29{,}002$
   Write the system in standard form.

$$\begin{array}{rcrcl} x & - & y & = & 145 \\ x & + & y & = & 28{,}365 \\ \hline 2x & & & = & 28{,}510 \\ & & x & = & 14{,}255 \end{array}$$

Substitute 14,255 for x in $x + y = 28{,}365$.

$14{,}255 + y = 28{,}365$

$y = 14{,}110$

Longs Peak is 14,255 feet, and Pikes Peak is 14,110 feet.

15. We are asked to find the distance between the cities. Let d be the distance between the cities and t the scheduled time of the flight. Make a chart.

	R	$\times$ T	$=$ D
Late:	300	$t+\frac{1}{2}$	d
On time:	350	t	d

Write the system of equations.

$300(t+\frac{1}{2}) = d$

$350t = d$

Substitute $350t$ for d in the first equation.

$300(t+\frac{1}{2}) = 350t$

$300t + 150 = 350t$

$150 = 50t$

$3 = t$

Substitute 3 for t in the equation $d = 350t$.

$d = 350(3) = 1050$

The distance between the cities is 1050 miles.

17. We are asked to find the speed of the boat in still water and the speed of the current. Let x be the speed of the boat in still water and y be the speed of the current. Make a chart.

	R	$\times$ T	$=$ D
Downstream:	$x+y$	$2\frac{1}{2}$	45
Upstream:	$x-y$	$3\frac{1}{4}$	39

Write the system of equations, using $2\frac{1}{2} = \frac{5}{2}$ and $3\frac{1}{4} = \frac{13}{4}$.

$\frac{5}{2}(x+y) = 45$

$\frac{13}{4}(x-y) = 39$

Multiply the first equation by $\frac{2}{5}$, and divide the second equation by $\frac{13}{4}$.

$$\begin{array}{rcrcl} x & + & y & = & 18 \\ x & - & y & = & 12 \\ \hline 2x & & & = & 30 \\ & & x & = & 15 \end{array}$$

Substitute 15 for x in the equation $x + y = 18$.

$15 + y = 18$

$y = 3$

The speed of the boat in still water is 15 miles per hour, and the speed of the current is 3 miles per hour.

19. We are asked to find the distance between Bill's home and office. Let d be the distance and t the time it takes at 40 miles per hour. Make a chart. Note that it takes Bill 10 minutes longer driving at 30 miles per hour and that 10 minutes is $\frac{1}{6}$ hour.

	R	$\times$ T	$=$ D
5 minutes early:	40	t	d
5 minutes late:	30	$t+\frac{1}{6}$	d

Write the system of equations.

$40t = d$

$30\left(t+\frac{1}{6}\right) = d$

Substitute $40t$ for d in the second equation.

$30\left(t+\frac{1}{6}\right)=40t$

$30t+5=40t$

$5=10t$

$\frac{5}{10}=t$

$\frac{1}{2}=t$

Substitute $\frac{1}{2}$ for t in the equation $d=40t$.

$d=40\left(\frac{1}{2}\right)=20$

The distance between Bill's home and office is 20 miles.

21. We are asked to find the amount invested at each rate. Let x be the amount invested at 6% and y the amount invested at 8%. Make a chart.

	P	$\times$ R	$=$ I
6%:	x	0.06	$0.06x$
8%:	y	0.08	$0.08y$

Write the system of equations

$x+y=20,000$

$0.06x+0.08y=1500$

Multiply the first equation by –0.06.

$$\begin{array}{rcrcr} -0.06x & - & 0.06y & = & -1200 \\ 0.06x & + & 0.08y & = & 1500 \\ \hline & & 0.02y & = & 300 \\ & & y & = & 15,000 \end{array}$$

Substitute 15,000 for y in the equation $x+y=20,000$.

$x+15,000=20,000$

$x=5000$

Fred invested $5000 at 6% and $15,000 at 8%.

23. We are asked to find the amount in the savings account. Let x be the amount in the savings account and y the amount in certificates of deposit. We make a chart.

	P	$\times$ R	$=$ I
Savings:	x	0.05	$0.05x$
CDs:	y	0.07	$0.07y$

Write the system of equations

$x+y=18,000$

$0.05x+0.07y=1100$

Multiply the first equation by –0.07.

$$\begin{array}{rcrcr} -0.07x & - & 0.07y & = & -1260 \\ 0.05x & + & 0.07y & = & 1100 \\ \hline -0.02x & & & = & -160 \\ & & x & = & 8000 \end{array}$$

Dominic has $8000 in his savings account.

25. Let x be the number of students enrolled in public institutions and y the number enrolled in private institutions. Write a system of equations.

$$\begin{array}{rcrcr} x & + & y & = & 15 \\ x & - & y & = & 8.4 \\ \hline 2x & & & = & 23.4 \\ & & x & = & 11.7 \end{array}$$

Substitute 11.7 for x in the equation $x+y=15$.

$11.7+y=15$

$y=3.3$

There were 11.7 million students enrolled in public institutions and 3.3 million enrolled in private institutions.

27. Let x be the number of Home Box Office subscribers and y the number of Showtime subscribers. Write a system of equations.

$$\begin{array}{rcrcr} x & + & y & = & 28,700,000 \\ x & - & y & = & 7300 \\ \hline 2x & & & = & 28,707,300 \\ & & x & = & 14,353,650 \end{array}$$

Substitute 14,353,650 for x in the equation $x+y=28,700,000$.

$14,353,650+y=28,700,000$

$y=14,346,350$

There were 14,353,650 Home Box Office subscribers and 14,346,350 Showtime subscribers.

29. Graph the line $x+2y=4$ as a dashed line since the inequality symbol is <. Test the point (0, 0). Since (0, 0) is a solution of the inequality $x+2y<4$, shade the region containing (0, 0).

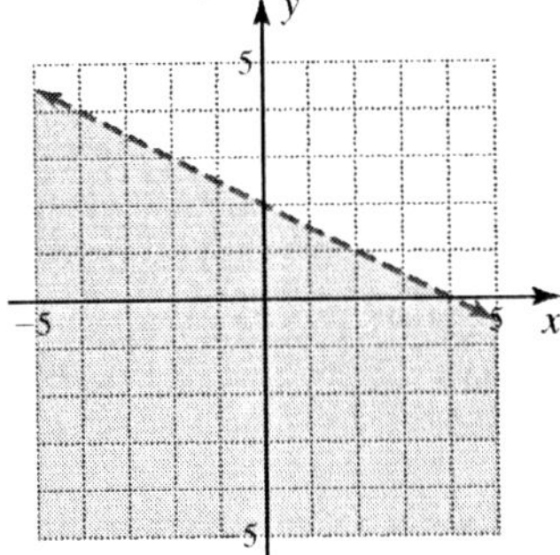

31. Graph the line $x=y$ as a dashed line since the inequality symbol is >. Test the point (1, 4). Since (1, 4) is not a solution of the inequality $x>y$, shade the region that does not contain (1, 4).

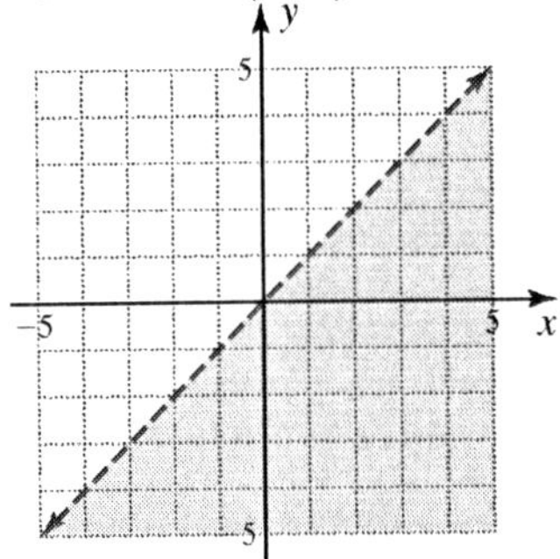

33. Answers may vary.

35. Answers may vary.

37. We are asked to find the numbers of nickels and dimes. Let n be the number of nickels and d the number of dimes. Translate the problem into a system of equations.

$0.05n+0.10d=2$

$n=2d$

Substitute $2d$ for n in the first equation.

$0.05(2d)+0.10d=2$

$0.10d+0.10d=2$

$0.20d=2$

$d=10$

Substitute 10 for d in the equation $n=2d$.

$n=2(10)=20$

Jill has 20 nickels and 10 dimes.

39. We are asked to find the speed of the plane in still air and the speed of the wind. Let x be the speed of the plane in still air and y be the speed of the wind. Make a chart.

	R	$\times$	T	$=$	D
With the wind:	$x+y$		3		1200
Against the wind:	$x-y$		4		1200

Write the system of equations.

$3(x+y)=1200$

$4(x-y)=1200$

Divide the first equation by 3, and divide the second equation by 4.

$$\begin{array}{rcrcl} x & + & y & = & 400 \\ x & - & y & = & 300 \\ \hline 2x & & & = & 700 \\ & & x & = & 350 \end{array}$$

Substitute 350 for x in the equation $x+y=400$.

$350+y=400$

$y=50$

The speed of the plane in still air is 350 miles per hour, and the speed of the wind is 50 miles per hour.

41. We are asked to find the number of votes each of the tickets received. Let x be the number received by the Democratic ticket and y be the number received by the Republican ticket. Write a system of equations.

$$\begin{array}{rcrcl} x & + & y & = & 84 \\ x & - & y & = & 6 \\ \hline 2x & & & = & 90 \\ & & x & = & 45 \end{array}$$

Substitute 45 for x in the equation $x+y=84$.

$45+y=84$

$y=39$

The Democratic ticket received 45 million votes, and the Republican ticket received 39 million votes.

8.5 Systems of Linear Inequalities

Problems 8.5

1. Since $x=2$ is a vertical line, $x \leq 2$ consists of the graph of the line $x=2$ and all points to the left. The condition $y \geq 0$ defines all points on the line $y=0$ (the x-axis) and above. The solution set is the set satisfying both conditions.

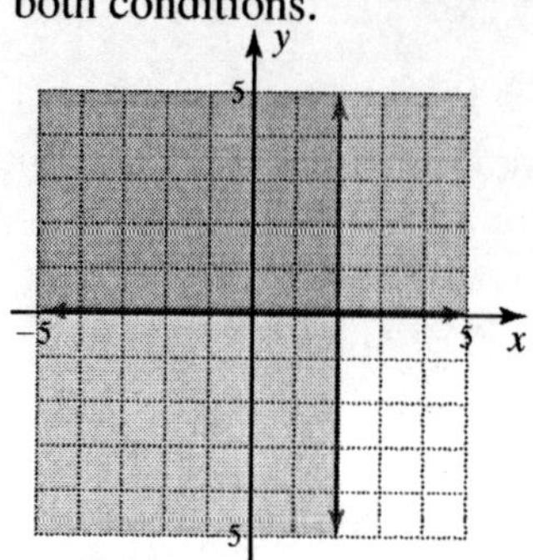

2. Graph the lines $2x+y=5$ (solid) and $x-y=3$ (dashed). Using (0, 0) as a test point, $2x+y \leq 5$ becomes $2 \cdot 0+0 \leq 5$, a true statement. So we shade the region containing (0, 0): the points on or below the line $2x+y=5$. The inequality $x-y<3$ is also satisfied by the test point (0, 0), so we shade the points above the line $x-y=3$. The solution set is the set satisfying both conditions.

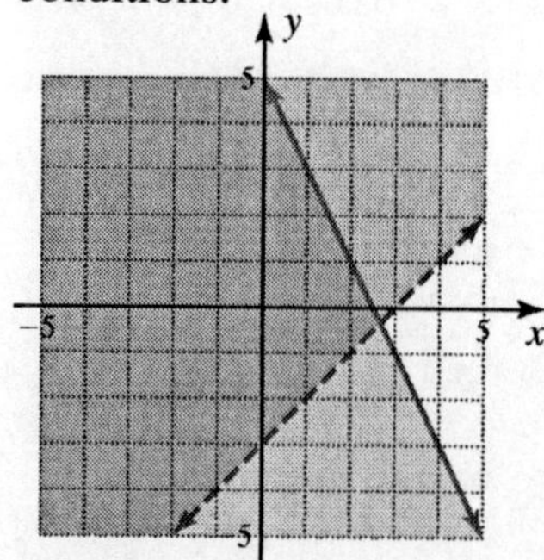

3. Graph the lines $y+x=3$ (solid) and $y-x=3$ (solid). Since the test point (0, 0) satisfies the inequality $y+x \leq 3$, we shade the region containing (0, 0): the points on or below the line $y+x=3$. Since the test point (0, 0) does not satisfy the inequality $y-x \geq 3$, we shade the region that does not contain (0, 0): the points on or above the line $y-x=3$. The solution set is the set satisfying both conditions.

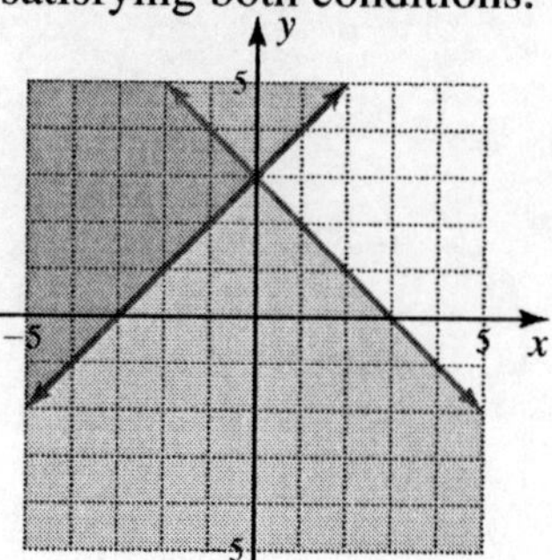

Exercises 8.5

1. Since $x=0$ is a vertical line, $x \geq 0$ consists of the graph of the line $x=0$ and all points to the right. The condition $y \leq 2$ defines all points on the line $y=2$ and below. The solution set is the set satisfying both conditions.

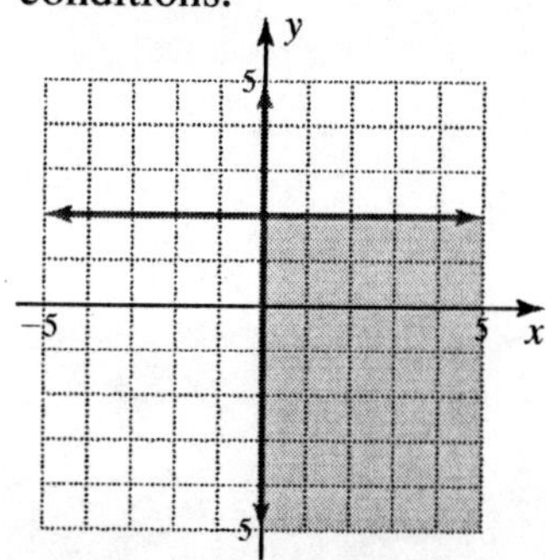

3. Since $x=-1$ is a vertical line, $x<-1$ consists all points to the left of the line $x=-1$. The condition $y>-2$ defines all points above the line $y=-2$. The solution set is the set satisfying both conditions.

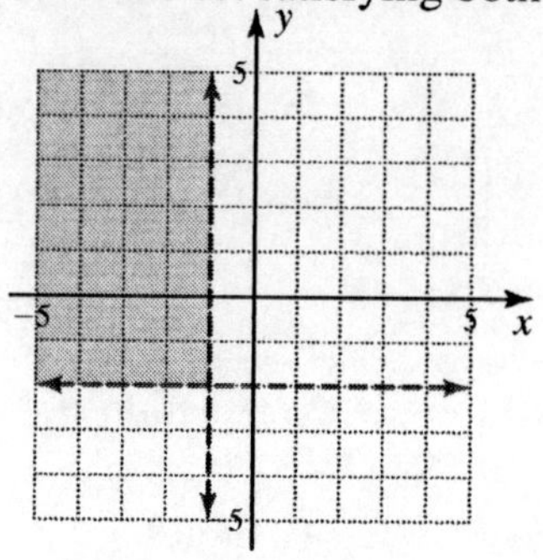

5. Graph the lines $x+2y=3$ (solid) and $x=y$ (dashed). Since the test point (0, 0) satisfies the inequality $x+2y\le 3$, we shade the region containing (0, 0): the points on or below the line $x+2y=3$. Since the test point (0, 1) satisfies the inequality $x<y$, we shade the region containing (0, 1): the points above the line $x=y$. The solution set is the set satisfying both conditions.

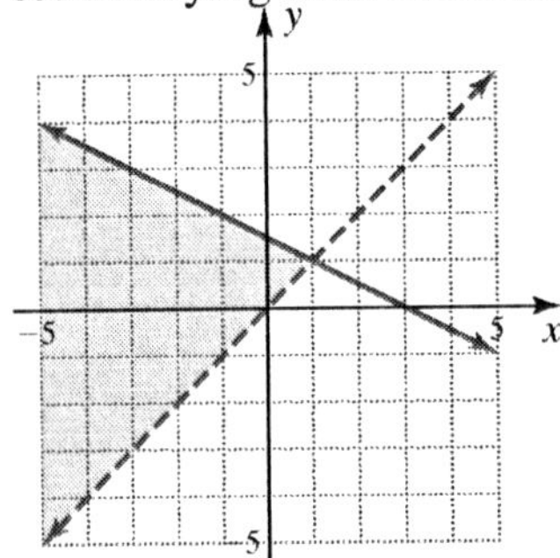

7. Graph the lines $4x-y=-1$ (dashed) and $-2x-y=-3$ (solid). Since the test point (0, 0) satisfies the inequality $4x-y>-1$, we shade the region containing (0, 0): the points below the line $4x-y=-1$. Since the test point (0, 0) does not satisfy the inequality $-2x-y\le -3$, we shade the region that does not contain (0, 0): the points on or above the line $-2x-y=-3$. The solution set is the set satisfying both conditions.

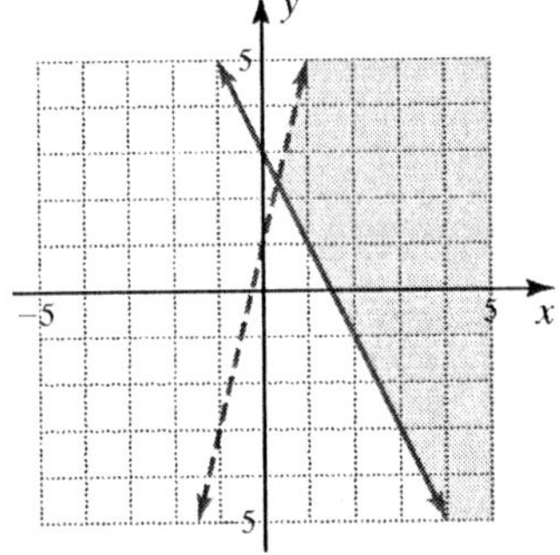

9. Graph the lines $-2x+y=3$ (dashed) and $5x-y=-10$ (solid). Since the test point (0, 0) does not satisfy the inequality $-2x+y>3$, we shade the region that does not contain (0, 0): the points above the line $-2x+y=3$. Since the test point (0, 0) does not satisfy the inequality $5x-y\le -10$, we shade the region that does not contain (0, 0): the points on or above the line $5x-y=-10$. The solution set is the set satisfying both conditions.

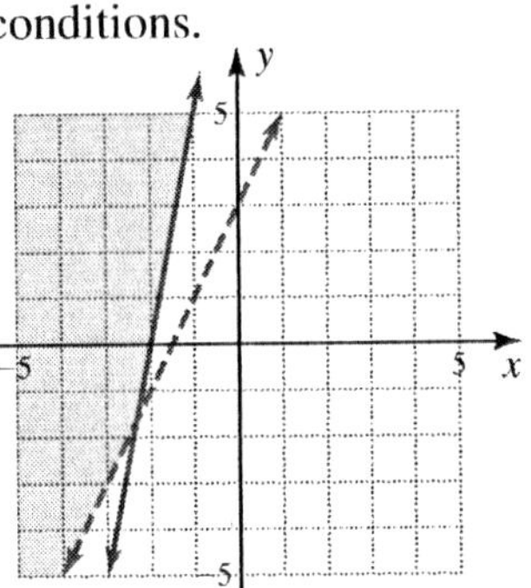

11. Graph the lines $2x-3y=5$ (dashed) and $x=y$ (solid). Since the test point (0, 0) satisfies the inequality $2x-3y<5$, we shade the region containing (0, 0): the points above the line $2x-3y=5$. Since the test point (0, 1) does not satisfy the inequality $x\ge y$, we shade the region that does not contain (0, 1): the points on or below the line $x=y$. The solution set is the set satisfying both conditions.

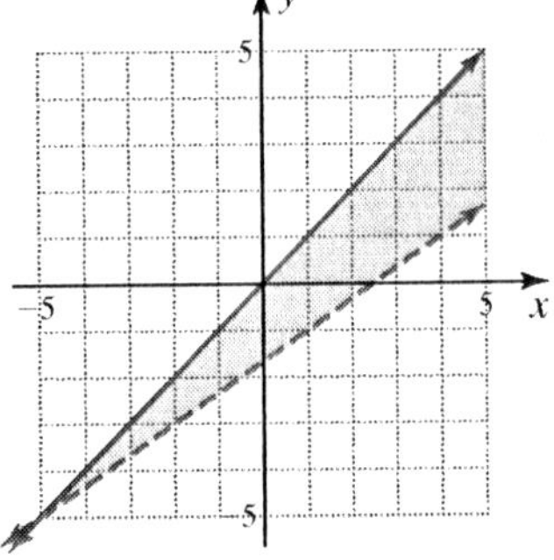

13. Graph the lines $x+3y=6$ (solid) and $x=y$ (dashed). Since the test point (0, 0) satisfies the inequality $x+3y\le 6$, we shade the region containing (0, 0): the points on or below the line $x+3y=6$. Since the test point (0, 1) does not satisfy the inequality $x>y$, we shade the region that does not contain (0, 1): the points below the line $x=y$. The solution set is the set satisfying both conditions.

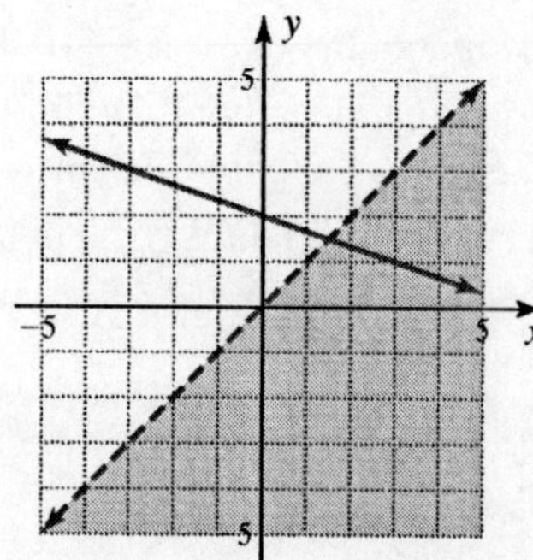

15. Graph the lines $3x + y = 6$ (dashed) and $x = y$ (slolid). Since the test point (0, 0) does not satisfy the inequality $3x + y > 6$, we shade the region that does not contain (0, 0): the points above the line $3x + y = 6$. Since the test point (0, 1) satisfies the inequality $x \leq y$, we shade the region containing (0, 1): the points on or above the line $x = y$. The solution set is the set satisfying both conditions.

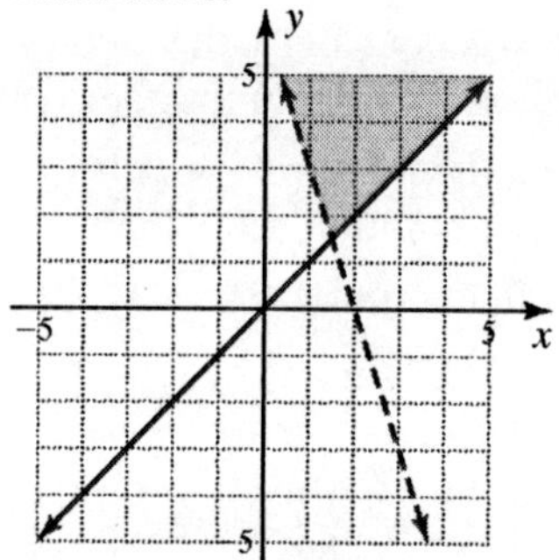

17. $\sqrt{64} = \sqrt{8^2} = 8$

19. Graph $p = -\frac{2a}{3} + 150$. Shade above the line. See graph below.

21. Graph $a = 10$ and $a = 70$. Shade between the lines. See graph below.

19–21.

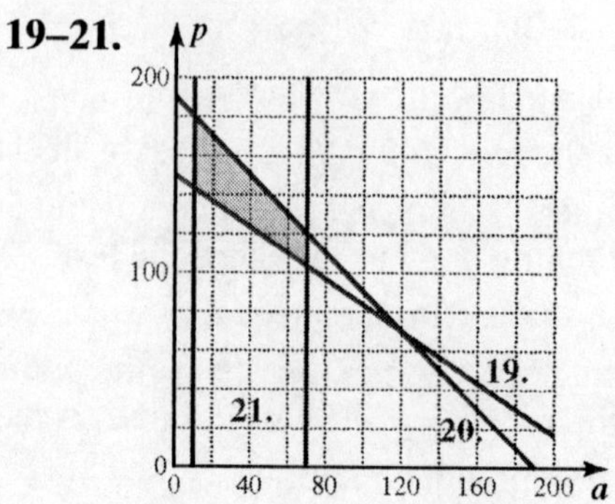

23. Answers may vary. Sample answer: the points on or to the right of the line $x = k$.

25. Since $x = 2$ is a vertical line, $x > 2$ consists all points to the right of the line $x = 2$. The condition $y < 3$ defines all points below the line $y = 3$. The solution set is the set satisfying both conditions.

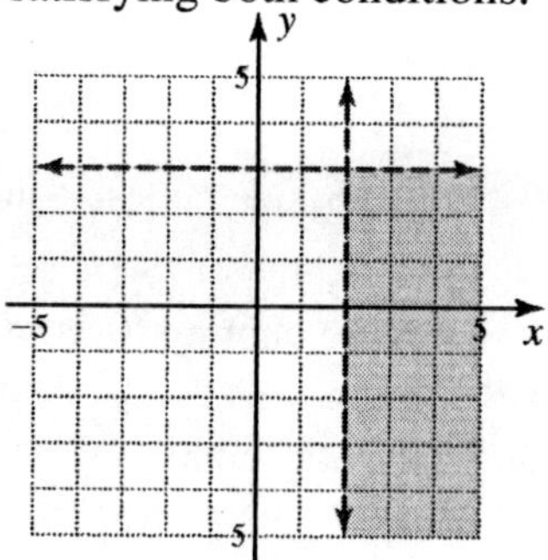

27. Graph the lines $3x - y = -1$ (dashed) and $x + 2y = 2$ (solid). Since the test point (0, 0) does not satisfy the inequality $3x - y < -1$, we shade the region that does not contain (0, 0): the points above the line $3x - y = -1$. Since the test point (0, 0) satisfies the inequality $x + 2y \leq 2$, we shade the region containing (0, 0): the points on or below the line $x + 2y = 2$. The solution set is the set satisfying both conditions.

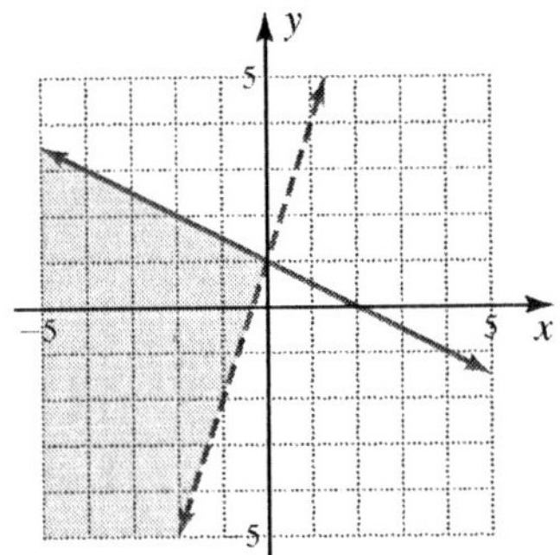

Review Exercises

1. a. Graph the equation $2x + y = 4$ using the points (0, 4) and (2, 0). Then graph the equation $y - 2x = 0$ using the points (0, 0) and (1, 2).

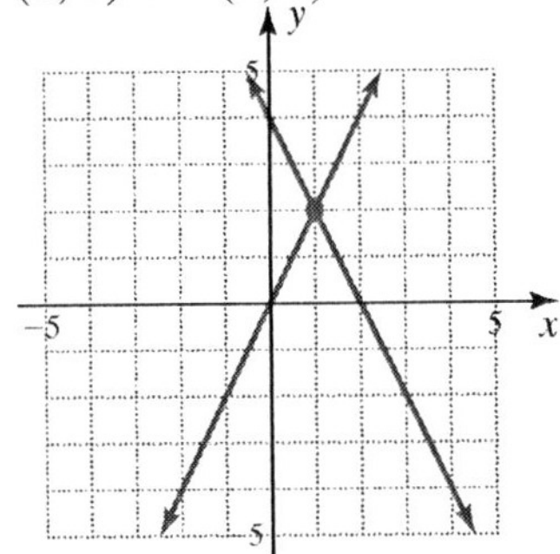

The system is consistent; the solution is (1, 2).

b. Graph the equation $x + y = 4$ using the points (0, 4) and (4, 0). Then graph the equation $y - x = 0$ using the points (0, 0) and (1, 1).

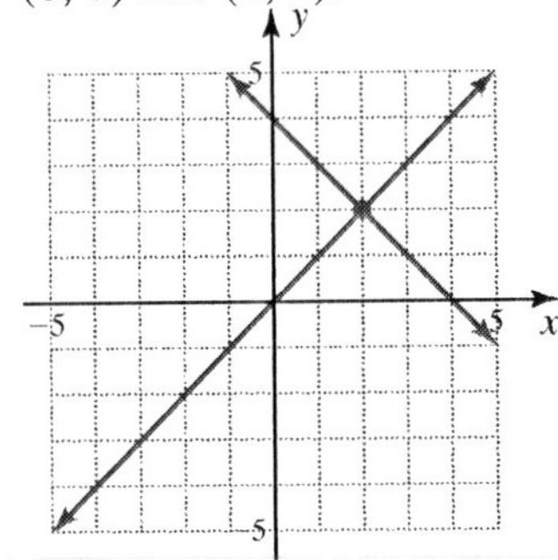

The system is consistent; the solution is (2, 2).

c. Graph the equation $x + y = 4$ using the points (0, 4) and (4, 0). Then graph the equation $y - 3x = 0$ using the points (0, 0) and (1, 3).

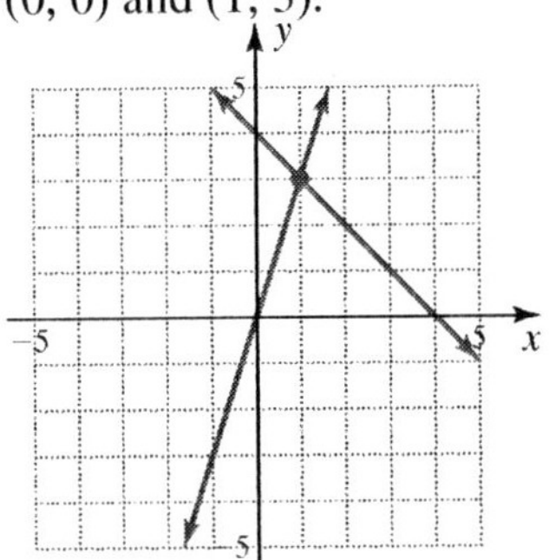

The system is consistent; the solution is (1, 3).

2. a. Graph the equation $y - 3x = 3$ using the points (0, 3) and (−1, 0). Then graph the equation $2y - 6x = 12$ using the points (−1, 3) and (−2, 0).

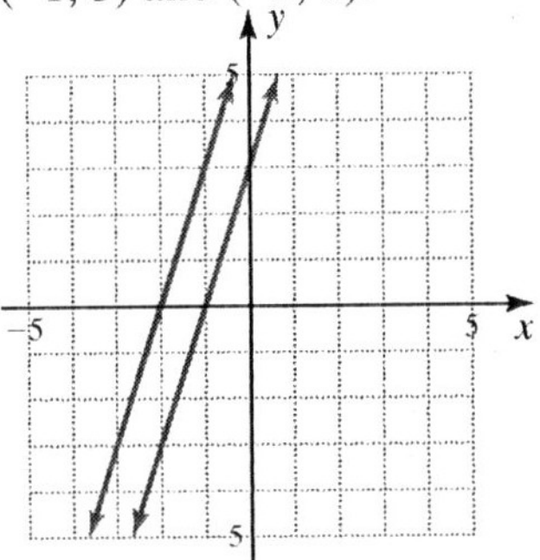

The system is inconsistent; there is no solution.

b. Graph the equation $y - 2x = 2$ using the points (0, 2) and (−1, 0). Then graph the equation $2y - 4x = 8$ using the points (0, 4) and (−2, 0).

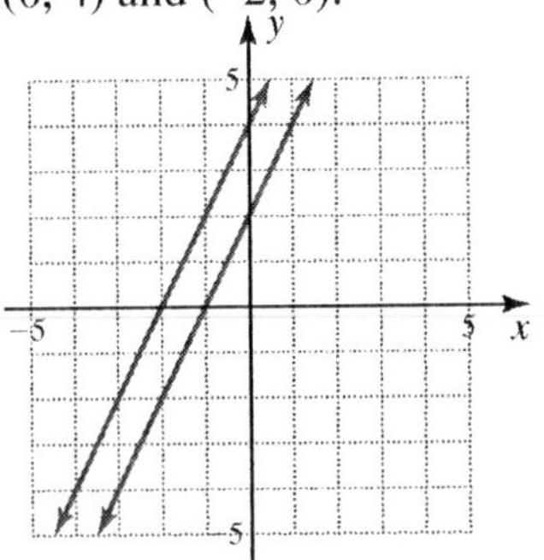

The system is inconsistent; there is no solution.

c. Graph the equation $y-3x=6$ using the points (–1, 3) and (–2, 0). Then graph the equation $2y-6x=6$ using the points (0, 3) and (–1, 0).

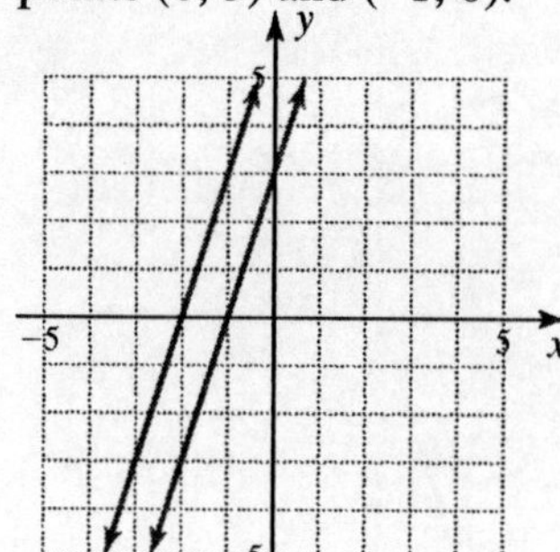

The system is inconsistent; there is no solution.

3. a. Solve the first equation for *x*: $x=5-4y$. Substitute $5-4y$ for *x* in $2x+8y=15$.

$$2(5-4y)+8y=15$$
$$10-8y+8y=15$$
$$10=15$$

The result, $10=15$, is never true. The system has no solution; it is inconsistent.

b. Solve the first equation for *x*: $x=6-3y$. Substitute $6-3y$ for *x* in $3x+9y=12$.

$$3(6-3y)+9y=12$$
$$18-9y+9y=12$$
$$18=12$$

The result, $18=12$, is never true. The system has no solution; it is inconsistent.

c. Solve the first equation for *x*: $x=5-4y$. Substitute $5-4y$ for *x* in $2x+13y=15$.

$$2(5-4y)+13y=15$$
$$10-8y+13y=15$$
$$5y=5$$
$$y=1$$

Substitute 1 for *y* in $x=5-4y$.

$$x=5-4(1)=1$$

The solution is (1, 1).

4. a. Solve the first equation for *x*: $x=5-4y$. Substitute $5-4y$ for *x* in $2x+8y=10$.

$$2(5-4y)+8y=10$$
$$10-8y+8y=10$$
$$10=10$$

The result, $10=10$, is always true. The system has infinitely many solutions; it is dependent.

b. Solve the first equation for *x*: $x=6-3y$. Substitute $6-3y$ for *x* in $3x+9y=18$.

$$3(6-3y)+9y=18$$
$$18-9y+9y=18$$
$$18=18$$

The result, $18=18$, is always true. The system has infinitely many solutions; it is dependent.

c. Solve the first equation for *x*: $x=5-4y$. Substitute $5-4y$ for *x* in $-2x-8y=-10$.

$$-2(5-4y)-8y=-10$$
$$-10+8y-8y=-10$$
$$-10=-10$$

The result, $-10=-10$, is always true. The system has infinitely many solutions; it is dependent.

5. a. Multiply the second equation by –2. Leave the first equation as is.

$$\begin{array}{rcrcr} 3x & + & 2y & = & 1 \\ -4x & - & 2y & = & 0 \\ \hline -x & & & = & 1 \\ & & x & = & -1 \end{array}$$

Substitute –1 for *x* in $2x+y=0$.

$$2(-1)+y=0$$
$$y=2$$

The solution of the system is (–1, 2).

b. Multiply the second equation by –2. Leave the first equation as is.

$$\begin{array}{rcrcr} 3x & + & 2y & = & 4 \\ -4x & - & 2y & = & -6 \\ \hline -x & & & = & -2 \\ & & x & = & 2 \end{array}$$

Substitute 2 for x in $2x + y = 3$.

$2(2) + y = 3$

$y = -1$

The solution of the system is (2, −1).

c. Multiply the second equation by −2. Leave the first equation as is.

$$\begin{aligned} 3x + 2y &= -7 \\ -4x - 2y &= 8 \\ \hline -x &= 1 \\ x &= -1 \end{aligned}$$

Substitute −1 for x in $2x + y = -4$.

$2(-1) + y = -4$

$y = -2$

The solution of the system is (−1, −2).

6. a. Multiply the first equation by 2. Leave the second equation as is.

$$\begin{aligned} 4x - 6y &= 12 \\ -4x + 6y &= -2 \\ \hline 0 &= 10 \end{aligned}$$

There is no solution; the system is inconsistent.

b. Multiply the first equation by 3. Leave the second equation as is.

$$\begin{aligned} 9x - 6y &= 24 \\ -9x + 6y &= -4 \\ \hline 0 &= 20 \end{aligned}$$

There is no solution; the system is inconsistent.

c. Leave both equations as is.

$$\begin{aligned} 3x - 5y &= 6 \\ -3x + 5y &= -12 \\ \hline 0 &= -6 \end{aligned}$$

There is no solution; the system is inconsistent.

7. a. Multiply the first equation by −2. Leave the second equation as is.

$$\begin{aligned} -6y - 4x &= -2 \\ 6y + 4x &= 2 \\ \hline 0 &= 0 \end{aligned}$$

There are infinitely many solutions; the system is dependent.

b. Multiply the first equation by −2 and rewrite in standard form.

$$\begin{aligned} -6x - 4y &= -2 \\ 6x + 4y &= 2 \\ \hline 0 &= 0 \end{aligned}$$

There are infinitely many solutions; the system is dependent.

c. Multiply the first equation by −2 and rewrite in standard form.

$$\begin{aligned} -8x - 6y &= 22 \\ 8x + 6y &= -22 \\ \hline 0 &= 0 \end{aligned}$$

There are infinitely many solutions; the system is dependent.

8. Let n be the number of nickels and d the number of dimes.

a. Translate the problem into a system of equations.

$0.05n + 0.10d = 3$

$d = n$

Substitute n for d in the first equation.

$0.05n + 0.10n = 3$

$0.15n = 3$

$n = 20$

Substitute 20 for n in the equation $d = n$: $d = 20$.

Desi has 20 nickels and 20 dimes.

b. Translate the problem into a system of equations.

$0.05n + 0.10d = 3$

$n = 4d$

Substitute $4d$ for n in the first equation.

$0.05(4d) + 0.10d = 3$

$0.20d + 0.10d = 3$

$0.30d = 3$

$d = 10$

Substitute 10 for d in the equation $n = 4d$.

$n = 4(10) = 40$

Desi has 40 nickels and 10 dimes.

c. Translate the problem into a system of equations.

$0.05n + 0.10d = 3$

$n = 10d$

Substitute $10d$ for n in the first equation.

$0.05(10d) + 0.10d = 3$

$0.50d + 0.10d = 3$

$0.60d = 3$

$d = 5$

Substitute 5 for d in the equation $n = 10d$.

$n = 10(5) = 50$

Desi has 50 nickels and 5 dimes.

9. Let x and y be the numbers.

a. Translate the problem into a system of equations.

$$\begin{array}{rcl} x + y &=& 180 \\ x - y &=& 40 \\ \hline 2x &=& 220 \\ x &=& 110 \end{array}$$

Substitute 110 for x in $x + y = 180$.

$110 + y = 180$

$y = 70$

The numbers are 70 and 110.

b. Translate the problem into a system of equations.

$$\begin{array}{rcl} x + y &=& 180 \\ x - y &=& 60 \\ \hline 2x &=& 240 \\ x &=& 120 \end{array}$$

Substitute 120 for x in $x + y = 180$.

$120 + y = 180$

$y = 60$

The numbers are 60 and 120.

c. Translate the problem into a system of equations.

$$\begin{array}{rcl} x + y &=& 180 \\ x - y &=& 80 \\ \hline 2x &=& 260 \\ x &=& 130 \end{array}$$

Substitute 130 for x in $x + y = 180$.

$130 + y = 180$

$y = 50$

The numbers are 50 and 130.

10. We are asked to find the speed of the plane in still air. Let x be the speed of the plane in still air and y be the speed of the wind.

a. Make a chart.

	R ×	T =	D
With the wind:	$x + y$	3	2400
Against the wind:	$x - y$	8	2400

Write the system of equations.

$3(x + y) = 2400$

$8(x - y) = 2400$

Divide the first equation by 3 and the second equation by 8.

$$\begin{array}{rcl} x + y &=& 800 \\ x - y &=& 300 \\ \hline 2x &=& 1100 \\ x &=& 550 \end{array}$$

The speed of the plane in still air was 550 miles per hour.

b. Make a chart.

	R ×	T =	D
With the wind:	$x + y$	3	2400
Against the wind:	$x - y$	10	2400

Write the system of equations.

$3(x + y) = 2400$

$10(x - y) = 2400$

Divide the first equation by 3 and the second equation by 10.

$$\begin{array}{rcl} x + y &=& 800 \\ x - y &=& 240 \\ \hline 2x &=& 1040 \\ x &=& 520 \end{array}$$

The speed of the plane in still air was 520 miles per hour.

c. Make a chart.

	R ×	T =	D
With the wind:	$x+y$	3	2400
Against the wind:	$x-y$	12	2400

Write the system of equations.

$3(x+y)=2400$

$12(x-y)=2400$

Divide the first equation by 3 and the second equation by 12.

$$\begin{array}{rcrcl} x & + & y & = & 800 \\ x & - & y & = & 200 \\ \hline 2x & & & = & 1000 \\ & & x & = & 500 \end{array}$$

The speed of the plane in still air was 500 miles per hour.

11. Let x be the amount in bonds and y the amount in certificates of deposit. Make a chart.

	P ×	R =	I
Bonds:	x	0.05	$0.05x$
CDs:	y	0.10	$0.10y$

a. Write the system of equations

$x+y=20,000$

$0.05x+0.10y=1750$

Multiply the first equation by −0.05.

$$\begin{array}{rcrcr} -0.05x & - & 0.05y & = & -1000 \\ 0.05x & + & 0.10y & = & 1750 \\ \hline & & 0.05y & = & 750 \\ & & y & = & 15,000 \end{array}$$

Substitute 15,000 for y in the equation $x+y=20,000$.

$x+15,000=20,000$

$x=5000$

The investor had $5000 in bonds and $15,000 in CDs.

b. Write the system of equations

$x+y=20,000$

$0.05x+0.10y=1150$

Multiply the first equation by −0.05.

$$\begin{array}{rcrcr} -0.05x & - & 0.05y & = & -1000 \\ 0.05x & + & 0.10y & = & 1150 \\ \hline & & 0.05y & = & 150 \\ & & y & = & 3000 \end{array}$$

Substitute 3000 for y in the equation $x+y=20,000$.

$x+3000=20,000$

$x=17,000$

The investor had $17,000 in bonds and $3,000 in CDs.

c. Write the system of equations

$x+y=20,000$

$0.05x+0.10y=1500$

Multiply the first equation by −0.05.

$$\begin{array}{rcrcr} -0.05x & - & 0.05y & = & -1000 \\ 0.05x & + & 0.10y & = & 1500 \\ \hline & & 0.05y & = & 500 \\ & & y & = & 10,000 \end{array}$$

Substitute 10,000 for y in the equation $x+y=20,000$.

$x+10,000=20,000$

$x=10,000$

The investor had $10,000 in bonds and $10,000 in CDs.

12. a. Since $x=4$ is a vertical line, $x>4$ consists all points to the right of the line $x=4$. The condition $y<-1$ defines all points below the line $y=-1$. The solution set is the set satisfying both conditions.

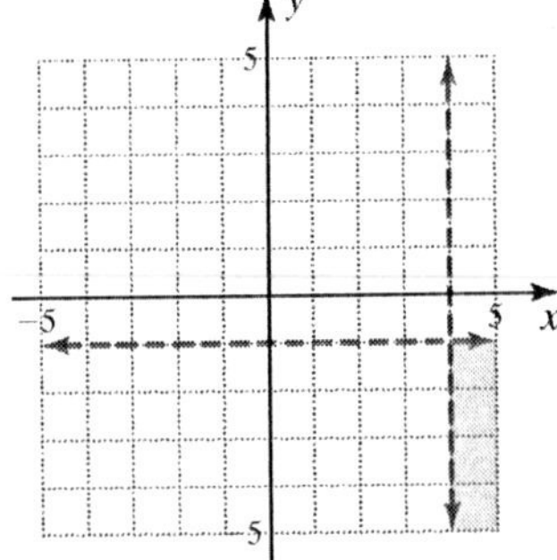

b. Graph the lines $x+y=3$ (dashed) and $x-y=4$ (dashed). Since the test point (0, 0) does not satisfy the inequality $x+y>3$, we shade the region that does not contain (0, 0): the points above the line $x+y=3$. Since the test point (0, 0) satisfies the inequality $x-y<4$, we shade the region containing (0, 0): the points above the line $x-y=4$. The solution set is the set satisfying both conditions.

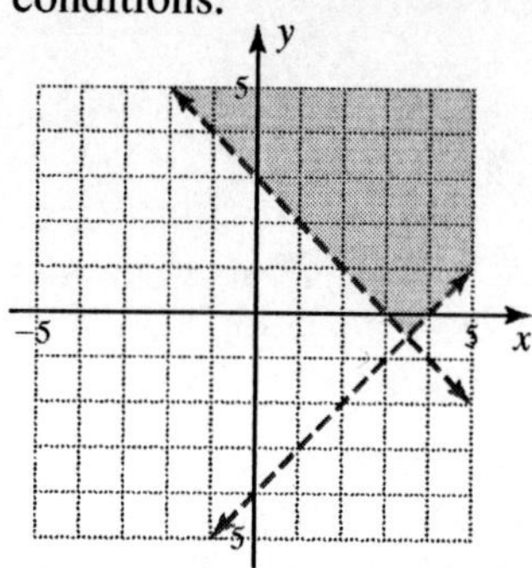

c. Graph the lines $2x+y=4$ (solid) and $x-2y=2$ (dashed). Since the test point (0, 0) satisfies the inequality $2x+y\le 4$, we shade the region containing (0, 0): the points on or below the line $2x+y=4$. Since the test point (0, 0) does not satisfy the inequality $x-2y>2$, we shade the region that does not contain (0, 0): the points below the line $x-2y=2$. The solution set is the set satisfying both conditions.

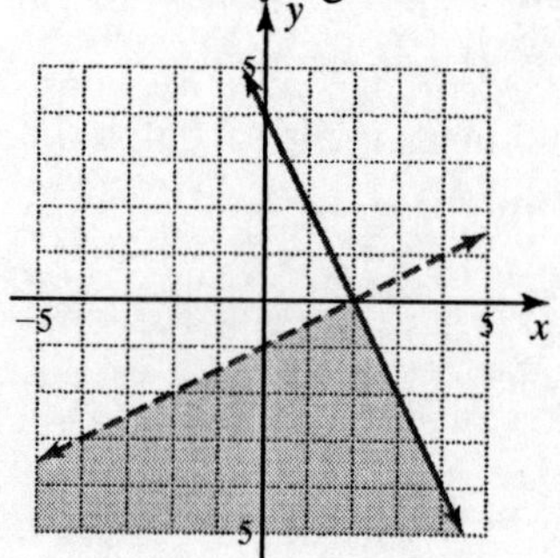

Cumulative Review Chapters 1–8

1. $-\frac{3}{7}+\left(-\frac{1}{6}\right)=-\frac{18}{42}+\left(-\frac{7}{42}\right)=-\frac{25}{42}$

2. $(-4)^4=(-4)(-4)(-4)(-4)=256$

3. $-\frac{1}{4}\div\left(-\frac{1}{8}\right)=-\frac{1}{4}\cdot\left(-\frac{8}{1}\right)=2$

4. $$\begin{aligned} y\div 2\cdot x-z&=24\div 2\cdot 6-3\\ &=12\cdot 6-3\\ &=72-3\\ &=69\end{aligned}$$

5. $$\begin{aligned} &2x-(x+4)-2(x+3)\\ &=2x-x-4-2x-6\\ &=2x-x-2x-4-6\\ &=-x-10\end{aligned}$$

6. $\frac{x-5y}{z}$

7. $$\begin{aligned} 3&=3(x-2)+3-2x\\ 3&=3x-6+3-2x\\ 3&=x-3\\ 6&=x \text{ or } x=6\end{aligned}$$

8. $$\begin{aligned} \frac{x}{4}-\frac{x}{9}&=5\\ 36\left(\frac{x}{4}-\frac{x}{9}\right)&=36(5)\\ 9x-4x&=180\\ 5x&=180\\ x&=36\end{aligned}$$

9. $$\begin{aligned} -\frac{x}{3}+\frac{x}{8}&\ge\frac{x-8}{8}\\ 24\left(-\frac{x}{3}+\frac{x}{8}\right)&\ge 24\left(\frac{x-8}{8}\right)\\ -8x+3x&\ge 3x-24\\ -8x&\ge -24\\ x&\le 3\end{aligned}$$

0 3

10. Starting at the origin, move 1 unit to the right and 2 units down.

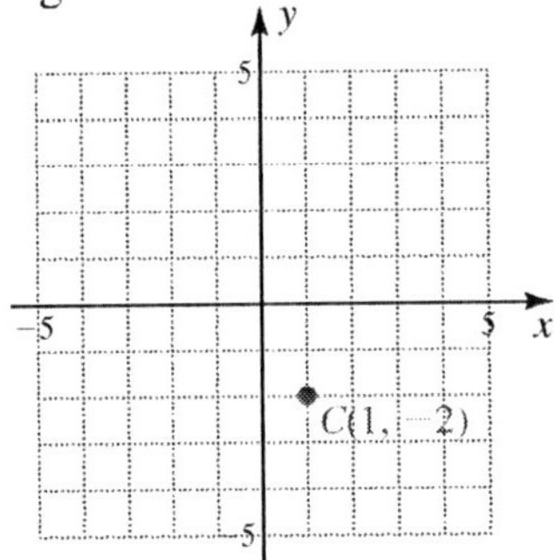

11. $4(-1)+3(-1)=-1$
$$-4-3=-1$$
$$-7=-1$$
No, $(-1, -1)$ is not a solution of the equation.

12. $3x-2(2)=-13$
$$3x-4=-13$$
$$3x=-9$$
$$x=-3$$

13. Let $x=0$. Then $y=1$. Graph (0, 1). Let $y=0$. Then $x=1$. Graph (1, 0). Draw the line through the points.

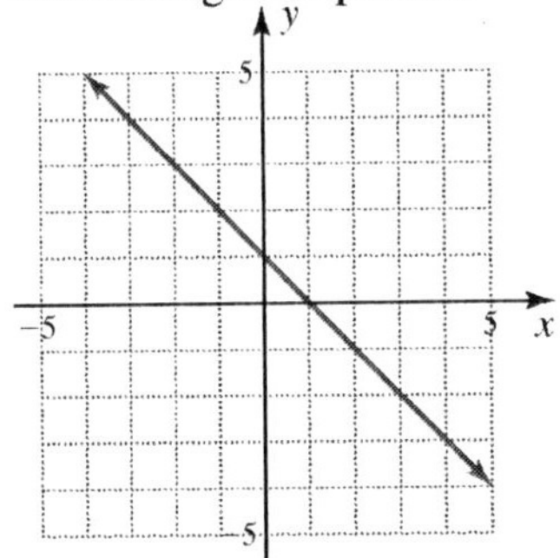

14. The graph of $2y+8=0$ is the horizontal line passing through (0, –4).

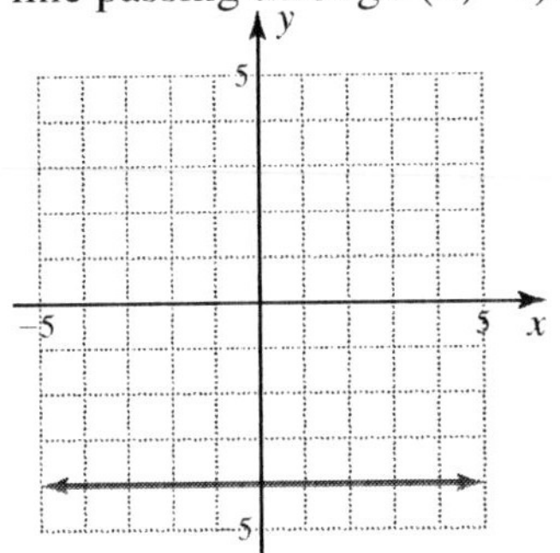

15. $m=\dfrac{-4-(-3)}{5-(-2)}=\dfrac{-1}{7}=-\dfrac{1}{7}$

16. $12x-4y=14$
$$-4y=-12x+14$$
$$y=3x-\frac{7}{2}$$
The slope is 3.

17. (1) $2y=-x+6$
$$y=-\frac{1}{2}x+3$$

(2) $-4x-8y=6$
$$-8y=4x+6$$
$$y=-\frac{1}{2}x-\frac{3}{4}$$

(3) $8y-4x=6$
$$8y=4x+6$$
$$y=\frac{1}{2}x+\frac{3}{4}$$

Lines (1) and (2) both have slope $-\frac{1}{2}$, so they are parallel.

18. $(2x^4y^{-3})^{-3}=2^{-3}x^{-12}y^9=\dfrac{y^9}{8x^{12}}$

19. $0.000035=3.5\times10^{-5}$

20. $\dfrac{5.46\times10^{-3}}{2.6\times10^4}=\dfrac{5.46}{2.6}\times10^{-3-4}=2.1\times10^{-7}$

21. $\left(6x^2-\dfrac{1}{4}\right)^2=(6x^2)^2-2(6x^2)\left(\dfrac{1}{4}\right)+\left(\dfrac{1}{4}\right)^2$
$$=36x^4-3x^2+\frac{1}{16}$$

22.

$$\begin{array}{r|l} & 2x^2 + x - 1 \\ \hline x-1 & 2x^3 - x^2 - 2x + 4 \\ & \underline{2x^3 - 2x^2} \\ & x^2 - 2x \\ & \underline{x^2 - x} \\ & -x + 4 \\ & \underline{-x + 1} \\ & 3 \end{array}$$

The quotient is $(2x^2 + x - 1)$ R 3.

23. $x^2 - 11x + 30 = (x-6)(x-5)$

24. $15x^2 - 29xy + 12y^2 = (3x - 4y)(5x - 3y)$

25. $81x^2 - 64y^2 = (9x)^2 - (8y)^2 = (9x + 8y)(9x - 8y)$

26. $-3x^4 + 12x^2 = -3x^2(x^2 - 4) = -3x^2(x^2 - 2^2) = -3x^2(x+2)(x-2)$

27. $4x^3 - 4x^2 - 8x = 4x(x^2 - x - 2) = 4x(x+1)(x-2)$

28. $4x^2 + 3x + 4x + 3 = x(4x+3) + 1(4x+3) = (4x+3)(x+1)$

29. $25kx^2 - 30kx + 9k = k(25x^2 - 30x + 9) = k[(5x)^2 - 2(5x)(3) + 3^2] = k(5x-3)^2$

30.

$$3x^2 + 4x = 15$$
$$3x^2 + 4x - 15 = 0$$
$$(3x - 5)(x + 3) = 0$$
$$3x - 5 = 0 \quad \text{or} \quad x + 3 = 0$$
$$3x = 5 \qquad x = -3$$
$$x = \frac{5}{3}$$

31. $\frac{4x}{3y} \cdot \frac{5y}{5y} = \frac{20xy}{15y^2}$

32. $\frac{-9(x^2 - y^2)}{3(x-y)} = \frac{-9(x+y)(x-y)}{3(x-y)} = -3(x+y)$

33. $\frac{x^2 - 4x - 12}{6 - x} = -\frac{(x-6)(x+2)}{x-6} = -(x+2)$

34. $(x-5)\cdot\dfrac{x-2}{x^2-25}=\dfrac{x-5}{1}\cdot\dfrac{x-2}{(x+5)(x-5)}=\dfrac{x-2}{x+5}$

35. $\dfrac{x+4}{x-4}\div\dfrac{x^2-16}{4-x}=\dfrac{x+4}{x-4}\cdot\dfrac{4-x}{(x+4)(x-4)}=-\dfrac{1}{x-4}$ or $\dfrac{1}{4-x}$

36. $\dfrac{9}{2(x-9)}+\dfrac{3}{2(x-9)}=\dfrac{12}{2(x-9)}=\dfrac{6}{x-9}$

37.
$$\begin{aligned}\frac{x+2}{x^2+x-6}-\frac{x+3}{x^2-4}&=\frac{x+2}{(x+3)(x-2)}-\frac{x+3}{(x+2)(x-2)}\\&=\frac{(x+2)(x+2)}{(x+3)(x+2)(x-2)}-\frac{(x+3)(x+3)}{(x+3)(x+2)(x-2)}\\&=\frac{(x^2+4x+4)-(x^2+6x+9)}{(x+3)(x+2)(x-2)}\\&=\frac{-2x-5}{(x+3)(x+2)(x-2)}\end{aligned}$$

38. $\dfrac{\frac{4}{x}-\frac{3}{4x}}{\frac{2}{3x}-\frac{1}{2x}}=\dfrac{12x\left(\frac{4}{x}-\frac{3}{4x}\right)}{12x\left(\frac{2}{3x}-\frac{1}{2x}\right)}=\dfrac{48-9}{8-6}=\dfrac{39}{2}$

39.
$$\begin{aligned}\frac{5x}{x-5}-2&=\frac{x}{x-5}\\(x-5)\left(\frac{5x}{x-5}-2\right)&=(x-5)\left(\frac{x}{x-5}\right)\\5x-2(x-5)&=x\\5x-2x+10&=x\\3x+10&=x\\2x&=-10\\x&=-5\end{aligned}$$

40.
$$\begin{aligned}\frac{x}{x^2-4}+\frac{2}{x-2}&=\frac{1}{x+2}\\(x+2)(x-2)\left(\frac{x}{(x+2)(x-2)}+\frac{2}{x-2}\right)&=(x+2)(x-2)\left(\frac{1}{x+2}\right)\\x+2(x+2)&=x-2\\x+2x+4&=x-2\\3x+4&=x-2\\2x&=-6\\x&=-3\end{aligned}$$

41.

$$\frac{x}{x+4}-\frac{1}{5}=\frac{-4}{x+4}$$

$$5(x+4)\left(\frac{x}{x+4}-\frac{1}{5}\right)=5(x+4)\left(\frac{-4}{x+4}\right)$$

$$5x-(x+4)=-20$$

$$5x-x-4=-20$$

$$4x=-16$$

$$x=-4$$

Since $x = -4$ makes the denominator $x + 4$ equal 0, −4 is an extraneous solution; there is no solution.

42.

$$2+\frac{8}{x-2}=\frac{32}{x^2-4}$$

$$(x+2)(x-2)\left(2+\frac{8}{x-2}\right)=(x+2)(x-2)\left(\frac{32}{(x+2)(x-2)}\right)$$

$$2(x+2)(x-2)+8(x+2)=32$$

$$2(x^2-4)+8(x+2)=32$$

$$2x^2-8+8x+16=32$$

$$2x^2+8x+8=32$$

$$2x^2+8x-24=0$$

$$x^2+4x-12=0$$

$$(x+6)(x-2)=0$$

$$x+6=0 \qquad x-2=0$$

$$x=-6 \qquad x=2$$

Since $x = 2$ makes the denominator $x - 2$ equal 0, 2 is an extraneous solution; −6 is the only solution.

43. Set up a proportion.

$$\frac{60}{4}=\frac{195}{x}$$

$$60x=780$$

$$x=13$$

The van needs 13 gallons of gas to travel 195 miles.

44.

$$\frac{x+1}{4}=\frac{9}{5}$$

$$5x+5=36$$

$$5x=31$$

$$x=\frac{31}{5}$$

45. Let t be the time it takes Sandra and Roger to paint the kitchen together.

$$\frac{t}{5}+\frac{t}{4}=1$$
$$20\left(\frac{t}{5}+\frac{t}{4}\right)=20(1)$$
$$4t+5t=20$$
$$9t=20$$
$$t=\frac{20}{9}$$

It takes $\frac{20}{9}$ or $2\frac{2}{9}$ hours for them to paint the kitchen together.

46.
$$y-y_1=m(x-x_1)$$
$$y-(-2)=-4[x-(-6)]$$
$$y+2=-4(x+6)$$
$$y+2=-4x-24$$
$$y=-4x-26$$

47. $y=mx+b$
$y=5x+2$

48. Graph the line $4x-3y=-12$ as a dashed line. Since the test point (0, 0) satisfies the inequality $4x-3y>-12$, shade the region containing (0, 0).

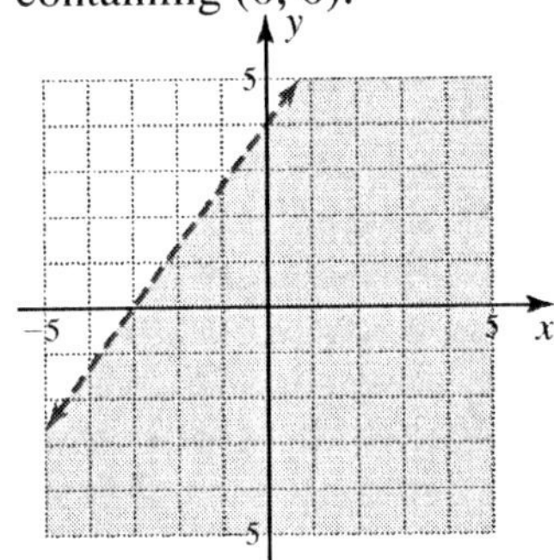

49. Graph the line $-y=-3x+6$ as a solid line. Since the test point (0, 0) satisfies the inequality $-y\le-3x+6$, shade the region containing (0, 0).

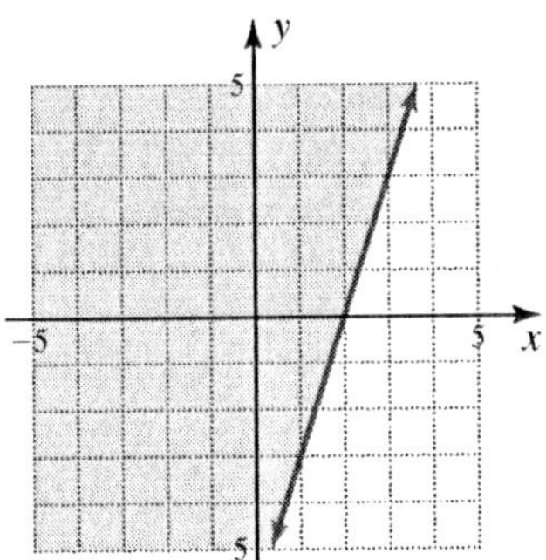

50. $P=kT$
Let $P=7$ and $T=350$.
$7=350k$
$\frac{7}{350}=k$
$\frac{1}{50}=k$

51. $P=\frac{k}{V}$
Let $P=1560$ and $V=4$.
$1560=\frac{k}{4}$
$6240=k$

52. Graph the equation $x+2y=6$ using the points (0, 3) and (2, 2). Then graph the equation $2y-x=-2$ using the points (0, −1) and (2, 0).

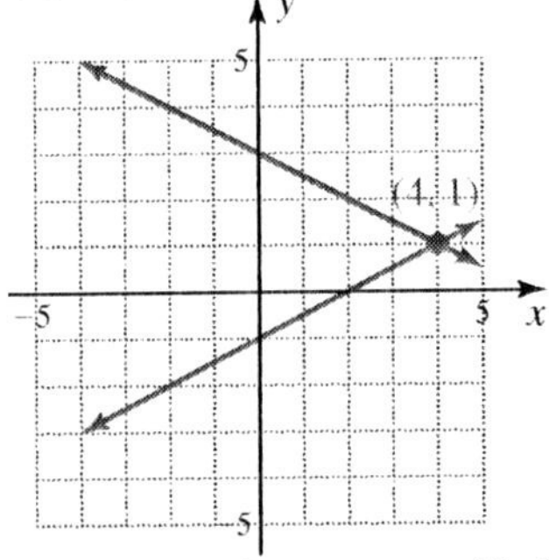

The graphs intersect at (4, 1), which is the solution of the system of equations.

53. Graph the equation $y-x=-1$ using the points (0, –1) and (1, 0). Then graph the equation $2y-2x=-4$ using the points (0, –2) and (2, 0).

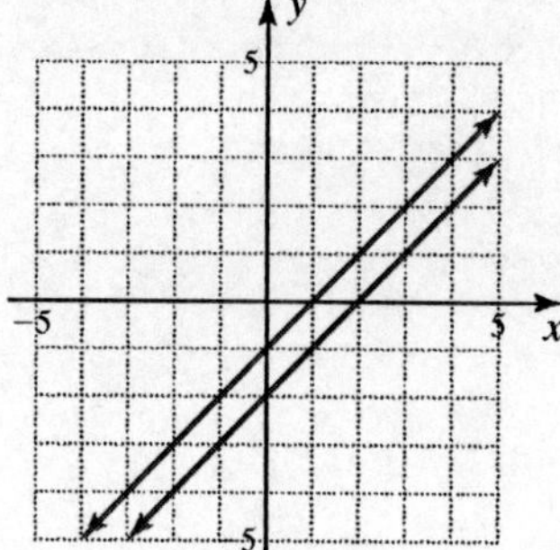

The lines are parallel. There is no solution; the system is inconsistent.

54. Solve the first equation, $x-4y=5$, for x: $x=4y+5$. Substitute $4y+5$ for x in the second equation, $-3x+12y=-17$.

$$-3(4y+5)+12y=-17$$
$$-12y-15+12y=-17$$
$$-15=-17$$

The result, $-15=-17$, is never true. The system is inconsistent; it has no solution.

55. Solve the first equation, $x-4y=-13$, for x: $x=4y-13$. Substitute $4y-13$ for x in the second equation, $3x-12y=-39$.

$$3(4y-13)-12y=-39$$
$$12y-39-12y=-39$$
$$-39=-39$$

The result, $-39=-39$, is always true. The system is dependent; it has infinitely many solutions.

56. Multiply the first equation by 2 and the second equation by –3.

$$\begin{array}{rcrcr} 6x & + & 8y & = & 14 \\ -6x & - & 9y & = & -15 \\ \hline & & -y & = & -1 \\ & & y & = & 1 \end{array}$$

Substitute 1 for y in $2x+3y=5$.

$$2x+3(1)=5$$
$$2x=2$$
$$x=1$$

The solution of the system is (1, 1).

57. Multiply the second equation by –1.

$$\begin{array}{rcrcr} 4x & - & 3y & = & 1 \\ -4x & + & 3y & = & 4 \\ \hline & & 0 & = & 5 \end{array}$$

The system is inconsistent; it has no solution.

58. Multiply the first equation by –2 and rewrite in standard form.

$$\begin{array}{rcrcr} 2x & - & 4y & = & 0 \\ -2x & + & 4y & = & 0 \\ \hline & & 0 & = & 0 \end{array}$$

The system is dependent; it has infinitely many solutions.

59. Let n be the number of nickels and d the number of dimes. Translate the problem into a system of equations.

$$0.05n+0.10d=4.20$$
$$d=3n$$

Substitute $3n$ for d in the first equation.

$$0.05n+0.10(3n)=4.20$$
$$0.05n+0.30n=4.20$$
$$0.35n=4.20$$
$$n=12$$

Substitute 12 for n in the equation $d=3n$.

$d=3(12)=36$

Kaye has 12 nickels and 36 dimes.

60. Let x and y be the numbers.

$$\begin{array}{rcrcr} x & + & y & = & 165 \\ x & - & y & = & 95 \\ \hline 2x & & & = & 260 \\ & & x & = & 130 \end{array}$$

Substitute 130 for x in $x+y=165$.

$$130+y=165$$
$$y=35$$

The numbers are 35 and 130.

Chapter 9 Roots and Radicals

9.1 Finding Roots

Problems 9.1

1. a. $\sqrt{169} = 13$

   b. $-\sqrt{64} = -8$

   c. $\sqrt{\frac{81}{25}} = \frac{9}{5}$

2. a. $\left(\sqrt{11}\right)^2 = 11$

   b. $\left(-\sqrt{36}\right)^2 = 36$

   c. $\left(\sqrt{x^2+7}\right)^2 = x^2 + 7$

3. a. $\frac{36}{25} = \left(\frac{6}{5}\right)^2$ is a perfect square, so $\sqrt{\frac{36}{25}} = \frac{6}{5}$ is a rational number.

   b. $81 = 9^2$ is a perfect square, so $-\sqrt{81} = -9$ is a rational number.

   c. 19 is not a perfect square, so $\sqrt{19} \approx 4.358898944$ is irrational.

   d. $\sqrt{-9}$ is not a real number, since there is no number that we can square and get -9 for an answer.

4. a. Since $3^4 = 81$, $\sqrt[4]{81} = 3$.

   b. There is no number whose fourth power is -16, so $\sqrt[4]{-16}$ is not a real number.

   c. Since $(-4)^3 = -64$, $\sqrt[3]{-64} = -4$.

   d. There is no number whose fourth power is -81, so $\sqrt[4]{-81}$ is not a real number.

5. $t = \sqrt{\frac{d}{16}} = \sqrt{\frac{81}{16}} = \frac{9}{4} = 2.25$ sec

Exercises 9.1

1. $\sqrt{25} = 5$

3. $-\sqrt{9} = -3$

5. $\sqrt{\frac{16}{9}} = \frac{4}{3}$

7. $-\sqrt{\frac{4}{81}} = -\frac{2}{9}$

9. $\sqrt{\frac{25}{81}} = \frac{5}{9}$
   $-\sqrt{\frac{25}{81}} = -\frac{5}{9}$

11. $\sqrt{\frac{49}{100}} = \frac{7}{10}$
    $-\sqrt{\frac{49}{100}} = -\frac{7}{10}$

13. $\left(\sqrt{5}\right)^2 = 5$

15. $\left(-\sqrt{11}\right)^2 = 11$

17. $\left(\sqrt{x^2+1}\right)^2 = x^2 + 1$

19. $\left(-\sqrt{3y^2+7}\right)^2 = 3y^2 + 7$

21. $36 = 6^2$ is a perfect square, so $\sqrt{36} = 6$ is a rational number.

23. $\sqrt{-100}$ is not a real number, since there is no number that we can square and get -100 for an answer.

25. $64 = 8^2$ is a perfect square, so $-\sqrt{64} = -8$ is a rational number.

27. $\frac{16}{9} = \left(\frac{4}{3}\right)^2$ is a perfect square, so $\sqrt{\frac{16}{9}} = \frac{4}{3}$ is a rational number.

29. 6 is not a perfect square, so $-\sqrt{6} \approx -2.449489743$ is irrational.

31. 2 is not a perfect square, so $-\sqrt{2} \approx -1.414213562$ is irrational.

33. Since $3^4 = 81$, $\sqrt[4]{81} = 3$.

35. Since $3^4 = 81$, $-\sqrt[4]{81} = -3$.

37. Since $4^3 = 64$, $-\sqrt[3]{64} = -4$.

39. Since $(-5)^3 = -125$, $-\sqrt[3]{-125} = -(-5) = 5$.

41. $t = \sqrt{\frac{d}{16}} = \sqrt{\frac{1600}{16}} = \frac{40}{4} = 10$ sec

43. $t = \sqrt{\frac{d}{5}} = \sqrt{\frac{100}{5}} = \sqrt{20} \approx 4.5$ sec

45.
$$\begin{aligned} c &= \sqrt{a^2 + b^2} \\ &= \sqrt{4^2 + 3^2} \\ &= \sqrt{16 + 9} \\ &= \sqrt{25} \\ &= 5 \text{ in.} \end{aligned}$$

47.
$$\begin{aligned} V_m &= 1.22\sqrt{a} \\ &= 1.22\sqrt{40{,}000} \\ &= 1.22(200) \\ &= 244 \text{ mi} \end{aligned}$$

49.
$$\begin{aligned} A &= s^2 \\ 121 &= s^2 \\ 11 &= s \end{aligned}$$
The room is 11 × 11 feet.

51.
$$\begin{aligned} \sqrt{b^2 - 4ac} &= \sqrt{(-4)^2 - 4(1)(3)} \\ &= \sqrt{16 - 12} \\ &= \sqrt{4} \\ &= 2 \end{aligned}$$

53.
$$\begin{aligned} \sqrt{b^2 - 4ac} &= \sqrt{5^2 - 4(2)(-3)} \\ &= \sqrt{25 + 24} \\ &= \sqrt{49} \\ &= 7 \end{aligned}$$

55. a. $26 - 25 = 1$

$\sqrt{25} = 5$

$\sqrt{26} \approx 5 + \frac{1}{11} = 5\frac{1}{11}$

$\sqrt{36} = 6$

$36 - 25 = 11$

b. $\sqrt{28} \approx 5\frac{1}{11} + \frac{2}{11} = 5\frac{3}{11}$

c. $\sqrt{30} \approx 5\frac{3}{11} + \frac{2}{11} = 5\frac{5}{11}$

57. One; answers may vary. Sample answer: The square root of a number is the positive square root of the number.

59. One

61. One

63. The variable a must be greater than or equal to 0 or else $\sqrt{a}$ is not a real number.

65. $\sqrt{\frac{16}{49}} = \frac{4}{7}$

67. $\sqrt{144} = 12$

69. $\left(\sqrt{17}\right)^2 = 17$

71. $\frac{49}{121}=\left(\frac{7}{11}\right)^2$ is a perfect square, so $-\sqrt{\frac{49}{121}}=-\frac{7}{11}$ is a rational number.

73. 15 is not a perfect square, so $\sqrt{15}\approx 3.872983346$ is irrational.

75. Since $(-5)^3=-125$, $\sqrt[3]{-125}=-5$.

77. Since $(-3)^3=-27$, $-\sqrt[3]{-27}=-(-3)=3$.

9.2 Multiplication and Division of Radicals

Problems 9.2

1. a. $\sqrt{72}=\sqrt{36\cdot 2}=\sqrt{36}\cdot\sqrt{2}=6\sqrt{2}$

b. $\sqrt{98}=\sqrt{49\cdot 2}=\sqrt{49}\cdot\sqrt{2}=7\sqrt{2}$

2. a. $\sqrt{5}\cdot\sqrt{7}=\sqrt{35}$

b. $\sqrt{40}\cdot\sqrt{10}=\sqrt{400}=20$

c. $\sqrt{6}\cdot\sqrt{x}=\sqrt{6x}$

3. a. $\sqrt{\frac{7}{16}}=\frac{\sqrt{7}}{\sqrt{16}}=\frac{\sqrt{7}}{4}$

b. $\frac{\sqrt{30}}{\sqrt{6}}=\sqrt{\frac{30}{6}}=\sqrt{5}$

c. $\frac{16\sqrt{30}}{8\sqrt{10}}=\frac{2\sqrt{30}}{\sqrt{10}}=2\sqrt{\frac{30}{10}}=2\sqrt{3}$

4. a. $\sqrt{81y^2}=\sqrt{81}\cdot\sqrt{y^2}=9y$

b. $\sqrt{9n^4}=\sqrt{9}\cdot\sqrt{n^4}=3n^2$

c.
$$\begin{aligned}\sqrt{72x^{10}}&=\sqrt{72}\cdot\sqrt{x^{10}}\\&=\sqrt{36\cdot 2}\cdot\sqrt{x^{10}}\\&=6\sqrt{2}\cdot x^5\\&=6x^5\sqrt{2}\end{aligned}$$

d. $\sqrt{y^{13}}=\sqrt{y^{12}\cdot y}=y^6\sqrt{y}$

5. a. $\sqrt[3]{16}=\sqrt[3]{8\cdot 2}=\sqrt[3]{2^3\cdot 2}=2\sqrt[3]{2}$

b. $\sqrt[4]{112}=\sqrt[4]{16\cdot 7}=\sqrt[4]{2^4\cdot 7}=2\sqrt[4]{7}$

c. $\sqrt[3]{\frac{64}{27}}=\frac{\sqrt[3]{64}}{\sqrt[3]{27}}=\frac{\sqrt[3]{4^3}}{\sqrt[3]{3^3}}=\frac{4}{3}$

Exercises 9.2

1. $\sqrt{45}=\sqrt{9\cdot 5}=\sqrt{9}\cdot\sqrt{5}=3\sqrt{5}$

3. $\sqrt{125}=\sqrt{25\cdot 5}=\sqrt{25}\cdot\sqrt{5}=5\sqrt{5}$

5. $\sqrt{180}=\sqrt{36\cdot 5}=\sqrt{36}\cdot\sqrt{5}=6\sqrt{5}$

7. $\sqrt{200}=\sqrt{100\cdot 2}=\sqrt{100}\cdot\sqrt{2}=10\sqrt{2}$

9. $\sqrt{384}=\sqrt{64\cdot 6}=\sqrt{64}\cdot\sqrt{6}=8\sqrt{6}$

11. $\sqrt{75}=\sqrt{25\cdot 3}=\sqrt{25}\cdot\sqrt{3}=5\sqrt{3}$

13. $\sqrt{600}=\sqrt{100\cdot 6}=\sqrt{100}\cdot\sqrt{6}=10\sqrt{6}$

15. $\sqrt{361}=19$

17. $\sqrt{700}=\sqrt{100\cdot 7}=\sqrt{100}\cdot\sqrt{7}=10\sqrt{7}$

19. $\sqrt{432}=\sqrt{144\cdot 3}=\sqrt{144}\cdot\sqrt{3}=12\sqrt{3}$

21. $\sqrt{3}\cdot\sqrt{5}=\sqrt{15}$

23. $\sqrt{27}\cdot\sqrt{3}=\sqrt{81}=9$

25. $\sqrt{7}\cdot\sqrt{7}=\sqrt{49}=7$

27. $\sqrt{3}\cdot\sqrt{x}=\sqrt{3x}$

29. $\sqrt{2a}\cdot\sqrt{18a}=\sqrt{36a^2}=6a$

31. $\sqrt{\frac{2}{25}}=\frac{\sqrt{2}}{\sqrt{25}}=\frac{\sqrt{2}}{5}$

33. $\frac{\sqrt{20}}{\sqrt{5}}=\sqrt{\frac{20}{5}}=\sqrt{4}=2$

35. $\frac{\sqrt{27}}{\sqrt{3}}=\sqrt{\frac{27}{3}}=\sqrt{9}=3$

37. $\frac{15\sqrt{30}}{3\sqrt{10}}=\frac{5\sqrt{30}}{\sqrt{10}}=5\sqrt{\frac{30}{10}}=5\sqrt{3}$

39. $\frac{18\sqrt{40}}{3\sqrt{10}}=\frac{6\sqrt{40}}{\sqrt{10}}=6\sqrt{\frac{40}{10}}=6\sqrt{4}=6\cdot 2=12$

41. $\sqrt{100a^2}=\sqrt{100}\cdot\sqrt{a^2}=10a$

43. $\sqrt{49a^4}=\sqrt{49}\cdot\sqrt{a^4}=7a^2$

45. $-\sqrt{32a^6}=-\sqrt{32}\cdot\sqrt{a^6}$
$=-\sqrt{16\cdot 2}\cdot\sqrt{a^6}$
$=-4\sqrt{2}\cdot a^3$
$=-4a^3\sqrt{2}$

47. $\sqrt{m^{13}}=\sqrt{m^{12}\cdot m}=m^6\sqrt{m}$

49. $-\sqrt{27m^{11}}=-\sqrt{27}\cdot\sqrt{m^{11}}$
$=-\sqrt{9\cdot 3}\cdot\sqrt{m^{10}\cdot m}$
$=-3\sqrt{3}\cdot m^5\sqrt{m}$
$=-3m^5\sqrt{3m}$

51. $\sqrt[3]{40}=\sqrt[3]{8\cdot 5}=\sqrt[3]{2^3\cdot 5}=2\sqrt[3]{5}$

53. $\sqrt[3]{-16}=\sqrt[3]{-8\cdot 2}=\sqrt[3]{(-2)^3\cdot 2}=-2\sqrt[3]{2}$

55. $\sqrt[3]{\frac{8}{27}}=\frac{\sqrt[3]{8}}{\sqrt[3]{27}}=\frac{\sqrt[3]{2^3}}{\sqrt[3]{3^3}}=\frac{2}{3}$

57. $\sqrt[4]{48}=\sqrt[4]{16\cdot 3}=\sqrt[4]{2^4\cdot 3}=2\sqrt[4]{3}$

59. $\sqrt[3]{\frac{64}{27}}=\frac{\sqrt[3]{64}}{\sqrt[3]{27}}=\frac{\sqrt[3]{4^3}}{\sqrt[3]{3^3}}=\frac{4}{3}$

61. $5x+7x=12x$

63. $9x^3+7x^3-2x^3=14x^3$

65. Answers may vary. Sample answer:
$(-2)^3=-8$ and $\sqrt[3]{-8}=-2$.

67. a. Irrational

b. Yes; answers may vary. Sample answer: p has no factors and so has no factors that are perfect squares.

69. $\sqrt[3]{135}=\sqrt[3]{27\cdot 5}=\sqrt[3]{3^3\cdot 5}=3\sqrt[3]{5}$

71. $\sqrt[3]{\frac{64}{27}}=\frac{\sqrt[3]{64}}{\sqrt[3]{27}}=\frac{\sqrt[3]{4^3}}{\sqrt[3]{3^3}}=\frac{4}{3}$

73. $\sqrt{100x^6}=\sqrt{100}\cdot\sqrt{x^6}=10x^3$

75. $\frac{\sqrt{28}}{\sqrt{14}}=\sqrt{\frac{28}{14}}=\sqrt{2}$

77. $\sqrt{\frac{4}{9}}=\frac{\sqrt{4}}{\sqrt{9}}=\frac{2}{3}$

79. $\sqrt{72}\cdot\sqrt{2}=\sqrt{72\cdot 2}=\sqrt{144}=12$

9.3 Addition and Subtraction of Radicals

Problems 9.3

1. a. $8\sqrt{5}+2\sqrt{5}=10\sqrt{5}$

b. $7\sqrt{6}-2\sqrt{6}=5\sqrt{6}$

2. a. $\sqrt{150}=\sqrt{25\cdot 6}=\sqrt{25}\cdot\sqrt{6}=5\sqrt{6}$
$\sqrt{24}=\sqrt{4\cdot 6}=\sqrt{4}\cdot\sqrt{6}=2\sqrt{6}$
$\sqrt{150}+\sqrt{24}=5\sqrt{6}+2\sqrt{6}=7\sqrt{6}$

b. $\sqrt{20}=\sqrt{4\cdot 5}=\sqrt{4}\cdot\sqrt{5}=2\sqrt{5}$
$\sqrt{80}=\sqrt{16\cdot 5}=\sqrt{16}\cdot\sqrt{5}=4\sqrt{5}$
$\sqrt{45}=\sqrt{9\cdot 5}=\sqrt{9}\cdot\sqrt{5}=3\sqrt{5}$
$\sqrt{20}+\sqrt{80}-\sqrt{45}=2\sqrt{5}+4\sqrt{5}-3\sqrt{5}$
$=3\sqrt{5}$

3. a. $\sqrt{3}\left(\sqrt{45}-\sqrt{2}\right)=\sqrt{3}\sqrt{45}-\sqrt{3}\sqrt{2}$
$=\sqrt{135}-\sqrt{6}$
$=\sqrt{9\cdot 15}-\sqrt{6}$
$=3\sqrt{15}-\sqrt{6}$

b. $\sqrt{5}\left(\sqrt{5}-\sqrt{3}\right)=\sqrt{5}\sqrt{5}-\sqrt{5}\sqrt{3}$
$=\sqrt{25}-\sqrt{15}$
$=5-\sqrt{15}$

4. $\sqrt{\frac{5}{12}}=\frac{\sqrt{5}}{\sqrt{12}}=\frac{\sqrt{5}\cdot\sqrt{3}}{\sqrt{12}\cdot\sqrt{3}}=\frac{\sqrt{15}}{\sqrt{36}}=\frac{\sqrt{15}}{6}$

5. $\sqrt{\frac{x^2}{50}}=\frac{\sqrt{x^2}}{\sqrt{50}}=\frac{\sqrt{x^2}\cdot\sqrt{2}}{\sqrt{50}\cdot\sqrt{2}}=\frac{\sqrt{x^2}\cdot\sqrt{2}}{\sqrt{100}}=\frac{x\sqrt{2}}{10}$

Exercises 9.3

1. $6\sqrt{7}+4\sqrt{7}=10\sqrt{7}$

3. $9\sqrt{13}-4\sqrt{13}=5\sqrt{13}$

5. $\sqrt{32}=\sqrt{16\cdot 2}=\sqrt{16}\cdot\sqrt{2}=4\sqrt{2}$
$\sqrt{50}=\sqrt{25\cdot 2}=\sqrt{25}\cdot\sqrt{2}=5\sqrt{2}$
$\sqrt{72}=\sqrt{36\cdot 2}=\sqrt{36}\cdot\sqrt{2}=6\sqrt{2}$
$\sqrt{32}+\sqrt{50}-\sqrt{72}=4\sqrt{2}+5\sqrt{2}-6\sqrt{2}$
$=3\sqrt{2}$

7. $\sqrt{162}=\sqrt{81\cdot 2}=\sqrt{81}\cdot\sqrt{2}=9\sqrt{2}$
$\sqrt{50}=\sqrt{25\cdot 2}=\sqrt{25}\cdot\sqrt{2}=5\sqrt{2}$
$\sqrt{200}=\sqrt{100\cdot 2}=\sqrt{100}\cdot\sqrt{2}=10\sqrt{2}$
$\sqrt{162}+\sqrt{50}-\sqrt{200}=9\sqrt{2}+5\sqrt{2}-10\sqrt{2}$
$=4\sqrt{2}$

9. $9\sqrt{48}=9\sqrt{16\cdot 3}=9\sqrt{16}\cdot\sqrt{3}=36\sqrt{3}$
$5\sqrt{27}=5\sqrt{9\cdot 3}=5\sqrt{9}\cdot\sqrt{3}=15\sqrt{3}$
$3\sqrt{12}=3\sqrt{4\cdot 3}=3\sqrt{4}\cdot\sqrt{3}=6\sqrt{3}$
$9\sqrt{48}-5\sqrt{27}+3\sqrt{12}=36\sqrt{3}-15\sqrt{3}+6\sqrt{3}$
$=27\sqrt{3}$

11. $3\sqrt{28}=3\sqrt{4\cdot 7}=3\sqrt{4}\cdot\sqrt{7}=6\sqrt{7}$
$2\sqrt{63}=2\sqrt{9\cdot 7}=2\sqrt{9}\cdot\sqrt{7}=6\sqrt{7}$
$5\sqrt{7}-3\sqrt{28}-2\sqrt{63}=5\sqrt{7}-6\sqrt{7}-6\sqrt{7}$
$=-7\sqrt{7}$

13. $8\sqrt{75}=8\sqrt{25\cdot 3}=8\sqrt{25}\cdot\sqrt{3}=40\sqrt{3}$
$2\sqrt{27}=2\sqrt{9\cdot 3}=2\sqrt{9}\cdot\sqrt{3}=6\sqrt{3}$
$-5\sqrt{3}+8\sqrt{75}-2\sqrt{27}=-5\sqrt{3}+40\sqrt{3}-6\sqrt{3}$
$=29\sqrt{3}$

15. $-3\sqrt{45}=-3\sqrt{9\cdot 5}=-3\sqrt{9}\cdot\sqrt{5}=-9\sqrt{5}$
$\sqrt{20}=\sqrt{4\cdot 5}=\sqrt{4}\cdot\sqrt{5}=2\sqrt{5}$
$-3\sqrt{45}+\sqrt{20}-\sqrt{5}=-9\sqrt{5}+2\sqrt{5}-\sqrt{5}$
$=-8\sqrt{5}$

17. $\sqrt{10}\left(\sqrt{20}-\sqrt{3}\right)=\sqrt{10}\sqrt{20}-\sqrt{10}\sqrt{3}$
$=\sqrt{200}-\sqrt{30}$
$=\sqrt{100\cdot 2}-\sqrt{30}$
$=10\sqrt{2}-\sqrt{30}$

19. $\sqrt{6}\left(\sqrt{14}+\sqrt{5}\right)=\sqrt{6}\sqrt{14}+\sqrt{6}\sqrt{5}$
$=\sqrt{84}+\sqrt{30}$
$=\sqrt{4\cdot 21}+\sqrt{30}$
$=2\sqrt{21}+\sqrt{30}$

21. $\sqrt{3}\left(\sqrt{3}-\sqrt{2}\right)=\sqrt{3}\sqrt{3}-\sqrt{3}\sqrt{2}$
$=\sqrt{9}-\sqrt{6}$
$=3-\sqrt{6}$

23. $\sqrt{5}\left(\sqrt{2}+\sqrt{5}\right)=\sqrt{5}\sqrt{2}+\sqrt{5}\sqrt{5}$
$=\sqrt{10}+\sqrt{25}$
$=\sqrt{10}+5$

25. $\sqrt{6}\left(\sqrt{2}-\sqrt{3}\right)=\sqrt{6}\sqrt{2}-\sqrt{6}\sqrt{3}$
$=\sqrt{12}-\sqrt{18}$
$=\sqrt{4\cdot 3}-\sqrt{9\cdot 2}$
$=2\sqrt{3}-3\sqrt{2}$

27. $2\left(\sqrt{2}-5\right)=2\sqrt{2}-10$

29. $\sqrt{2}\left(\sqrt{6}-3\right)=\sqrt{2}\sqrt{6}-3\sqrt{2}$
$=\sqrt{12}-3\sqrt{2}$
$=\sqrt{4\cdot 3}-3\sqrt{2}$
$=2\sqrt{3}-3\sqrt{2}$

31. $\frac{3}{\sqrt{6}}=\frac{3\cdot\sqrt{6}}{\sqrt{6}\cdot\sqrt{6}}=\frac{3\sqrt{6}}{\sqrt{36}}=\frac{3\sqrt{6}}{6}=\frac{\sqrt{6}}{2}$

33. $\frac{-10}{\sqrt{5}}=\frac{-10\cdot\sqrt{5}}{\sqrt{5}\cdot\sqrt{5}}=\frac{-10\sqrt{5}}{\sqrt{25}}=\frac{-10\sqrt{5}}{5}=-2\sqrt{5}$

35. $\frac{\sqrt{8}}{\sqrt{2}}=\frac{\sqrt{8}\cdot\sqrt{2}}{\sqrt{2}\cdot\sqrt{2}}=\frac{\sqrt{16}}{\sqrt{4}}=\frac{4}{2}=2$

37. $\frac{-\sqrt{2}}{\sqrt{5}}=\frac{-\sqrt{2}\cdot\sqrt{5}}{\sqrt{5}\cdot\sqrt{5}}=\frac{-\sqrt{10}}{\sqrt{25}}=\frac{-\sqrt{10}}{5}$

39. $\frac{\sqrt{2}}{\sqrt{8}}=\frac{\sqrt{2}\cdot\sqrt{2}}{\sqrt{8}\cdot\sqrt{2}}=\frac{\sqrt{4}}{\sqrt{16}}=\frac{2}{4}=\frac{1}{2}$

41. $\frac{\sqrt{x^2}}{\sqrt{18}}=\frac{\sqrt{x^2}\cdot\sqrt{2}}{\sqrt{18}\cdot\sqrt{2}}=\frac{x\sqrt{2}}{\sqrt{36}}=\frac{x\sqrt{2}}{6}$

43. $\frac{\sqrt{a^2}}{\sqrt{b}}=\frac{\sqrt{a^2}\cdot\sqrt{b}}{\sqrt{b}\cdot\sqrt{b}}=\frac{a\sqrt{b}}{\sqrt{b^2}}=\frac{a\sqrt{b}}{b}$

45. $\sqrt{\frac{3}{10}}=\frac{\sqrt{3}}{\sqrt{10}}=\frac{\sqrt{3}\cdot\sqrt{10}}{\sqrt{10}\cdot\sqrt{10}}=\frac{\sqrt{30}}{\sqrt{100}}=\frac{\sqrt{30}}{10}$

47. $\sqrt{\frac{x^2}{32}}=\frac{\sqrt{x^2}}{\sqrt{32}}=\frac{\sqrt{x^2}\cdot\sqrt{2}}{\sqrt{32}\cdot\sqrt{2}}=\frac{x\sqrt{2}}{\sqrt{64}}=\frac{x\sqrt{2}}{8}$

49. $\sqrt{\frac{x^4}{20}}=\frac{\sqrt{x^4}}{\sqrt{20}}=\frac{\sqrt{x^4}\cdot\sqrt{5}}{\sqrt{20}\cdot\sqrt{5}}=\frac{x^2\sqrt{5}}{\sqrt{100}}=\frac{x^2\sqrt{5}}{10}$

51. $(x+3)(x-3)=x^2-3x+3x-9=x^2-9$

53. $\frac{6x+12}{3}=\frac{6x}{3}+\frac{12}{3}=2x+4$

55. **45.** $\sqrt{\frac{3}{10}}=\sqrt{\frac{3\cdot 10}{10\cdot 10}}=\sqrt{\frac{30}{100}}=\frac{\sqrt{30}}{10}$

46. $\sqrt{\frac{2}{27}}=\sqrt{\frac{2\cdot 3}{27\cdot 3}}=\sqrt{\frac{6}{81}}=\frac{\sqrt{6}}{9}$

47. $\sqrt{\frac{x^2}{32}}=\sqrt{\frac{x^2\cdot 2}{32\cdot 2}}=\sqrt{\frac{2x^2}{64}}=\frac{x\sqrt{2}}{8}$

48. $\sqrt{\frac{x}{18}}=\sqrt{\frac{x\cdot 2}{18\cdot 2}}=\sqrt{\frac{2x}{36}}=\frac{\sqrt{2x}}{6}$

49. $\sqrt{\frac{x^4}{20}}=\sqrt{\frac{x^4\cdot 5}{20\cdot 5}}=\sqrt{\frac{5x^4}{100}}=\frac{x^2\sqrt{5}}{10}$

50. $\sqrt{\frac{x^6}{72}}=\sqrt{\frac{x^6\cdot 2}{72\cdot 2}}=\sqrt{\frac{2x^6}{144}}=\frac{x^3\sqrt{2}}{12}$

57. Answers may vary. Sample answer:
$\sqrt{\frac{3}{8}}=\sqrt{\frac{3\cdot 2}{8\cdot 2}}=\sqrt{\frac{6}{16}}=\frac{\sqrt{6}}{\sqrt{16}}=\frac{\sqrt{6}}{4}$

59. $8\sqrt{3}+5\sqrt{3}=13\sqrt{3}$

61. $\sqrt{18}=\sqrt{9\cdot 2}=\sqrt{9}\cdot\sqrt{2}=3\sqrt{2}$
$\sqrt{18}+3\sqrt{2}=3\sqrt{2}+3\sqrt{2}=6\sqrt{2}$

63. $7\sqrt{18}=7\sqrt{9\cdot 2}=7\sqrt{9}\cdot\sqrt{2}=21\sqrt{2}$
$7\sqrt{8}+7\sqrt{4\cdot 2}=7\sqrt{4}\cdot\sqrt{2}=14\sqrt{2}$
$7\sqrt{18}+5\sqrt{2}-7\sqrt{8}=21\sqrt{2}+5\sqrt{2}-14\sqrt{2}$
$=12\sqrt{2}$

65. $\sqrt{3}\left(5-\sqrt{3}\right)=5\sqrt{3}-\sqrt{3}\sqrt{3}=5\sqrt{3}-\sqrt{9}=5\sqrt{3}-3$

67. $\dfrac{3}{\sqrt{7}}=\dfrac{3\cdot\sqrt{7}}{\sqrt{7}\cdot\sqrt{7}}=\dfrac{3\sqrt{7}}{\sqrt{49}}=\dfrac{3\sqrt{7}}{7}$

69. $\dfrac{\sqrt{x}}{\sqrt{2}}=\dfrac{\sqrt{x}\cdot\sqrt{2}}{\sqrt{2}\cdot\sqrt{2}}=\dfrac{\sqrt{2x}}{\sqrt{4}}=\dfrac{\sqrt{2x}}{2}$

9.4 Simplifying Radicals

Problems 9.4

1. a. $\sqrt{60+4}-\sqrt{9}=\sqrt{64}-\sqrt{9}=8-3=5$

b. $8\sqrt{10}-\sqrt{2}\sqrt{5}=8\sqrt{10}-\sqrt{10}=(8-1)\sqrt{10}=7\sqrt{10}$

c. $\dfrac{\sqrt{10x^3}}{\sqrt{5x^2}}=\sqrt{\dfrac{10x^3}{5x^2}}=\sqrt{2x}$

2. a. $\dfrac{\sqrt[3]{108}}{\sqrt[3]{2}}=\sqrt[3]{54}=\sqrt[3]{27\cdot 2}=3\sqrt[3]{2}$

b. $\dfrac{7}{\sqrt[3]{9}}=\dfrac{7\cdot\sqrt[3]{3}}{\sqrt[3]{9}\cdot\sqrt[3]{3}}=\dfrac{7\sqrt[3]{3}}{\sqrt[3]{27}}=\dfrac{7\sqrt[3]{3}}{3}$

3. a.
$$\begin{aligned}\left(\sqrt{3}+5\sqrt{2}\right)\left(\sqrt{3}-4\sqrt{2}\right)&=\sqrt{3}\cdot\sqrt{3}+\sqrt{3}\left(-4\sqrt{2}\right)+5\sqrt{2}\left(\sqrt{3}\right)+5\sqrt{2}\left(-4\sqrt{2}\right)\\&=3-4\sqrt{6}+5\sqrt{6}-20\cdot 2\\&=3+\sqrt{6}-40\\&=-37+\sqrt{6}\end{aligned}$$

b. $\left(2+3\sqrt{5}\right)\left(2-3\sqrt{5}\right)=(2)^2-\left(3\sqrt{5}\right)^2=4-\left[(3)^2\left(\sqrt{5}\right)^2\right]=4-(9)(5)=4-45=-41$

4. a. $\dfrac{5}{\sqrt{7}+1}=\dfrac{5\cdot\left(\sqrt{7}-1\right)}{\left(\sqrt{7}+1\right)\left(\sqrt{7}-1\right)}=\dfrac{5\sqrt{7}-5}{\left(\sqrt{7}\right)^2-(1)^2}=\dfrac{5\sqrt{7}-5}{7-1}=\dfrac{5\sqrt{7}-5}{6}$

b. $\dfrac{5}{\sqrt{6}-\sqrt{2}}=\dfrac{5\left(\sqrt{6}+\sqrt{2}\right)}{\left(\sqrt{6}-\sqrt{2}\right)\left(\sqrt{6}+\sqrt{2}\right)}=\dfrac{5\sqrt{6}+5\sqrt{2}}{\left(\sqrt{6}\right)^2-\left(\sqrt{2}\right)^2}=\dfrac{5\sqrt{6}+5\sqrt{2}}{6-2}=\dfrac{5\sqrt{6}+5\sqrt{2}}{4}$

5. a. $\dfrac{-9+\sqrt{18}}{3}=\dfrac{-9+\sqrt{9\cdot 2}}{3}$
$=\dfrac{-9+3\sqrt{2}}{3}$
$=\dfrac{3\left(-3+\sqrt{2}\right)}{3}$
$=-3+\sqrt{2}$

b. $\dfrac{-12+\sqrt{24}}{6}=\dfrac{-12+\sqrt{4\cdot 6}}{6}$
$=\dfrac{-12+2\sqrt{6}}{6}$
$=\dfrac{2\left(-6+\sqrt{6}\right)}{6}$
$=\dfrac{-6+\sqrt{6}}{3}$

Exercises 9.4

1. $\sqrt{36}+\sqrt{100}=6+10=16$

3. $\sqrt{144+25}=\sqrt{169}=13$

5. $\sqrt{4}-\sqrt{36}=2-6=-4$

7. $\sqrt{13^2-12^2}=\sqrt{169-144}=\sqrt{25}=5$

9. $15\sqrt{10}+\sqrt{90}=15\sqrt{10}+\sqrt{9\cdot 10}$
$=15\sqrt{10}+3\sqrt{10}$
$=(15+3)\sqrt{10}$
$=18\sqrt{10}$

11. $14\sqrt{11}-\sqrt{44}=14\sqrt{11}-\sqrt{4\cdot 11}$
$=14\sqrt{11}-2\sqrt{11}$
$=(14-2)\sqrt{11}$
$=12\sqrt{11}$

13. $\sqrt[3]{54}-\sqrt[3]{8}=\sqrt[3]{27\cdot 2}-\sqrt[3]{8}=3\sqrt[3]{2}-2$

15. $5\sqrt[3]{16}-3\sqrt[3]{54}=5\sqrt[3]{8\cdot 2}-3\sqrt[3]{27\cdot 2}$
$=5\cdot 2\sqrt[3]{2}-3\cdot 3\sqrt[3]{2}$
$=10\sqrt[3]{2}-9\sqrt[3]{2}$
$=(10-9)\sqrt[3]{2}$
$=\sqrt[3]{2}$

17. $\sqrt{\dfrac{9x^2}{x}}=\sqrt{9x}=3\sqrt{x}$

19. $\sqrt{\dfrac{81y^7}{16y^5}}=\sqrt{\dfrac{81y^2}{16}}=\dfrac{9y}{4}$

21. $\sqrt{\dfrac{64a^4b^6}{3ab^4}}=\sqrt{\dfrac{64a^3b^2}{3}}$
$=\dfrac{\sqrt{64a^3b^2}}{\sqrt{3}}$
$=\dfrac{\sqrt{64a^3b^2}\cdot\sqrt{3}}{\sqrt{3}\cdot\sqrt{3}}$
$=\dfrac{8ab\sqrt{3a}}{3}$

23. $\sqrt[3]{\dfrac{8b^6c^{10}}{27bc}}=\sqrt[3]{\dfrac{8b^5c^9}{27}}=\dfrac{\sqrt[3]{8b^5c^9}}{\sqrt[3]{27}}=\dfrac{2bc^3\sqrt[3]{b^2}}{3}$

25. $\dfrac{\sqrt[3]{500}}{\sqrt[3]{2}}=\sqrt[3]{\dfrac{500}{2}}=\sqrt[3]{250}=\sqrt[3]{125\cdot 2}=5\sqrt[3]{2}$

27. $\dfrac{6}{\sqrt[3]{9}}=\dfrac{6\cdot\sqrt[3]{3}}{\sqrt[3]{9}\cdot\sqrt[3]{3}}=\dfrac{6\sqrt[3]{3}}{\sqrt[3]{27}}=\dfrac{6\sqrt[3]{3}}{3}=2\sqrt[3]{3}$

29. $\left(\sqrt{3}+6\sqrt{5}\right)\left(\sqrt{3}-4\sqrt{5}\right)=\sqrt{3}\cdot\sqrt{3}+\sqrt{3}\left(-4\sqrt{5}\right)+6\sqrt{5}\left(\sqrt{3}\right)+6\sqrt{5}\left(-4\sqrt{5}\right)$
$=3-4\sqrt{15}+6\sqrt{15}-24\cdot 5$
$=3+2\sqrt{15}-120$
$=2\sqrt{15}-117$

31. $\left(\sqrt{2}+3\sqrt{3}\right)\left(\sqrt{2}+3\sqrt{3}\right)=\sqrt{2}\cdot\sqrt{2}+\sqrt{2}\left(3\sqrt{3}\right)+3\sqrt{3}\left(\sqrt{2}\right)+3\sqrt{3}\left(3\sqrt{3}\right)$
$=2+3\sqrt{6}+3\sqrt{6}+9\cdot 3$
$=2+6\sqrt{6}+27$
$=6\sqrt{6}+29$

33. $\left(5\sqrt{2}-3\sqrt{3}\right)\left(5\sqrt{2}-\sqrt{3}\right)=5\sqrt{2}\left(5\sqrt{2}\right)+5\sqrt{2}\left(-\sqrt{3}\right)+\left(-3\sqrt{3}\right)\left(5\sqrt{2}\right)+\left(-3\sqrt{3}\right)\left(-\sqrt{3}\right)$
$=25\cdot 2-5\sqrt{6}-15\sqrt{6}+3\cdot 3$
$=50-20\sqrt{6}+9$
$=59-20\sqrt{6}$

35. $\left(\sqrt{13}+2\sqrt{2}\right)\left(\sqrt{13}-2\sqrt{2}\right)=\left(\sqrt{13}\right)^2-\left(2\sqrt{2}\right)^2=13-\left[(2)^2\left(\sqrt{2}\right)^2\right]=13-(4)(2)=13-8=5$

37. $\dfrac{3}{\sqrt{2}+1}=\dfrac{3\cdot\left(\sqrt{2}-1\right)}{\left(\sqrt{2}+1\right)\left(\sqrt{2}-1\right)}=\dfrac{3\sqrt{2}-3}{\left(\sqrt{2}\right)^2-(1)^2}=\dfrac{3\sqrt{2}-3}{2-1}=\dfrac{3\sqrt{2}-3}{1}=3\sqrt{2}-3$

39. $\dfrac{4}{\sqrt{7}-1}=\dfrac{4\cdot\left(\sqrt{7}+1\right)}{\left(\sqrt{7}-1\right)\left(\sqrt{7}+1\right)}=\dfrac{4\sqrt{7}+4}{\left(\sqrt{7}\right)^2-(1)^2}=\dfrac{4\sqrt{7}+4}{7-1}=\dfrac{4\sqrt{7}+4}{6}=\dfrac{2\left(2\sqrt{7}+2\right)}{6}=\dfrac{2\sqrt{7}+2}{3}$

41. $\dfrac{\sqrt{2}}{2+\sqrt{3}}=\dfrac{\sqrt{2}\cdot\left(2-\sqrt{3}\right)}{\left(2+\sqrt{3}\right)\left(2-\sqrt{3}\right)}=\dfrac{2\sqrt{2}-\sqrt{6}}{(2)^2-\left(\sqrt{3}\right)^2}=\dfrac{2\sqrt{2}-\sqrt{6}}{4-3}=\dfrac{2\sqrt{2}-\sqrt{6}}{1}=2\sqrt{2}-\sqrt{6}$

43. $\dfrac{\sqrt{5}}{2-\sqrt{3}}=\dfrac{\sqrt{5}\cdot\left(2+\sqrt{3}\right)}{\left(2-\sqrt{3}\right)\left(2+\sqrt{3}\right)}=\dfrac{2\sqrt{5}+\sqrt{15}}{(2)^2-\left(\sqrt{3}\right)^2}=\dfrac{2\sqrt{5}+\sqrt{15}}{4-3}=\dfrac{2\sqrt{5}+\sqrt{15}}{1}=2\sqrt{5}+\sqrt{15}$

45. $\dfrac{\sqrt{5}}{\sqrt{2}+\sqrt{3}}=\dfrac{\sqrt{5}\cdot\left(\sqrt{2}-\sqrt{3}\right)}{\left(\sqrt{2}+\sqrt{3}\right)\left(\sqrt{2}-\sqrt{3}\right)}=\dfrac{\sqrt{10}-\sqrt{15}}{\left(\sqrt{2}\right)^2-\left(\sqrt{3}\right)^2}=\dfrac{\sqrt{10}-\sqrt{15}}{2-3}=\dfrac{\sqrt{10}-\sqrt{15}}{-1}=-\sqrt{10}+\sqrt{15}$

47. $\dfrac{\sqrt{3}}{\sqrt{5}-\sqrt{2}}=\dfrac{\sqrt{3}\cdot\left(\sqrt{5}+\sqrt{2}\right)}{\left(\sqrt{5}-\sqrt{2}\right)\left(\sqrt{5}+\sqrt{2}\right)}=\dfrac{\sqrt{15}+\sqrt{6}}{\left(\sqrt{5}\right)^2-\left(\sqrt{2}\right)^2}=\dfrac{\sqrt{15}+\sqrt{6}}{5-2}=\dfrac{\sqrt{15}+\sqrt{6}}{3}$

49. $\dfrac{\sqrt{3}+\sqrt{2}}{\sqrt{3}-\sqrt{2}}=\dfrac{(\sqrt{3}+\sqrt{2})(\sqrt{3}+\sqrt{2})}{(\sqrt{3}-\sqrt{2})(\sqrt{3}+\sqrt{2})}$

$=\dfrac{(\sqrt{3})^2+2(\sqrt{3})(\sqrt{2})+(\sqrt{2})^2}{(\sqrt{3})^2-(\sqrt{2})^2}$

$=\dfrac{3+2\sqrt{6}+2}{3-2}$

$=\dfrac{5+2\sqrt{6}}{1}$

$=5+2\sqrt{6}$

51. $\dfrac{-8+\sqrt{16}}{2}=\dfrac{-8+4}{2}=\dfrac{-4}{2}=-2$

53. $\dfrac{-6-\sqrt{4}}{6}=\dfrac{-6-2}{6}=\dfrac{-8}{6}=\dfrac{-4}{3}$

55. $\dfrac{2+2\sqrt{3}}{6}=\dfrac{2(1+\sqrt{3})}{6}=\dfrac{1+\sqrt{3}}{3}$

57. $\dfrac{-2+2\sqrt{23}}{4}=\dfrac{2(-1+\sqrt{23})}{4}=\dfrac{-1+\sqrt{23}}{2}$

59. $\dfrac{-6+3\sqrt{10}}{9}=\dfrac{3(-2+\sqrt{10})}{9}=\dfrac{-2+\sqrt{10}}{3}$

61. $\dfrac{-8+\sqrt{28}}{6}=\dfrac{-8+\sqrt{4\cdot 7}}{6}$

$=\dfrac{-8+2\sqrt{7}}{6}$

$=\dfrac{2(-4+\sqrt{7})}{6}$

$=\dfrac{-4+\sqrt{7}}{3}$

63. $\dfrac{-9+\sqrt{243}}{6}=\dfrac{-9+\sqrt{81\cdot 3}}{6}$

$=\dfrac{-9+9\sqrt{3}}{6}$

$=\dfrac{3(-3+3\sqrt{3})}{6}$

$=\dfrac{-3+3\sqrt{3}}{2}$

65. $\left(\sqrt{x-1}\right)^2=x-1$

67. $\left(2\sqrt{x}\right)^2=(2)^2\left(\sqrt{x}\right)^2=4x$

69. $\left(\sqrt{x^2+2x+1}\right)^2=x^2+2x+1$

71. $x^2-3x=x(x-3)$

73. $x^2-3x+2=(x-2)(x-1)$

75. $\sqrt{\dfrac{2}{\gamma}}\sqrt{\dfrac{P_2-P_1}{P_1}}=\sqrt{\dfrac{2(P_2-P_1)}{\gamma P_1}}$

77. Answers may vary.

79. No, because there will still be a radical in the denominator. You should multiply by $\sqrt{2}-1$.

81. $\dfrac{-4+\sqrt{28}}{2}=\dfrac{-4+\sqrt{4\cdot 7}}{2}$

$=\dfrac{-4+2\sqrt{7}}{2}$

$=\dfrac{2(-2+\sqrt{7})}{2}$

$=-2+\sqrt{7}$

83. $\frac{3}{\sqrt{2}+2}=\frac{3\cdot(\sqrt{2}-2)}{(\sqrt{2}+2)(\sqrt{2}-2)}=\frac{3\sqrt{2}-6}{(\sqrt{2})^2-(2)^2}=\frac{3\sqrt{2}-6}{2-4}=\frac{3\sqrt{2}-6}{-2}=\frac{6-3\sqrt{2}}{2}$

85. $\frac{3}{\sqrt[3]{2}}=\frac{3\sqrt[3]{4}}{\sqrt[3]{2}\cdot\sqrt[3]{4}}=\frac{3\sqrt[3]{4}}{\sqrt[3]{8}}=\frac{3\sqrt[3]{4}}{2}$

87. $\frac{\sqrt[3]{500}}{\sqrt[3]{2}}=\sqrt[3]{\frac{500}{2}}=\sqrt[3]{250}=\sqrt[3]{125\cdot 2}=5\sqrt[3]{2}$

89. $\frac{\sqrt[3]{16a^5}}{\sqrt[3]{2a^3}}=\sqrt[3]{\frac{16a^5}{2a^3}}=\sqrt[3]{8a^2}=2\sqrt[3]{a^2}$

91. $(\sqrt{5}+2\sqrt{3})(\sqrt{5}-3\sqrt{3})=(\sqrt{5})^2+\sqrt{5}(-3\sqrt{3})+2\sqrt{3}(\sqrt{5})+2\sqrt{3}(-3\sqrt{3})$

$=5-3\sqrt{15}+2\sqrt{15}-6\cdot 3$

$=5-\sqrt{15}-18$

$=-13-\sqrt{15}$

9.5 Applications

Problems 9.5

1. a. $\sqrt{x+1}=5$

$(\sqrt{x+1})^2=5^2$

$x+1=25$

$x=24$

Check: $\sqrt{24+1}\stackrel{?}{=}5$

$\sqrt{25}\stackrel{?}{=}5$

$5=5$ True

24 is the only solution.

b. $\sqrt{x+1}=x+1$

$(\sqrt{x+1})^2=(x+1)^2$

$x+1=x^2+2x+1$

$0=x^2+x$

$0=(x+1)x$

$x+1=0$ or $x=0$

$x=-1$

Check: If $x=-1$, $\sqrt{-1+1} \stackrel{?}{=} -1+1$
$$\sqrt{0} \stackrel{?}{=} 0$$
$$0=0 \quad \text{True}$$
If $x=0$, $\sqrt{0+1} \stackrel{?}{=} 0+1$
$$\sqrt{1} \stackrel{?}{=} 1$$
$$1=1 \quad \text{True}$$
-1 and 0 are solutions.

2. $\sqrt{x+3}-x=-3$
$$\sqrt{x+3}=x-3$$
$$\left(\sqrt{x+3}\right)^2=(x-3)^2$$
$$x+3=x^2-6x+9$$
$$0=x^2-7x+6$$
$$0=(x-1)(x-6)$$
$x-1=0$ or $x-6=0$
$x=1$ $\quad x=6$

Check: If $x=1$, $\sqrt{1+3}-1 \stackrel{?}{=} -3$
$$\sqrt{4}-1 \stackrel{?}{=} -3$$
$$2-1 \stackrel{?}{=} -3 \quad \text{False}$$
If $x=6$, $\sqrt{6+3}-6 \stackrel{?}{=} -3$
$$\sqrt{9}-6 \stackrel{?}{=} -3$$
$$3-6 \stackrel{?}{=} -3$$
$$-3=-3 \quad \text{True}$$
6 is the only solution.

3. a. $\sqrt{x+4}=\sqrt{2x+1}$
$$\left(\sqrt{x+4}\right)^2=\left(\sqrt{2x+1}\right)^2$$
$$x+4=2x+1$$
$$4=x+1$$
$$3=x$$
Check: $\sqrt{3+4} \stackrel{?}{=} \sqrt{2\cdot 3+1}$
$$\sqrt{7}=\sqrt{7} \quad \text{True}$$
The solution is 3.

b. $\sqrt{x+3}-3\sqrt{2x-11}=0$
$$\sqrt{x+3}=3\sqrt{2x-11}$$
$$\left(\sqrt{x+3}\right)^2=\left(3\sqrt{2x-11}\right)^2$$
$$x+3=3^2(2x-11)$$
$$x+3=9(2x-11)$$
$$x+3=18x-99$$
$$3=17x-99$$
$$102=17x$$
$$6=x$$
Check: $\sqrt{6+3}-3\sqrt{2\cdot 6-11} \stackrel{?}{=} 0$
$$\sqrt{9}-3\sqrt{1} \stackrel{?}{=} 0$$
$$3-3=0 \quad \text{True}$$
The solution is 6.

4. $\sqrt{t+5}=4$
$$\left(\sqrt{t+5}\right)^2=4^2$$
$$t+5=16$$
$$t=11$$
11 years after 2000, that is, in the year 2011.

Exercises 9.5

1. $\sqrt{x}=4$
$$\left(\sqrt{x}\right)^2=4^2$$
$$x=16$$
Check: $\sqrt{16} \stackrel{?}{=} 4$
$$4=4 \quad \text{True}$$
The solution is $x=16$.

3. $\sqrt{x-1}=-2$
$$\left(\sqrt{x-1}\right)^2=(-2)^2$$
$$x-1=4$$
$$x=5$$
Check: $\sqrt{5-1} \stackrel{?}{=} -2$
$$\sqrt{4}=-2 \quad \text{False}$$
There is no solution.

5. $\sqrt{y}-2=0$

$\sqrt{y}=2$

$\left(\sqrt{y}\right)^2=2^2$

$y=4$

Check: $\sqrt{4}-2\stackrel{?}{=}0$

$2-2=0$ True

The solution is $y = 4$.

7. $\sqrt{y+1}-3=0$

$\sqrt{y+1}=3$

$\left(\sqrt{y+1}\right)^2=3^2$

$y+1=9$

$y=8$

Check: $\sqrt{8+1}-3\stackrel{?}{=}0$

$\sqrt{9}-3\stackrel{?}{=}0$

$3-3=0$ True

The solution is $y = 8$.

9. $\sqrt{x+1}=x-5$

$\left(\sqrt{x+1}\right)^2=(x-5)^2$

$x+1=x^2-10x+25$

$0=x^2-11x+24$

$0=(x-8)(x-3)$

$x-8=0$ or $x-3=0$

$x=8$ $\quad x=3$

Check: If $x = 8$, $\sqrt{8+1}\stackrel{?}{=}8-5$

$\sqrt{9}=3$ True

If $x = 3$, $\sqrt{3+1}\stackrel{?}{=}3-5$

$\sqrt{4}\stackrel{?}{=}-2$

$2=-2$ False

The solution is $x = 8$.

11. $\sqrt{x+4}=x+2$

$\left(\sqrt{x+4}\right)^2=(x+2)^2$

$x+4=x^2+4x+4$

$0=x^2+3x$

$0=x(x+3)$

$x=0$ or $x+3=0$

$x=-3$

Check: If $x = 0$, $\sqrt{0+4}\stackrel{?}{=}0+2$

$\sqrt{4}=2$ True

If $x = -3$, $\sqrt{-3+4}\stackrel{?}{=}-3+2$

$\sqrt{1}\stackrel{?}{=}-1$ False

The solution is $x = 0$.

13. $\sqrt{x-1}-x=-3$

$\sqrt{x-1}=x-3$

$\left(\sqrt{x-1}\right)^2=(x-3)^2$

$x-1=x^2-6x+9$

$0=x^2-7x+10$

$0=(x-5)(x-2)$

$x-5=0$ or $x-2=0$

$x=5$ $\quad x=2$

Check: If $x = 5$, $\sqrt{5-1}-5\stackrel{?}{=}-3$

$\sqrt{4}-5\stackrel{?}{=}-3$

$2-5\stackrel{?}{=}-3$ True

If $x = 2$, $\sqrt{2-1}-2\stackrel{?}{=}-3$

$\sqrt{1}-2\stackrel{?}{=}-3$

$1-2\stackrel{?}{=}-3$ False

The solution is $x = 5$.

15. $y-10-\sqrt{5y}=0$

$y-10=\sqrt{5y}$

$(y-10)^2=\left(\sqrt{5y}\right)^2$

$y^2-20y+100=5y$

$y^2-25y+100=0$

$(y-20)(y-5)=0$

$y-20=0$ or $y-5=0$

$y=20$ $\quad y=5$

Check: If $y = 20$,

$20-10-\sqrt{5\cdot 20}\stackrel{?}{=}0$

$10-\sqrt{100}\stackrel{?}{=}0$

$10-10\stackrel{?}{=}0$ True

If $y = 5$, $5-10-\sqrt{5\cdot 5}\stackrel{?}{=}0$

$-5-\sqrt{25}\stackrel{?}{=}0$

$-5-5\stackrel{?}{=}0$ False

The solution is $y = 20$.

17. $\sqrt{y}+20=y$

$\sqrt{y}=y-20$

$\left(\sqrt{y}\right)^2=(y-20)^2$

$y=y^2-40y+400$

$0=y^2-41y+400$

$0=(y-25)(y-16)$

$y-25=0$ or $y-16=0$

$y=25$ $y=16$

Check: If $y = 25$, $\sqrt{25}+20 \stackrel{?}{=} 25$

$5+20=25$ True

If $y = 16$, $\sqrt{16}+20 \stackrel{?}{=} 16$

$4+20 \stackrel{?}{=} 16$ False

The solution is $y = 25$.

19. $4\sqrt{y}=y+3$

$\left(4\sqrt{y}\right)^2=(y+3)^2$

$4^2(y)=y^2+6y+9$

$16y=y^2+6y+9$

$0=y^2-10y+9$

$0=(y-9)(y-1)$

$y-9=0$ or $y-1=0$

$y=9$ $y=1$

Check: If $y=9$, $4\sqrt{9} \stackrel{?}{=} 9+3$

$4\cdot 3=12$ True

If $y = 1$, $4\sqrt{1} \stackrel{?}{=} 1+3$

$4\cdot 1=4$ True

The solutions are $y = 9$ and $y = 1$.

21. $\sqrt{y+3}=\sqrt{2y-3}$

$\left(\sqrt{y+3}\right)^2=\left(\sqrt{2y-3}\right)^2$

$y+3=2y-3$

$3=y-3$

$6=y$

Check: $\sqrt{6+3} \stackrel{?}{=} \sqrt{2\cdot 6-3}$

$\sqrt{9} \stackrel{?}{=} \sqrt{9}$ True

The solution is $y = 6$.

23. $\sqrt{3x+1}=\sqrt{2x+6}$

$\left(\sqrt{3x+1}\right)^2=\left(\sqrt{2x+6}\right)^2$

$3x+1=2x+6$

$x+1=6$

$x=5$

Check: $\sqrt{3\cdot 5+1} \stackrel{?}{=} \sqrt{2\cdot 5+6}$

$\sqrt{15+1} \stackrel{?}{=} \sqrt{10+6}$

$\sqrt{16} \stackrel{?}{=} \sqrt{16}$ True

The solution is $x = 5$.

25. $2\sqrt{x+5}=\sqrt{8x+4}$

$\left(2\sqrt{x+5}\right)^2=\left(\sqrt{8x+4}\right)^2$

$2^2(x+5)=8x+4$

$4(x+5)=8x+4$

$4x+20=8x+4$

$20=4x+4$

$16=4x$

$4=x$

Check: If $x = 4$, $2\sqrt{4+5} \stackrel{?}{=} \sqrt{8\cdot 4+4}$

$2\sqrt{9} \stackrel{?}{=} \sqrt{36}$

$2\cdot 3 \stackrel{?}{=} 6$ True

The solution is $x = 4$.

27. $\sqrt{4x-1}-\sqrt{x+10}=0$

$\sqrt{4x-1}=\sqrt{x+10}$

$\left(\sqrt{4x-1}\right)^2=\left(\sqrt{x+10}\right)^2$

$4x-1=x+10$

$3x-1=10$

$3x=11$

$x=\frac{11}{3}$

Check: If $x=\frac{11}{3}$,

$\sqrt{4\left(\frac{11}{3}\right)-1}-\sqrt{\frac{11}{3}+10} \stackrel{?}{=} 0$

$\sqrt{\frac{44}{3}-\frac{3}{3}}-\sqrt{\frac{11}{3}+\frac{30}{3}} \stackrel{?}{=} 0$

$\sqrt{\frac{41}{3}}-\sqrt{\frac{41}{3}} \stackrel{?}{=} 0$ True

The solution is $x=\frac{11}{3}$.

29. $\sqrt{3y-2}-\sqrt{2y+3}=0$

$$\sqrt{3y-2}=\sqrt{2y+3}$$
$$\left(\sqrt{3y-2}\right)^2=\left(\sqrt{2y+3}\right)^2$$
$$3y-2=2y+3$$
$$y-2=3$$
$$y=5$$

Check: If $y = 5$,

$$\sqrt{3\cdot 5-2}-\sqrt{2\cdot 5+3}\overset{?}{=}0$$
$$\sqrt{15-2}-\sqrt{10+3}\overset{?}{=}0$$
$$\sqrt{13}-\sqrt{13}\overset{?}{=}0 \quad \text{True}$$

The solution is $y = 5$.

31.

$$r=\sqrt{\frac{S}{4\pi}}$$
$$2=\sqrt{\frac{S}{4\pi}}$$
$$2^2=\left(\sqrt{\frac{S}{4\pi}}\right)^2$$
$$4=\frac{S}{4\pi}$$
$$16\pi=S$$
$$16\cdot 3.14\approx S$$
$$S=50.24 \text{ sq ft}$$

33.

$$t=\sqrt{\frac{d}{16}}$$
$$3=\sqrt{\frac{d}{16}}$$
$$3^2=\left(\sqrt{\frac{d}{16}}\right)^2$$
$$9=\frac{d}{16}$$
$$144 \text{ ft}=d$$

35.

$$t=2\pi\sqrt{\frac{L}{32}}$$
$$2\approx 2\cdot\frac{22}{7}\sqrt{\frac{L}{32}}$$
$$2\approx\frac{44}{7}\sqrt{\frac{L}{32}}$$
$$\frac{7}{22}\approx\sqrt{\frac{L}{32}}$$
$$\left(\frac{7}{22}\right)^2\approx\left(\sqrt{\frac{L}{32}}\right)^2$$
$$\frac{49}{484}\approx\frac{L}{32}$$
$$L\approx 3.24 \text{ ft}$$

37.

$$A=\sqrt{1+0.04y}$$
$$1.2=\sqrt{1+0.04y}$$
$$(1.2)^2=\left(\sqrt{1+0.04y}\right)^2$$
$$1.44=1+0.04y$$
$$0.44=0.04y$$
$$11=y$$

11 years after 1989, that is, in the year 2000.

39. $\sqrt{49}=7$

41. $\sqrt{\frac{5}{16}}=\frac{\sqrt{5}}{\sqrt{16}}=\frac{\sqrt{5}}{4}$

43.

$$\sqrt[3]{x}=3$$
$$\left(\sqrt[3]{x}\right)^3=3^3$$
$$x=27$$

45.

$$\sqrt[3]{x+1}=2$$
$$\left(\sqrt[3]{x+1}\right)^3=2^3$$
$$x+1=8$$
$$x=7$$

47.

$$\sqrt[4]{x}=2$$
$$\left(\sqrt[4]{x}\right)^4=2^4$$
$$x=16$$

49. $\sqrt[4]{x-1}=-2$

$\left(\sqrt[4]{x-1}\right)^4=(-2)^4$

$x-1=16$

$x=17$

Check: $\sqrt[4]{17-1} \stackrel{?}{=} -2$

$\sqrt[4]{16} \stackrel{?}{=} -2$

$2 \stackrel{?}{=} -2$ False

There is no solution.

51. Answers may vary. Sample answer: The first step is to isolate the radical, so $\sqrt{x}=-3$. Since there is no real number whose square root is negative, there is no real-number solution.

53. Answers may vary.

55. $C=\sqrt{0.3p+1}$

$2=\sqrt{0.3p+1}$

$2^2=\left(\sqrt{0.3p+1}\right)^2$

$4=0.3p+1$

$3=0.3p$

$p=10$ thousand

57. $\sqrt{3y+10}=\sqrt{y+14}$

$\left(\sqrt{3y+10}\right)^2=\left(\sqrt{y+14}\right)^2$

$3y+10=y+14$

$2y+10=14$

$2y=4$

$y=2$

Check: $\sqrt{3\cdot 2+10} \stackrel{?}{=} \sqrt{2+14}$

$\sqrt{16} \stackrel{?}{=} \sqrt{16}$ True

The solution is $y = 2$.

59. $\sqrt{5x+1}-\sqrt{x+9}=0$

$\sqrt{5x+1}=\sqrt{x+9}$

$\left(\sqrt{5x+1}\right)^2=\left(\sqrt{x+9}\right)^2$

$5x+1=x+9$

$4x+1=9$

$4x=8$

$x=2$

Check: $\sqrt{5\cdot 2+1}-\sqrt{2+9} \stackrel{?}{=} 0$

$\sqrt{11}-\sqrt{11} \stackrel{?}{=} 0$ True

The solution is $x = 2$.

61. $\sqrt{x+2}-x=-4$

$\sqrt{x+2}=x-4$

$\left(\sqrt{x+2}\right)^2=(x-4)^2$

$x+2=x^2-8x+16$

$0=x^2-9x+14$

$0=(x-7)(x-2)$

$x-7=0$ or $x-2=0$

$x=7$ $\quad$ $x=2$

Check: If $x = 7$, $\sqrt{7+2}-7 \stackrel{?}{=} -4$

$\sqrt{9}-7 \stackrel{?}{=} -4$

$3-7 \stackrel{?}{=} -4$ True

If $x = 2$, $\sqrt{2+2}-2 \stackrel{?}{=} -4$

$\sqrt{4}-2 \stackrel{?}{=} -4$

$2-2 \stackrel{?}{=} -4$ False

The solution is $x = 7$.

63. $\sqrt{x-1}=-2$

$\left(\sqrt{x-1}\right)^2=(-2)^2$

$x-1=4$

$x=5$

Check: $\sqrt{5-1} \stackrel{?}{=} -2$

$\sqrt{4} \stackrel{?}{=} -2$ False

There is no solution.

65. $\sqrt{x-2} = x-2$

$\left(\sqrt{x-2}\right)^2 = (x-2)^2$

$x-2 = x^2 - 4x + 4$

$0 = x^2 - 5x + 6$

$0 = (x-2)(x-3)$

$x-2=0$ or $x-3=0$

$x=2$ $\quad x=3$

Check: If $x = 2$, $\sqrt{2-2} \stackrel{?}{=} 2-2$

$\sqrt{0} \stackrel{?}{=} 0$ True

If $x = 3$, $\sqrt{3-2} \stackrel{?}{=} 3-2$

$\sqrt{1} \stackrel{?}{=} 1$ True

The solutions are $x = 2$ and $x = 3$.

Review Exercises

1. a. $\sqrt{81} = 9$

b. $\sqrt{-64}$ is not a real number, since there is no number that we can square and get –64 for an answer.

c. $\sqrt{\frac{36}{25}} = \frac{6}{5}$

2. a. $-\sqrt{36} = -6$

b. $-\sqrt{\frac{64}{25}} = -\frac{8}{5}$

c. $\sqrt{-\frac{9}{4}}$ is not a real number, since there is no number that we can square and get $-\frac{9}{4}$ for an answer.

3. a. $\left(\sqrt{8}\right)^2 = 8$

b. $\left(\sqrt{25}\right)^2 = 25$

c. $\left(\sqrt{17}\right)^2 = 17$

4. a. $\left(-\sqrt{36}\right)^2 = 36$

b. $\left(-\sqrt{17}\right)^2 = 17$

c. $\left(-\sqrt{64}\right) = 64$

5. a. $\left(\sqrt{x^2+1}\right)^2 = x^2 + 1$

b. $\left(\sqrt{x^2+4}\right)^2 = x^2 + 4$

c. $\left(-\sqrt{x^2+5}\right)^2 = x^2 + 5$

6. a. 11 is not a perfect square, so $\sqrt{11} \approx 3.3166$ is irrational.

b. $25 = 5^2$ is a perfect square, so $-\sqrt{25} = -5$ is a rational number.

c. $\sqrt{-9}$ is not a real number, since there is no number that we can square and get –9 for an answer.

7. a. $\frac{9}{4} = \left(\frac{3}{2}\right)^2$ is a perfect square, so $\sqrt{\frac{9}{4}} = \frac{3}{2}$ is a rational number.

b. $-\sqrt{\frac{9}{4}} = -\frac{3}{2}$ is a rational number.

c. $\sqrt{-\frac{9}{4}}$ is not a real number, since there is no number that we can square and get $-\frac{9}{4}$ for an answer.

8. a. $\sqrt[3]{64} = \sqrt[3]{4^3} = 4$

b. $\sqrt[3]{-8} = \sqrt[3]{(-2)^3} = -2$

c. $-\sqrt[4]{81} = -\sqrt[4]{3^4} = -3$

9. a. $\sqrt[4]{16} = \sqrt[4]{2^4} = 2$

b. $-\sqrt[4]{16} = -\sqrt[4]{2^4} = -2$

c. $\sqrt[4]{-16}$ is not a real number, since there is no number whose fourth power is -16.

10. $t = \sqrt{\dfrac{d}{16}}$

a. $t = \sqrt{\dfrac{121}{16}} = \dfrac{11}{4} = 2\dfrac{3}{4}$ sec

b. $t = \sqrt{\dfrac{144}{16}} = \dfrac{12}{4} = 3$ sec

c. $t = \sqrt{\dfrac{169}{16}} = \dfrac{13}{4} = 3\dfrac{1}{4}$ sec

11. a. $\sqrt{32} = \sqrt{16 \cdot 2} = \sqrt{16} \cdot \sqrt{2} = 4\sqrt{2}$

b. $\sqrt{48} = \sqrt{16 \cdot 3} = \sqrt{16} \cdot \sqrt{3} = 4\sqrt{3}$

c. $\sqrt{196} = 14$

12. a. $\sqrt{3} \cdot \sqrt{7} = \sqrt{21}$

b. $\sqrt{12} \cdot \sqrt{3} = \sqrt{36} = 6$

c. $\sqrt{5}\sqrt{y} = \sqrt{5y}$

13. a. $\sqrt{\dfrac{3}{16}} = \dfrac{\sqrt{3}}{\sqrt{16}} = \dfrac{\sqrt{3}}{4}$

b. $\sqrt{\dfrac{5}{36}} = \dfrac{\sqrt{5}}{\sqrt{36}} = \dfrac{\sqrt{5}}{6}$

c. $\sqrt{\dfrac{9}{4}} = \dfrac{3}{2}$

14. a. $\dfrac{\sqrt{8}}{\sqrt{2}} = \sqrt{\dfrac{8}{2}} = \sqrt{4} = 2$

b. $\dfrac{\sqrt{21}}{\sqrt{3}} = \sqrt{\dfrac{21}{3}} = \sqrt{7}$

c. $\dfrac{6\sqrt{50}}{2\sqrt{10}} = 3\sqrt{\dfrac{50}{10}} = 3\sqrt{5}$

15. a. $\sqrt{36x^2} = \sqrt{36} \cdot \sqrt{x^2} = 6x$

b. $\sqrt{100y^4} = \sqrt{100} \cdot \sqrt{y^4} = 10y^2$

c. $\sqrt{81n^8} = \sqrt{81} \cdot \sqrt{n^8} = 9n^4$

16. a.
$$\begin{aligned}\sqrt{72y^{10}} &= \sqrt{36 \cdot 2y^{10}} \\ &= \sqrt{36} \cdot \sqrt{2} \cdot \sqrt{y^{10}} \\ &= 6y^5\sqrt{2}\end{aligned}$$

b.
$$\begin{aligned}\sqrt{147z^8} &= \sqrt{49 \cdot 3z^8} \\ &= \sqrt{49} \cdot \sqrt{3} \cdot \sqrt{z^8} \\ &= 7z^4\sqrt{3}\end{aligned}$$

c.
$$\begin{aligned}\sqrt{48x^{12}} &= \sqrt{16 \cdot 3x^{12}} \\ &= \sqrt{16} \cdot \sqrt{3} \cdot \sqrt{x^{12}} \\ &= 4x^6\sqrt{3}\end{aligned}$$

17. a. $\sqrt{y^{15}} = \sqrt{y^{14} \cdot y} = y^7\sqrt{y}$

b. $\sqrt{y^{13}} = \sqrt{y^{12} \cdot y} = y^6\sqrt{y}$

c. $\sqrt{50n^7} = \sqrt{25 \cdot 2 \cdot n^6 \cdot n} = 5n^3\sqrt{2n}$

18. a. $\sqrt[3]{24} = \sqrt[3]{8 \cdot 3} = \sqrt[3]{2^3 \cdot 3} = 2\sqrt[3]{3}$

b. $\sqrt[3]{\dfrac{8}{27}} = \dfrac{\sqrt[3]{8}}{\sqrt[3]{27}} = \dfrac{\sqrt[3]{2^3}}{\sqrt[3]{3^3}} = \dfrac{2}{3}$

c. $\sqrt[3]{-\frac{125}{64}} = \sqrt[3]{\left(-\frac{5}{4}\right)^3} = -\frac{5}{4}$

19. a. $\sqrt[4]{81} = \sqrt[4]{3^4} = 3$

b. $\sqrt[4]{48} = \sqrt[4]{16 \cdot 3} = \sqrt[4]{2^4 \cdot 3} = 2\sqrt[4]{3}$

c. $\sqrt[4]{80} = \sqrt[4]{16 \cdot 5} = \sqrt[4]{2^4 \cdot 5} = 2\sqrt[4]{5}$

20. a. $7\sqrt{3} + 8\sqrt{3} = 15\sqrt{3}$

b. $\sqrt{32} + 5\sqrt{2} = \sqrt{16 \cdot 2} + 5\sqrt{2}$
$= \sqrt{16} \cdot \sqrt{2} + 5\sqrt{2}$
$= 4\sqrt{2} + 5\sqrt{2}$
$= 9\sqrt{2}$

c. $\sqrt{12} + \sqrt{48} = \sqrt{4 \cdot 3} + \sqrt{16 \cdot 3}$
$= \sqrt{4} \cdot \sqrt{3} + \sqrt{16} \cdot \sqrt{3}$
$= 2\sqrt{3} + 4\sqrt{3}$
$= 6\sqrt{3}$

21. a. $9\sqrt{11} - 6\sqrt{11} = 3\sqrt{11}$

b. $\sqrt{50} - 4\sqrt{2} = \sqrt{25 \cdot 2} - 4\sqrt{2}$
$= \sqrt{25} \cdot \sqrt{2} - 4\sqrt{2}$
$= 5\sqrt{2} - 4\sqrt{2}$
$= \sqrt{2}$

c. $\sqrt{108} - \sqrt{75} = \sqrt{36 \cdot 3} - \sqrt{25 \cdot 3}$
$= \sqrt{36} \cdot \sqrt{3} - \sqrt{25} \cdot \sqrt{3}$
$= 6\sqrt{3} - 5\sqrt{3}$
$= \sqrt{3}$

22. a. $\sqrt{3}\left(\sqrt{20} - \sqrt{2}\right) = \sqrt{3}\sqrt{20} - \sqrt{3}\sqrt{2}$
$= \sqrt{60} - \sqrt{6}$
$= \sqrt{4 \cdot 15} - \sqrt{6}$
$= \sqrt{4} \cdot \sqrt{15} - \sqrt{6}$
$= 2\sqrt{15} - \sqrt{6}$

b. $\sqrt{5}\left(\sqrt{5} - \sqrt{3}\right) = \sqrt{5}\sqrt{5} - \sqrt{5}\sqrt{3}$
$= \sqrt{25} - \sqrt{15}$
$= 5 - \sqrt{15}$

c. $\sqrt{7}\left(\sqrt{7} - \sqrt{98}\right) = \sqrt{7}\sqrt{7} - \sqrt{7}\sqrt{98}$
$= \sqrt{49} - \sqrt{686}$
$= 7 - \sqrt{49 \cdot 14}$
$= 7 - \sqrt{49} \cdot \sqrt{14}$
$= 7 - 7\sqrt{14}$

23. a. $\sqrt{\frac{5}{8}} = \frac{\sqrt{5}}{\sqrt{8}} = \frac{\sqrt{5} \cdot \sqrt{2}}{\sqrt{8} \cdot \sqrt{2}} = \frac{\sqrt{10}}{\sqrt{16}} = \frac{\sqrt{10}}{4}$

b. $\sqrt{\frac{x^2}{50}} = \frac{\sqrt{x^2}}{\sqrt{50}} = \frac{\sqrt{x^2} \cdot \sqrt{2}}{\sqrt{50} \cdot \sqrt{2}} = \frac{x\sqrt{2}}{\sqrt{100}} = \frac{x\sqrt{2}}{10}$

c. $\sqrt{\frac{y^2}{27}} = \frac{\sqrt{y^2}}{\sqrt{27}} = \frac{\sqrt{y^2} \cdot \sqrt{3}}{\sqrt{27} \cdot \sqrt{3}} = \frac{y\sqrt{3}}{\sqrt{81}} = \frac{y\sqrt{3}}{9}$

24. a. $\sqrt{32+4} - \sqrt{9} = \sqrt{36} - \sqrt{9} = 6 - 3 = 3$

b. $\sqrt{18+7} - \sqrt{4} = \sqrt{25} - \sqrt{4} = 5 - 2 = 3$

c. $\sqrt{60+4} - \sqrt{16} = \sqrt{64} - \sqrt{16} = 8 - 4 = 4$

25. a. $8\sqrt{15} - \sqrt{3} \cdot \sqrt{5} = 8\sqrt{15} - \sqrt{15} = 7\sqrt{15}$

b. $7\sqrt{6} - \sqrt{2} \cdot \sqrt{3} = 7\sqrt{6} - \sqrt{6} = 6\sqrt{6}$

c. $9\sqrt{14} - 2\sqrt{7} \cdot \sqrt{2} = 9\sqrt{14} - 2\sqrt{14} = 7\sqrt{14}$

26. a. $\frac{\sqrt[3]{162}}{\sqrt[3]{2}} = \sqrt[3]{81} = \sqrt[3]{27 \cdot 3} = 3\sqrt[3]{3}$

b. $\frac{\sqrt[3]{135}}{\sqrt[3]{5}} = \sqrt[3]{27} = 3$

c. $\frac{\sqrt[3]{192}}{\sqrt[3]{24}} = \sqrt[3]{8} = 2$

27. a. $\frac{7}{\sqrt[3]{4}} = \frac{7 \cdot \sqrt[3]{2}}{\sqrt[3]{4} \cdot \sqrt[3]{2}} = \frac{7\sqrt[3]{2}}{\sqrt[3]{8}} = \frac{7\sqrt[3]{2}}{2}$

b. $\frac{5}{\sqrt[3]{9}}=\frac{5\cdot\sqrt[3]{3}}{\sqrt[3]{9}\cdot\sqrt[3]{3}}=\frac{5\sqrt[3]{3}}{\sqrt[3]{27}}=\frac{5\sqrt[3]{3}}{3}$

c. $\frac{9}{\sqrt[3]{25}}=\frac{9\cdot\sqrt[3]{5}}{\sqrt[3]{25}\cdot\sqrt[3]{5}}=\frac{9\sqrt[3]{5}}{\sqrt[3]{125}}=\frac{9\sqrt[3]{5}}{5}$

28. a.
$$\begin{aligned}\left(\sqrt{3}+3\sqrt{2}\right)\left(\sqrt{3}-5\sqrt{2}\right)&=\sqrt{3}\cdot\sqrt{3}+\sqrt{3}\left(-5\sqrt{2}\right)+3\sqrt{2}\cdot\sqrt{3}+3\sqrt{2}\left(-5\sqrt{2}\right)\\&=3-5\sqrt{6}+3\sqrt{6}-15\cdot 2\\&=3-2\sqrt{6}-30\\&=-27-2\sqrt{6}\end{aligned}$$

b.
$$\begin{aligned}\left(\sqrt{7}+3\sqrt{5}\right)\left(\sqrt{7}-2\sqrt{5}\right)&=\sqrt{7}\cdot\sqrt{7}+\sqrt{7}\left(-2\sqrt{5}\right)+3\sqrt{5}\cdot\sqrt{7}+3\sqrt{5}\left(-2\sqrt{5}\right)\\&=7-2\sqrt{35}+3\sqrt{35}-6\cdot 5\\&=7+\sqrt{35}-30\\&=-23+\sqrt{35}\end{aligned}$$

29. a. $\left(\sqrt{7}+2\sqrt{3}\right)\left(\sqrt{7}-2\sqrt{3}\right)=\left(\sqrt{7}\right)^2-\left(2\sqrt{3}\right)^2=7-\left[2^2\left(\sqrt{3}\right)^2\right]=7-4\cdot 3=7-12=-5$

b. $\left(\sqrt{11}+3\sqrt{5}\right)\left(\sqrt{11}-3\sqrt{5}\right)=\left(\sqrt{11}\right)^2-\left(3\sqrt{5}\right)^2=11-\left[3^2\left(\sqrt{5}\right)^2\right]=11-9\cdot 5=11-45=-34$

30. a. $\frac{3}{\sqrt{3}+1}=\frac{3\cdot\left(\sqrt{3}-1\right)}{\left(\sqrt{3}+1\right)\left(\sqrt{3}-1\right)}=\frac{3\sqrt{3}-3}{\left(\sqrt{3}\right)^2-1^2}=\frac{3\sqrt{3}-3}{3-1}=\frac{3\sqrt{3}-3}{2}$

b. $\frac{5}{\sqrt{2}-1}=\frac{5\cdot\left(\sqrt{2}+1\right)}{\left(\sqrt{2}-1\right)\left(\sqrt{2}+1\right)}=\frac{5\sqrt{2}+5}{\left(\sqrt{2}\right)^2-1^2}=\frac{5\sqrt{2}+5}{2-1}=\frac{5\sqrt{2}+5}{1}=5\sqrt{2}+5$

31. a. $\frac{7}{\sqrt{3}-\sqrt{2}}=\frac{7\cdot\left(\sqrt{3}+\sqrt{2}\right)}{\left(\sqrt{3}-\sqrt{2}\right)\left(\sqrt{3}+\sqrt{2}\right)}=\frac{7\sqrt{3}+7\sqrt{2}}{\left(\sqrt{3}\right)^2-\left(\sqrt{2}\right)^2}=\frac{7\sqrt{3}+7\sqrt{2}}{3-2}=\frac{7\sqrt{3}+7\sqrt{2}}{1}=7\sqrt{3}+7\sqrt{2}$

b. $\frac{2}{\sqrt{5}-\sqrt{2}}=\frac{2\cdot\left(\sqrt{5}+\sqrt{2}\right)}{\left(\sqrt{5}-\sqrt{2}\right)\left(\sqrt{5}+\sqrt{2}\right)}=\frac{2\sqrt{5}+2\sqrt{2}}{\left(\sqrt{5}\right)^2-\left(\sqrt{2}\right)^2}=\frac{2\sqrt{5}+2\sqrt{2}}{5-2}=\frac{2\sqrt{5}+2\sqrt{2}}{3}$

32. a. $\frac{-8+\sqrt{8}}{2}=\frac{-8+\sqrt{4\cdot 2}}{2}=\frac{-8+2\sqrt{2}}{2}=\frac{2\left(-4+\sqrt{2}\right)}{2}=-4+\sqrt{2}$

b. $\frac{-16+\sqrt{12}}{4} = \frac{-16+\sqrt{4\cdot 3}}{4}$
$= \frac{-16+2\sqrt{3}}{4}$
$= \frac{2\left(-8+\sqrt{3}\right)}{4}$
$= \frac{-8+\sqrt{3}}{2}$

33. a. $\sqrt{x+2} = 3$
$\left(\sqrt{x+2}\right)^2 = 3^2$
$x+2 = 9$
$x = 7$
Check: $\sqrt{7+2} \overset{?}{=} 3$
$\sqrt{9} = 3$ True
The solution is $x = 7$.

b. $\sqrt{x-2} = -2$
$\left(\sqrt{x-2}\right)^2 = (-2)^2$
$x-2 = 4$
$x = 6$
Check: $\sqrt{6-2} \overset{?}{=} -2$
$\sqrt{4} \overset{?}{=} -2$ False
There is no solution.

34. a. $\sqrt{x+5} = x-1$
$\left(\sqrt{x+5}\right)^2 = (x-1)^2$
$x+5 = x^2 - 2x + 1$
$0 = x^2 - 3x - 4$
$0 = (x-4)(x+1)$
$x-4 = 0$ or $x+1 = 0$
$x = 4$ $\quad$ $x = -1$
Check: If $x = 4$, $\sqrt{4+5} \overset{?}{=} 4-1$
$\sqrt{9} = 3$ True
If $x = -1$, $\sqrt{-1+5} \overset{?}{=} -1-1$
$\sqrt{4} \overset{?}{=} -2$ False
The solution is $x = 4$.

b. $\sqrt{x+10} = x-2$
$\left(\sqrt{x+10}\right)^2 = (x-2)^2$
$x+10 = x^2 - 4x + 4$
$0 = x^2 - 5x - 6$
$0 = (x-6)(x+1)$
$x-6 = 0$ or $x+1 = 0$
$x = 6$ $\quad$ $x = -1$
Check: If $x = 6$, $\sqrt{6+10} \overset{?}{=} 6-2$
$\sqrt{16} = 4$ True
If $x = -1$, $\sqrt{-1+10} \overset{?}{=} -1-2$
$\sqrt{9} \overset{?}{=} -3$ False
The solution is $x = 6$.

35. a. $\sqrt{x+4} - x = -2$
$\sqrt{x+4} = x-2$
$\left(\sqrt{x+4}\right)^2 = (x-2)^2$
$x+4 = x^2 - 4x + 4$
$0 = x^2 - 5x$
$0 = x(x-5)$
$x = 0$ or $x-5 = 0$
$x = 5$
Check: If $x = 0$, $\sqrt{0+4} - 0 \overset{?}{=} -2$
$\sqrt{4} \overset{?}{=} -2$ False
If $x = 5$, $\sqrt{5+4} - 5 \overset{?}{=} -2$
$\sqrt{9} - 5 \overset{?}{=} -2$
$3 - 5 \overset{?}{=} -2$ True
The solution is $x = 5$.

b. $\sqrt{x+2} - x = -4$
$\sqrt{x+2} = x-4$
$\left(\sqrt{x+2}\right)^2 = (x-4)^2$
$x+2 = x^2 - 8x + 16$
$0 = x^2 - 9x + 14$
$0 = (x-7)(x-2)$
$x-7 = 0$ or $x-2 = 0$
$x = 7$ $\quad$ $x = 2$

Check: If $x = 7$, $\sqrt{7+2} - 7 \stackrel{?}{=} -4$

$\sqrt{9} - 7 \stackrel{?}{=} -4$

$3 - 7 \stackrel{?}{=} -4$ True

If $x = 2$, $\sqrt{2+2} - 2 \stackrel{?}{=} -4$

$\sqrt{4} - 2 \stackrel{?}{=} -4$

$2 - 2 \stackrel{?}{=} -4$ False

The solution is $x = 7$.

36. a.

$$\sqrt{y+5} = \sqrt{3y-3}$$
$$\left(\sqrt{y+5}\right)^2 = \left(\sqrt{3y-3}\right)^2$$
$$y+5 = 3y-3$$
$$5 = 2y-3$$
$$8 = 2y$$
$$4 = y$$

Check: $\sqrt{4+5} \stackrel{?}{=} \sqrt{3\cdot 4-3}$

$\sqrt{9} \stackrel{?}{=} \sqrt{12-3}$

$\sqrt{9} \stackrel{?}{=} \sqrt{9}$ True

The solution is $y = 4$.

b.

$$\sqrt{y+5} = \sqrt{2y+5}$$
$$\left(\sqrt{y+5}\right)^2 = \left(\sqrt{2y+5}\right)^2$$
$$y+5 = 2y+5$$
$$5 = y+5$$
$$0 = y$$

Check: $\sqrt{0+5} \stackrel{?}{=} \sqrt{2\cdot 0+5}$

$\sqrt{5} \stackrel{?}{=} \sqrt{5}$ True

The solution is $y = 0$.

37. a.

$$\sqrt{y+8} - 3\sqrt{2y-1} = 0$$
$$\sqrt{y+8} = 3\sqrt{2y-1}$$
$$\left(\sqrt{y+8}\right)^2 = \left(3\sqrt{2y-1}\right)^2$$
$$y+8 = 3^2(2y-1)$$
$$y+8 = 9(2y-1)$$
$$y+8 = 18y-9$$
$$8 = 17y-9$$
$$17 = 17y$$
$$1 = y$$

Check: $\sqrt{1+8} - 3\sqrt{2\cdot 1-1} \stackrel{?}{=} 0$

$\sqrt{9} - 3\sqrt{1} \stackrel{?}{=} 0$

$3 - 3 = 0$ True

The solution is $y = 1$.

b.

$$\sqrt{y+9} - 3\sqrt{2y+1} = 0$$
$$\sqrt{y+9} = 3\sqrt{2y+1}$$
$$\left(\sqrt{y+9}\right)^2 = \left(3\sqrt{2y+1}\right)^2$$
$$y+9 = 3^2(2y+1)$$
$$y+9 = 9(2y+1)$$
$$y+9 = 18y+9$$
$$9 = 17y+9$$
$$0 = 17y$$
$$0 = y$$

Check: $\sqrt{0+9} - 3\sqrt{2\cdot 0+1} \stackrel{?}{=} 0$

$\sqrt{9} - 3\sqrt{1} \stackrel{?}{=} 0$

$3 - 3 \stackrel{?}{=} 0$ True

The solution is $y = 0$.

38. a.

$$C = \sqrt{0.2x+1}$$
$$3 = \sqrt{0.2x+1}$$
$$3^2 = \left(\sqrt{0.2x+1}\right)^2$$
$$9 = 0.2x+1$$
$$8 = 0.2x$$

$x = 40$ thousand or 40,000

b. $C=\sqrt{0.2x+1}$
$$7=\sqrt{0.2x+1}$$
$$7^2=\left(\sqrt{0.2x+1}\right)^2$$
$$49=0.2x+1$$
$$48=0.2x$$
$x=240$ thousand or 240,000

Cumulative Review Chapters 1–9

1. $9=3\cdot 3$
$8=2\cdot 2\cdot 2$
$\text{LCD}=2\cdot 2\cdot 2\cdot 3\cdot 3=72$
$$-\frac{2}{9}=-\frac{2\cdot 8}{9\cdot 8}=-\frac{16}{72}$$
$$-\frac{1}{8}=-\frac{1\cdot 9}{8\cdot 9}=-\frac{9}{72}$$
$$-\frac{2}{9}+\left(-\frac{1}{8}\right)=-\frac{16}{72}+\left(-\frac{9}{72}\right)=-\frac{25}{72}$$

2. $(-3)^4=(-3)\cdot(-3)\cdot(-3)\cdot(-3)=81$

3. $-\frac{1}{6}\div\left(-\frac{7}{12}\right)=-\frac{1}{\cancel{6}_1}\cdot\left(-\frac{\cancel{12}^2}{7}\right)=\frac{2}{7}$

4. $y\div 5\cdot x-z=50\div 5\cdot 5-3$
$=10\cdot 5-3$
$=50-3$
$=47$

5. $2x-(x+4)-2(x+3)$
$=2x-x-4-2x-6$
$=(2x-x-2x)+(-4-6)$
$=-x+(-10)$ or $-x-10$

6. The quotient of $(m+3n)$ and p is $\frac{m+3n}{p}$.

7.
$$2=5(x-3)+1-4x$$
$$2=5x-15+1-4x$$
$$2=(5x-4x)+(-15+1)$$
$$2=x+(-14)$$
$$2+14=x+(-14)+14$$
$$16=x$$

8.
$$\frac{x}{2}-\frac{x}{3}=1$$
$$6\cdot\frac{x}{2}-6\cdot\frac{x}{3}=6\cdot 1$$
$$3x-2x=6$$
$$x=6$$

9.
$$-\frac{x}{6}+\frac{x}{2}\le\frac{x-2}{2}$$
$$6\cdot\left(-\frac{x}{6}\right)+6\cdot\frac{x}{2}\le 6\left(\frac{x-2}{2}\right)$$
$$-x+3x\le 3(x-2)$$
$$2x\le 3x-6$$
$$2x-3x\le 3x-3x-6$$
$$-x\le -6$$
$$x\ge 6$$

0 6

10. Start at the origin. To reach the point $(-1,-3)$, go 1 unit to the left and 3 units down.

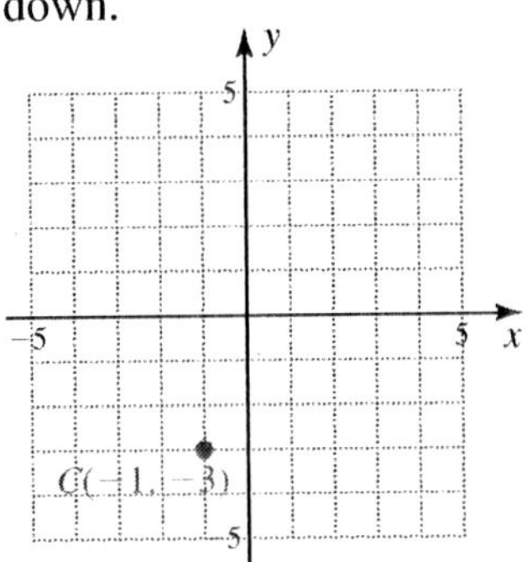

11. $4x-y=-1$
$4(-1)-3=-1$
$-4-3=-1$ not true
$(-1, 3)$ is not a solution of $4x-y=-1$.

12.
$$2x-4y=-10$$
$$2x-4(3)=-10$$
$$2x-12=-10$$
$$2x-12+12=-10+12$$
$$2x=2$$
$$x=1$$

13. $5x+y=5$
Let $x=0$. $5(0)+y=5$
$y=5$
$(0, 5)$ is the y-intercept.

Let $y = 0$. $5x + 0 = 5$

$$5x = 5$$
$$x = 1$$

(1, 0) is the x-intercept. Graph the points (0, 5) and (1, 0) and connect them with a line.

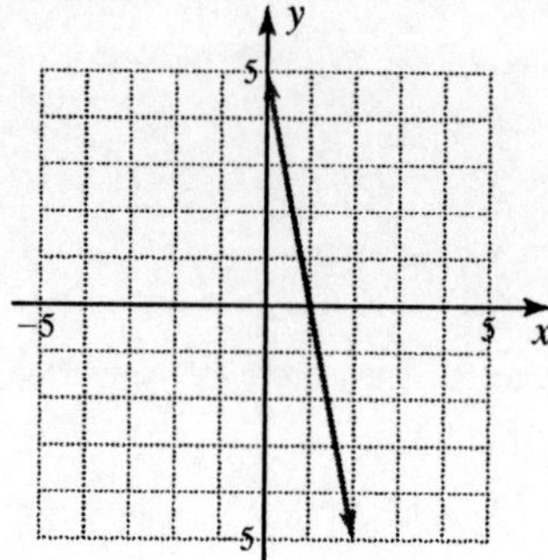

14. $4y - 8 = 0$

$$4y = 8$$
$$y = 2$$

The graph of $y = 2$ is a horizontal line crossing the y-axis at 2.

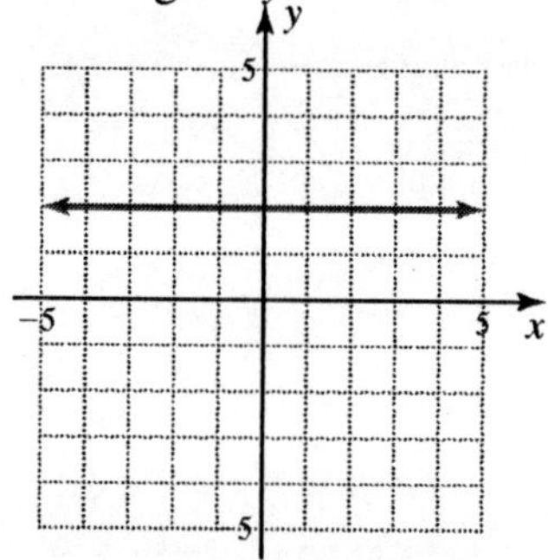

15. $(x_1, y_1) = (0, 4)$

$(x_2, y_2) = (6, 1)$

$$m = \frac{1-4}{6-0} = \frac{-3}{6} = -\frac{1}{2}$$

16. $3x - 3y = -8$

$$-3y = -3x - 8$$
$$y = x + \frac{8}{3}$$

Since the coefficient of x is 1, the slope is 1.

17. **(1)** $15x - 12y = 4$

$$-12y = -15x + 4$$
$$y = \frac{15}{12}x - \frac{4}{12}$$
$$y = \frac{5}{4}x - \frac{1}{3}$$

(2) $12y + 15x = 4$

$$12y = -15x + 4$$
$$y = -\frac{15}{12}x + \frac{4}{12}$$
$$y = -\frac{5}{4}x + \frac{1}{3}$$

(3) $-4y = -5x + 4$

$$y = \frac{5}{4}x - 1$$

The lines in equations (1) and (3) have the same slope, $\frac{5}{4}$.

18. $(2x^4y^{-3})^{-4} = 2^{-4}x^{-16}y^{12}$

$$= \frac{y^{12}}{2^4x^{16}}$$
$$= \frac{y^{12}}{16x^{16}}$$

19. $0.00000025 = 2.5 \times 10^{-7}$

20. $2.72 \div 1.6 = 1.7$

$10^{-4} \div 10^4 = 10^{-4-4} = 10^{-8}$

$(2.72 \times 10^{-4}) \div (1.6 \times 10^4) = 1.7 \times 10^{-8}$

21. $\left(4x^2 - \frac{1}{2}\right)^2 = (4x^2)^2 - 2 \cdot \frac{1}{2} \cdot 4x^2 + \left(\frac{1}{2}\right)^2$

$$= 16x^4 - 4x^2 + \frac{1}{4}$$

22.
$$\begin{array}{r} 2x^2 - x - 2 \\ x-3\overline{\big) \; 2x^3 - 7x^2 \;\; + x + 9} \\ \underline{(-)2x^3 - 6x^2} \qquad\quad \\ -x^2 \;\; + x \\ \underline{(-)-x^2 + 3x} \\ -2x + 9 \\ \underline{(-)-2x + 6} \\ 3 \end{array}$$

$(2x^3 - 7x^2 + x + 9) \div (x - 3)$
$= (2x^2 - x - 3)$ R 3

23. $x^2 - 4x + 3$
Factors: $-1, -3$
Sum: -4
$x^2 - 4x + 3 = (x - 1)(x - 3)$

24. $9x^2 - 27xy + 20y^2$
Key number: $9 \cdot 20 = 180$
$= 9x^2 - 12xy - 15xy + 20y^2$
$= (9x^2 - 12xy) - (15xy - 20y^2)$
$= 3x(3x - 4y) - 5y(3x - 4y)$
$= (3x - 5y)(3x - 4y)$

25. $4x^2 - 25y^2 = (2x)^2 - (5y)^2$
$= (2x + 5y)(2x - 5y)$

26. $-5x^4 + 5x^2 = -5x^2(x^2 - 1)$
$= -5x^2(x + 1)(x - 1)$

27. $4x^3 - 8x^2 - 12x = 4x(x^2 - 2x - 3)$
$= 4x(x + 1)(x - 3)$

28. $3x^2 + 4x + 9x + 12 = (3x^2 + 9x) + (4x + 12)$
$= 3x(x + 3) + 4(x + 3)$
$= (3x + 4)(x + 3)$

29. $16kx^2 + 8kx + k = k(16x^2 + 8x + 1)$
$= k(4x + 1)(4x + 1)$
$= k(4x + 1)^2$

30. $4x^2 + 17x = 15$
$4x^2 + 17x - 15 = 0$
$(4x - 3)(x + 5) = 0$
$4x - 3 = 0$ or $x + 5 = 0$
$4x = 3$ $\quad x = -5$
$x = \dfrac{3}{4}$

31. $15y^2 = 5y(3y)$
$\dfrac{2x}{5y} = \dfrac{2x(3y)}{5y(3y)} = \dfrac{6xy}{15y^2}$

32. $\dfrac{-16(x^2 - y^2)}{4(x - y)} = \dfrac{-4 \cdot \overset{1}{\cancel{4}}(x + y)\overset{1}{\cancel{(x - y)}}}{\underset{1}{\cancel{4}}\underset{1}{\cancel{(x - y)}}}$

$= -4(x + y)$

33. $x^2 + 4x - 21 = (x + 7)(x - 3)$
$3 - x = -1(x - 3)$
$\dfrac{x^2 + 4x - 21}{3 - x} = \dfrac{(x + 7)\overset{1}{\cancel{(x - 3)}}}{-1\underset{1}{\cancel{(x - 3)}}} = -(x + 7)$

34. $x^2 - 64 = (x + 8)(x - 8)$
$(x - 8) \cdot \dfrac{x + 4}{x^2 - 64} = \dfrac{\overset{1}{\cancel{x - 8}}}{1} \cdot \dfrac{x + 4}{(x + 8)\underset{1}{\cancel{(x - 8)}}}$

$= \dfrac{x + 4}{x + 8}$

35. $\frac{x+5}{x-5} \div \frac{x^2-25}{5-x} = \frac{x+5}{x-5} \cdot \frac{5-x}{x^2-25}$

$= \frac{\overset{1}{\cancel{x+5}}}{\underset{1}{\cancel{x-5}}} \cdot \frac{\overset{-1}{\cancel{5-x}}}{\underset{1}{\cancel{(x+5)}}(x-5)}$

$= \frac{-1}{x-5}$

$= \frac{(-1)(-1)}{(-1)(x-5)}$

$= \frac{1}{-x+5}$

$= \frac{1}{5-x}$

36. $\frac{7}{3(x+5)} + \frac{5}{3(x+5)} = \frac{7+5}{3(x+5)}$

$= \frac{\overset{4}{\cancel{12}}}{\underset{1}{\cancel{3}}(x+5)}$

$= \frac{4}{x+5}$

37. $x^2 + x - 20 = (x+5)(x-4)$

$x^2 - 16 = (x-4)(x+4)$

$\text{LCD} = (x+5)(x-4)(x+4)$

$\frac{x+4}{x^2+x-20} - \frac{x+5}{x^2-16} = \frac{(x+4)(x+4)}{(x+5)(x-4)(x+4)} - \frac{(x+5)(x+5)}{(x+5)(x-4)(x+4)}$

$= \frac{(x^2+8x+16)-(x^2+10x+25)}{(x+5)(x-4)(x+4)}$

$= \frac{x^2+8x+16-x^2-10x-25}{(x+5)(x-4)(x+4)}$

$= \frac{-2x-9}{(x+5)(x-4)(x+4)}$

38. $\frac{\frac{2}{3x}+\frac{1}{2x}}{\frac{1}{x}+\frac{1}{4x}} = \frac{12x \cdot \left(\frac{2}{3x}+\frac{1}{2x}\right)}{12x \cdot \left(\frac{1}{x}+\frac{1}{4x}\right)} = \frac{12x \cdot \frac{2}{3x} + 12x \cdot \frac{1}{2x}}{12x \cdot \frac{1}{x} + 12x \cdot \frac{1}{4x}} = \frac{8+6}{12+3} = \frac{14}{15}$

39.
$$\frac{4x}{x-4}+1=\frac{3x}{x-4}$$
$$(x-4)\cdot\frac{4x}{x-4}+1(x-4)=(x-4)\cdot\frac{3x}{x-4}$$
$$4x+x-4=3x$$
$$5x-4=3x$$
$$-4=-2x$$
$$2=x$$

40.
$$\frac{x}{x^2-4}+\frac{2}{x-2}=\frac{1}{x+2}$$
$$\frac{x}{(x+2)(x-2)}+\frac{2}{x-2}=\frac{1}{x+2}$$
$$(x+2)(x-2)\cdot\frac{x}{(x+2)(x-2)}+(x+2)(x-2)\cdot\frac{2}{x-2}=(x+2)(x-2)\cdot\frac{1}{x+2}$$
$$x+2(x+2)=x-2$$
$$x+2x+4=x-2$$
$$3x+4=x-2$$
$$2x=-6$$
$$x=-3$$

41.
$$\frac{x}{x+6}-\frac{1}{7}=\frac{-6}{x+6}$$
$$7(x+6)\cdot\frac{x}{x+6}-\frac{1}{7}\cdot 7(x+6)=\frac{-6}{x+6}\cdot 7(x+6)$$
$$7x-x-6=-42$$
$$6x-6=-42$$
$$6x=-36$$
$$x=-6$$

Check: $\frac{-6}{-6+6}-\frac{1}{7}=\frac{-6}{-6+6}$

$$\frac{-6}{0}-\frac{1}{7}=\frac{-6}{0}$$

Two of the terms are not defined.
The equation has no solution.

42.
$$1+\frac{2}{x-5}=\frac{20}{x^2-25}$$
$$1+\frac{2}{x-5}=\frac{20}{(x+5)(x-5)}$$
$$(x+5)(x-5)\cdot 1+(x+5)(x-5)\cdot\frac{2}{x-5}=(x+5)(x-5)\cdot\frac{20}{(x+5)(x-5)}$$
$$x^2-25+2(x+5)=20$$
$$x^2-25+2x+10=20$$
$$x^2+2x-15=20$$
$$x^2+2x-35=0$$
$$(x+7)(x-5)=0$$
$x+7=0$ or $x-5=0$
$x=-7$ $\quad x=5$

Since $x = 5$ makes the denominator $x - 5$ equal to zero, the only possible solution is $x = -7$.

43. Let g be the gallons needed.
$$\frac{100}{4}=\frac{625}{g}$$
$$4g\cdot\frac{100}{4}=4g\cdot\frac{625}{g}$$
$$100g=2500$$
$$g=25$$
25 gallons will be needed.

44.
$$\frac{x-5}{5}=\frac{5}{4}$$
$$4(x-5)=25$$
$$4x-20=25$$
$$4x=45$$
$$x=\frac{45}{4}$$

45. Let h be the number of hours it takes Janet and James to paint the kitchen together.
$$\frac{1}{3}+\frac{1}{4}=\frac{1}{h}$$
$$12h\cdot\frac{1}{3}+12h\cdot\frac{1}{4}=12h\cdot\frac{1}{h}$$
$$4h+3h=12$$
$$7h=12$$
$$h=\frac{12}{7}=1\frac{5}{7}\text{ hr}$$

46. $y-4=5(x-6)$
$y-4=5x-30$
$y=5x-26$

47. $m=4$, $b=3$
$y=4x+3$

48. First graph the boundary line $x-5y=-5$.
When $x=0$, $-5y=-5$, and $y=1$
When $y=0$, $x=-5$
Join the points (0, 1) and (−5, 0) with a dashed line.
Test the point (0, 0):
$0-5\cdot 0<-5$
$0<-5$ False
(0, 0) is not part of the solution.

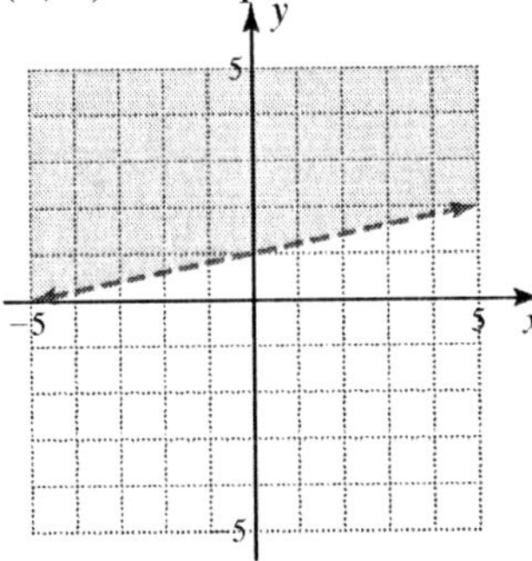

49. First graph the line $-y=-5x-5$.
When $x=0$, $-y=-5\cdot 0-5$
$-y=-5$
$y=5$
When $y=0$, $0=-5x-5$
$5=-5x$
$-1=x$
Join the points (0, 5) and (−1, 0) with a solid line. Test the point (0, 0):
$0\ge -5\cdot 0-5$
$0\ge -5$ True
(0, 0) is part of the solution.

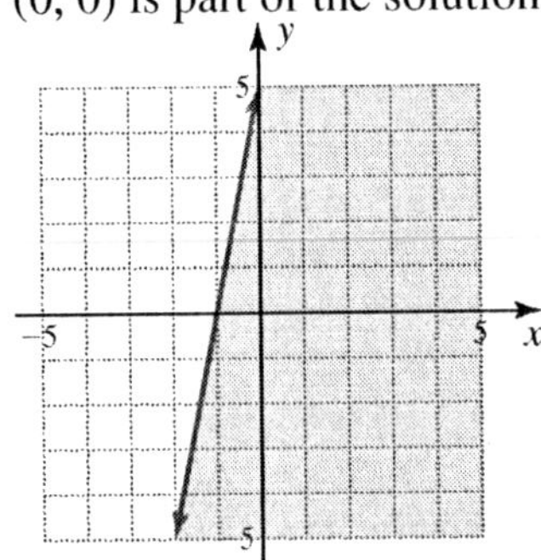

50. $P=kT$
$3=k(240)$
$\frac{3}{240}=k$
$\frac{1}{80}=k$

51. $P=\frac{k}{V}$
$1800=\frac{k}{6}$
$10{,}800=k$

52. $x+4y=16$

x	y
0	4
16	0

$4y-x=12$

x	y
0	3
−12	0

Graph the lines. They intersect at $\left(2, \frac{7}{2}\right)$.

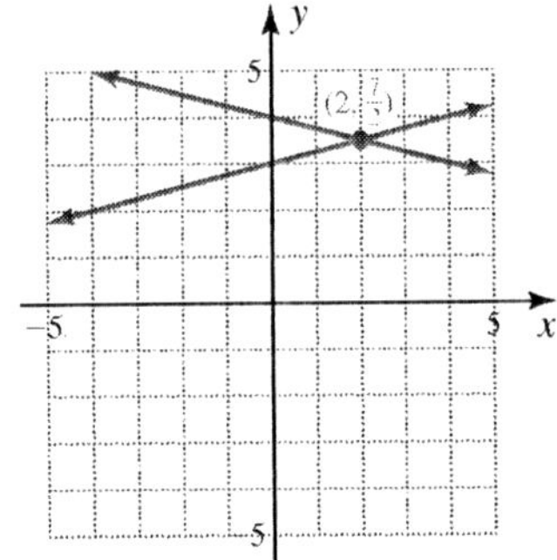

53. $y+3x=-3$

x	y
0	−3
−1	0

$2y+6x=-12$

x	y
0	−6
−2	0

Graph the lines. They appear to be parallel.

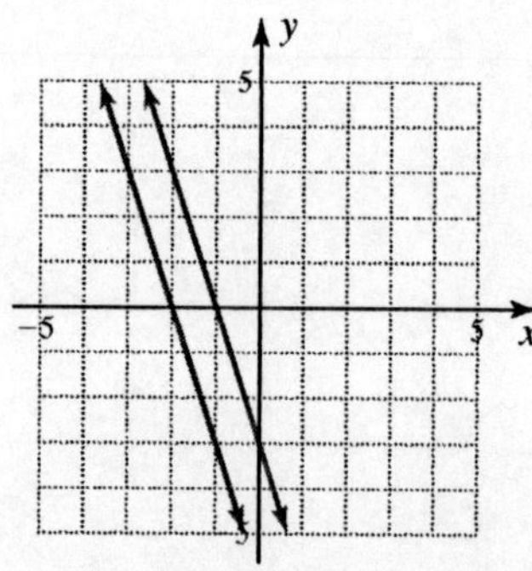

Notice that $2y + 6x = -12$ can be written as $y + 3x = -6$, and the other line is $y + 3x = -3$. Both equations cannot be true at the same time, and their graphs cannot intersect. There is no solution.

54. $x+4y=-18$
$$x=-4y-18$$
$$-2x-8y=32$$
$$-2(-4y-18)-8y=32$$
$$8y+36-8y=32$$
$$36=32 \quad \text{contradiction}$$
There is no solution. The system is inconsistent.

55. $x+3y=10$
$$x=-3y+10$$
$$-2x-6y=-20$$
$$-2(-3y+10)-6y=-20$$
$$6y-20-6y=-20$$
$$-20=-20$$
The equations are dependent. There are infinitely many solutions.

56. $x-2y=3$
$$x=2y+3$$
$$2x-y=0$$
$$2(2y+3)-y=0$$
$$4y+6-y=0$$
$$3y+6=0$$
$$3y=-6$$
$$y=-2$$
$$x-2y=3$$
$$x-2(-2)=3$$
$$x+4=3$$
$$x=-1$$
The solution is (–1, –2).

57. $5x+4y=-18$ $-10x-8y=3$
$$\begin{aligned} 10x+8y&=-36\\ -10x-8y&=3\\ \hline 0&=-33 \end{aligned}$$
There is no solution. The system is inconsistent.

58. $4y+3x=11$
$$6x+8y=22$$
The second equation is a constant multiple of the first one. The system is dependent. There are infinitely many solutions.

59. Let n be the number of nickels and d the number of dimes.
$$350=5n+10d$$
$$3n=d$$
$$350=5n+10(3n)$$
$$350=5n+30n$$
$$350=35n$$
$$10=n$$
$d = 3(10) = 30$
Sara has 10 nickels and 30 dimes.

60. Let x and y be the two numbers.
$$\begin{aligned} x+y&=215\\ x-y&=85\\ \hline 2x&=300\\ x&=150 \end{aligned}$$
$$x+y=215$$
$$150+y=215$$
$$y=65$$
The numbers are 150 and 65.

61. Since $(-4)^3=-64$, $\sqrt[3]{-64}=-4$.

62. $\sqrt[7]{(-3)^7}=-3$

63. $\sqrt{\dfrac{7}{243}}=\dfrac{\sqrt{7}}{\sqrt{243}}=\dfrac{\sqrt{7}\cdot\sqrt{3}}{\sqrt{243}\cdot\sqrt{3}}=\dfrac{\sqrt{21}}{\sqrt{729}}=\dfrac{\sqrt{21}}{27}$

64.
$$\begin{aligned} \sqrt[3]{128a^8b^9}&=\sqrt[3]{64\cdot 2\cdot a^6\cdot a^2\cdot b^9}\\ &=\sqrt[3]{4^3\cdot 2(a^2)^3\cdot a^2(b^3)^3}\\ &=4a^2b^3\sqrt[3]{2a^2} \end{aligned}$$

65. $\sqrt{48}=\sqrt{16\cdot 3}=\sqrt{16}\cdot\sqrt{3}=4\sqrt{3}$
$\sqrt{12}=\sqrt{4\cdot 3}=\sqrt{4}\cdot\sqrt{3}=2\sqrt{3}$
$\sqrt{48}+\sqrt{12}=4\sqrt{3}+2\sqrt{3}=6\sqrt{3}$

66.
$$\begin{aligned}\sqrt[3]{2x}\left(\sqrt[3]{4x^2}-\sqrt[3]{81x}\right)&=\sqrt[3]{2x}\cdot\sqrt[3]{4x^2}-\sqrt[3]{2x}\cdot\sqrt[3]{81x}\\&=\sqrt[3]{8x^3}-\sqrt[3]{162x^2}\\&=2x-\sqrt[3]{27\cdot 6x^2}\\&=2x-\sqrt[3]{3^3\cdot 6x^2}\\&=2x-3\sqrt[3]{6x^2}\end{aligned}$$

67. $\dfrac{\sqrt{3}}{\sqrt{2p}}=\dfrac{\sqrt{3}\cdot\sqrt{2p}}{\sqrt{2p}\cdot\sqrt{2p}}=\dfrac{\sqrt{6p}}{2p}$

68.
$$\begin{aligned}\left(\sqrt{125}+\sqrt{343}\right)\left(\sqrt{245}+\sqrt{175}\right)&=\sqrt{125}\cdot\sqrt{245}+\sqrt{125}\cdot\sqrt{175}+\sqrt{343}\cdot\sqrt{245}+\sqrt{343}\cdot\sqrt{175}\\&=\sqrt{30{,}625}+\sqrt{21{,}875}+\sqrt{84{,}035}+\sqrt{60{,}025}\\&=175+\sqrt{625\cdot 35}+\sqrt{2401\cdot 35}+245\\&=420+25\sqrt{35}+49\sqrt{35}\\&=420+74\sqrt{35}\end{aligned}$$

69. $\dfrac{\sqrt{x}}{\sqrt{x}-\sqrt{5}}=\dfrac{\sqrt{x}\cdot\left(\sqrt{x}+\sqrt{5}\right)}{\left(\sqrt{x}-\sqrt{5}\right)\left(\sqrt{x}+\sqrt{5}\right)}=\dfrac{\left(\sqrt{x}\right)^2+\sqrt{5x}}{\left(\sqrt{x}\right)^2-\left(\sqrt{5}\right)^2}=\dfrac{x+\sqrt{5x}}{x-5}$

70. $\dfrac{6+\sqrt{18}}{3}=\dfrac{6+\sqrt{9\cdot 2}}{3}=\dfrac{6+3\sqrt{2}}{3}=\dfrac{3\left(2+\sqrt{2}\right)}{3}=2+\sqrt{2}$

71.
$$\begin{aligned}\sqrt{x+3}&=-2\\\left(\sqrt{x+3}\right)^2&=(-2)^2\\x+3&=4\\x&=1\end{aligned}$$
Check: $\sqrt{1+3}\stackrel{?}{=}-2$
$\sqrt{4}\stackrel{?}{=}-2$ False
There is no real-number solution.

72. $\sqrt{x-6}-x=-6$

$$\sqrt{x-6}=x-6$$
$$\left(\sqrt{x-6}\right)^2=(x-6)^2$$
$$x-6=x^2-12x+36$$
$$0=x^2-13x+42$$
$$0=(x-7)(x-6)$$
$$x-7=0 \quad \text{or} \quad x-6=0$$
$$x=7 \qquad\qquad x=6$$

Check: If $x = 7$, $\sqrt{7-6}-7 \stackrel{?}{=} -6$

$$\sqrt{1}-7 \stackrel{?}{=} -6$$
$$1-7 \stackrel{?}{=} -6 \quad \text{True}$$

If $x = 6$, $\sqrt{6-6}-6 \stackrel{?}{=} -6$

$$\sqrt{0}-6 \stackrel{?}{=} -6$$
$$-6 \stackrel{?}{=} -6 \quad \text{True}$$

The solutions are $x = 6$ and $x = 7$.

Chapter 10 Quadratic Equations

10.1 Solving Quadratic Equations by the Square Root Property

Problems 10.1

1. a.
$$\begin{aligned} x^2 &= 81 \\ x &= \pm\sqrt{81} \\ x &= \pm 9 \end{aligned}$$

b.
$$\begin{aligned} x^2 - 1 &= 0 \\ x^2 &= 1 \\ x &= \pm\sqrt{1} \\ x &= \pm 1 \end{aligned}$$

c.
$$\begin{aligned} x^2 &= 13 \\ x &= \pm\sqrt{13} \end{aligned}$$

2.
$$\begin{aligned} 9x^2 - 16 &= 0 \\ 9x^2 &= 16 \\ x^2 &= \frac{16}{9} \\ x &= \pm\sqrt{\frac{16}{9}} \\ x &= \pm\frac{4}{3} \end{aligned}$$

3. a.
$$\begin{aligned} 49x^2 - 3 &= 0 \\ 49x^2 &= 3 \\ x^2 &= \frac{3}{49} \\ x &= \pm\sqrt{\frac{3}{49}} \\ x &= \pm\frac{\sqrt{3}}{7} \end{aligned}$$

b.
$$\begin{aligned} 10x^2 + 9 &= 0 \\ 10x^2 &= -9 \\ x^2 &= -\frac{9}{10} \end{aligned}$$

$x^2 \geq 0,$ so this equation has no real-number solution.

4. a.
$$\begin{aligned} (x+6)^2 &= 36 \\ x+6 &= \pm\sqrt{36} \\ x+6 &= \pm 6 \\ x &= -6 \pm 6 \\ x = -6+6 \quad &\text{or} \quad x = -6-6 \\ x = 0 \quad &\text{or} \quad x = -12 \end{aligned}$$

b.
$$\begin{aligned} (x+2)^2 - 9 &= 0 \\ (x+2)^2 &= 9 \\ x+2 &= \pm\sqrt{9} \\ x+2 &= \pm 3 \\ x &= -2 \pm 3 \\ x = -2+3 \quad &\text{or} \quad x = -2-3 \\ x = 1 \quad &\text{or} \quad x = -5 \end{aligned}$$

c.
$$\begin{aligned} 16(x+1)^2 - 5 &= 0 \\ 16(x+1)^2 &= 5 \\ (x+1)^2 &= \frac{5}{16} \\ x+1 &= \pm\sqrt{\frac{5}{16}} \\ x &= -1 \pm \frac{\sqrt{5}}{4} \\ x = -1 + \frac{\sqrt{5}}{4} \quad &\text{or} \quad x = -1 - \frac{\sqrt{5}}{4} \\ x = \frac{-4+\sqrt{5}}{4} \quad &\text{or} \quad x = \frac{-4-\sqrt{5}}{4} \end{aligned}$$

5.
$$\begin{aligned} 16(x-3)^2 + 7 &= 0 \\ 16(x-3)^2 &= -7 \\ (x-3)^2 &= -\frac{7}{16} \end{aligned}$$

$(x-3)^2 \geq 0,$ so this equation has no real-number solution.

6.
$$\begin{aligned} 2(3x-2)^2 &= 36 \\ (3x-2)^2 &= 18 \\ 3x-2 &= \pm\sqrt{18} \\ 3x &= 2 \pm 3\sqrt{2} \\ x &= \frac{2 \pm 3\sqrt{2}}{3} \end{aligned}$$

7. $\text{BMI} = \frac{705W}{H^2}$

$30 = \frac{705(180)}{H^2}$

$30H^2 = 705(180)$

$H^2 = 705 \cdot 6 = 4230$

$H = \sqrt{4230} \approx 65$ inches

Exercises 10.1

1. $x^2 = 100$

$x = \pm\sqrt{100}$

$x = \pm 10$

3. $x^2 = 0$

$x = \pm\sqrt{0}$

$x = 0$

5. $y^2 = -4$

$y^2 \geq 0,$ so this equation has no real-number solution.

7. $x^2 = 7$

$x = \pm\sqrt{7}$

9. $x^2 - 9 = 0$

$x^2 = 9$

$x = \pm\sqrt{9} = \pm 3$

11. $x^2 - 3 = 0$

$x^2 = 3$

$x = \pm\sqrt{3}$

13. $25x^2 - 1 = 0$

$25x^2 = 1$

$x^2 = \frac{1}{25}$

$x = \sqrt{\frac{1}{25}} = \pm\frac{1}{5}$

15. $100x^2 - 49 = 0$

$100x^2 = 49$

$x^2 = \frac{49}{100}$

$x = \pm\sqrt{\frac{49}{100}}$

$x = \pm\frac{7}{10}$

17. $25y^2 - 17 = 0$

$25y^2 = 17$

$y^2 = \frac{17}{25}$

$y = \pm\sqrt{\frac{17}{25}} = \pm\frac{\sqrt{17}}{5}$

19. $25x^2 + 3 = 0$

$25x^2 = -3$

$x^2 = -\frac{3}{25}$

$x^2 \geq 0,$ so this equation has no real-number solution.

21. $(x+1)^2 = 81$

$x + 1 = \pm\sqrt{81}$

$x = -1 \pm 9$

$x = -1 + 9$ or $x = -1 - 9$

$x = 8$ or $x = -10$

23. $(x-2)^2 = 36$

$x - 2 = \pm\sqrt{36}$

$x = 2 \pm 6$

$x = 2 + 6$ or $x = 2 - 6$

$x = 8$ or $x = -4$

25. $(z-4)^2 = -25$

$(z-4)^2 \geq 0,$ so this equation has no real-number solution.

27. $(x-9)^2=81$
$$x-9=\pm\sqrt{81}$$
$$x=9\pm 9$$
$$x=9+9 \quad \text{or} \quad x=9-9$$
$$x=18 \quad \text{or} \quad x=0$$

29. $(x+4)^2=16$
$$x+4=\pm\sqrt{16}$$
$$x=-4\pm 4$$
$$x=-4+4 \quad \text{or} \quad x=-4-4$$
$$x=0 \quad \text{or} \quad x=-8$$

31. $25(x+1)^2-1=0$
$$25(x+1)^2=1$$
$$(x+1)^2=\frac{1}{25}$$
$$x+1=\pm\sqrt{\frac{1}{25}}$$
$$x=-1\pm\frac{1}{5}$$
$$x=-1+\frac{1}{5} \quad \text{or} \quad x=-1-\frac{1}{5}$$
$$x=-\frac{4}{5} \quad \text{or} \quad x=-\frac{6}{5}$$

33. $36(x-3)^2-49=0$
$$36(x-3)^2=49$$
$$(x-3)^2=\frac{49}{36}$$
$$x-3=\pm\sqrt{\frac{49}{36}}$$
$$x=3\pm\frac{7}{6}$$
$$x=3+\frac{7}{6} \quad \text{or} \quad x=3-\frac{7}{6}$$
$$x=\frac{25}{6} \quad \text{or} \quad x=\frac{11}{6}$$

35. $4(x+1)^2-25=0$
$$4(x+1)^2=25$$
$$(x+1)^2=\frac{25}{4}$$
$$x+1=\pm\sqrt{\frac{25}{4}}$$
$$x=-1\pm\frac{5}{2}$$
$$x=-1+\frac{5}{2} \quad \text{or} \quad x=-1-\frac{5}{2}$$
$$x=\frac{3}{2} \quad \text{or} \quad x=-\frac{7}{2}$$

37. $9(x-1)^2-5=0$
$$9(x-1)^2=5$$
$$(x-1)^2=\frac{5}{9}$$
$$x-1=\pm\sqrt{\frac{5}{9}}$$
$$x=1\pm\frac{\sqrt{5}}{3}$$
$$x=\frac{3\pm\sqrt{5}}{3}$$

39. $16(x+1)^2+1=0$
$$16(x+1)^2=-1$$
$$(x+1)^2=-\frac{1}{16}$$

$(x+1)^2\geq 0,$ so this equation has no real-number solution.

41. $x^2=\frac{1}{81}$
$$x=\pm\sqrt{\frac{1}{81}}$$
$$x=\pm\frac{1}{9}$$

43. $x^2 - \frac{1}{16} = 0$

$x^2 = \frac{1}{16}$

$x = \pm\sqrt{\frac{1}{16}} = \pm\frac{1}{4}$

45. $6x^2 - 24 = 0$

$6x^2 = 24$

$x^2 = 4$

$x = \pm\sqrt{4} = \pm 2$

47. $2(v+1)^2 - 18 = 0$

$2(v+1)^2 = 18$

$(v+1)^2 = 9$

$v+1 = \pm\sqrt{9}$

$v = -1 \pm 3$

$v = -1+3$ or $v = -1-3$

$v = 2$ or $v = -4$

49. $8(x-1)^2 - 18 = 0$

$8(x-1)^2 = 18$

$(x-1)^2 = \frac{18}{8}$

$(x-1)^2 = \frac{9}{4}$

$x - 1 = \pm\sqrt{\frac{9}{4}}$

$x = 1 \pm \frac{3}{2}$

$x = 1 + \frac{3}{2}$ or $x = 1 - \frac{3}{2}$

$x = \frac{5}{2}$ or $x = -\frac{1}{2}$

51. $4(2y-3)^2 = 32$

$(2y-3)^2 = 8$

$2y - 3 = \pm\sqrt{8}$

$2y = 3 \pm 2\sqrt{2}$

$y = \frac{3 \pm 2\sqrt{2}}{2}$

53. $8(2x-3)^2 - 64 = 0$

$8(2x-3)^2 = 64$

$(2x-3)^2 = 8$

$2x - 3 = \pm\sqrt{8}$

$2x = 3 \pm 2\sqrt{2}$

$x = \frac{3 \pm 2\sqrt{2}}{2}$

55. $3\left(\frac{1}{2}x+1\right)^2 = 54$

$\left(\frac{1}{2}x+1\right)^2 = 18$

$\frac{1}{2}x + 1 = \pm\sqrt{18}$

$\frac{1}{2}x = -1 \pm 3\sqrt{2}$

$x = -2 \pm 6\sqrt{2}$

57. $2\left(\frac{1}{3}x-1\right)^2 - 40 = 0$

$2\left(\frac{1}{3}x-1\right)^2 = 40$

$\left(\frac{1}{3}x-1\right)^2 = 20$

$\frac{1}{3}x - 1 = \pm\sqrt{20}$

$\frac{1}{3}x = 1 \pm 2\sqrt{5}$

$x = 3 \pm 6\sqrt{5}$

59. $5\left(\frac{1}{2}y-1\right)^2 + 60 = 0$

$5\left(\frac{1}{2}y-1\right)^2 = -60$

$\left(\frac{1}{2}y-1\right)^2 = -12$

$\left(\frac{1}{2}y-1\right)^2 \geq 0$, so this equation has no real-number solution.

61. $(x+7)^2 = (x+7)(x+7) = x^2 + 14x + 49$

63. $(x-3)^2 = (x-3)(x-3) = x^2 - 6x + 9$

65.
$$A = \pi r^2$$
$$25\pi = \pi r^2$$
$$25 = r^2$$
$$5 = r$$
The radius is 5 inches.

67.
$$A = 4\pi r^2$$
$$49\pi = 4\pi r^2$$
$$49 = 4r^2$$
$$\frac{49}{4} = r^2$$
$$r = \sqrt{\frac{49}{4}} = \frac{7}{2} = 3\frac{1}{2}$$
The radius is $3\frac{1}{2}$ feet.

69. Sample answer: $x^2 + 6 = 0$
$$x^2 = -6$$
Since x^2 is nonnegative and $-6 < 0$, $x^2 + 6 = 0$ has no real-number solution.

71. If $X^2 = A$ has no real-number solution, $A < 0$.

73. If $X^2 = A$ has two solutions, $A > 0$.

75. If $X^2 = A$ and A is a positive perfect square, $X^2 = A$ has rational solutions.

77. $3(x-1)^2 - 24 = 0$
$$3(x-1)^2 = 24$$
$$(x-1)^2 = 8$$
$$x - 1 = \pm\sqrt{8}$$
$$x = 1 \pm 2\sqrt{2}$$

79. $(x+2)^2 - 9 = 0$
$$(x+2)^2 = 9$$
$$x + 2 = \pm\sqrt{9}$$
$$x = -2 \pm 3$$
$x = -2 + 3$ or $x = -2 - 3$
$x = 1$ or $x = -5$

81. $49x^2 - 3 = 0$
$$49x^2 = 3$$
$$x^2 = \frac{3}{49}$$
$$x = \pm\sqrt{\frac{3}{49}} = \pm\frac{\sqrt{3}}{7}$$

83. $9x^2 - 16 = 0$
$$9x^2 = 16$$
$$x^2 = \frac{16}{9}$$
$$x = \pm\sqrt{\frac{16}{9}} = \pm\frac{4}{3}$$

85. $x^2 - 1 = 0$
$$x^2 = 1$$
$$x = \pm\sqrt{1} = \pm 1$$

10.2 Solving Quadratic Equations by Completing the Square

Problems 10.2

1. a. $(x+11)^2 = x^2 + 22x + \square$
$$\left(\frac{22}{2}\right)^2 = 11^2 = 121$$

b. $\left(x - \frac{1}{4}\right)^2 = x^2 - \frac{1}{2}x + \square$
$$\left(\frac{-\frac{1}{2}}{2}\right) = \left(-\frac{1}{4}\right)^2 = \frac{1}{16}$$

2. a. $x^2 - 12x + \square = (\ )^2$

$\left(-\frac{12}{2}\right)^2 = (-6)^2$, so $(\ )^2 = (x-6)^2$.

b. $x^2 + 5x + \square = (\ )^2$

$\left(\frac{5}{2}\right)^2$, so $(\ )^2 = \left(x + \frac{5}{2}\right)^2$.

3. $4x^2 - 24x + 27 = 0$

$$\begin{aligned}
4x^2 - 24x &= -27 \\
x^2 - 6x &= -\frac{27}{4} \\
x^2 - 6x + (-3)^2 &= -\frac{27}{4} + (-3)^2 \\
(x-3)^2 &= \frac{9}{4} \\
x - 3 &= \pm\sqrt{\frac{9}{4}} \\
x &= 3 \pm \frac{3}{2}
\end{aligned}$$

$x = 3 + \frac{3}{2}$ or $x = 3 - \frac{3}{2}$

$x = \frac{9}{2}$ or $x = \frac{3}{2}$

4. $4x^2 + 24x + 31 = 0$

$$\begin{aligned}
4x^2 + 24x &= -31 \\
x^2 + 6x &= -\frac{31}{4} \\
x^2 + 6x + 3^2 &= -\frac{31}{4} + 3^2 \\
(x+3)^2 &= \frac{5}{4} \\
x + 3 &= \pm\sqrt{\frac{5}{4}} \\
x &= -3 \pm \frac{\sqrt{5}}{2} \\
x &= \frac{-6 \pm \sqrt{5}}{2}
\end{aligned}$$

Exercises 10.2

1. $x^2 + 18x + \square$

$\left(\frac{18}{2}\right)^2 = 9^2 = 81$

3. $x^2 - 16x + \square$

$\left(\frac{-16}{2}\right) = (-8)^2 = 64$

5. $x^2 + 7x + \square$

$\left(\frac{7}{2}\right)^2 = \frac{49}{4}$

7. $x^2 - 3x + \square$

$\left(\frac{-3}{2}\right)^2 = \frac{9}{4}$

9. $x^2 + x + \square$

$\left(\frac{1}{2}\right)^2 = \frac{1}{4}$

11. $x^2 + 4x + \square = (\ )^2$

$\left(\frac{4}{2}\right) = 2^2 = 4$, so $(\ )^2 = (x+2)^2$.

13. $x^2 + 3x + \square = (\ )^2$

$\left(\frac{3}{2}\right)^2 = \frac{9}{4}$, so $(\ )^2 = \left(x + \frac{3}{2}\right)^2$.

15. $x^2 - 6x + \square = (\ )^2$

$\left(\frac{-6}{2}\right)^2 = (-3)^2 = 9$, so $(\ )^2 = (x-3)^2$.

17. $x^2 - 5x + \square = (\ )^2$

$\left(\frac{-5}{2}\right)^2 = \frac{25}{4}$, so $(\ )^2 = \left(x - \frac{5}{2}\right)^2$.

19. $x^2 - \frac{3}{2}x + \square = (\ )^2$

$\left(\frac{-\frac{3}{2}}{2}\right) = \left(-\frac{3}{4}\right)^2 = \frac{9}{16}$, so $(\ )^2 = \left(x - \frac{3}{4}\right)^2$.

21.
$$\begin{aligned} x^2 + 2x + 7 &= 0 \\ x^2 + 2x &= -7 \\ x^2 + 2x + 1^2 &= -7 + 1^2 \\ (x+1)^2 &= -6 \end{aligned}$$

$(x+1)^2 \geq 0$, so this equation has no real-number solution.

23.
$$\begin{aligned} x^2 + x - 1 &= 0 \\ x^2 + x &= 1 \\ x^2 + x + \left(\frac{1}{2}\right)^2 &= 1 + \left(\frac{1}{2}\right)^2 \\ \left(x + \frac{1}{2}\right)^2 &= \frac{5}{4} \\ x + \frac{1}{2} &= \pm\sqrt{\frac{5}{4}} \\ x &= -\frac{1}{2} \pm \frac{\sqrt{5}}{2} \\ x &= \frac{-1 \pm \sqrt{5}}{2} \end{aligned}$$

25.
$$\begin{aligned} x^2 + 3x - 1 &= 0 \\ x^2 + 3x &= 1 \\ x^2 + 3x + \left(\frac{3}{2}\right)^2 &= 1 + \left(\frac{3}{2}\right)^2 \\ \left(x + \frac{3}{2}\right)^2 &= \frac{13}{4} \\ x + \frac{3}{2} &= \pm\sqrt{\frac{13}{4}} \\ x &= -\frac{3}{2} \pm \frac{\sqrt{13}}{2} \\ x &= \frac{-3 \pm \sqrt{13}}{2} \end{aligned}$$

27.
$$\begin{aligned} x^2 - 3x - 3 &= 0 \\ x^2 - 3x &= 3 \\ x^2 - 3x + \left(-\frac{3}{2}\right)^2 &= 3 + \left(-\frac{3}{2}\right)^2 \\ \left(x - \frac{3}{2}\right)^2 &= \frac{21}{4} \\ x - \frac{3}{2} &= \pm\sqrt{\frac{21}{4}} \\ x &= \frac{3}{2} \pm \frac{\sqrt{21}}{2} \\ x &= \frac{3 \pm \sqrt{21}}{2} \end{aligned}$$

29.
$$\begin{aligned} 4x^2 + 4x - 3 &= 0 \\ 4x^2 + 4x &= 3 \\ x^2 + x &= \frac{3}{4} \\ x^2 + x + \left(\frac{1}{2}\right)^2 &= \frac{3}{4} + \left(\frac{1}{2}\right)^2 \\ \left(x + \frac{1}{2}\right)^2 &= 1 \\ x + \frac{1}{2} &= \pm\sqrt{1} \\ x &= -\frac{1}{2} \pm 1 \end{aligned}$$

$x = -\frac{1}{2} + 1$ or $x = -\frac{1}{2} - 1$

$x = \frac{1}{2}$ or $x = -\frac{3}{2}$

31.
$$\begin{aligned}4x^2-16x&=15\\x^2-4x&=\frac{15}{4}\\x^2-4x+(-2)^2&=\frac{15}{4}+(-2)^2\\(x-2)^2&=\frac{31}{4}\\x-2&=\pm\sqrt{\frac{31}{4}}\\x&=2\pm\frac{\sqrt{31}}{2}\\x&=\frac{4\pm\sqrt{31}}{2}\end{aligned}$$

33.
$$\begin{aligned}4x^2-7&=4x\\4x^2-4x&=7\\x^2-x&=\frac{7}{4}\\x^2-x+\left(-\frac{1}{2}\right)^2&=\frac{7}{4}+\left(-\frac{1}{2}\right)^2\\\left(x-\frac{1}{2}\right)^2&=2\\x-\frac{1}{2}&=\pm\sqrt{2}\\x&=\frac{1}{2}\pm\sqrt{2}\\x&=\frac{1\pm2\sqrt{2}}{2}\end{aligned}$$

35.
$$\begin{aligned}2x^2+1&=4x\\2x^2-4x&=-1\\x^2-2x&=-\frac{1}{2}\\x^2-2x+(-1)^2&=-\frac{1}{2}+(-1)^2\\(x-1)^2&=\frac{1}{2}\\x&=1\pm\sqrt{\frac{1}{2}}\\x&=1\pm\frac{\sqrt{2}}{2}\\x&=\frac{2\pm\sqrt{2}}{2}\end{aligned}$$

37.
$$\begin{aligned}(x+3)(x-2)&=-4\\x^2+x-6&=-4\\x^2+x&=2\\x^2+x+\left(\frac{1}{2}\right)^2&=2+\left(\frac{1}{2}\right)^2\\\left(x+\frac{1}{2}\right)^2&=\frac{9}{4}\\x+\frac{1}{2}&=\pm\sqrt{\frac{9}{4}}\\x&=-\frac{1}{2}\pm\frac{3}{2}\end{aligned}$$
$$x=-\frac{1}{2}+\frac{3}{2}\quad\text{or}\quad x=-\frac{1}{2}-\frac{3}{2}$$
$$x=1\quad\text{or}\quad x=-2$$

39.
$$\begin{aligned}2x(x+5)-1&=0\\2x^2+10x&=1\\x^2+5x&=\frac{1}{2}\\x^2+5x+\left(\frac{5}{2}\right)^2&=\frac{1}{2}+\left(\frac{5}{2}\right)^2\\\left(x+\frac{5}{2}\right)^2&=\frac{27}{4}\\x+\frac{5}{2}&=\pm\sqrt{\frac{27}{4}}\\x&=-\frac{5}{2}\pm\frac{3\sqrt{3}}{2}\\x&=\frac{-5\pm3\sqrt{3}}{2}\end{aligned}$$

41.
$$\begin{aligned}10x^2+5x&=12\\10x^2+5x-12&=0\end{aligned}$$

43.
$$\begin{aligned}9x&=x^2\\-x^2+9x&=0\\x^2-9x&=0\end{aligned}$$

45. $$\frac{x}{4} = \frac{3}{5} - \frac{x^2}{2}$$
$$\frac{x^2}{2} + \frac{x}{4} - \frac{3}{5} = 0$$
$$10x^2 + 5x - 12 = 0$$

47. a. $\overline{C} = x^2 - 4x + 6$

$\overline{C} = x^2 - 4x + 4 + 2$

$\overline{C} = (x-2)^2 + 2$

$x - 2 = 0$ if $x = 2$, so 2000 units should be produced.

b. When $x = 2$, $\overline{C} = \$2$.

49. $B = 20t^2 - 120t + 200$

$B = 20(t^2 - 6t + 10)$

$B = 20(t^2 - 6t + 9 + 1)$

$B = 20(t-3)^2 + 20$

The lowest number of bacteria occurs when $t - 3 = 0$ or $t = 3$ days.

51. $x^3 + 2x^2 = 5$ cannot be solved by completing the square because x is cubed in the equation.

53. Sample answer:

$x^2 - 6x + \square$

$\left(\frac{-6}{2}\right)^2 = (-3)^2 = 9$, so $x^2 - 6x + 9$.

55. $\left(x - \frac{1}{4}\right)^2 = x^2 - \frac{1}{2}x + \square$

$\left(-\frac{1}{4}\right)^2 = \frac{1}{16}$

57. $x^2 + 5x + \square = (\ )^2$

$\left(\frac{5}{2}\right)^2 = \frac{25}{4}$, so $(\ )^2 = \left(x + \frac{5}{2}\right)^2$.

59. $4x^2 + 24x + 31 = 0$
$$4x^2 + 24x = -31$$
$$x^2 + 6x = -\frac{31}{4}$$
$$x^2 + 6x + 3^2 = -\frac{31}{4} + 3^2$$
$$(x+3)^2 = \frac{5}{4}$$
$$x + 3 = \pm\sqrt{\frac{5}{4}}$$
$$x = -3 \pm \frac{\sqrt{5}}{2}$$
$$x = \frac{-6 \pm \sqrt{5}}{2}$$

10.3 Solving Quadratic Equations by the Quadratic Formula

Problems 10.3

1. $3x^2 + 2x - 5 = 0$
$$x = \frac{-2 \pm \sqrt{2^2 - 4(3)(-5)}}{2(3)}$$
$$x = \frac{-2 \pm \sqrt{4 + 60}}{6}$$
$$x = \frac{-2 \pm \sqrt{64}}{6}$$
$$x = \frac{-2 \pm 8}{6}$$
$$x = \frac{-2 + 8}{6} = \frac{6}{6} = 1 \text{ or}$$
$$x = \frac{-2 - 8}{6} = \frac{-10}{6} = -\frac{5}{3}$$

2. $x^2 = 4x + 4$

$x^2 - 4x - 4 = 0$

$x = \dfrac{-(-4) \pm \sqrt{(-4)^2 - 4(1)(-4)}}{2(1)}$

$x = \dfrac{4 \pm \sqrt{16+16}}{2}$

$x = \dfrac{4 \pm \sqrt{32}}{2} = \dfrac{4 \pm 4\sqrt{2}}{2} = 2 \pm 2\sqrt{2}$

3. $6x = x^2$

$0 = x^2 - 6x$

$x = \dfrac{-(-6) \pm \sqrt{(-6)^2 - 4(1)(0)}}{2(1)}$

$x = \dfrac{6 \pm \sqrt{36}}{2}$

$x = \dfrac{6 \pm 6}{2}$

$x = \dfrac{6+6}{2} = 6$ or $x = \dfrac{6-6}{2} = 0$

4. $\dfrac{x^2}{4} - \dfrac{3}{8}x = \dfrac{1}{4}$

$2x^2 - 3x - 2 = 0$

$x = \dfrac{-(-3) \pm \sqrt{(-3)^2 - 4(2)(-2)}}{2(2)}$

$x = \dfrac{3 \pm \sqrt{9+16}}{4}$

$x = \dfrac{3 \pm \sqrt{25}}{4} = \dfrac{3 \pm 5}{4}$

$x = \dfrac{3+5}{4} = 2$ or $x = \dfrac{3-5}{4} = -\dfrac{1}{2}$

5. $3x^2 + 2x = -1$

$3x^2 + 2x + 1 = 0$

$x = \dfrac{-2 \pm \sqrt{2^2 - 4(3)(1)}}{2(3)}$

$x = \dfrac{-2 \pm \sqrt{4-12}}{6} = \dfrac{-2 \pm \sqrt{-8}}{6}$

$\sqrt{-8}$ is not a real number, so this equation has no real-number solution.

Exercises 10.3

1. $x^2 + 3x + 2 = 0$

$x = \dfrac{-3 \pm \sqrt{3^2 - 4(1)(2)}}{2(1)}$

$x = \dfrac{-3 \pm \sqrt{9-8}}{2}$

$x = \dfrac{-3 \pm \sqrt{1}}{2} = \dfrac{-3 \pm 1}{2}$

$x = \dfrac{-3+1}{2} = -1$ or $x = \dfrac{-3-1}{2} = -2$

3. $x^2 + x - 2 = 0$

$x = \dfrac{-1 \pm \sqrt{1^2 - 4(1)(-2)}}{2(1)}$

$x = \dfrac{-1 \pm \sqrt{1+8}}{2}$

$x = \dfrac{-1 \pm \sqrt{9}}{2} = \dfrac{-1 \pm 3}{2}$

$x = \dfrac{-1+3}{2} = 1$ or $x = \dfrac{-1-3}{2} = -2$

5. $2x^2 + x - 2 = 0$

$x = \dfrac{-1 \pm \sqrt{1^2 - 4(2)(-2)}}{2(2)}$

$x = \dfrac{-1 \pm \sqrt{1+16}}{4}$

$x = \dfrac{-1 \pm \sqrt{17}}{4}$

7. $3x^2 + x = 2$

$3x^2 + x - 2 = 0$

$x = \dfrac{-1 \pm \sqrt{1^2 - 4(3)(-2)}}{2(3)}$

$x = \dfrac{-1 \pm \sqrt{1+24}}{6}$

$x = \dfrac{-1 \pm \sqrt{25}}{6} = \dfrac{-1 \pm 5}{6}$

$x = \dfrac{-1+5}{6} = \dfrac{2}{3}$ or $x = \dfrac{-1-5}{6} = -1$

9. $2x^2+7x=-6$

$2x^2+7x+6=0$

$x=\dfrac{-7\pm\sqrt{7^2-4(2)(6)}}{2(2)}$

$x=\dfrac{-7\pm\sqrt{49-48}}{4}$

$x=\dfrac{-7\pm\sqrt{1}}{4}=\dfrac{-7\pm1}{4}$

$x=\dfrac{-7+1}{4}=-\dfrac{3}{2}$ or $x=\dfrac{-7-1}{4}=-2$

11. $7x^2=12x-5$

$7x^2-12x+5=0$

$x=\dfrac{-(-12)\pm\sqrt{(-12)^2-4(7)(5)}}{2(7)}$

$x=\dfrac{12\pm\sqrt{144-140}}{14}$

$x=\dfrac{12\pm\sqrt{4}}{14}=\dfrac{12\pm2}{14}$

$x=\dfrac{12+2}{14}=1$ or $x=\dfrac{12-2}{14}=\dfrac{5}{7}$

13. $5x^2=11x-4$

$5x^2-11x+4=0$

$x=\dfrac{-(-11)\pm\sqrt{(-11)^2-4(5)(4)}}{2(5)}$

$x=\dfrac{11\pm\sqrt{121-80}}{10}$

$x=\dfrac{11\pm\sqrt{41}}{10}$

15. $\dfrac{x^2}{5}-\dfrac{x}{2}=-\dfrac{3}{10}$

$2x^2-5x=-3$

$2x^2-5x+3=0$

$x=\dfrac{-(-5)\pm\sqrt{(-5)^2-4(2)(3)}}{2(2)}$

$x=\dfrac{5\pm\sqrt{25-24}}{4}$

$x=\dfrac{5\pm\sqrt{1}}{4}=\dfrac{5\pm1}{4}$

$x=\dfrac{5+1}{4}=\dfrac{3}{2}$ or $x=\dfrac{5-1}{4}=1$

17. $\dfrac{x^2}{2}=\dfrac{3x}{4}-\dfrac{1}{8}$

$4x^2=6x-1$

$4x^2-6x+1=0$

$x=\dfrac{-(-6)\pm\sqrt{(-6)^2-4(4)(1)}}{2(4)}$

$x=\dfrac{6\pm\sqrt{36-16}}{8}$

$x=\dfrac{6\pm\sqrt{20}}{8}=\dfrac{6\pm2\sqrt{5}}{8}=\dfrac{3\pm\sqrt{5}}{4}$

19. $\dfrac{x^2}{8}=-\dfrac{x}{4}-\dfrac{1}{8}$

$x^2=-2x-1$

$x^2+2x+1=0$

$x=\dfrac{-2\pm\sqrt{2^2-4(1)(1)}}{2(1)}$

$x=\dfrac{-2\pm\sqrt{4-4}}{2}$

$x=\dfrac{-2+0}{2}=-1$

21. $6x = 4x^2 + 1$

$-4x^2 + 6x - 1 = 0$

$4x^2 - 6x + 1 = 0$

$x = \dfrac{-(-6) \pm \sqrt{(-6)^2 - 4(4)(1)}}{2(4)}$

$x = \dfrac{6 \pm \sqrt{36 - 16}}{8}$

$x = \dfrac{6 \pm 2\sqrt{5}}{8} = \dfrac{3 \pm \sqrt{5}}{4}$

23. $3x = 1 - 3x^2$

$3x^2 + 3x - 1 = 0$

$x = \dfrac{-3 \pm \sqrt{3^2 - 4(3)(-1)}}{2(3)}$

$x = \dfrac{-3 \pm \sqrt{9 + 12}}{6}$

$x = \dfrac{-3 \pm \sqrt{21}}{6}$

25. $x(x+2) = 2x(x+1) - 4$

$x^2 + 2x = 2x^2 + 2x - 4$

$0 = x^2 - 4$

$x = \dfrac{-0 \pm \sqrt{0^2 - 4(1)(-4)}}{2(1)}$

$x = \dfrac{\pm\sqrt{16}}{2} = \pm\dfrac{4}{2} = \pm 2$

27. $6x(x+5) = (x+15)^2$

$6x^2 + 30x = x^2 + 30x + 225$

$5x^2 - 225 = 0$

$x^2 - 45 = 0$

$x = \dfrac{-0 \pm \sqrt{0^2 - 4(1)(-45)}}{2(1)}$

$x = \dfrac{\pm\sqrt{180}}{2} = \pm\dfrac{6\sqrt{5}}{2} = \pm 3\sqrt{5}$

29. $(x-2)^2 = 4x(x-1)$

$x^2 - 4x + 4 = 4x^2 - 4x$

$0 = 3x^2 - 4$

$x = \dfrac{-0 \pm \sqrt{0^2 - 4(3)(-4)}}{2(3)}$

$x = \pm\dfrac{\sqrt{48}}{6} = \pm\dfrac{4\sqrt{3}}{6} = \pm\dfrac{2\sqrt{3}}{3}$

31. $(-2)^2 = (-1)^2(2)^2 = 4$

33. $(-1)^2 = (-1)(-1) = 1$

35. $ax^2 + bx + c = 0$ becomes $4a^2x^2 + 4abx + 4ac = 0$ if each term is multiplied by $4a$.

37. $4a^2x^2 + 4abx = -4ac$ becomes $4a^2x^2 + 4abx + b^2 = b^2 - 4ac$ if b^2 is added on both sides.

39. $(2ax + b)^2 = b^2 - 4ac$ becomes $2ax + b = \pm\sqrt{b^2 - 4ac}$ if the square root of each side is taken.

41. $2ax = -b \pm \sqrt{b^2 - 4ac}$ becomes $x = \dfrac{-b \pm \sqrt{b^2 - 4ac}}{2a}$ if each side is divided by $2a$.

43. $x^2 - 17 = 0$

$x^2 = 17$

$x = \pm\sqrt{17}$

45. $x^2 + 4x = -4$

$x^2 + 4x + 4 = 0$

$(x+2)^2 = 0$

$x + 2 = 0$

$x = -2$

47. $y^2 - y = 0$

$y(y-1) = 0$

$y = 0$ or $y - 1 = 0$

$y = 0$ or $y = 1$

49. $x^2 + 5x + 6 = 0$

$x = \dfrac{-5 \pm \sqrt{5^2 - 4(1)(6)}}{2(1)}$

$x = \dfrac{-5 \pm \sqrt{25 - 24}}{2}$

$x = \dfrac{-5 \pm \sqrt{1}}{2} = \dfrac{-5 \pm 1}{2}$

$x = \dfrac{-5+1}{2} = -2$ or $x = \dfrac{-5-1}{2} = -3$

51. $(z+2)(z+4) = 8$

$z^2 + 6z + 8 = 8$

$z^2 + 6z = 0$

$z(z+6) = 0$

$z = 0$ or $z + 6 = 0$

$z = 0$ or $z = -6$

53. $y^2 = 1 - \dfrac{8}{3}y$

$y^2 + \dfrac{8}{3}y - 1 = 0$

$3y^2 + 8y - 3 = 0$

$x = \dfrac{-8 \pm \sqrt{8^2 - 4(3)(-3)}}{2(3)}$

$x = \dfrac{-8 \pm \sqrt{64 + 36}}{6}$

$x = \dfrac{-8 \pm \sqrt{100}}{6} = \dfrac{-8 \pm 10}{6}$

$x = \dfrac{-8+10}{6} = \dfrac{1}{3}$ or $x = \dfrac{-8-10}{6} = -3$

55. $3z^2 + 1 = z$

$3z^2 - z + 1 = 0$

$z = \dfrac{-(-1) \pm \sqrt{(-1)^2 - 4(3)(1)}}{2(3)}$

$z = \dfrac{1 \pm \sqrt{1-12}}{6}$

$z = \dfrac{1 \pm \sqrt{-11}}{6}$

$\sqrt{-11}$ is not a real number, so this equation has no real-number solution.

57. If $D = 0$, $x = -\dfrac{b}{2a}$, so $ax^2 + bx + c = 0$ has one rational solution.

59. If $D < 0$, $\sqrt{D}$ is not a real number, so $ax^2 + bx + c = 0$ has no solutions.

61. $3x^2 + 2x = -1$

$3x^2 + 2x + 1 = 0$

$x = \dfrac{-2 \pm \sqrt{2^2 - 4(3)(1)}}{2(3)}$

$x = \dfrac{-2 \pm \sqrt{4-12}}{6}$

$x = \dfrac{-2 \pm \sqrt{-8}}{6}$

$\sqrt{-8}$ is not a real number, so this equation has no real-number solution.

63. $x^2 = 4x + 4$

$x^2 - 4x - 4 = 0$

$x = \dfrac{-(-4) \pm \sqrt{(-4)^2 - 4(1)(-4)}}{2(1)}$

$x = \dfrac{4 \pm \sqrt{16+16}}{2}$

$x = \dfrac{4 \pm \sqrt{32}}{2} = \dfrac{4 \pm 4\sqrt{2}}{2} = 2 \pm 2\sqrt{2}$

65. $\frac{x^2}{4}-\frac{3}{8}x=\frac{1}{4}$

$2x^2-3x=2$

$2x^2-3x-2=0$

$x=\frac{-(-3)\pm\sqrt{(-3)^2-4(2)(-2)}}{2(2)}$

$x=\frac{3\pm\sqrt{9+16}}{4}$

$x=\frac{3\pm\sqrt{25}}{4}$

$x=\frac{3\pm 5}{4}$

$x=\frac{3+5}{4}=2$ or $x=\frac{3-5}{4}=-\frac{1}{2}$

10.4 Graphing Quadratic Equations

Problems 10.4

1. $y=-2x^2$

x	y
0	0
1	−2
−1	−2

2. $-y=-x^2-1$

x	y
0	−1
1	−2
−1	−2

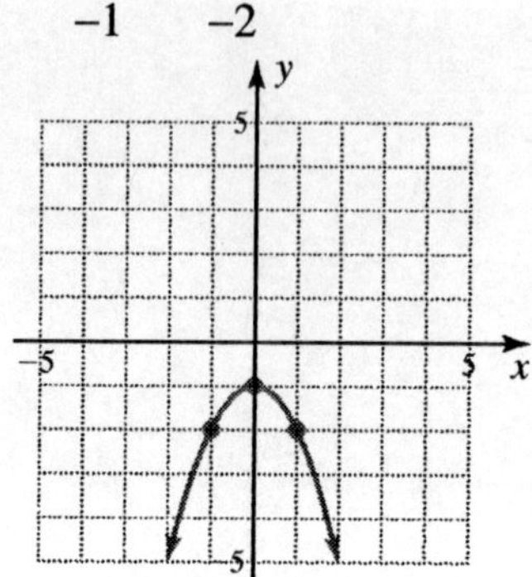

3. $y=-(x-2)^2-1$

x	y
1	−2
2	−1
3	−2

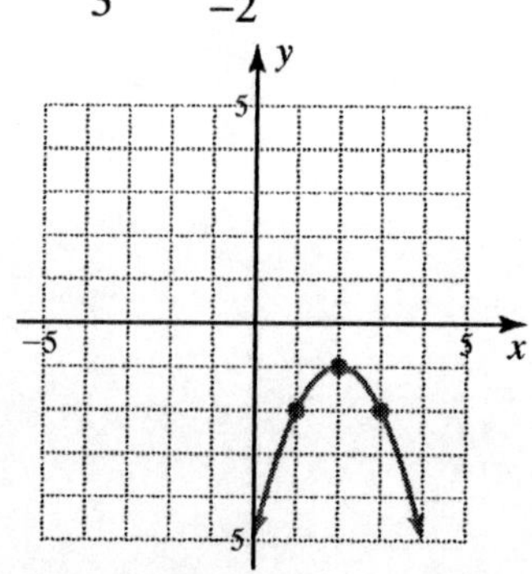

4. $y=-x^2-2x+8$

y-int: $-0-0+8=8$

$0=-x^2-2x+8$

$0=x^2+2x-8$

$0=(x+4)(x-2)$

$x=-4$ or $x=2$

x-int: −4, 2

Vertex: $x = \dfrac{-4+2}{2} = -1$

$y = -(-1)^2 - 2(-1) + 8 = 9$

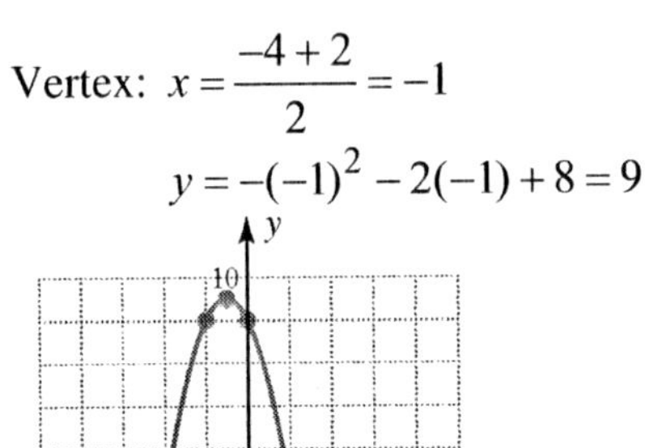

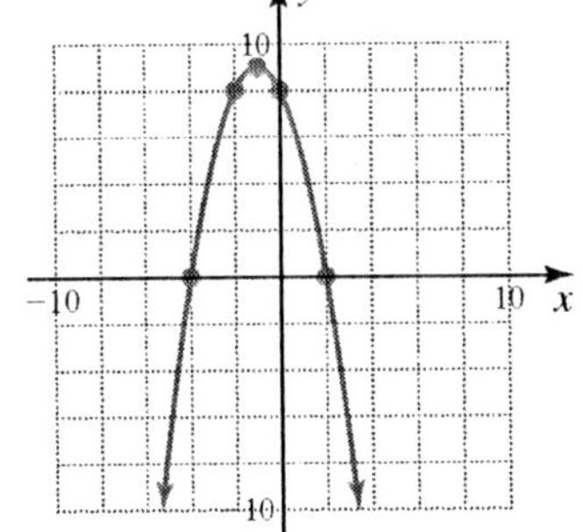

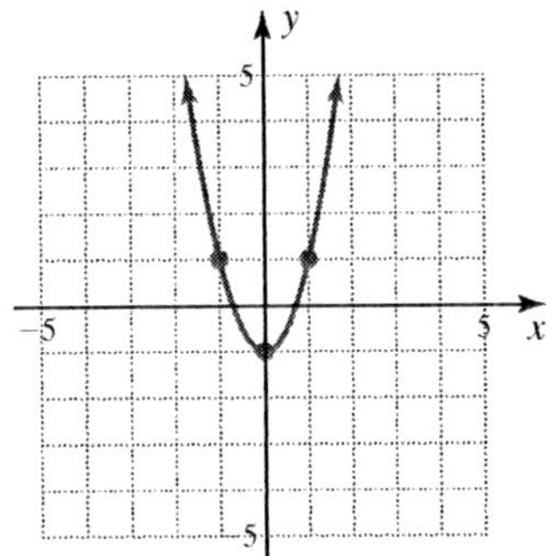

Exercises 10.4

1. $y = 2x^2$

x	y
0	0
1	2
−1	2

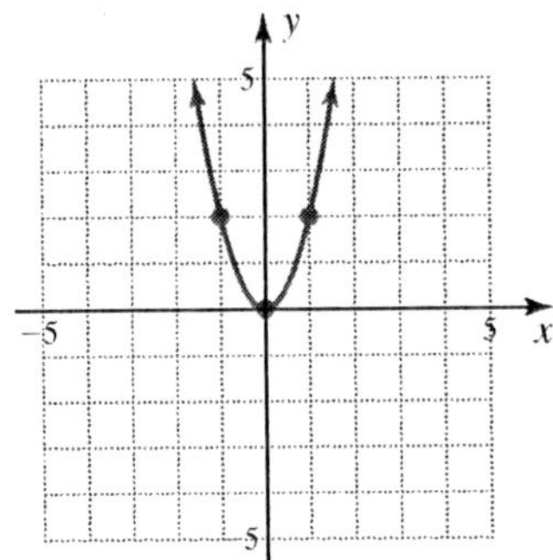

3. $y = 2x^2 - 1$

x	y
0	−1
1	1
−1	1

5. $y = -2x^2 + 2$

x	y
0	2
1	0
−1	0

7. $y = -2x^2 - 2$

x	y
0	−2
1	−4
−1	−4

9. $y = (x-2)^2$

x	y
1	1
2	0
3	1

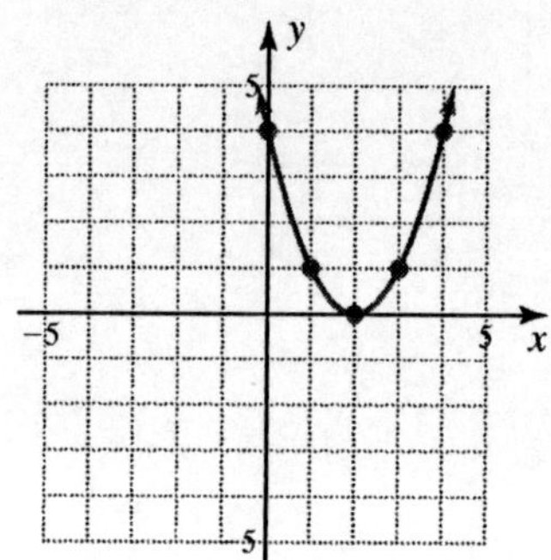

11. $y = (x-2)^2 - 2$

x	y
0	2
1	−1
2	−2
3	−1
4	2

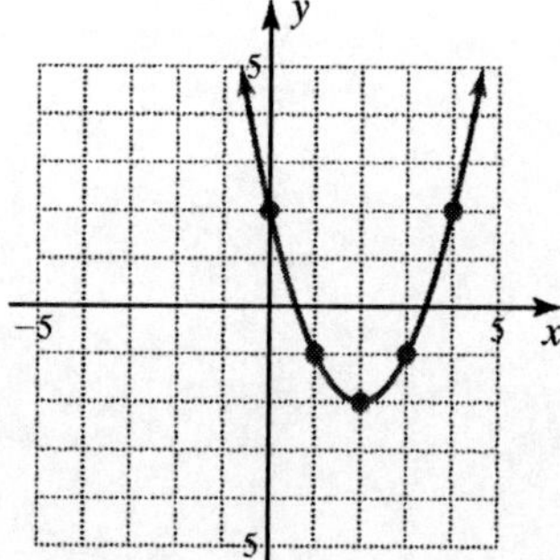

13. $y = -(x-2)^2$

x	y
1	−1
2	0
3	−1

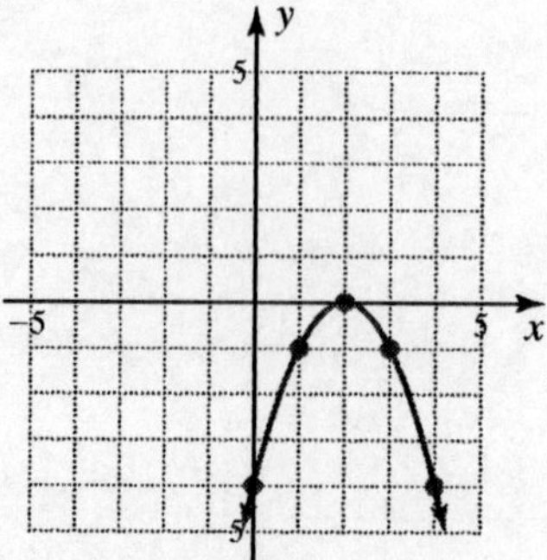

15. $y = -(x-2)^2 - 2$

x	y
1	−3
2	−2
3	−3

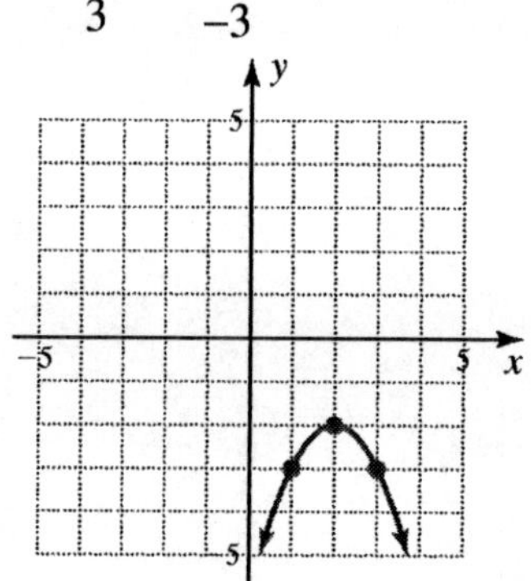

17. $y = x^2 + 4x + 3$

$0 = x^2 + 4x + 3$

$0 = (x+1)(x+3)$

$x = -1$ or $x = -3$

x-int: $(-1, 0)$, $(-3, 0)$

y-int: $0 + 0 + 3 = 3$; $(0, 3)$

vertex: $x = \dfrac{-1+(-3)}{2} = -\dfrac{4}{2} = -2$

$y = (-2)^2 + 4(-2) + 3 = 4 - 8 + 3 = -1$

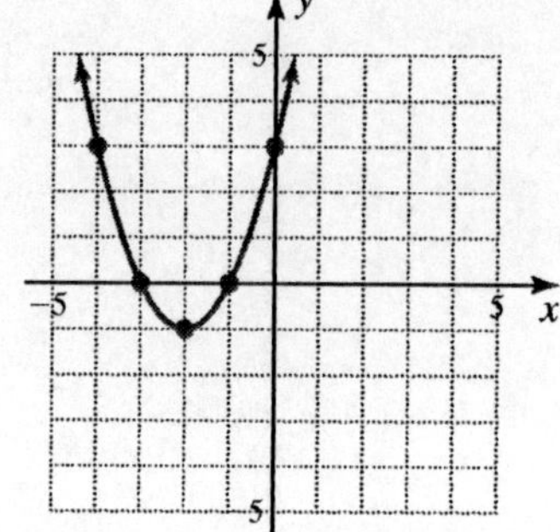

19. $y = x^2 + 2x - 3$
$0 = x^2 + 2x - 3$
$0 = (x+3)(x-1)$
$x = -3$ or $x = 1$
x-int: $(-3, 0)$, $(1, 0)$
y-int: $0 + 0 - 3 = -3$; $(0, -3)$
vertex: $x = \frac{-3+1}{2} = \frac{-2}{2} = -1$
$y = (-1)^2 + 2(-1) - 3 = 1 - 2 - 3 = -4$

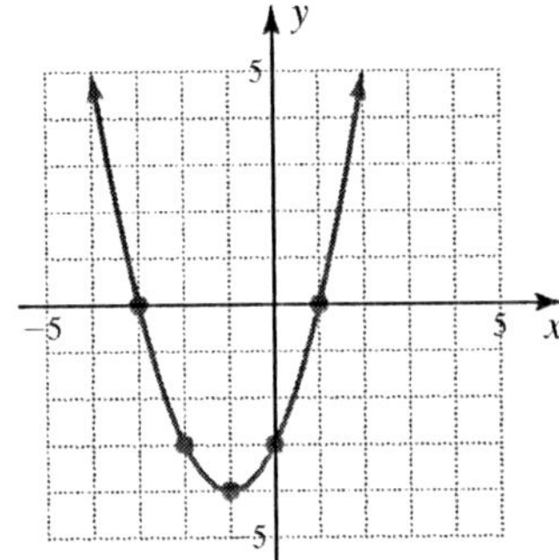

21. $y = -x^2 + 4x - 3$
$0 = -x^2 + 4x - 3$
$0 = x^2 - 4x + 3$
$0 = (x-3)(x-1)$
$x = 3$ or $x = 1$
x-int: $(3, 0)$, $(1, 0)$
y-int: $0 + 0 - 3 = -3$; $(0, -3)$
vertex: $x = \frac{3+1}{2} = \frac{4}{2} = 2$
$y = -2^2 + 4(2) - 3 = -4 + 8 - 3 = 1$

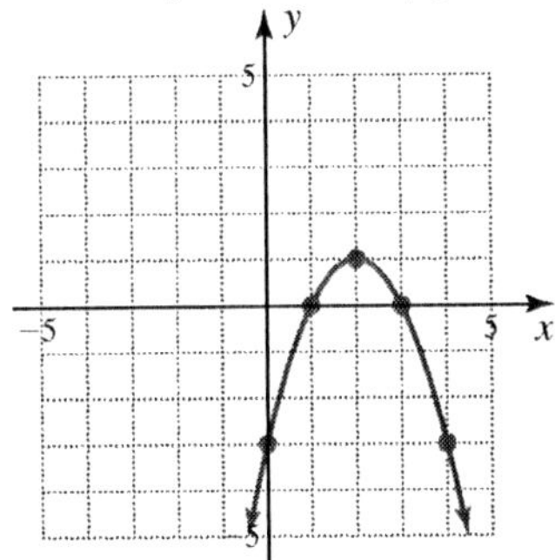

23. On the horizontal axis, 10 corresponds to 10,000 km/hr. The point on the graph is (10, 40), so the trip will take about 40 hours.

25. On the horizontal axis, 6.4 corresponds to 6400 km/hr. The point on the graph is about (6.4, 60), so the trip took about 60 hours.

27. $3^2 + 4^2 = 9 + 16 = 25$

29. $\frac{120}{0.003} = \frac{120,000}{3} = 40,000$

31. $P_T = (60 - x)(x - 20)$
$0 = (60 - x)(x - 20)$
$x = 60$ or $x = 20$
vertex: $x = \frac{60+20}{2} = \frac{80}{2} = 40$
A price of \$40 would maximize profits.

33. $P_T = (60 - 40)(40 - 20)$
$P_T = (20)(20) = 400$
The maximum profits will be \$400.

35. The graph of $y = ax^2$ opens upward if $a > 0$.

37. The graph of $y = ax^2$ "stretches" upward if a is a positive integer and a increases.

39. The graph of $y = ax^2$ is shifted upward if the positive constant k is added to y. If k is negative, the graph of y is shifted downward.

41. $y = -(x-2)^2 - 1$

x	y
1	−2
2	−1
3	−2

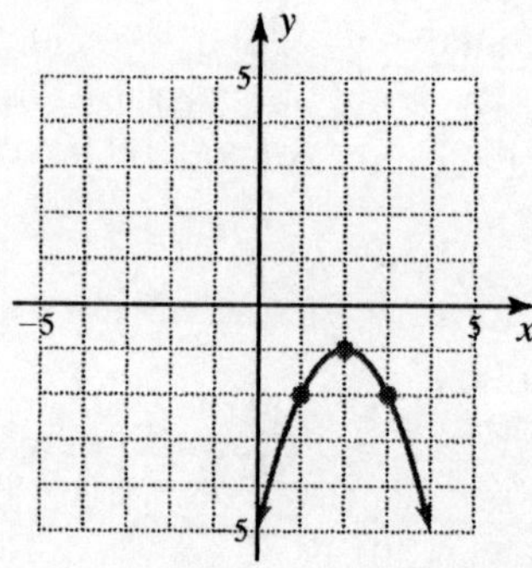

43. $y = -2x^2$

x	y
0	0
1	−2
−1	−2

45. $y = -x^2 - 2x + 3$

$0 = -x^2 - 2x + 3$

$0 = x^2 + 2x - 3$

$0 = (x+3)(x-1)$

$x = -3$ or $x = 1$

x-int: (−3, 0), (1, 0)

y-int: 0 − 0 + 3 = 3; (0, 3)

vertex:

$$x = \frac{-3+1}{2} = \frac{-2}{2} = -1$$

$$y = -(-1)^2 - 2(-1) + 3 = -1 + 2 + 3 = 4$$

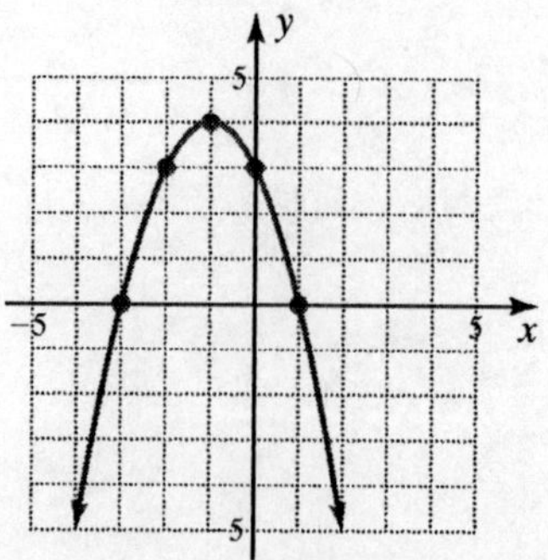

10.5 The Pythagorean Theorem and Other Applications

Problems 10.5

1. $5^2 + 14^2 = L^2$

$25 + 196 = L^2$

$221 = L^2$

$\pm\sqrt{221} = L$

Length is positive, so the length of the ladder is $\sqrt{221}$ ft.

2. $P = 0.003v^2$

$7.5 = 0.003v^2$

$\frac{7.5}{0.003} = v^2$

$2500 = v^2$

$\sqrt{2500} = v$

$\pm 50 = v$

Speed is positive, so the wind speed is 50 mi/hr.

3. $d = 5t^2 + v_0 t$

$180 = 5t^2 + (0)t$

$180 = 5t^2$

$36 = t^2$

$t^2 = 36$

$t = \pm\sqrt{36} = \pm 6$

Time is positive, so it takes 6 seconds for the object to hit the ground.

4. $R = C$

$$4x = x^2 - 3x + 6$$
$$0 = x^2 - 7x + 6$$
$$0 = (x-6)(x-1)$$
$$x - 6 = 0 \quad \text{or} \quad x - 1 = 0$$
$$x = 6 \quad \text{or} \quad x = 1$$

The company breaks even if it produces either 1000 or 6000 items.

Exercises 10.5

1. $a^2 + b^2 = h^2$

$$12^2 + b^2 = 13^2$$
$$144 + b^2 = 169$$
$$b^2 = 25$$
$$b = \sqrt{25} = 5$$

3. $a^2 + b^2 = h^2$

$$5^2 + b^2 = 15^2$$
$$25 + b^2 = 225$$
$$b^2 = 200$$
$$b = \sqrt{200} = 10\sqrt{2}$$

5. $h^2 = a^2 + b^2$

$$h^2 = \left(\sqrt{6}\right)^2 + 3^2$$
$$h^2 = 6 + 9 = 15$$
$$h = \sqrt{15}$$

7. $h^2 = a^2 + b^2$

$$h^2 = \left(\sqrt{5}\right)^2 + 2^2$$
$$h^2 = 5 + 4 = 9$$
$$h = \sqrt{9} = 3$$

9. $a^2 + b^2 = h^2$

$$3^2 + b^2 = \left(\sqrt{13}\right)^2$$
$$9 + b^2 = 13$$
$$b^2 = 4$$
$$b = \sqrt{4} = 2$$

11. $30^2 + 40^2 = L^2$

$$900 + 1600 = L^2$$
$$2500 = L^2$$
$$\pm\sqrt{2500} = L$$
$$\pm 50 = L$$

Since length is positive, the length of the wire is 50 ft.

13. $P = 0.003v^2$

$$30 = 0.003v^2$$
$$\frac{30}{0.003} = v^2$$
$$10{,}000 = v^2$$
$$\pm\sqrt{10{,}000} = v$$
$$\pm 100 = v$$

Speed is positive, so the wind speed is 100 mi/hr.

15. $d = 5t^2 + v_0 t$

$$320 = 5t^2 + (0)t$$
$$320 = 5t^2$$
$$64 = t^2$$
$$\pm\sqrt{64} = t$$
$$\pm 8 = t$$

Time is positive, so it takes 8 seconds for the object to hit the ground.

17. $d = 5t^2 + v_0 t$

$$8 = 5t^2 + 3t$$
$$0 = 5t^2 + 3t - 8$$
$$t = \frac{-3 \pm \sqrt{3^2 - 4(5)(-8)}}{2(5)}$$
$$t = \frac{-3 \pm \sqrt{9 + 160}}{10}$$
$$t = \frac{-3 \pm \sqrt{169}}{10} = \frac{-3 \pm 13}{10}$$
$$t = \frac{-3 + 13}{10} = 1 \text{ or } t = \frac{-3 - 13}{10} = -1.6$$

Time is positive, so it takes 1 second for the object to travel 8 meters.

19. $R = C$

$$5x = x^2 - x + 9$$
$$0 = x^2 - 6x + 9$$
$$0 = (x-3)^2$$
$$0 = x - 3$$
$$3 = x$$

The company must produce 3000 units to break even.

21. $A = WL$

$$2(2\cdot 3) = (2+x)(3+x)$$
$$12 = 6 + 5x + x^2$$
$$0 = x^2 + 5x - 6$$
$$0 = (x+6)(x-1)$$
$$x + 6 = 0 \quad \text{or} \quad x - 1 = 0$$
$$x = -6 \quad \text{or} \quad x = 1$$

The sides were increased, so the new dimensions are 3 feet by 4 feet.

23. $a^2 + b^2 = h^2$

$$a^2 + (a+14)^2 = (a+16)^2$$
$$a^2 + a^2 + 28a + 196 = a^2 + 32a + 256$$
$$a^2 - 4a - 60 = 0$$
$$(a-10)(a+6) = 0$$
$$a - 10 = 0 \quad \text{or} \quad a + 6 = 0$$
$$a = 10 \quad \text{or} \quad a = -6$$

$a > 0$, so a = 10 cm, b = 24 cm, and c = 26 cm.

25. $A + 20 = 32$

$$A + 20 - 20 = 32 - 20$$
$$A = 12$$

27. $18 + B = 30$

$$18 - 18 + B = 30 - 18$$
$$B = 12$$

29. $E = mc^2$

$$mc^2 = E$$
$$c^2 = \frac{E}{m}$$
$$c = \pm\sqrt{\frac{E}{m}} = \pm\frac{\sqrt{Em}}{m}$$

$c > 0$, so $c = \dfrac{\sqrt{Em}}{m}$.

31. $F = \dfrac{GMm}{r^2}$

$$r^2F = GMm$$
$$r^2 = \frac{GMm}{F}$$
$$r = \pm\sqrt{\frac{GMm}{F}} = \pm\frac{\sqrt{FGMm}}{F}$$

$r > 0$, so $r = \dfrac{\sqrt{FGMm}}{F}$.

33. $I = kP^2$

$$kP^2 = I$$
$$P^2 = \frac{I}{k}$$
$$P = \pm\sqrt{\frac{I}{k}} = \pm\frac{\sqrt{Ik}}{k}$$

$P > 0$, so $P = \dfrac{\sqrt{Ik}}{k}$.

35. Sample answer: The negative answer was discarded because speed is positive.

37. $R = C$

$$4x = x^2 - 3x + 6$$
$$0 = x^2 - 7x + 6$$
$$0 = (x-6)(x-1)$$
$$x - 6 = 0 \quad \text{or} \quad x - 1 = 0$$
$$x = 6 \quad \text{or} \quad x = 1$$

The company must produce 100 or 600 items to break even.

39.
$$P = 0.003v^2$$
$$0.003v^2 = P$$
$$0.003v^2 = 7.5$$
$$v^2 = \frac{7.5}{0.003}$$
$$v^2 = 2500$$
$$v = \pm\sqrt{2500} = \pm 50$$
Speed is positive, so the wind speed is 50 mi/hr.

10.6 Functions

Problems 10.6

1. $T = \{(6, -4), (4, -6), (-1, 3)\}$
The domain D is the set of all possible x-values. Thus $D = \{6, 4, -1\}$.
The range R is the set of all possible y-values.
Thus $R = \{-4, -6, 3\}$.

2. $\left\{(x, y) \middle| y = \frac{1}{x+1}\right\}$
The domain is the set of real numbers except -1, since $\frac{1}{-1+1} = \frac{1}{0}$ which is undefined.
The range is set of real numbers except 0, since $y = \frac{1}{x+1}$ is never 0.

3. a. $\{(5, 6), (6, 5), (6, 6), (7, 6)\}$
The relation is not a function because the two ordered pairs (6, 5) and (6, 6) have the same first coordinate.

b. $\{(x, y) | y = 3x + 2\}$
The relation is a function because if x is any real number, the expression $y = 3x + 2$ yields one and only one value, so there will never be two ordered pairs with the same first coordinate.

4. Since $f(x) = 4x - 5$,

a. $f(3) = 4 \cdot 3 - 5 = 12 - 5 = 7$

b. $f(4) = 4 \cdot 4 - 5 = 16 - 5 = 11$

c. $f(3) + f(4) = 7 + 11 = 18$

d.
$$\begin{aligned} f(x-1) &= 4(x-1) - 5 \\ &= 4x - 4 - 5 \\ &= 4x - 9 \end{aligned}$$

5. Since $g(x) = x^3 - 3x^2 + 2x - 1$,

a.
$$\begin{aligned} g(3) &= 3^3 - 3(3)^2 + 2(3) - 1 \\ &= 27 - 27 + 6 - 1 \\ &= 5 \end{aligned}$$

b.
$$\begin{aligned} g(-2) &= (-2)^3 - 3(-2)^2 + 2(-2) - 1 \\ &= -8 - 12 - 4 - 1 \\ &= -25 \end{aligned}$$

c. $g(3) - g(-2) = 5 - (-25) = 30$

6. (3, 6), (4, 8), (1.1, 2.2), $\left(\frac{3}{5}, \frac{6}{5}\right)$
Each y is 2 times x, so $f(x) = 2x$. Thus (_, 10), (_, 4.4), and (6, _) are (5, 10), (2.2, 4.4), and (6, 12).

7. $L(a) = -\frac{2}{3}a + 150$

a. $L(21) = -\frac{2}{3}(21) + 150 = -14 + 150 = 136$

b. $L(33) = -\frac{2}{3}(33) + 150 = -22 + 150 = 128$

Exercises 10.6

1. $\{(1, 2), (2, 3), (3, 4)\}$
The domain D is the set of all possible x-values. Thus $D = \{1, 2, 3\}$.
The range R is the set of all possible y-values.
Thus $R = \{2, 3, 4\}$.

3. $\{(1, 1), (2, 2), (3, 3)\}$
The domain D is the set of all possible x-values. Thus $D = \{1, 2, 3\}$.
The range R is the set of all possible y-values. Thus $R = \{1, 2, 3\}$.

5. $\{(x, y)|y = 3x\}$
The domain is the set of all real numbers.
The range is the set of all real numbers.

7. $\{(x, y)|y = x + 1\}$
The domain is the set of all real numbers.
The range is the set of all real numbers.

9. $\{(x, y) \mid y = x^2\}$
The domain is the set of all real numbers.
Since $y = x^2 \geq 0$ for all x, the range is the set of all nonnegative real numbers.

11. $\{(x, y) \mid y^2 = x\}$
Since $x = y^2 \geq 0$ for all y, the domain is the set of all nonnegative real numbers. The range is the set of all real numbers.

13. $\left\{(x, y) \middle| y = \dfrac{1}{x-3}\right\}$
The domain is the set of real numbers except 3, since $\dfrac{1}{3-3} = \dfrac{1}{0}$ which is undefined.
The range is the set of real numbers except 0, since $y = \dfrac{1}{x-3}$ is never 0.

15. $\{(x, y)|y = 2x$, x an integer between -1 and 2, inclusive$\}$
The domain consists of the integers -1, 0, 1, and 2: $D = \{-1, 0, 1, 2\}$.
Multiply each element of the domain by 2 to obtain the elements of the range. The range is $R = \{-2, 0, 2, 4\}$.
The ordered pairs are $(-1, -2)$, $(0, 0)$, $(1, 2)$, and $(2, 4)$.

17. $\{(x, y)|y = 2x - 3$, x an integer between 0 and 4, inclusive$\}$
The domain consists of the integers 0, 1, 2, 3, 4: $D = \{0, 1, 2, 3, 4\}$.
Use $y = 2x - 3$ to obtain the elements of the range. The range is $R = \{-3, -1, 1, 3, 5\}$.
The ordered pairs are $(0, -3)$, $(1, -1)$, $(2, 1)$, $(3, 3)$, and $(4, 5)$.

19. $\left\{(x, y) \middle| y = \sqrt{x},\ x = 0, 1, 4, 9, 16, \text{ or } 25\right\}$
The domain is $D = \{0, 1, 4, 9, 16, 25\}$.
Find the square root of each element of the domain to obtain the elements of the range.
The range is $R = \{0, 1, 2, 3, 4, 5\}$.
The ordered pairs are $(0, 0)$, $(1, 1)$, $(4, 2)$, $(9, 3)$, $(16, 4)$, and $(25, 5)$.

21. $\{(x, y)|y > x$, x and y positive integers less than $5\}$
Since $x < y$ and $y < 5$, x must be less than 4, but greater than 0.
The domain is $D = \{1, 2, 3\}$. Since $y > x$ and $x > 0$, y must be less than 5, but greater than 1. The range is $R = \{2, 3, 4\}$.
The ordered pairs are $(1, 2)$, $(1, 3)$, $(1, 4)$, $(2, 3)$, $(2, 4)$, and $(3, 4)$.

23. $\{(0, 1), (1, 2), (2, 3)\}$
The relation is a function because there is one y-value for each x-value.

25. $\{(-1, 1), (-2, 2), (-3, 3)\}$
The relation is a function because there is one y-value for each x-value.

27. $\{(x, y)|y = 5x + 6\}$
The relation is a function because if x is any real number, the expression $y = 5x + 6$ yields one and only one value, so there will never be two ordered pairs with the same first coordinate.

29. $\{(x, y) \mid x = y^2\}$
The relation is not a function because there are two values of y for each positive value of x.

31. Since $f(x) = 3x + 1$,

a. $f(0) = 3(0) + 1 = 0 + 1 = 1$

b. $f(2) = 3(2) + 1 = 6 + 1 = 7$

c. $f(-2) = 3(-2) + 1 = -6 + 1 = -5$

33. Since $F(x) = \sqrt{x-1}$,

a. $F(1) = \sqrt{1-1} = \sqrt{0} = 0$

b. $F(5) = \sqrt{5-1} = \sqrt{4} = 2$

c. $F(26) = \sqrt{26-1} = \sqrt{25} = 5$

35. Since $f(x) = 3x + 1$,

a. $f(x + h) = 3(x + h) + 1 = 3x + 3h + 1$

b. $\begin{aligned} f(x+h) - f(x) &= 3x + 3h + 1 - (3x+1) \\ &= 3x + 3h + 1 - 3x - 1 \\ &= 3h \end{aligned}$

c. $\dfrac{f(x+h) - f(x)}{h} = \dfrac{3h}{h} = 3,\ h \neq 0$

37. $\left(\frac{1}{2}, \frac{1}{4}\right)$, (1.2, 1.44), (5, 25), (7, 49)

Each y is the square of x, so $y = g(x) = x^2$.

Thus $\left(\frac{1}{4}, \_\right)$, (2.1, _), and (_, 64) are $\left(\frac{1}{4}, \frac{1}{16}\right)$, (2.1, 4.41), and (±8, 64).

39. Since $g(x) = 2x^3 + x^2 - 3x + 1$,

a. $\begin{aligned} g(0) &= 2(0)^3 + 0^2 - 3(0) + 1 \\ &= 0 + 0 - 0 + 1 \\ &= 1 \end{aligned}$

b. $\begin{aligned} g(-2) &= 2(-2)^3 + (-2)^2 - 3(-2) + 1 \\ &= -16 + 4 + 6 + 1 \\ &= -5 \end{aligned}$

c. $\begin{aligned} g(2) &= 2(2)^3 + 2^2 - 3(2) + 1 \\ &= 16 + 4 - 6 + 1 \\ &= 15 \end{aligned}$

41. $U(a) = -a + 190$

a. $U(50) = -50 + 190 = 140$

b. $U(60) = -60 + 190 = 130$

43. $w(h) = 5h - 190$

a. $\begin{aligned} w(70) &= 5(70) - 190 \\ &= 350 - 190 \\ &= 160 \text{ lb} \end{aligned}$

b. $\begin{aligned} 200 &= 5h - 190 \\ 390 &= 5h \\ 5h &= 390 \\ h &= 78 \text{ in.} \end{aligned}$

45. $P(d) = 63.9d$

a. $P(10) = 63.9(10) = 639 \text{ lb/ft}^2$

b. $P(100) = 63.9(100) = 6390 \text{ lb/ft}^2$

47. $S(t) = \frac{1}{2}gt^2$

a. $S(3) = \frac{1}{2}(32)(3)^2 = 144 \text{ ft}$

b. $S(5) = \frac{1}{2}(32)(5)^2 = 400 \text{ ft}$

49. $E(v) = \frac{1}{2}mv^2$

$\dfrac{36 \text{ km}}{1 \text{ hr}} = \dfrac{36{,}000 \text{ m}}{3600 \text{ s}} = 10 \text{ m/s}$

$E(10) = \frac{1}{2}(1200)(10)^2 = 60{,}000 \text{ joules}$

51. Any vertical line $x = a$, $-\infty < a < \infty$, intersects the graph of $y = |x|$ at one and only one point, so $y = |x|$ is a function.

53. Any vertical line $x = a$, $-\infty < a < \infty$, intersects the graph of $y = -x^2 + 3$ at one and only one point, so $y = -x^2 + 3$ is a function.

55. Sample answer: No; for example, $x = y^2$ is a relation, but not a function, since there are two y-values for every positive x-value.

57. Sample answer: The domain of a function defined by a set of ordered pairs is the set of x-values of the ordered pairs.

59. {(–5, 5), (–6, 6), (–7, 7)}
The domain D is the set of all possible x-values. Thus $D = \{-5, -6, -7\}$.
The range R is the set of all possible y-values. Thus $R = \{5, 6, 7\}$.

61. $\left\{(x, y) \middle| y = \frac{1}{x-3}\right\}$
The domain is the set of real numbers except 3, since $\frac{1}{3-3} = \frac{1}{0}$ which is undefined.
The range is the set of real numbers except 0, since $y = \frac{1}{x-3}$ is never 0.

63. {(–4, 5), (–5, 5), (–6, 5)}
The relation is a function because there is one y-value for each x-value.

65. $\{(x, y) | y = 2x + 3\}$
The relation is a function because if x is any real number, the expression $y = 2x + 3$ yields one and only one value, so there will never be two ordered pairs with the same first coordinate.

67. $\{(x, y) \mid x = y^2\}$
The relation is not a function because there are two values of y for each positive value of x.

69. Since $g(x) = x^3 - 1$,
$g(-1) = (-1)^3 - 1 = -1 - 1 = -2.$

Review Exercises

1. a. $x^2 = 1$
$x = \pm\sqrt{1} = \pm 1$

b. $x^2 = 100$
$x = \pm\sqrt{100} = \pm 10$

c. $x^2 = 81$
$v = \pm\sqrt{81} = \pm 9$

2. a. $16x^2 - 25 = 0$
$16x^2 = 25$
$x^2 = \frac{25}{16}$
$x = \pm\sqrt{\frac{25}{16}} = \pm\frac{5}{4}$

b. $25x^2 - 9 = 0$
$25x^2 = 9$
$x^2 = \frac{9}{25}$
$x = \pm\sqrt{\frac{9}{25}} = \pm\frac{3}{5}$

c. $64x^2 - 25 = 0$
$64x^2 = 25$
$x^2 = \frac{25}{64}$
$x = \pm\sqrt{\frac{25}{64}} = \pm\frac{5}{8}$

3. a. $7x^2 + 36 = 0$
$7x^2 = -36$
$x^2 = -\frac{36}{7}$
Since $x^2 \geq 0$, this equation has no real-number solution.

b. $8x^2 + 49 = 0$
$8x^2 = -49$
$x^2 = -\frac{49}{8}$
Since $x^2 \geq 0$, this equation has no real-number solution.

c. $3x^2 + 81 = 0$

$3x^2 = -81$

$x^2 = -\frac{81}{3} = -27$

Since $x^2 \geq 0$, this equation has no real-number solution.

4. a. $49(x+1)^2 - 3 = 0$

$49(x+1)^2 = 3$

$(x+1)^2 = \frac{3}{49}$

$x + 1 = \pm\sqrt{\frac{3}{49}} = \pm\frac{\sqrt{3}}{7}$

$x = -1 \pm \frac{\sqrt{3}}{7}$

$x = \frac{-7 \pm \sqrt{3}}{7}$

b. $25(x+2)^2 - 2 = 0$

$25(x+2)^2 = 2$

$(x+2)^2 = \frac{2}{25}$

$x + 2 = \pm\sqrt{\frac{2}{25}} = \pm\frac{\sqrt{2}}{5}$

$x = -2 \pm \frac{\sqrt{2}}{5}$

$x = \frac{-10 \pm \sqrt{2}}{5}$

c. $16(x+1)^2 - 5 = 0$

$16(x+1)^2 = 5$

$(x+1)^2 = \frac{5}{16}$

$x + 1 = \pm\sqrt{\frac{5}{16}} = \pm\frac{\sqrt{5}}{4}$

$x = -1 \pm \frac{\sqrt{5}}{4}$

$x = \frac{-4 \pm \sqrt{5}}{4}$

5. a. $(x+3)^2 = x^2 + 6x + \square$

$3^2 = 9$

b. $(x+7)^2 = x^2 + 14x + \square$

$7^2 = 49$

c. $(x+6)^2 = x^2 + 12x + \square$

$6^2 = 36$

6. a. $x^2 - 6x + \square = (\ )^2$

$\left(\frac{-6}{2}\right)^2 = (-3)^2 = 9$, so $(\ )^2 = (x-3)^2$.

b. $x^2 - 10x + \square = (\ )^2$

$\left(\frac{-10}{2}\right)^2 = (-5)^2 = 25$, so

$(\ )^2 = (x-5)^2$.

c. $x^2 - 12x + \square = (\ )^2$

$\left(-\frac{12}{2}\right)^2 = (-6)^2 = 36$, so

$(\ )^2 = (x-6)^2$.

7. a. $7x^2 - 14x = -4$

$7(x^2 - 2x) = -4$

$x^2 - 2x = -\frac{4}{7}$

The number is 7.

b. $6x^2 - 18x = -2$

$6(x^2 - 3x) = -2$

$x^2 - 3x = -\frac{2}{6} = -\frac{1}{3}$

The number is 6.

c. $5x^2 - 15x = -3$

$5(x^2 - 3x) = -3$

$x^2 - 3x = -\frac{3}{5}$

The number is 5.

8. a. $x^2 - 4x = -4$

$\left(\frac{-4}{2}\right)^2 = (-2)^2 = 4$

The number is 4.

b. $x^2 - 6x = -9$

$\left(\frac{-6}{2}\right)^2 = (-3)^2 = 9$

The number is 9.

c. $x^2 - 12x = -3$

$\left(\frac{-12}{2}\right)^2 = (-6)^2 = 36$

The number is 36.

9. a. $dx^2 + ex + f = 0,\ d \neq 0$

$$x = \frac{-e \pm \sqrt{e^2 - 4(d)(f)}}{2(d)}$$

$$x = \frac{-e \pm \sqrt{e^2 - 4df}}{2d}$$

b. $gx^2 + hx + i = 0,\ g \neq 0$

$$x = \frac{-h \pm \sqrt{h^2 - 4(g)(i)}}{2(g)}$$

$$x = \frac{-h \pm \sqrt{h^2 - 4gi}}{2g}$$

c. $jx^2 + kx + m = 0,\ j \neq 0$

$$x = \frac{-k \pm \sqrt{k^2 - 4(j)(m)}}{2(j)}$$

$$x = \frac{-k \pm \sqrt{k^2 - 4jm}}{2j}$$

10. a. $2x^2 - x - 1 = 0$

$$x = \frac{-(-1) \pm \sqrt{(-1)^2 - 4(2)(-1)}}{2(2)}$$

$$x = \frac{1 \pm \sqrt{1+8}}{4}$$

$$x = \frac{1 \pm \sqrt{9}}{4} = \frac{1 \pm 3}{4}$$

$$x = \frac{1+3}{4} = 1 \text{ or } x = \frac{1-3}{4} = -\frac{1}{2}$$

b. $2x^2 - 2x - 5 = 0$

$$x = \frac{-(-2) \pm \sqrt{(-2)^2 - 4(2)(-5)}}{2(2)}$$

$$x = \frac{2 \pm \sqrt{4+40}}{4}$$

$$x = \frac{2 \pm 2\sqrt{11}}{4} = \frac{1 \pm \sqrt{11}}{2}$$

c. $2x^2 - 3x - 3 = 0$

$$x = \frac{-(-3) \pm \sqrt{(-3)^2 - 4(2)(-3)}}{2(2)}$$

$$x = \frac{3 \pm \sqrt{9+24}}{4}$$

$$x = \frac{3 \pm \sqrt{33}}{4}$$

11. a. $3x^2 - x = 1$

$3x^2 - x - 1 = 0$

$$x = \frac{-(-1) \pm \sqrt{(-1)^2 - 4(3)(-1)}}{2(3)}$$

$$x = \frac{1 \pm \sqrt{1+12}}{6}$$

$$x = \frac{1 \pm \sqrt{13}}{6}$$

b. $3x^2 - 2x = 2$

$3x^2 - 2x - 2 = 0$

$x = \dfrac{-(-2) \pm \sqrt{(-2)^2 - 4(3)(-2)}}{2(3)}$

$x = \dfrac{2 \pm \sqrt{4 + 24}}{6}$

$x = \dfrac{2 \pm \sqrt{28}}{6}$

$x = \dfrac{2 \pm 2\sqrt{7}}{6} = \dfrac{1 \pm \sqrt{7}}{3}$

c. $3x^2 - 3x = 2$

$3x^2 - 3x - 2 = 0$

$x = \dfrac{-(-3) \pm \sqrt{(-3)^2 - 4(3)(-2)}}{2(3)}$

$x = \dfrac{3 \pm \sqrt{9 + 24}}{6}$

$x = \dfrac{3 \pm \sqrt{33}}{6}$

12. a. $9x = x^2$

$0 = x^2 - 9x$

$0 = x(x - 9)$

$x = 0$ or $x - 9 = 0$

$x = 0$ or $x = 9$

b. $4x = x^2$

$0 = x^2 - 4x$

$0 = x(x - 4)$

$x = 0$ or $x - 4 = 0$

$x = 0$ or $x = 4$

c. $25x = x^2$

$0 = x^2 - 25x$

$0 = x(x - 25)$

$x = 0$ or $x - 25 = 0$

$x = 0$ or $x = 25$

13. a. $\dfrac{x^2}{9} - x = -\dfrac{4}{9}$

$x^2 - 9x = -4$

$x^2 - 9x + 4 = 0$

$x = \dfrac{-(-9) \pm \sqrt{(-9)^2 - 4(1)(4)}}{2(1)}$

$x = \dfrac{9 \pm \sqrt{81 - 16}}{2}$

$x = \dfrac{9 \pm \sqrt{65}}{2}$

b. $\dfrac{x^2}{5} - \dfrac{x}{2} = \dfrac{3}{10}$

$2x^2 - 5x = 3$

$2x^2 - 5x - 3 = 0$

$x = \dfrac{-(-5) \pm \sqrt{(-5)^2 - 4(2)(-3)}}{2(2)}$

$x = \dfrac{5 \pm \sqrt{25 + 24}}{4}$

$x = \dfrac{5 \pm \sqrt{49}}{4} = \dfrac{5 \pm 7}{4}$

$x = \dfrac{5 + 7}{4} = 3$ or $x = \dfrac{5 - 7}{4} = -\dfrac{1}{2}$

c. $\dfrac{x^2}{3} + \dfrac{x}{6} = -\dfrac{1}{2}$

$2x^2 + x = -3$

$2x^2 + x + 3 = 0$

$x = \dfrac{-1 \pm \sqrt{1^2 - 4(2)(3)}}{2(2)}$

$x = \dfrac{-1 \pm \sqrt{1 - 24}}{4}$

$x = \dfrac{-1 \pm \sqrt{-23}}{4}$

Since $\sqrt{-23}$ is not a real number, this equation has no real-number solution.

14. See below for graphs.

a. $y = x^2 + 1$

x	y
0	1
1	2
–1	2

b. $y = x^2 + 2$

x	y
0	2
1	3
–1	3

c. $y = x^2 + 3$

x	y
0	3
1	4
–1	4

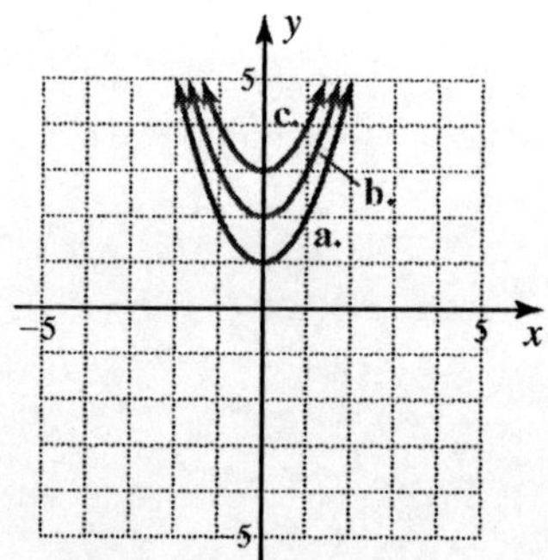

15. See below for graph.

a. $y = -x^2 - 1$

x	y
0	–1
1	–2
–1	–2

b. $y = -x^2 - 2$

x	y
0	–2
1	–3
–1	–3

c. $y = -x^2 - 3$

x	y
0	–3
1	–4
–1	–4

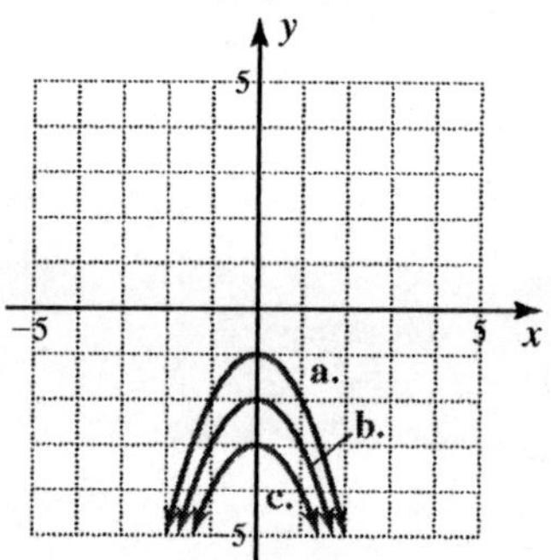

16. See below for graph.

a. $y = -(x-2)^2$

x	y
1	–1
2	0
3	–1

b. $y = -(x-3)^2$

x	y
2	–1
3	0
4	–1

c. $y = -(x-4)^2$

x	y
3	−1
4	0
5	−1

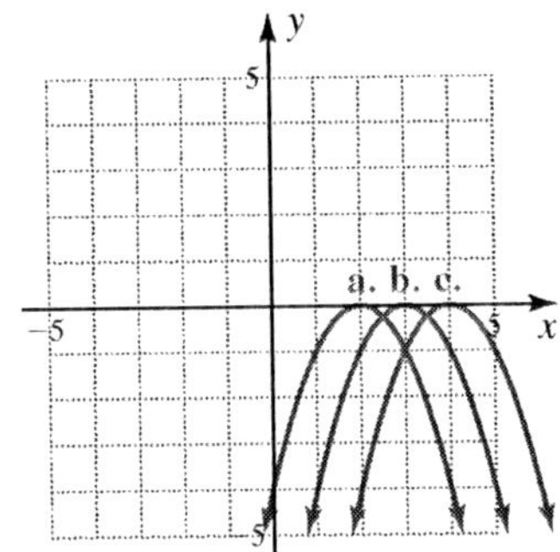

17. See below for graph.

a. $y = (x-2)^2 + 1$

x	y
1	2
2	1
3	2

b. $y = (x-2)^2 + 2$

x	y
1	3
2	2
3	3

c. $y = (x-2)^2 + 3$

x	y
1	4
2	3
3	4

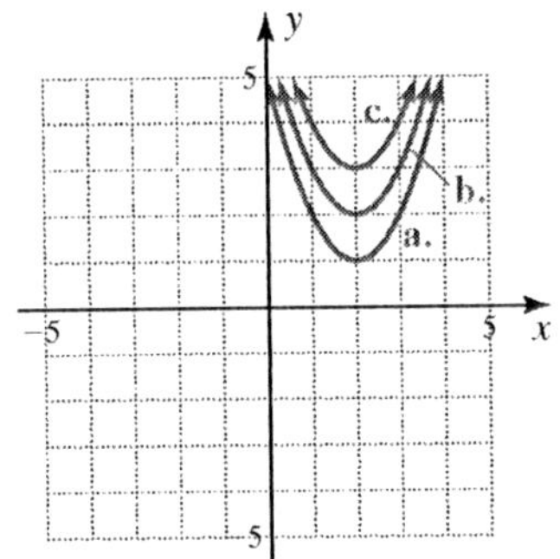

18. a. $y = -(x-2)^2 - 1$

x	y
1	−2
2	−1
3	−2

b. $y = -(x-2)^2 - 2$

x	y
1	−3
2	−2
3	−3

c. $y = -(x-2)^2 - 3$

x	y
1	−4
2	−3
3	−4

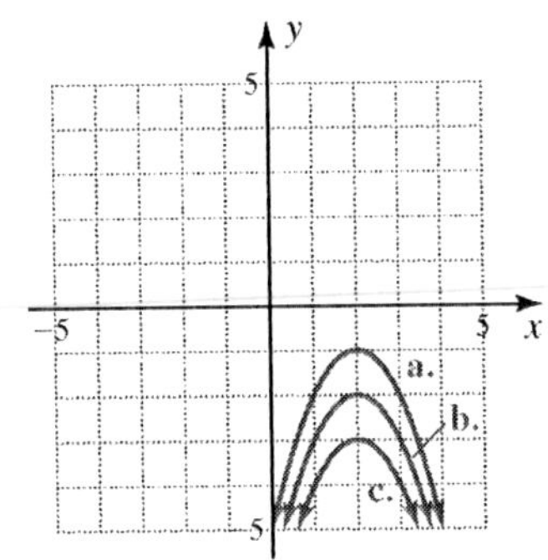

19. a. $h^2 = a^2 + b^2$
$h^2 = 5^2 + 12^2$
$h^2 = 25 + 144 = 169$
$h = \sqrt{169} = 13$ in.

b. $h^2 = a^2 + b^2$
$h^2 = 2^2 + 3^2$
$h^2 = 4 + 9 = 13$
$h = \sqrt{13}$ in.

c. $h^2 = a^2 + b^2$
$h^2 = 4^2 + 5^2$
$h^2 = 16 + 25 = 41$
$h = \sqrt{41}$ in.

20. $d = 5t^2 + v_0 t$

a. $125 = 5t^2 + (0)t$
$125 = 5t^2$
$25 = t^2$
$\pm 5 = t$
$t > 0$, so $t = 5$ seconds.

b. $245 = 5t^2$
$49 = t^2$
$\pm 7 = t$
$t > 0$, so $t = 7$ seconds.

c. $320 = 5t^2$
$64 = t^2$
$\pm 8 = t$
$t > 0$, so $t = 8$ seconds.

21. a. $\{(-3, 1), (-4, 1), (-5, 2)\}$
The domain D is the set of all possible x-values. Thus $D = \{-3, -4, -5\}$.
The range R is the set of all possible y-values. Thus $R = \{1, 2\}$.

b. $\{(2, -4), (-1, 3), (-1, 4\}$
The domain D is the set of all possible x-values. Thus $D = \{-1, 2\}$.
The range R is the set of all possible y-values. Thus $R = \{-4, 3, 4\}$.

c. $\{(-1, 2), (-1, 3), (-1, 4)\}$
The domain D is the set of all possible x-values. Thus $D = \{-1\}$.
The range R is the set of all possible y-values. Thus $R = \{2, 3, 4\}$.

22. a. $\{(x, y) | y = 2x - 3\}$
The domain is the set of all real numbers. The range is the set of all real numbers.

b. $\{(x, y) | y = -3x - 2\}$
The domain is the set of all real numbers. The range is the set of all real numbers.

c. $\{(x, y) \mid y = x^2\}$
The domain is the set of all real numbers. Since $y = x^2 \geq 0$ for all x, the range is the set of all nonnegative real numbers.

23. a. $\{(-3, 1), (-4, 1), (-5, 2)\}$
The relation is a function because there is one y-value for each x-value.

b. $\{(2, -4), (-1, 3), (-1, 4)\}$
The relation is not a function because the two ordered pairs $(-1, 3)$ and $(-1, 4)$ have the same first coordinate.

c. $\{(-1, 2), (-1, 3), (-1, 4)\}$
The relation is not a function because all three ordered pairs have the same first coordinate.

24. Since $f(x) = x^3 - 2x^2 + x - 1$,

a. $f(2) = 2^3 - 2(2)^2 + 2 - 1$
$= 8 - 8 + 2 - 1$
$= 1$

b. $f(-2) = (-2)^3 - 2(-2)^2 + (-2) - 1$
$= -8 - 8 - 2 - 1$
$= -19$

c. $f(1) = 1^3 - 2(1)^2 + 1 - 1$
$= 1 - 2 + 1 - 1$
$= -1$

25. $P(n) = 25 - 0.3n$

a. $P(10) = 25 - 0.3(10) = 25 - 3 = \$22$

b. $P(20) = 25 - 0.3(20) = 25 - 6 = \$19$

c. $P(30) = 25 - 0.3(30) = 25 - 9 = \$16$

Cumulative Review Chapters 1–10

1. $-\frac{3}{8}+\left(-\frac{1}{7}\right)=-\frac{21}{56}-\frac{8}{56}=-\frac{29}{56}$

2. $(-4)^4=(-4)(-4)(-4)(-4)=256$

3. $-\frac{3}{8}\div\left(-\frac{5}{24}\right)=\frac{3}{8}\times\frac{24}{5}=\frac{3}{1}\times\frac{3}{5}=\frac{9}{5}$

4.
$$\begin{aligned} y\div 4\cdot x-z &= 16\div 4\cdot 4-2\\ &=4\cdot 4-2\\ &=16-2\\ &=14\end{aligned}$$

5.
$$\begin{aligned} 2x-(x+4)-2(x+3) &= 2x-x-4-2x-6\\ &=-x-10\end{aligned}$$

6. $\frac{x-5y}{z}$

7.
$$\begin{aligned} 5&=3(x-1)+5-2x\\ 0&=3x-3-2x\\ 0&=x-3\\ 3&=x\end{aligned}$$

8.
$$\begin{aligned} \frac{x}{8}-\frac{x}{9}&=1\\ 9x-8x&=72\\ x&=72\end{aligned}$$

9.
$$\begin{aligned} -\frac{x}{4}+\frac{x}{8}&\geq\frac{x-8}{8}\\ -2x+x&\geq x-8\\ -x&\geq x-8\\ 8&\geq 2x\\ 4&\geq x\\ x&\leq 4\end{aligned}$$

0 4

10. To graph $C(-1, -2)$, start at the origin, go 1 unit left and 2 units down.

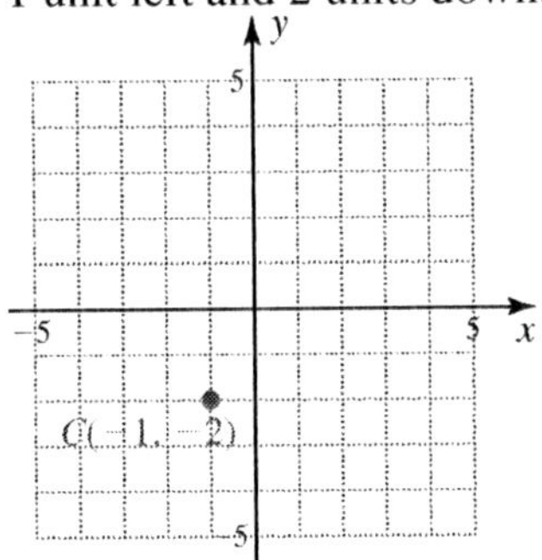

11.
$$\begin{aligned} 2x-y&=-1\\ 2(-2)-(-3)&\stackrel{?}{=}-1\\ -4+3&\stackrel{?}{=}-1\\ -1&=-1\end{aligned}$$

$(-2, -3)$ is a solution of $2x - y = -1$.

12.
$$\begin{aligned} 2x-3y&=-10\\ 2x-3(2)&=-10\\ 2x-6&=-10\\ 2x&=-4\\ x&=-2\end{aligned}$$

13. $2x + y = 4$

y-int: $2(0)+y=4$

$y=4$

x-int: $2x+0=4$

$2x=4$

$x=2$

Graph (0, 4) and (2, 0) and draw a straight line through these points.

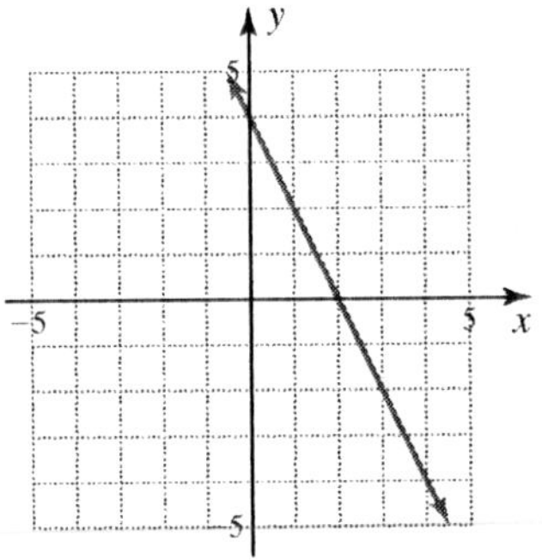

14. $2y-8=0$
$2y=8$
$y=4$
The graph is the horizontal line $y = 4$.

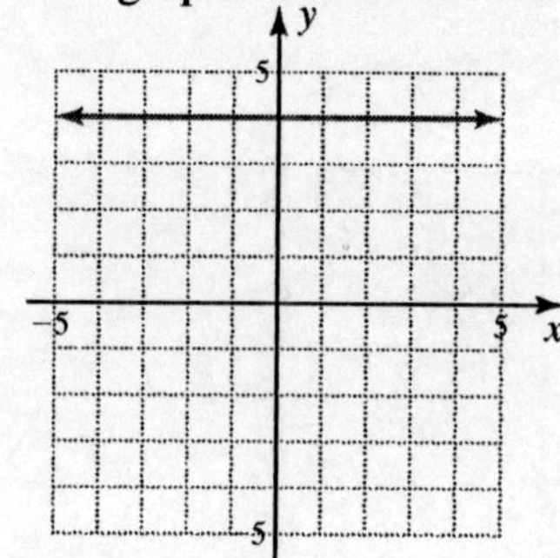

15. $m=\dfrac{6-(-8)}{6-(-5)}=\dfrac{14}{11}$

16. $8x-4y=-13$
$-4y=-8x-13$
$y=2x+\dfrac{13}{4}$
$y=mx+b$
$m = 2$, so the slope is 2.

17. **(1)** $8y+12x=7$
$8y=-12x+7$
$y=-\dfrac{3}{2}x+\dfrac{7}{8}$

(2) $12x-8y=7$
$-8y=-12x+7$
$y=\dfrac{3}{2}x-\dfrac{7}{8}$

(3) $-2y=-3x+7$
$y=\dfrac{3}{2}x-\dfrac{7}{2}$

(2) and (3) have the same slope, $\dfrac{3}{2}$.

18. $(3x^3y^{-5})^4=3^4x^{3\cdot4}y^{-5\cdot4}=\dfrac{81x^{12}}{y^{20}}$

19. $0.0050=5.0\times10^{-3}$
Move the decimal point 3 places to the right and multiply by 10^{-3}.

20. $1.65\times10^{-4}\div1.1\times10^3=\dfrac{1.65\times10^{-4}}{1.1\times10^3}$
$=\dfrac{1.65}{1.1}\times10^{-4-3}$
$=1.5\times10^{-7}$

21. $\left(2x^2-\dfrac{1}{5}\right)^2=\left(2x^2-\dfrac{1}{5}\right)\left(2x^2-\dfrac{1}{2}\right)$
$=4x^4-\dfrac{4}{5}x^2+\dfrac{1}{25}$

22.
$$\begin{array}{r} 2x^2\;-x-1\text{ R }1 \\ x+3\overline{\smash{)}\,2x^3+5x^2-4x-2} \\ \underline{2x^3+6x^2\qquad\qquad} \\ 0\;-x^2-4x\quad \\ \underline{-x^2-3x\quad} \\ 0\;\;-x-2 \\ \underline{-x-3} \\ 0+1 \end{array}$$

23. $15=-5(-3)$ and $-5+(-3)=-8$, so
$x^2-8x+15=(x-5)(x-3)$.

24. $12x^2-25xy+12y^2$
$=12x^2-16xy-9xy+12y^2$
$=4x(3x-4y)-3y(3x-4y)$
$=(4x-3y)(3x-4y)$

25. $25x^2 - 36y^2 = (5x)^2 - (6y)^2$
$= (5x + 6y)(5x - 6y)$

26. $-5x^4 + 80x^2 = -5x^2(x^2 - 16)$
$= -5x^2(x + 4)(x - 4)$

27. $3x^3 - 3x^2 - 18x = 3x(x^2 - x - 6)$
$= 3x(x + 2)(x - 3)$

28. $4x^2 + 12x + 5x + 15 = 4x(x + 3) + 5(x + 3)$
$= (x + 3)(4x + 5)$

29. $25kx^2 + 20kx + 4k = k(25x^2 + 20x + 4)$
$= k[(5x)^2 + 2 \cdot 5x \cdot 2 + 2^2]$
$= k(5x + 2)^2$

30. $3x^2 + 13x = 10$
$3x^2 + 13x - 10 = 0$
$(3x - 2)(x + 5) = 0$
$3x - 2 = 0$ or $x + 5 = 0$
$3x = 2$ $\quad x = -5$
$x = \frac{2}{3}$

31. $\frac{6x}{5y} = \frac{6x}{5y} \cdot \frac{2y^2}{2y^2} = \frac{12xy^2}{10y^3}$

32. $\frac{-4(x^2 - y^2)}{4(x - y)} = \frac{-(x + y)(x - y)}{x - y} = -(x + y)$

33. $\frac{x^2 - 4x - 21}{7 - x} = \frac{(x - 7)(x + 3)}{-(x - 7)} = -(x + 3)$

34. $(x - 4) \cdot \frac{x - 3}{x^2 - 16} = \frac{(x - 4)(x - 3)}{(x + 4)(x - 4)} = \frac{x - 3}{x + 4}$

35. $\frac{x + 2}{x - 2} \div \frac{x^2 - 4}{2 - x} = \frac{x + 2}{x - 2} \times \frac{2 - x}{x^2 - 4}$
$= \frac{-(x + 2)(x - 2)}{(x - 2)(x + 2)(x - 2)}$
$= -\frac{1}{x - 2}$
$= \frac{1}{2 - x}$

36. $\frac{5}{4(x + 1)} + \frac{7}{4(x + 1)} = \frac{5 + 7}{4(x + 1)}$
$= \frac{12}{4(x + 1)}$
$= \frac{3}{x + 1}$

37. $\frac{x + 7}{x^2 + x - 56} - \frac{x + 8}{x^2 - 49}$
$= \frac{x + 7}{(x - 7)(x + 8)} - \frac{x + 8}{(x + 7)(x - 7)}$
$= \frac{(x + 7)^2}{(x + 7)(x - 7)(x + 8)} - \frac{(x + 8)^2}{(x + 7)(x - 7)(x + 8)}$
$= \frac{x^2 + 14x + 49 - x^2 - 16x - 64}{(x + 8)(x + 7)(x - 7)}$
$= \frac{-2x - 15}{(x + 8)(x + 7)(x - 7)}$

38. $\frac{\frac{1}{x} + \frac{3}{2x}}{\frac{2}{3x} - \frac{1}{4x}} = \frac{\frac{2+3}{2x}}{\frac{8-3}{12x}} = \frac{\frac{5}{2x}}{\frac{5}{12x}} = \frac{5}{2x} \cdot \frac{12x}{5} = 6$

39. $\frac{x}{x - 3} + 6 = \frac{5x}{x - 3}$
$x + 6(x - 3) = 5x$
$x + 6x - 18 = 5x$
$2x = 18$
$x = 9$

40. $\frac{x}{x^2 - 16} + \frac{4}{x - 4} = \frac{1}{x + 4}$
$x + 4(x + 4) = x - 4$
$x + 4x + 16 = x - 4$
$4x = -20$
$x = -5$

41. $\frac{x}{x+4}-\frac{1}{5}=\frac{-4}{x+4}$
$x-\frac{x+4}{5}=-4$
$5x-x-4=-20$
$4x=-16$
$x=-4$
Since $x+4\neq 0$, $x\neq -4$, so the equation has no solution.

42. $2+\frac{8}{x-2}=\frac{32}{x^2-4}$
$2(x^2-4)+8(x+2)=32$
$2x^2-8+8x+16=32$
$2x^2+8x-24=0$
$x^2+4x-12=0$
$(x+6)(x-2)=0$
$x+6=0$ or $x-2=0$
$x=-6$ or $x=2$
Since $x-2\neq 0$, $x\neq 2$, so $x=-6$ is the only solution.

43. $\frac{x}{725}=\frac{10}{250}$
$250x=7250$
$x=29$ gallons

44. $\frac{x-8}{4}=\frac{6}{5}$
$x-8=\frac{24}{5}$
$x=\frac{24}{5}+8$
$x=\frac{24+40}{5}$
$x=\frac{64}{5}$

45. t = total time to paint

Janet does $\frac{1}{4}$ of the job in 1 hr. James does $\frac{1}{5}$ of the job in 1 hr. Together they do $\frac{1}{t}$ of the job in 1 hr.

$\frac{1}{4}+\frac{1}{5}=\frac{1}{t}$
$\frac{5+4}{20}=\frac{1}{t}$
$\frac{9}{20}=\frac{1}{t}$
$t=\frac{20}{9}=2\frac{2}{9}$ hr

46. $y-y_1=m(x-x_1)$
$y+4=6(x+2)$
$y=6x+12-4$
$y=6x+8$

47. $y=mx+b$
$y=-2x-5$

48. $3x-2y<-6$
$-2y<-3x-6$
$y>\frac{3}{2}x+3$

Graph the dashed line $y=\frac{3}{2}x+3$.

Check the point (0, 0).

$0>\frac{3}{2}(0)+3$
$0>0+3$
$0>3$ False

Shade the region above the line $y=\frac{3}{2}x+3$.

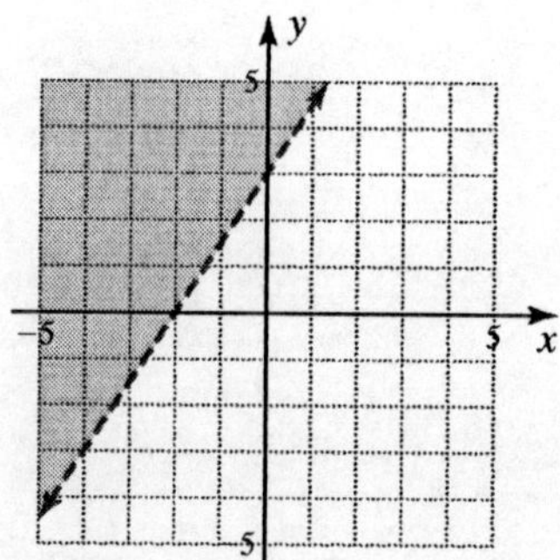

49. $-y\geq -5x+5$
$y\leq 5x-5$

Graph the solid line $y=5x-5$.
Check the point (0, 0).
$0\leq 5(0)-5$
$0\leq 0-5$
$0\leq -5$ False

Shade the region below the line $y = 5x - 5$.

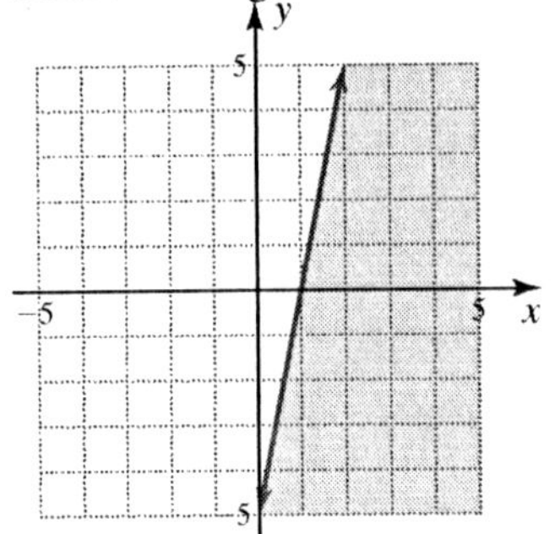

50. $$P = kT$$
$$7 = k(490)$$
$$\frac{7}{490} = k$$
$$\frac{1}{70} = k$$

51. $$P = \frac{k}{V}$$
$$1960 = \frac{k}{7}$$
$$13,720 = k$$

52. $$x + 3y = -12$$
$$3y = -x - 12$$
$$y = -\frac{1}{3}x - 4$$

$$2y - x = -2$$
$$2y = x - 2$$
$$y = \frac{1}{2}x - 1$$

$$-\frac{1}{3}x - 4 = \frac{1}{2}x - 1$$
$$-\frac{1}{3}x - \frac{1}{2}x = -1 + 4$$
$$-\frac{5}{6}x = 3$$
$$x = -\frac{18}{5}$$

$$y = \frac{1}{2}\left(-\frac{18}{5}\right) - 1 = -\frac{9}{5} - 1 = -\frac{14}{5}$$

The graphs intersect at $\left(-\frac{18}{5}, -\frac{14}{5}\right)$.

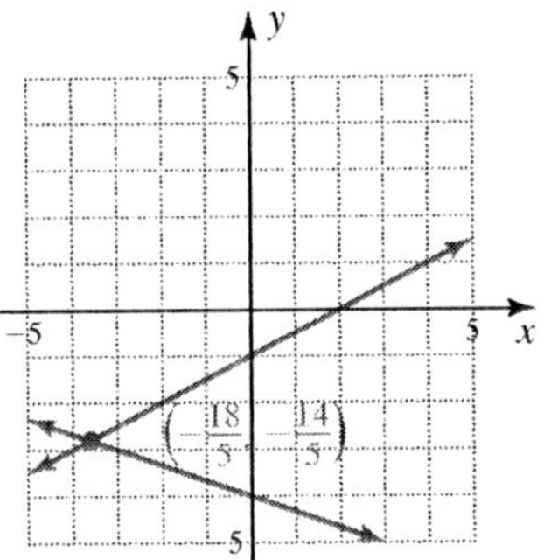

53. $$y + 3x = 3$$
$$y = -3x + 3$$

$$3y + 9x = 18$$
$$3y = -9x + 18$$
$$y = -3x + 6$$

The lines are parallel, so the system has no solution.

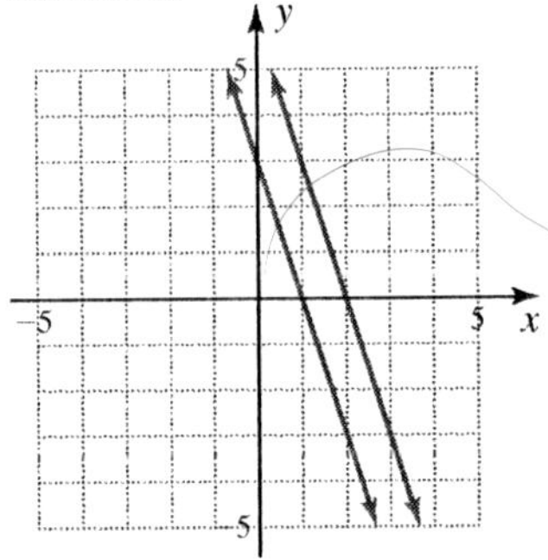

54. $$x + 3y = 10$$
$$x = -3y + 10$$

$$4x + 12y = 42$$
$$4(-3y + 10) + 12y = 42$$
$$-12y + 40 + 12y = 42$$
$$40 = 42$$

The equations are inconsistent, so there is no solution.

55. $$x - 3y = 3$$
$$x = 3y + 3$$

$$-3x + 9y = -9$$
$$-3(3y + 3) + 9y = -9$$
$$-9y - 9 + 9y = -9$$
$$-9 = -9$$

The equations are dependent, so there are infinitely many solutions.

56. Multiply the first equation by 2 and the second equation by –3.

$$\begin{aligned} 6x-8y&=6 \\ -6x+9y&=-9 \\ \hline y&=-3 \end{aligned}$$

Substitute –3 for y in $2x-3y=3$.
The solution is (–3, –3).

57. $4x+3y=18$
$8x+6y=-3$
Multiply the first equation by 2 and the second equation by –1.

$$\begin{aligned} 8x+6y&=36 \\ -8x-6y&=3 \\ \hline 0&=39 \end{aligned}$$

The equations are inconsistent, so there is no solution.

58. $3y+x=-13$
$-4x-12y=52$
Multiply the first equation by 4 and add to the second equation.

$$\begin{aligned} 4x+12y&=-52 \\ -4x-12y&=52 \\ \hline 0&=0 \end{aligned}$$

The equations are dependent, so there are infinitely many solutions.

59. n = nickels
d = dimes
$d = 2n$

$$\begin{aligned} 0.05n+0.1d&=2.50 \\ 0.05n+0.2n&=2.50 \\ 0.25n&=2.50 \\ n&=10 \end{aligned}$$

$d = 2n = 20$
Kaye has 10 nickels and 20 dimes.

60. $x+y=170$

$$\begin{aligned} x+y&=170 \\ x-y&=110 \\ \hline 2x&=280 \\ x&=140 \end{aligned}$$

$$\begin{aligned} 140+y&=170 \\ y&=30 \end{aligned}$$

The numbers are 30 and 140.

61. $\sqrt[5]{243}=\sqrt[5]{3^5}=3$

62. $\sqrt[4]{(-5)^4}=\sqrt[4]{625}=\sqrt[4]{5^4}=5$

63. $\sqrt{\dfrac{5}{243}}=\dfrac{\sqrt{5}}{\sqrt{3\cdot 9^2}}=\dfrac{\sqrt{5}}{9\sqrt{3}}=\dfrac{\sqrt{15}}{27}$

64. $\dfrac{\sqrt{2}}{\sqrt{7j}}=\dfrac{\sqrt{2}}{\sqrt{7j}}\cdot\dfrac{\sqrt{7j}}{\sqrt{7j}}=\dfrac{\sqrt{14j}}{7j}$

65.

$$\begin{aligned} \sqrt{32}+\sqrt{18}&=\sqrt{4^2\cdot 2}+\sqrt{3^2\cdot 2} \\ &=4\sqrt{2}+3\sqrt{2} \\ &=7\sqrt{2} \end{aligned}$$

66.

$$\begin{aligned} \sqrt[3]{3x}\left(\sqrt[3]{9x^2}-\sqrt[3]{16x}\right)&=\sqrt[3]{27x^3}-\sqrt[3]{48x^2} \\ &=\sqrt[3]{3^3x^3}-\sqrt[3]{2^3\cdot 6x^2} \\ &=3x-2\sqrt[3]{6x^2} \end{aligned}$$

67.

$$\begin{aligned} &\left(\sqrt{125}+\sqrt{63}\right)\left(\sqrt{245}+\sqrt{175}\right) \\ &=\left(\sqrt{5^2\cdot 5}+\sqrt{3^2\cdot 7}\right)\left(\sqrt{7^2\cdot 5}+\sqrt{5^2\cdot 7}\right) \\ &=\left(5\sqrt{5}+3\sqrt{7}\right)\left(7\sqrt{5}+5\sqrt{7}\right) \\ &=35\cdot 5+25\sqrt{35}+21\sqrt{35}+15\cdot 7 \\ &=280+46\sqrt{35} \end{aligned}$$

68. $\dfrac{20-\sqrt{32}}{4}=\dfrac{20-4\sqrt{2}}{4}=5-\sqrt{2}$

69.

$$\begin{aligned} \sqrt{x+4}&=-3 \\ x+4&=9 \\ x&=5 \end{aligned}$$

Check: $\sqrt{5+4}=\sqrt{9}=3\neq -3$
The equation has no real-number solution.

70. $\sqrt{x-4}-x=-4$

$$\sqrt{x-4}=x-4$$
$$\left(\sqrt{x-4}\right)^2=(x-4)^2$$
$$x-4=x^2-8x+16$$
$$0=x^2-9x+20$$
$$0=(x-4)(x-5)$$
$$x-4=0 \quad\text{or}\quad x-5=0$$
$$x=4 \quad\text{or}\quad x=5$$

71. $9x^2-4=0$

$$9x^2=4$$
$$x^2=\frac{4}{9}$$
$$x=\pm\sqrt{\frac{4}{9}}=\pm\frac{2}{3}$$

72. $64x^2+9=0$

$$64x^2=-9$$
$$x^2=-\frac{9}{64}$$

$x^2\geq 0,$ so the equation has no real-number solution.

73. $36(x-4)^2-7=0$

$$36(x-4)^2=7$$
$$(x-4)^2=\frac{7}{36}$$
$$x-4=\pm\sqrt{\frac{7}{36}}=\pm\frac{\sqrt{7}}{6}$$
$$x=4\pm\frac{\sqrt{7}}{6}$$

74. $(x+2)^2=x^2+4x+$ ___

$$\left(\frac{4}{2}\right)^2=2^2=4$$

The missing term is 4.

75. $4x^2+8x=14$

$$4(x^2+2x)=14$$
$$x^2+2x=\frac{14}{4}=\frac{7}{2}$$

Each term must be divided by 4.

76. $x^2+4x=5$

$$\left(\frac{4}{2}\right)^2=2^2=4$$
$$x^2+4x+4=5+4$$

The number 4 must be added to both sides.

77. $jx^2+kx+m=0$

$$x=\frac{-k\pm\sqrt{k^2-4(j)(m)}}{2(j)}$$
$$x=\frac{-k\pm\sqrt{k^2-4jm}}{2j}$$

78. $2x^2+3x-9=0$

$$x=\frac{-3\pm\sqrt{3^2-4(2)(-9)}}{2(2)}$$
$$x=\frac{-3\pm\sqrt{9+72}}{4}$$
$$x=\frac{-3\pm\sqrt{81}}{4}=\frac{-3\pm 9}{4}$$
$$x=\frac{-3+9}{4}=\frac{3}{2} \text{ or } x=\frac{-3-9}{4}=-3$$

79. $16x=x^2$

$$0=x^2-16x$$
$$0=x(x-16)$$
$$x=0 \quad\text{or}\quad x-16=0$$
$$x=0 \quad\text{or}\quad x=16$$

80. $y = -(x+3)^2 + 2$

x	y
−5	−2
−4	1
−3	2
−2	1
−1	−2

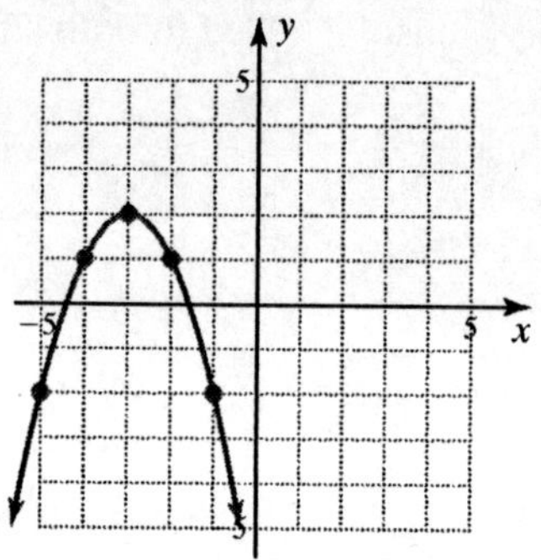

81. $d = 5t^2 + v_0 t$

$20 = 5t^2 + (0)t$

$20 = 5t^2$

$4 = t^2$

$\pm 2 = t$

$t > 0$, so $t = 2$ seconds.